Nanoalloys: From Theory to Application

University of Birmingham, UK
3–5 September 2007

FARADAY DISCUSSIONS
Volume 138, 2008

RSCPublishing

The Faraday Division of the Royal Society of Chemistry, previously the Faraday Society, founded in 1903 to promote the study of sciences lying between Chemistry, Physics and Biology.

EDITORIAL STAFF

Editor
Philip Earis

Assistant editor
Madelaine Chapman

Publishing assistant
Rachel Dilworth

Team leader, serials production
Joanna Stevens

Technical editor
Helen Lunn

Publisher
Janet Dean

Faraday Discussions (Print ISSN 1359-6640, Electronic ISSN 1364-5498) is published 3 times a year by the Royal Society of Chemistry, Thomas Graham House, Science Park, Milton Road, Cambridge, UK CB4 0WF.
Volume 138 ISBN: 0 85404 119 2
　　　　　ISBN-13: 978 0 85404 119 0

2008 annual subscription price: print+electronic £519, US $1,033; electronic only £467, US $929. Customers in Canada will be subject to a surcharge to cover GST. Customers in the EU subscribing to the electronic version only will be charged VAT. All orders, with cheques made payable to the Royal Society of Chemistry, should be sent to RSC Distribution Services, c/o Portland Customer Services, Commerce Way, Colchester, Essex, UK CO2 8HP.
Tel +44 (0) 1206 226050;
E-mail sales@rscdistribution.org

If you take an institutional subscription to any RSC journal you are entitled to free, site-wide web access to that journal. You can arrange access *via* Internet Protocol (IP) address at www.rsc.org/ip. Customers should make payments by cheque in sterling payable on a UK clearing bank or in US dollars payable on a US clearing bank. Periodicals postage is paid at Rahway, NJ and at additional mailing offices. Airfreight and mailing in the USA by Mercury Airfreight International Ltd., 365 Blair Road, Avenel, NJ 07001, USA.

US Postmaster: send address changes to *Faraday Discussions*, c/o Mercury Airfreight International Ltd., 365 Blair Road, Avenel, NJ 07001. All despatches outside the UK by Consolidated Airfreight.

PRINTED IN THE UK

Faraday Discussions documents a long-established series of *Faraday Discussion* meetings which provide a unique international forum for the exchange of views and newly acquired results in developing areas of physical chemistry, biophysical chemistry and chemical physics.

Nanoalloys: From Theory to Application

Faraday Discussions

www.rsc.org/faraday_d

A General Discussion on Nanoalloys: From Theory to Application was held at University of Birmingham, UK on 3rd, 4th and 5th September 2007.

RSC Publishing is a not-for-profit publisher and a division of the Royal Society of Chemistry. Any surplus made is used to support charitable activities aimed at advancing the chemical sciences. Full details are available from www.rsc.org

CONTENTS

ISSN 1359-6640; ISBN 0-85404-119-2
ISBN-978-085404-119-0

Faraday Discussions — Vol 138

Nanoalloys: From Theory to Application

RSCPublishing

Cover
See Nele Veldeman, Ewald Janssens, Klavs Hansen, Jorg De Haeck, Roger E. Silverans and Peter Lievens, *Faraday Discuss.*, 2008, **138**, 147–162.
Visualisation of a transition metal dopant encapsulated in a 16 atom gold cluster cation: smashing alloy clusters with laser light evidences enhanced stability of closed shell species.

Image reproduced by permission of Professor Peter Lievens, from *Faraday Discuss.*, 2008, **138**, 147.

PREFACE

INTRODUCTORY LECTURE

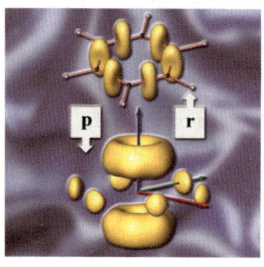

PAPERS AND DISCUSSIONS

CONCLUDING REMARKS

ADDITIONAL INFORMATION

Preface

Roy L. Johnston*[a] and Riccardo Ferrando*[b]

First published as an Advance Article on the web 30th November 2007
DOI: 10.1039/b717388c

The range of properties of metallic systems can be greatly extended by taking mixtures of elements to generate intermetallic compounds and alloys. The rich diversity of compositions, structures and properties of metallic alloys has led to widespread applications in electronics, engineering and catalysis. Recently, the desire to fabricate materials with well-defined, controllable properties and structures (on the nanometre scale), coupled with the flexibility afforded by intermetallic materials, has generated considerable interest in bimetallic and multimetallic alloy clusters—the so-called "nanoalloys". In addition to experimental studies (synthesis, characterisation and property measurement), nanoalloys are attracting increasing attention from the point of view of theory and simulation.

One of the major reasons for the tremendous growth of interest in nanoalloys is the fact that their chemical and physical properties can be tuned by varying the composition and type and degree of chemical ordering, as well as the size of the clusters. Nanoalloys often display structures and properties which are distinct from those of the pure elemental clusters and they may also display properties which are distinct from the corresponding bulk alloys, due to finite size effects. The structures, compositions and degree of segregation or mixing of the surfaces of nanoalloys are also critically important in many applications, especially catalysis.

The idea for a conference on nanoalloys, arose from discussions between the two of us going back to early 2004. The interdisciplinary nature of research into nanoalloys, ranging across Chemistry, Physics and Materials Science, the sheer volume of output (which we became acutely aware of when writing a review of the field), and the wide range of actual and potential applications suggested that it would be timely to organise the first international meeting devoted to nanoalloys. We also thought that the unique format of a Faraday Discussion would be the best forum to ensure lively discussion and the identification of important areas for future research. We proposed to hold the meeting in Birmingham. As we are both theoreticians, we then decided it would probably be a good idea to get some experimentalists involved, so we enlisted the help of Claude Henry, Sarah Horswell, Brian Johnson and Peter Lievens and then set about inviting speakers.

The meeting comprised four sessions which were intended to represent the four main strands of current research in nanoalloys. The first session was devoted to theory and simulation (with talks by Fortunelli, Rossi, Hou, Calvo, Leiva and Lequien); the second session concentrated on electronic, optical and magnetic properties (with talks by Broyer, Veldeman, El-Shall, Pastor and Mottet); the third session was concerned with catalysis (with talks by Hutchings, Caps, Sermon, Russell, Tromp and Thomas); the fourth session dealt with structural studies (with talks by Kiely, Mejia-Rosales, Li, Hampe and Marsault). These sessions followed the stimulating and thought-provoking Introductory Lecture by Prof. Julius Jellinek (Argonne National Laboratory, USA). The Concluding Remarks were made by Prof. Brian Johnson in his eloquent and inimitable style.

The meeting also included a lively poster session with 23 posters, the vast majority of which were of high quality. The Organising Committee awarded the Skinner prize

[a] *School of Chemistry, University of Birmingham, Edgbaston, Birmingham, UK B15 2TT.*
  E-mail: r.l.johnston@bham.ac.uk
[b] *Dipartimento di Fisica, Università di Genova, Via Dodecaneso 33, I16146 Genova, Italy.*
  E-mail: ferrando@fisica.unige.it

Plate 1 Participants at the Faraday Discussion 138 Meeting: Nanoalloys—From Theory to Application, held at University of Birmingham, 3–5 September 2007. Photograph courtesy of Merlin Fox.

(for best student poster) jointly to Giovanni Barcaro (CNR, Pisa, Italy) and Alvaro Mayoral-Garcia (University of Birmingham).

Past President of the Faraday Division and invited speaker, Prof. Sir John Meurig Thomas, delivered an illuminating and entertaining after-dinner speech, referring to the history of research in Chemistry and Physics in Birmingham (going back to the work of Priestley) and to the links between Michael Faraday and the topic of the Discussion. It was pleasing to note the number of times that Faraday's name came up in the presentations, and, continuing this theme, the conference photograph (Plate 1) was taken in front of the statue "Faraday" by Edoardo Paolozzi, near the University's West Gate.

We believe that our initial aims of outlining the state of the field and highlighting growth areas were successfully achieved and we hope that the reader will find the following papers and the associated discussion to be as interesting, useful and thought provoking as did those of us who attended the meeting.

It remains for us to say that we enjoyed organising this conference and that we are most grateful to our co-organisers and to the RSC staff who made the organisation much easier. Of course we wish to thank all of the contributors for delivering their work against a tight deadline and all participants for submitting their questions, replies and comments. We acknowledge financial support for the meeting from Johnson Matthey plc and the University of Birmingham Collaborative Network in Nanotechnology.

Roy L. Johnston (*Chair, co-editor*),

Riccardo Ferrando (*co-editor*)

Nanoalloys: tuning properties and characteristics through size and composition

Julius Jellinek*

Received 24th December 2007, Accepted 3rd January 2008
First published as an Advance Article on the web 5th February 2008
DOI: 10.1039/b800086g

A brief sketch of the history of metals and alloys is followed by examples illustrating the current status of the field of nanoalloys and a discussion of our results on the characterization of structural and dynamical (thermal) properties of Ni–Al bimetallic clusters.

1 Introduction

1.1 Metals and alloys: a brief historical sketch

This Faraday Discussion Meeting is the first in the series of Faraday Discussions devoted to the subject of nanoalloys. It also is the first major forum on the field of nanoalloys as such. Having this Discussion attests to the degree of interest in nanoalloy systems, both as complex and fascinating research subjects and as objects with a unique potential for profoundly affecting our lives. In that, nanoalloys are the present and, even more so, the future inheritors of the special role metals and alloys have played throughout the history of mankind.

Over millenia, the history of civilization was related to and defined by the history of metals and alloys. Different periods of civilization were labeled by the name of the metal that played the central role in their development. The Bronze Age, the origins of which date back to the 3rd millennium BC, was followed by the Iron Age, which started in the 12th century BC and which, in large measure, continues even today. The Industrial Revolution of the 18th century could not have happened without iron and its alloys. The societies that discovered and used metals and alloys first were more advanced, more prosperous, and more powerful than their neighbors. The initial uses of metals in jewelry, ornaments, objects of art, and as coinage were followed by utilization in domestic tools and appliances and, eventually, in means of production and warfare.

The history of the discovery of metals and alloys and of how to produce and perfect them is indeed fascinating.[1,2] By the end of the 17th century, only 12 metals were known to mankind. Twelve more were added to the list in the 18th century. The 19th century turned out to be a particularly fertile one with 41 new metals discovered. By the end of the 20th century, the count of metallic elements in the Periodic Table stood at 84.

Seven metals carry the distinction of being "metals of antiquity" as they were discovered and used in ancient Mesopotamia, Egypt, Greece, Rome, the Far East, and Africa. These are Au (discovered *ca.* 6000 BC), Cu (*ca.* 4200 BC), Ag (*ca.* 4000 BC), Pb (*ca.* 3500 BC), Sn (*ca.* 1750 BC), Fe (*ca.* 1500 BC), and Hg (*ca.* 750 BC). (These dates vary somewhat depending on the source.)

Chemical Sciences and Engineering Division, Argonne National Laboratory, Argonne, Illinois 60439, USA. E-mail: jellinek@anl.gov

The first alloy known to mankind was bronze (90% Cu and 10% Sn). By some accounts, it was the Sumerians who discovered it in the 3rd millenium BC, although the historians are not unanimous on this. The Bronze Age emerged in different parts of the world at different times. In Great Britain, for example, it is associated with the period between *ca.* 2100 and 700 BC. By some claims, the earliest use of steel ($\sim 99\%$ Fe and $\sim 1\%$ C) was by the Persians in the 16th century BC. Other early alloys were electrum (50% Au and 50% Ag), 18 karat yellow gold (75% Au, 12.5% Ag, and 12.5% Cu), and sterling silver (92.5% Ag and 7.5% Cu). The smiths learned that mixing metals and adding other elements to them was a way to improve their strength, ductility, malleability, and other characteristics. Today, virtually all uses of metals are in an alloyed form.

1.2 Marrying nano and alloys

The young and rapidly developing fields of nanoscience and nanotechnology have added a new dimension to the old, but never aging, subject of metals and alloys. The unprecedented degree of novelty and unique potential for principally new technologies, which one can anticipate from combining size and composition effects, have captured the imagination of many researchers and triggered vigorous activity. The expectations are hardly exaggerated. The unusual properties of even pure (*i.e.*, single-component) metal nanosystems are well documented. (Here we use a definition of nanosystems and nanophenomena that does not invoke a quantitative specification of sizes, as that specification kept changing over the years. For our purposes, the nanoscale is that range of sizes over which the properties, or phenomena, show variations with the system size. By this definition, the linear dimensions of nanosytems may vary from a few subnanometers to hundreds of nanometers).

Small or medium size atomic clusters of nominally metallic elements may lack altogether the attributes (or their analogs) that are usually associated with the bulk metallic state. These attributes then emerge as the clusters grow in size (see, *e.g.*, refs. 3 and 4 and the citations therein). Metal clusters and nanoparticles have also been shown to possess structural, electronic, dielectric, magnetic, optical, and chemical properties that are different from those of the corresponding bulk metals.[5] All these properties exhibit strong size-variations, which, as a rule, are nonmonotonic over the range of small sizes (the range of emergent phenomena), and which become monotonic and smooth in the range of larger sizes (the range of scaling behaviour). It is not surprising then that combining two or more metallic elements in clusters (we use this term for smaller size nanosytems), or in larger nanoparticles, leads to an even higher degree of novelty and complexity. The central goal of the field of nanoalloys is to explore, understand, and characterize the rich variety and variability of the alloy properties at the nanoscale as a function of size and composition, both elemental and percentile. Efforts to achieve this goal are well underway. In the next section we discuss a number of examples that illustrate the status of the field.

2 Overview

2.1 Structural properties

One of the added complexities of nanoalloys is that they form many more equilibrium structural forms than their single-component counterparts. The manifolds of the extra (so-called homotopic) conformations originate from the various possible distributions of the different types of constituent atoms between the sites of a given geometric (isomeric) structure (see refs. 6 and 7 and the citations therein). Fig. 1 illustrates some of the possibilities. The component elements can form solid solutions with different degrees of mixing or can segregate into parts with plane-like boundaries, core-shell arrangements, and even onion-like multishell structures.

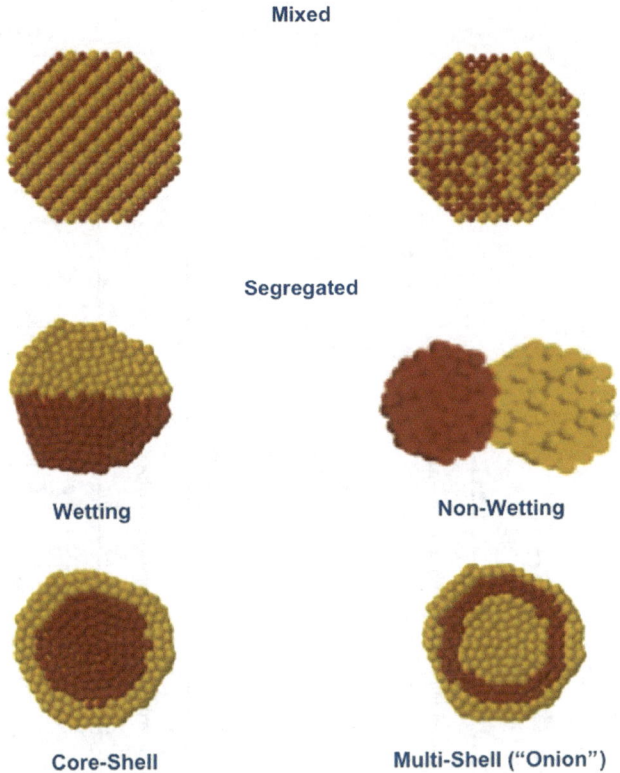

Mixed

Segregated

Wetting   **Non-Wetting**

Core-Shell   **Multi-Shell ("Onion")**

Fig. 1   Different structural patterns (cross sections) of bimetallic nanoalloys.[8]

Fig. 2 displays scannning transmission electron microscopy (STEM) images of $Au_{0.5}/Pd_{0.5}$ nanoparticles with a three-shell onion structure. Such onion conformations actually were first predicted computationally.[9] We return to a more detailed discussion of the isomeric and homotopic forms of bimetallic particles in section 3.

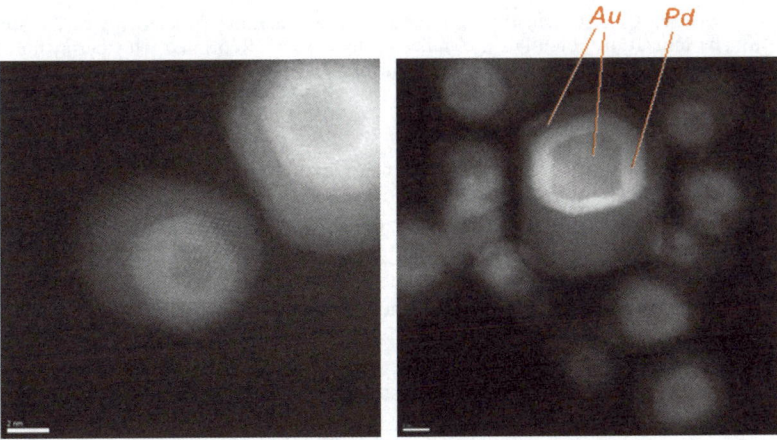

Fig. 2   STEM images of $Au_{0.5}/Pd_{0.5}$ nanoparticles. The scale in the lower left corner of the images corresponds to 2 nm. (Courtesy of M. José-Yacamán.)

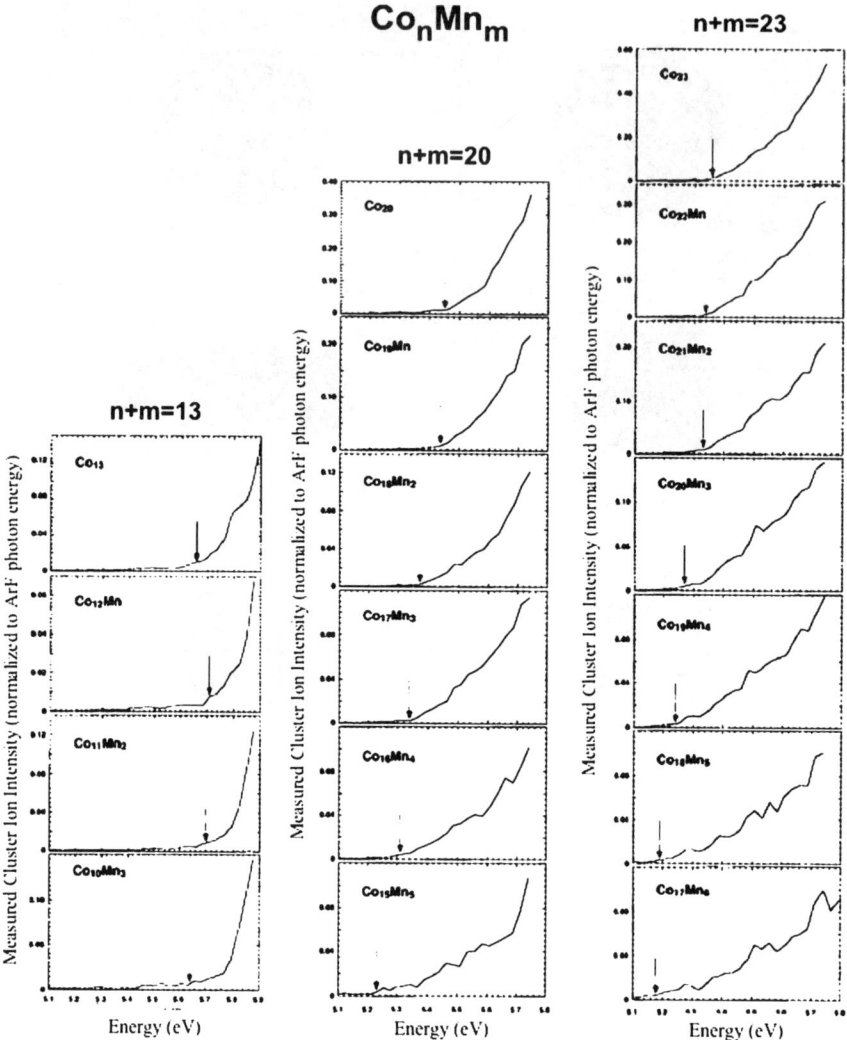

Fig. 3 Measured Co_nMn_m cluster ion intensities as a function of the energy of the ionizing photon. The ionization potentials are indicated by arrows. Reprinted with permission from ref. 14; copyright 1999 by the American Chemical Society.

2.2 Electronic features

The electronic features of nanoalloys were examined both experimentally and computationally. The properties explored include degree of stability, orbital character of charge density, charge transfer between the component elements, density of electronic states, HOMO–LUMO gaps, and ionization potentials (IPs) (see, *e.g.*, refs. 10–16 and the citations therein). The IPs were measured as a function of size and composition for a variety of bimetallic clusters which include Co_nAl_m,[10] Co_nNa_m,[11] Al_nCs_m,[12] Co_nV_m,[13] and Co_nMn_m.[14] Typical experimental data are shown in Fig. 3, which displays measured cluster ion intensities as a function of the energy of the ionizing photon for three different sizes and various compositions of Co_nMn_m. The ion emergence energies (indicated by arrows) are the ionization potentials. The overall trends are a reduction in the IP value with an increase in cluster size (although atomic clusters may, in general, exhibit nonmonotonic size-variations of

Fig. 4 The Lycurgus cup in reflected (left) and transmitted (right) light.[17]

IP), and a decrease in the IP value with an increase in the Mn concentration (although the first substitution of a Co atom by a Mn atom in Co_{13} results in a slight increase of the IP).

2.3 Optical properties

Fig. 4 shows two images of the Lycurgus cup (British Museum), a masterpiece of Roman glass craftsmanship, which dates back to *ca.* 4th century AD.[17] The cup depicts a scene from the myth of Lycurgus, a king of the Thracians around 800 BC, who was entrapped by branches of a vine as punishment for his evil behaviour. It appears greenish in reflected light and reddish in transmitted light. Analytical transmission electron microscopy (TEM) identified $Ag_{0.7}Au_{0.3}$ particles of a few tens of nanometers embedded in the glass of the cup[18] (see Fig. 5). It is the optical activity of these particles that is responsible for the cup's colours. The first to identify

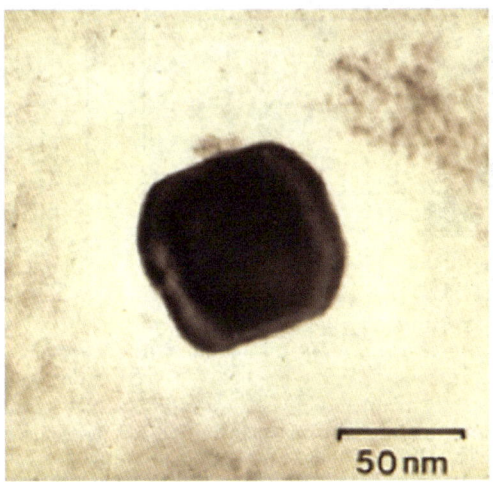

Fig. 5 TEM image of a $Ag_{0.7}/Au_{0.3}$ nanoparticle in a sample of the Lycurgus cup. Reprinted with permission from ref. 18; copyright 1990 by Wiley-Blackwell.

Fig. 6 Michael Faraday and a display of his colloidal solutions of gold at the Royal Institution.[20]

finely divided metal particles (specifically those of gold) as the source of the colour of colloidal solutions was Michael Faraday[19] (see Fig. 6). It is only befitting to have a Faraday Discussion Meeting on a subject touched, as were so many others, by the genius of the man.

An example of a modern exploration of optical properties of nanoalloys is shown in Fig. 7. The figure depicts scanning tunneling microscopy (STM) images and photon emission spectra of Au–Ag bimetallic particles supported by a thin Al_2O_3 film deposited on a NiAl(110) surface.[21] The particles are assembled through simultaneous and successive deposition of different amounts of Au and Ag atoms on the film, which leads, respectively, to mixed (cases a–d of the figure) and core-shell (cases e and f) structures with different Au–Ag percentile composition. Mie plasmon excitations of individual particles are generated through electron injection via an STM tip, and the radiative decay of the excitations is measured. The mixed particles display a single peak in their emission spectra. The wavelength of the maximum of the peak and its width increase with the increase of the Au content. The spectra of the core-shell particles possess two peaks, one of which originates from the core and the other from the shell. The intensities of the peaks depend on the fraction of the

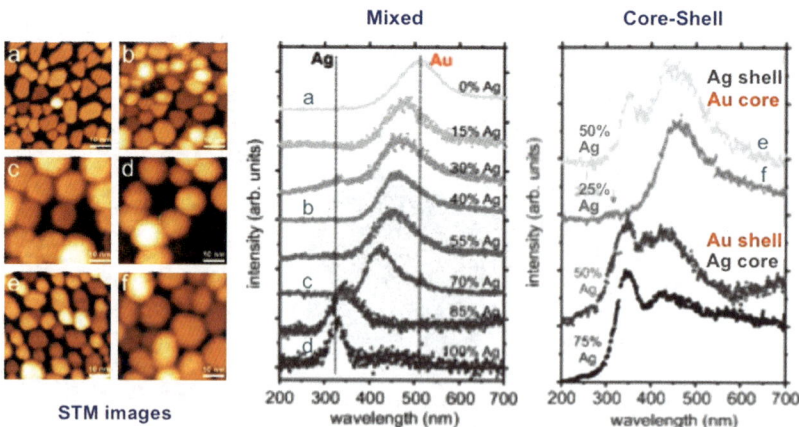

Fig. 7 STM images and measured photon emission spectra of individual Ag–Au nanoparticles supported on Al_2O_3/NiAl(110). The scale in the lower right corner of the images corresponds to 10 nm. Cases a–d are those of mixed particles, whereas e and f of particles with core-shell structures. Reprinted with permission from ref. 21; copyright 2005 by the American Physical Society.

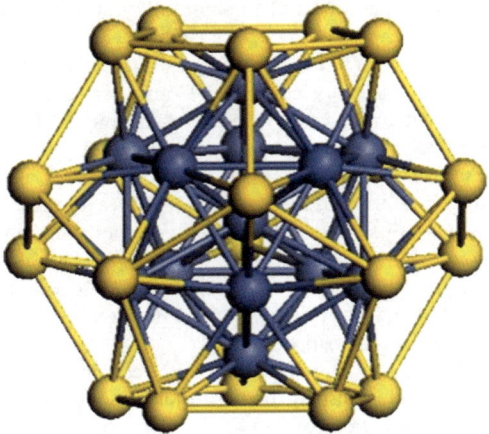

Fig. 8 Computed lowest energy structures of $[Mn_{13}@Au_{20}]^-$ and $[Co_{13}@Au_{20}]^-$.[28] The Mn and Co cores are depicted in blue, the Au cage in yellow.

material in the core and in the shell. The separation between the core and the shell peaks increases as the thickness of the shell decreases. The peak that corresponds to Ag shifts a bit to the blue, whereas the one that originates from Au moves a bit to the red. In general, the separation between the peaks is a little bit larger when the core is formed by Ag and the shell by Au. The reasons underlying these observations are discussed in ref. 21.

2.4 Magnetic properties

Magnetic properties of bimetallic clusters have been the subject of a number of experimental[16,22–24] and theoretical[23,25–28] studies. Atomic clusters of even single metals often exhibit magnetic attributes that are very different from those of the corresponding bulk metals and that change substantially, and not necessarily monotonically, especially in the range of small to medium size clusters (*cf. e.g.*, ref. 23 and the citations therein). Adding another metal leads only to further broadening of the spectrum of magnetic features. The outcome depends on the

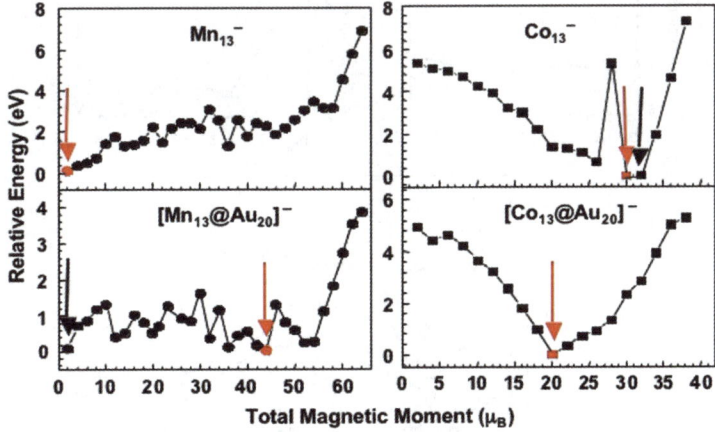

Fig. 9 Energies of Mn_{13}^-, Co_{13}^-, $[Mn_{13}@Au_{20}]^-$, and $[Co_{13}@Au_{20}]^-$ in different spin states referred to the energy of the corresponding most stable state (shown in red).[28] The black arrows indicate states that are energetically very close to the lowest energy state.

elemental and percentile composition of the nanoalloy, as well as on its isomeric and homotopic forms.

One can ask, for example: What is the effect of encapsulation, or "coating", of a metal cluster or nanoparticle by a cage of another metal? This question is particularly intriguing when the coating metal is nonmagnetic in bulk quantities. Fig. 8 shows the optimized structures of anionic $Mn_{13}@Au_{20}$ and $Co_{13}@Au_{20}$.[28] The most stable conformations of these bimetallic clusters, as well as of the bare anionic Mn_{13} and Co_{13}, are close to icosahedra. The relative energies of their different magnetic states are shown in Fig. 9. The magnetic moment of the most stable state of Mn_{13}^- is 2 μ_B whereas that of $[Mn_{13}@Au_{20}]^-$ is 44 μ_B. One notices, though, that another state of $[Mn_{13}@Au_{20}]^-$ with a magnetic moment of only 2 μ_B has just a slightly higher energy. Co_{13}^- forms two close low-energy states with magnetic moments of 30 μ_B and 32 μ_B, respectively. The magnetic moment of the distinct lowest energy state of $[Co_{13}@Au_{20}]^-$ is 20 μ_B. Gold coating of magnetic clusters can thus lead to both enhancement, as in the case of Mn_{13}^-, and attenuation, as in the case of Co_{13}^-, of their magnetic moments.

2.5 Chemical reactivity and catalysis

Single-component metal clusters exhibit striking size-dependent variations in their chemical reactivity.[29,30] Addition of a few atoms, or at times even just one, to a cluster may change its reactivity by orders of magnitude. For example, the room-temperature rate constant for D_2 adsorption on Ni_{10} is about a factor of 30 larger than the corresponding rate constant for Ni_9.[30] It is only natural that the variety of the chemical properties of nanoalloys is even richer since it is affected not only by their size, but also by their elemental and percentile composition. Studies, dominantly experimental at present, confirm this expectation. They were performed for $Co_{n-m}Al_m$,[31] $Nb_{n-m}Al_m$,[31] $Co_{n-m}V_m$,[31] and $Co_{n-m}Na_m$[11] clusters reacting with H_2; $Co_{n-m}Mn_m$ clusters reacting with N_2;[14] and other nanoalloy systems reacting with various molecules (see, e.g., ref. 8 and the citations therein).

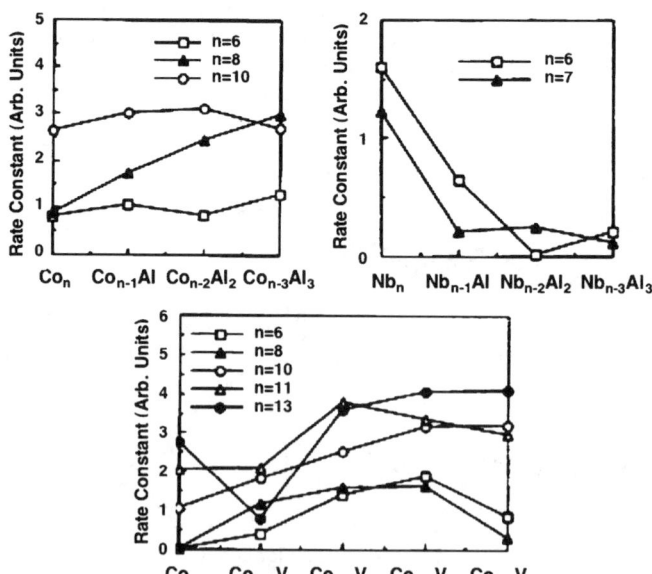

Fig. 10 Measured room-temperature rate constants for H_2 adsorption on $Co_{n-m}Al_m$, $Nb_{n-m}Al_m$, and $Co_{n-m}V_m$. Reprinted with permission from ref. 31; copyright 1991 by Springer.

Fig. 10 shows measured room-temperature rate constants for the first three cases mentioned above as a function of the cluster size and composition.[31] The reactivity of the 6- and 10-atom $Co_{n-m}Al_m$ changes only a little and nonmonotonically with the increase of the number of Al atoms. In distinct difference, the reactivity of the 8-atom $Co_{n-m}Al_m$ increases quite substantially and essentially linearly with its Al content. Both the 6- and the 7-atom Nb_n show a dramatic decrease in their reactivity as atoms in them get replaced by Al atoms. The reactivities of $Co_{n-m}V_m$, $n = 6, 8, 10, 11$, and 13, exhibit noticeable and, in general, nonmonotonic variations as the number of V atoms in them is increased. Especially prominent is the decrease in the reactivity of Co_{13} as one of its atoms gets replaced by a V atom.

The size and composition dependence of the chemical reactivity of nanoalloys made them particularly suitable as candidates for catalysts with fine-tuned activity, selectivity, and resistivity to poisoning. Nanoalloy catalysis is one of the most active and technologically promising areas of nanoscience. The variety of metal pairs explored so far is already quite broad and includes Co–Ni (for growth of carbon nanotubes); Co–Pt and Ru–Pt (for fuel cell electrocatalysis); Pd–Pt (for use in catalytic converters); Pd–Au (for formation of H_2O_2 from H_2 and O_2; CO and alcohol reduction; and synthesis of vinyl acetate); Ru–Pd (for single-step hydrogenation); as well as Ni–Pd, Ni–Pt, Ni–Au, Cu–Pd, Cu–Pt, Pt–Au, Mo–Pt, Rh–Pt, and Re–Ir. Many of these have been found to possess superior catalytic properties. The optimal size, or range of sizes, and the optimal percentile composition of the nanoalloy catalysts are usually related. The studies also point to the important role of the support and/or environment of the catalytic particles. Further details and references to original work can be found in a recent review.[8]

3 Nanoalloys: a challenge in complexity

3.1 General

The brief overview and illustrations of the preceding section furnish just a glimpse of the degree of complexity posed by nanoalloys as research subjects. Indeed, nanoalloys combine and amplify the vast array of size effects with the broad range of composition effects, giving rise to a plethora of new properties and characteristics that could not be obtained by varying either the size of pure metallic systems or the composition of bulk alloys alone. As mentioned, the large manifolds of the possible equilibrium geometric (isomeric) structures of pure nanosystems get greatly expanded by the various homotopic forms associated with each isomer. The complexity of the size-driven changes in electronic structure and electronic features becomes even more diverse when one combines atoms of different metals. The added complexity is also reflected in the optical, magnetic, and chemical properties of nanoalloys.

What defines the degree of stability or the energy ordering of the different structural forms in nanoalloys? What drives the propensity of the component elements in a nanoalloy to mix or to segregate? What is an appropriate quantitative measure of the degree of mixing or segregation? Can one introduce concept-based structural classification schemes? In general, can we devise analysis and characterization frameworks capable of addressing the full complexity of nanoalloys as a function of their size and composition? Or, alternatively, can we formulate a hierarchical approach based on various degrees of coarse-graining and detail of description that would address the complexity of nanoalloys at the level relevant within a given scientific context or technological application? It is these questions that we should keep in mind and pursue with the goal of identifying and formulating common trends and features, fundamental relationships, characteristic labels and descriptors, general guiding principles and rules, and possibly even universal laws. A good starting point for such a pursuit is bimetallic clusters of various sizes and compositions because they are amenable to high-level and detailed exploration and

Table 1 Parameters of the Gupta-type potential for Ni–Al systems[37]

Parameter	Ni–Ni	Al–Al	Ni–Al
A/eV	0.0376	0.1221	0.0563
ξ/eV	1.070	1.316	1.2349
p	16.999	8.612	14.997
q	1.189	2.516	1.2823
r_0/Å	2.4911	2.8637	2.5222

characterization. Below we discuss some examples of such exploration based on our theoretical/computational studies.

3.2 Understanding structural complexity

We illustrate and analyze this complexity using the case of 13-atom Ni–Al clusters as a paradigm.[6,7,32–35] The interatomic interactions are described by the so-called Gupta-type potential[36]

$$U = \sum_i \left\{ \sum_j A_{\alpha\beta} e^{-p_{\alpha\beta}(r_{ij}/r_o^{\alpha\beta}-1)} - \left[\sum_j \xi_{\alpha\beta}^2 e^{-2q_{\alpha\beta}(r_{ij}/r_o^{\alpha\beta}-1)} \right]^{\frac{1}{2}} \right\}, \qquad (1)$$

where U is the system configurational energy, α and β denote the constituent elements, i and j label the atoms, r_{ij} is the distance between the i-th and the j-th atoms, and the values of the five parameters A, ξ, p, q, and r_o for each of the three possible pairs of atoms (Ni–Ni, Al–Al, and Ni–Al) are listed in Table 1. (For the genesis of the Gupta-type potential see ref. 38).

Equilibrium structures of the pure Ni_{13} and Al_{13} and of Ni_nAl_m for all (n, m) pairs such that $n + m = 13$ were obtained using a combination of simulated cooling and gradient-based techniques. The structural relaxations were carried out over all degrees of freedom, and normal mode analysis was performed for every stationary configuration obtained to separate the (globally or locally) stable equilibrium structural forms from those that constitute transition state conformations. For the pure clusters, a large number of configurations generated along a high-energy trajectory (the clusters were melted) were utilized as initial guess structures. Fig. 11 shows the six lowest energy isomers of Ni_{13} and Al_{13} of the total of more than 100 obtained for each (for details see refs. 6, 7, and 32). A feature to notice is

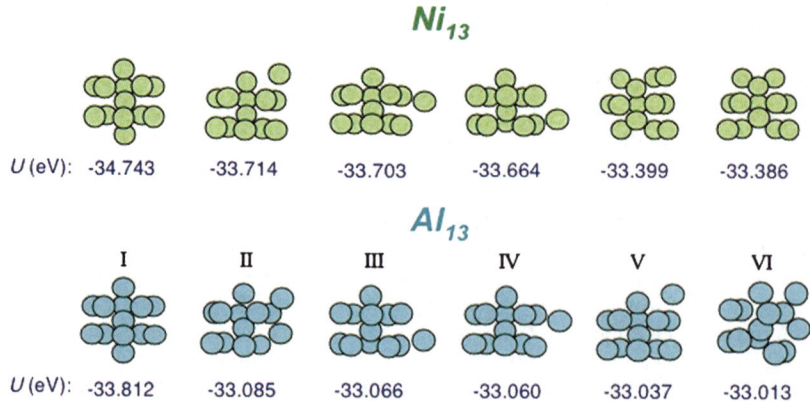

Fig. 11 The first six lowest energy isomers of the pure Ni_{13} and Al_{13} clusters as defined by the Gupta-type potential (*cf.* refs. 7 and 32). The six isomers of Al_{13} were used as template isomeric forms for Ni_nAl_m 13-mers. See the text for details.

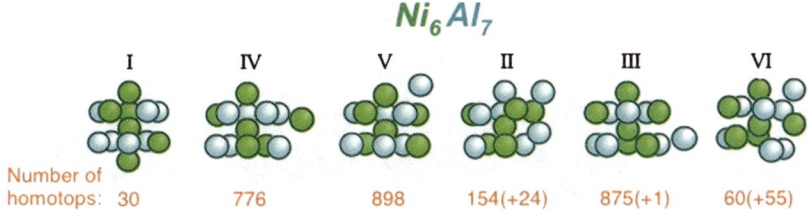

Ni_6Al_7

	I	IV	V	II	III	VI
Number of homotops:	30	776	898	154(+24)	875(+1)	60(+55)

Fig. 12 Energy ordering (from left to right) of the six isomers of Ni_6Al_7 considered. The Ni atoms are depicted in green, the Al atoms in blue. Each isomer is represented by its most stable homotope. The total number of homotopes of each isomer is indicated. The numbers in parentheses represent additional stationary homotopic structures that are transition state configurations, rather than stable homotopes. The Roman enumeration of the isomers corresponds to that in Fig. 11.

that although the icosahedron is the most stable form of both clusters, the energy ordering of their higher energy isomers is different. For Ni_nAl_m, we considered as isomeric templates the six structures of Al_{13} shown in Fig. 11, and for each template obtained all the corresponding homotopic forms. This was accomplished by performing all the possible replacements of n Al atoms by Ni atoms and reoptimizing the cluster structures (for details see refs. 6, 7, and 32). The energy ordering of the isomers of Ni_6Al_7, as represented by their corresponding lowest energy homotopes, is shown from left to right in Fig. 12. As is clear from the figure, this ordering is different from that of the isomers of the pure Al_{13}. In fact, as illustrated in Fig. 13, the energy ordering of the isomers depends on the (n, m) content of the cluster.

Fig. 12 also lists the number of distinct homotopes associated with the different isomers of Ni_6Al_7. In general, the total number $N(n, m)$ of homotopes associated with each isomeric form of a two-component system that contains n atoms of one type and m of the other is

$$N(n, m) = \frac{(n + m)!}{n!\, m!}. \tag{2}$$

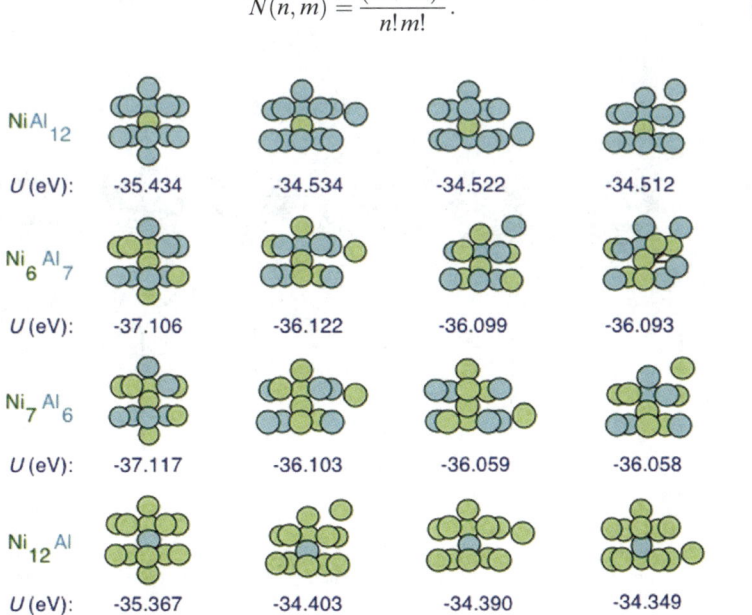

$NiAl_{12}$				
U (eV):	-35.434	-34.534	-34.522	-34.512
Ni_6Al_7				
U (eV):	-37.106	-36.122	-36.099	-36.093
Ni_7Al_6				
U (eV):	-37.117	-36.103	-36.059	-36.058
$Ni_{12}Al$				
U (eV):	-35.367	-34.403	-34.390	-34.349

Fig. 13 The first four (of the six considered) isomers of $NiAl_{12}$, Ni_6Al_7, Ni_7Al_6, and $Ni_{12}Al$ as represented by their corresponding most stable homotopes. The energies of the clusters are indicated. The Ni atoms are depicted in green, the Al atoms in blue.

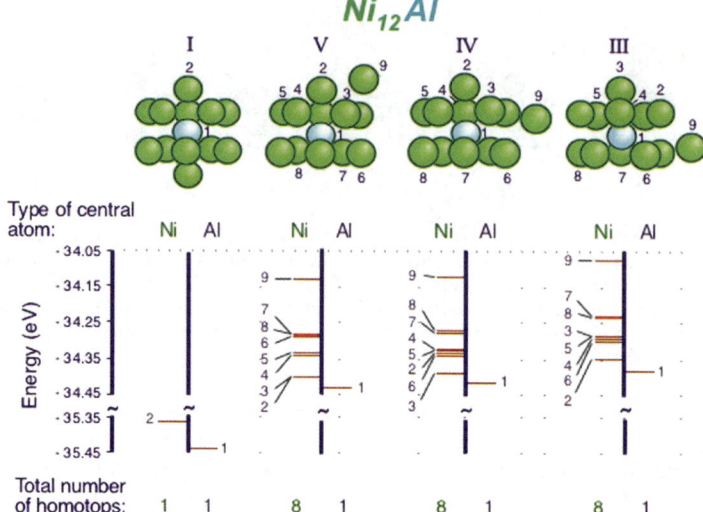

Fig. 14 Spectra of homotope energies for four isomers of $Ni_{12}Al$.[6,7] For each isomer, the homotopes are partitioned into two classes defined by the type (Ni or Al) of the central atom. The number of homotopes in each class is shown. The labels of the homotope energy levels indicate the position of the Al atom in the corresponding homotopes. The Ni atoms are depicted in green, the Al atoms in blue. The isomers are represented by their most stable homotopes. The Roman enumeration of the isomers corresponds to that in Fig. 11. Isomeric forms II and VI do not survive for $Ni_{12}Al$.

For the case of $n = 6$ and $m = 7$, $N(n, m) = 1716$. For isomers with point group symmetry, the number of distinct (*i.e.*, inequivalent) homotopes is smaller than that defined by eqn (2). But even with this reduction, the increase in the structural complexity of nanoalloys, as compared to that of their pure counterparts, is immense.

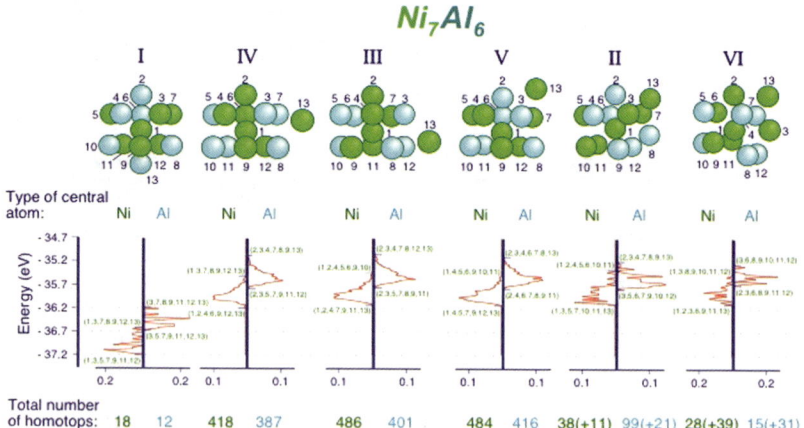

Fig. 15 The same as Fig. 14, but for six isomers of Ni_7Al_{16} (*cf.* refs. 6 and 7). Because of the large number of homotopes associated with each isomer, normalized distributions of the homotope energy levels, rather than the levels themselves, are shown. The seven numbers labelling the lowest and the highest energy levels within each homotopic class indicate the positions of the Ni atoms in the corresponding homotopes. The numbers in parentheses correspond to the additional stationary homotopic structures that are transition state configurations, rather than stable homotopes.

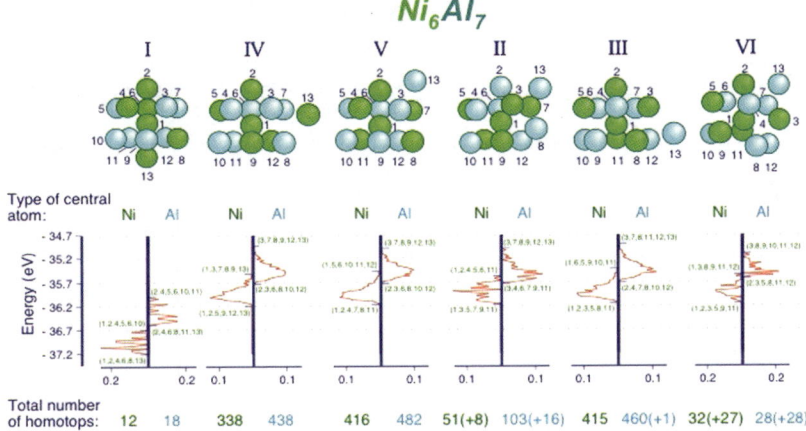

Fig. 16 The same as Fig. 15, but for six isomers of Ni_6Al_7 (*cf.* refs. 6 and 7). The six numbers labelling the lowest and the highest energy levels within each homotopic class indicate the positions of the Ni atoms in the corresponding homotopes.

Can one identify general trends in the spectrum of configurational energies of the isomeric and homotopic forms of a bimetallic cluster as a function of its composition? The answer to this question is yes. One such trend is the systematic change in the energy patterns of the homotopes associated with the individual isomers. The other is the systematic change in the gap between the energy ranges spanned by the homotopes of the most stable isomer, on the one hand, and those spanned by the homotopes of the higher energy isomers, on the other. The computations predict that the icosahedron is the most stable isomeric form of all 13-atom Ni_nAl_m clusters irrespective of their composition[6,7] (the term "icosahedron" is used here to represent a packing pattern, rather than a structure with a precisely icosahedral symmetry).

The two trends in the composition dependence can be identified through inspection of Fig. 14–18, which depict the spectra of isomer and homotope energies of the Ni_nAl_m 13-mer for various (n, m) values. Note that the homotope energy spectra of the individual isomers are partitioned in these figures into two classes defined by the

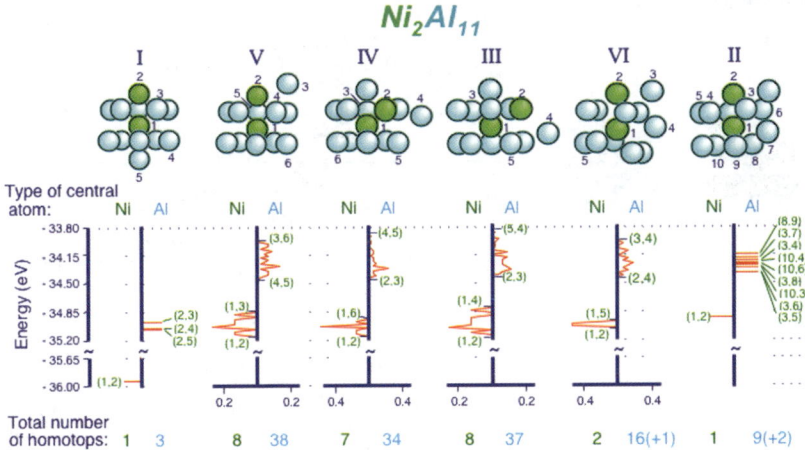

Fig. 17 The same as Fig. 14 and 15, but for the six isomers of Ni_2Al_{11}. The two numbers labelling the individual homotope energy levels, or the lowest and the highest energy levels within the homotopic classes, indicate the positions of the two Ni atoms in the corresponding homotopes.

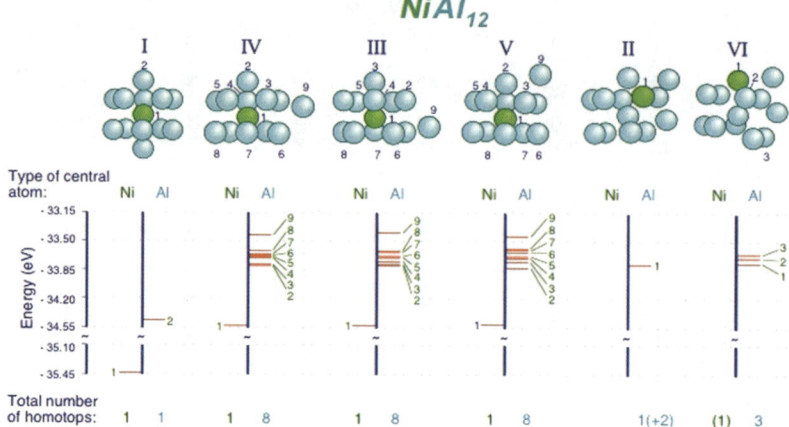

Fig. 18 The same as Fig. 14, but for six isomers of NiAl$_{12}$ (*cf.* refs. 6 and 7). The numbers labelling the homotope energy levels indicate the position of the Ni atom in the corresponding homotopes. The numbers in parentheses correspond to the additional stationary homotopic structures that are transition state configurations, rather than stable homotopes.

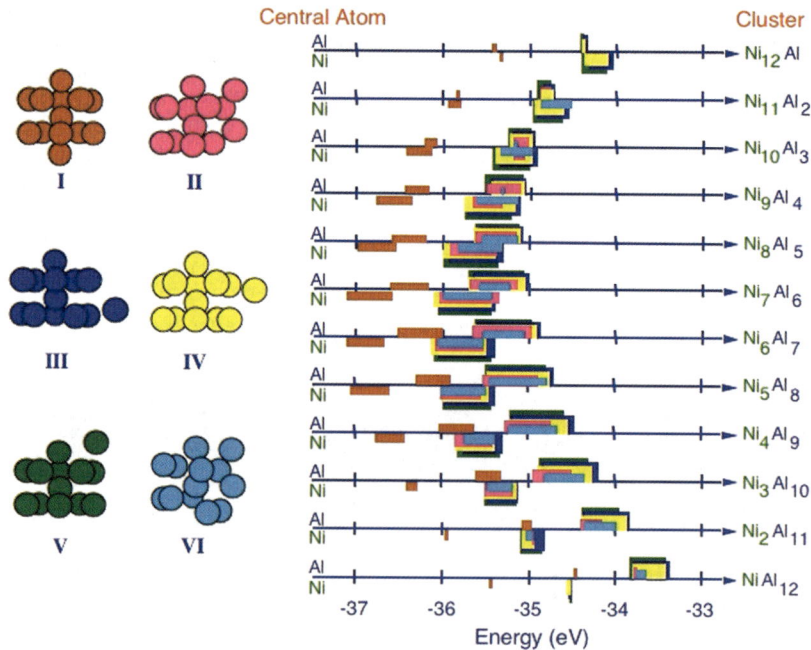

Fig. 19 The spans and locations of the energy ranges associated with each of the two classes of homotopes as defined by the type (Ni or Al) of the central atom, for six isomers and all possible compositions of the Ni$_n$Al$_m$ 13-mer. The colours establish a correlation between the graphs of the energy ranges and the isomeric forms. The labelling of the isomers by Roman numerals corresponds to that in Fig. 11. Notice that one or both homotopic class(es) of an isomer may not be supported by all compositions. Neither homotopic class of isomers II and VI survives for Ni$_{12}$Al (*cf.* Fig. 14), the homotopic class of isomer VI with Al in the center does not survive for Ni$_9$Al$_4$, and the homotopic classes of isomers II and VI with Ni in the center do not survive for NiAl$_{12}$ (*cf.* Fig. 18).

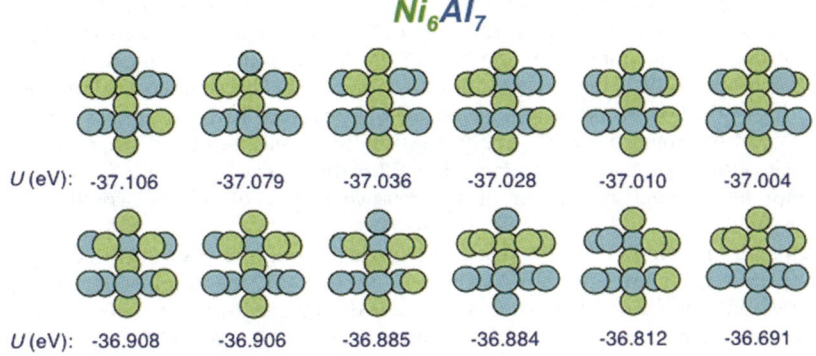

Ni₆Al₇

U (eV): -37.106   -37.079   -37.036   -37.028   -37.010   -37.004

U (eV): -36.908   -36.906   -36.885   -36.884   -36.812   -36.691

Fig. 20 The 12 homotopes of the icosahedral Ni_6Al_7 with Ni in the center in the order of increasing energy. The Ni atoms are depicted in green, the Al atoms in blue.

type (Ni or Al) of the central, or most coordinated, atom. First, one notices that when the number of Ni atoms in the cluster dominates, the energies of the two homotopic classes of each individual isomer are close. An increase in the fraction of the Al atoms, however, leads to an increase in the bifurcation of the energies of the two homotopic classes and the development of a gap between them. Second, in $Ni_{12}Al$ the homotope energies of the most stable, icosahedral isomer are separated by a large gap from the homotope energies of the higher energy isomers; the latter exhibit similar patterns of homotope energies. As the fraction of the Al atoms in the cluster increases, this gap gradually gets reduced. The cumulative effect of these two trends is a gradual closure of the gap and development of an overlap between the energy ranges spanned by the class of icosahedral homotopes with Al in the center, on the one hand, and the classes of homotopes of higher energy isomers with Ni in the center, on the other. The two trends and their consequence are clearly seen in Fig. 19, which shows the energy ranges spanned by the two homotopic classes corresponding to each of the six isomers and all possible compositions of the Ni_nAl_m 13-mer.

Can one complement the above characterization of the composition-driven changes in the patterns of the energy ranges spanned by the two homotopic classes associated with each isomer by an understanding of what defines the energy ordering of the homotopes within each individual class? A hint at an answer to this question is

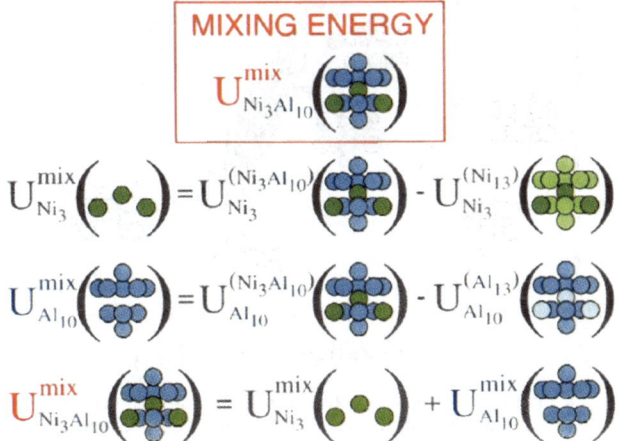

Fig. 21 Pictorial "cluster equation" definition of the mixing energy. See the text for details.

given by Fig. 20, which shows all twelve homotopes of the icosahedral Ni_6Al_7 with Ni in the center (*cf.* Fig. 16) arranged in the order of increasing energy. In the homotope of the lowest energy, the Ni and the Al atoms are intermixed, whereas in the homotope of the highest energy they are segregated. This suggests that the energy ordering of the homotopes within a class may be governed by the degree of mixing of the Ni and Al atoms. To verify this suggestion, one needs a quantitative measure of the degree of mixing; as is clear from Fig. 20, with the exception of a few cases, an attempt at a visual evaluation of the relative degree of mixing would not be successful.

A pictorial definition of the degree of mixing, using the case of Ni_3Al_{10} as an example, is given in Fig. 21. As defined by eqn (1), the total configurational (interaction) energy of the cluster is the sum of energies of its constituent atoms. As a consequence, the total configurational energy of a bimetallic cluster can be represented as the sum of energies of its two single-component units; this is achieved by partitioning the external sum over i in eqn (1) into two sums, each performed only over the atoms of the same type. In the case of Ni_3Al_{10}, these units are Ni_3 and Al_{10}. One can define the energies of these units in Ni_3Al_{10} and in their respective pure 13-atom clusters (*i.e.*, of the Ni_3 unit in Ni_{13} and of the Al_{10} unit in Al_{13}), the configurations of which are chosen to coincide with that of Ni_3Al_{10}. The differences between the energies of each of these units in the two-component cluster, on the one hand, and in their respective pure clusters, on the other, are their mixing energies. The sum of the mixing energies of the component units is the total mixing energy of the bimetallic cluster (for formal expressions see refs. 6 and 7). Thus, the mixing energy is that energy which is released (or consumed) when one takes out two units of different metals from their corresponding pure clusters, each of which has the same total number of atoms and the same structure as the bimetallic one, and combines them to form the bimetallic cluster. One can then define a mixing coefficient M,[6,7]

$$M = \frac{U_{mix}}{U} 100\%, \tag{3}$$

where U_{mix} and U are the mixing and the total energy, respectively. It is this mixing coefficient that serves as a quantitative measure of the degree of mixing.

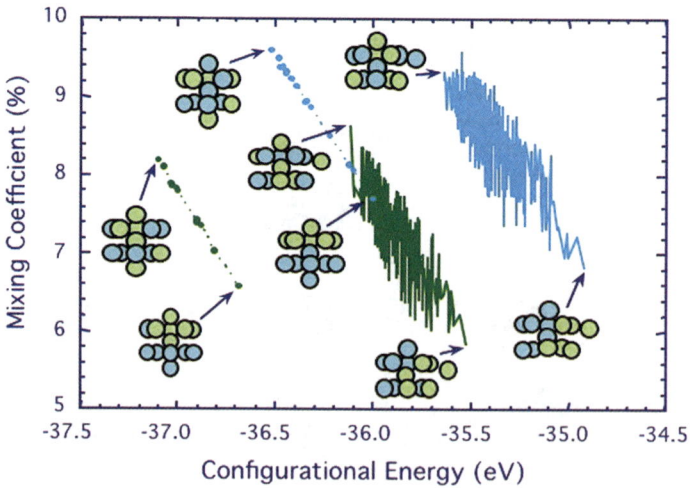

Fig. 22 Graphs of mixing coefficient as a function of configurational energy for homotopic classes with Ni in the center (green) and Al in the center (blue) for the first two isomers of Ni_6Al_7.[6,7,34] The structures shown are those of the lowest energy and the highest energy homotopes within each class. The Ni atoms are depicted in green, the Al atoms in blue.

Fig. 22 displays the graphs of the mixing coefficient as a function of the configurational energy for the homotopic classes associated with the first two isomers of Ni_6Al_7 (cf. Fig. 16). The two graphs that correspond to the homotopic classes of the first (icosahedral) isomer show a monotonic (essentially linear) dependence between the energy and the mixing coefficient. The same type of behaviour, only globally rather than locally, is also exhibited by the graphs of the homotopic classes of the second isomer. This strict or global correlation between the mixing coefficient and the configurational energy is a signature of the fact that the energy ordering of the structures that belong to the same homotopic class is driven by the degree of their mixing (or segregation). (The homotopic classes are defined by the commonality of size, composition, isomeric form, and the type of the central, or most coordinated, atom.) In the case of the Ni–Al clusters considered, the higher the degree of mixing of a homotope, the more stable it is relative to other homotopes within its class. We have verified that this correlation holds true for all (n, m) compositions.

The graphs for the other higher energy isomers of Ni_6Al_7 (not shown) are very similar to those for the second isomer in Fig. 22, and they also show local oscillations superimposed on the overall near-linear trend. The reason for these local oscillations is the following. The energy gaps between neighbouring structures in the homotopic classes of the higher energy (less symmetric) isomers are very small (cf. the number of homotopes and the extents of the energy ranges indicated for these classes in Fig. 16). The energy changes caused by the small relaxations in the interatomic distances as a consequence of permutations of unlike atoms are comparable with, or even larger than, these gaps. As a result, even minor relaxations may perturb the local ordering of the homotopes.

As mentioned, computations with the Gupta-type potential, eqn (1), predict that the most stable isomeric form of the 13-atom Ni_nAl_m clusters is an icosahedron. The

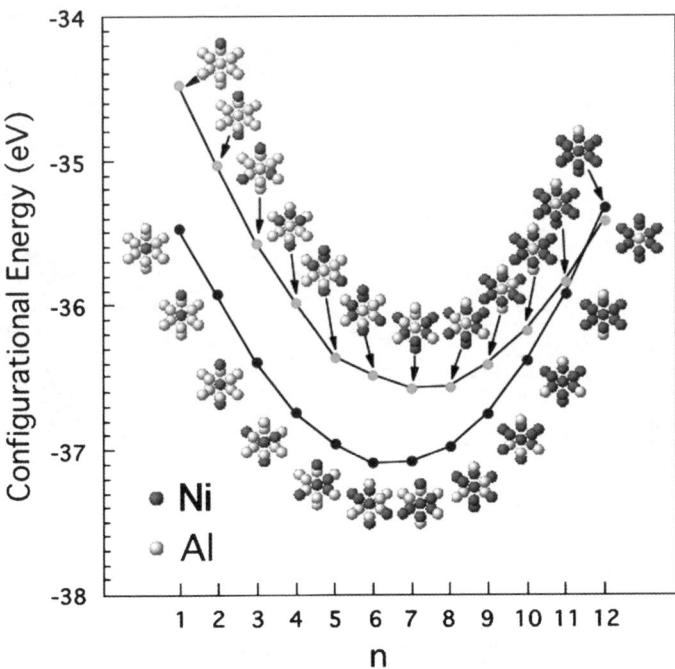

Fig. 23 Computed configurational energies of the most stable homotopes with Ni in the center and Al in the center for the icosahedral isomer of the 13-atom Ni_nAl_m clusters as a function of the number of Ni atoms.[39] The computations were performed with the Gupta-type potential (see the text for details). The lines are drawn to guide the eye.

most stable homotopes of this isomer with Ni in the center and Al in the center are shown as a function of the number of Ni atoms in Fig. 23.[39] As is clear from the figure, except for the case of $Ni_{12}Al$, the most stable configurations of the 13-atom Ni_nAl_m have Ni in their centers. For $Ni_{12}Al$, the Gupta-type potential predicts the homotope with Al in the center to be just a bit more stable than the homotope with Ni in the center. A DFT computation[40] also defines the icosahedral homotopes of $Ni_{12}Al$ with Ni and Al in the center to be very close in energy, but reverses the order of their stability.

These theoretical predictions have recently been verified experimentally.[39] The experiments capitalized on the fact that the N_2 molecule binds to nickel, but not to aluminum. Thus, achieving saturation coverage of Ni–Al clusters by N_2 and counting the number of the adsorbed molecules, one can map out the number of the exposed Ni atoms. The saturation coverages measured for the Ni_nAl_m 13-mers are shown in Fig. 24. One notices that for all compositions there is a magenta bar

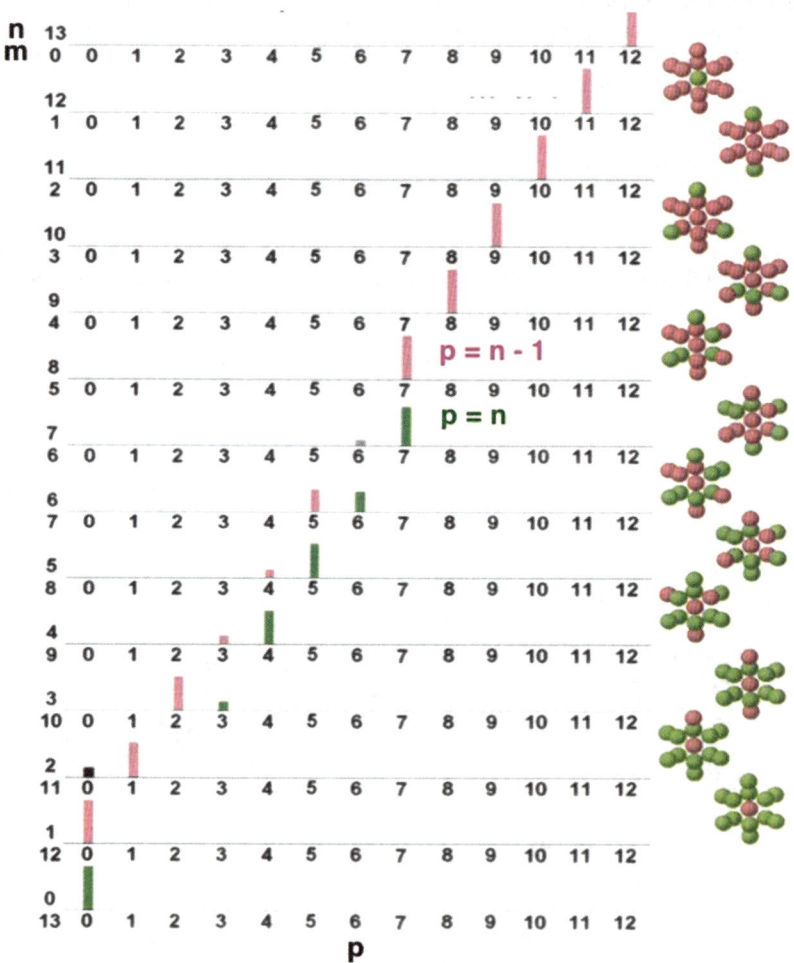

Fig. 24 Measured relative intensities (bars) of the saturation coverages p of 13-atom Ni_nAl_m clusters by Ni_2 molecules.[39] The magenta bars correspond to adsorption of $p = n - 1$ molecules, the green bars to adsorption of $p = n$ molecules. The depicted structures are those of the most stable forms of the clusters for each composition as obtained from the Gupta-type potential (cf. Fig. 23). The Ni atoms are shown in magenta, the Al atoms in green. See the text for details.

that represents adsorption of $n-1$ N_2 molecules. This is consistent with the prediction that in the most stable, icosahedral structure of the clusters the central site is occupied by a Ni atom.

Because of the close energetic proximity of the icosahedral forms of $Ni_{12}Al$ with Ni and Al in the center, one might expect the measurements on this cluster to detect the adsorption of 12 N_2 molecules as well. The data, however, do not support this expectation. A possible reason for this is that the entropic effect of the 12 available surface sites for the single Al atom, *versus* the single central site, tilts the energetically competitive nature of the icosahedral homotopes with Ni and Al in the center in favour of the homotope with Ni in the center.

The experiments, however, do show an additional pattern of adsorption of n N_2 molecules (the green bars in Fig. 24) for clusters with composition close to 50/50%. This finding can be explained by the fact that the configurational energies in this composition range are the lowest (*cf.* Fig. 19 and 23), and the icosahedral homotopes with Al in the center also become accessible under the experimental conditions (for further details see ref. 39).

A similar agreement between the structures predicted in computations with the Gupta-type potential and the N_2 saturation levels measured in experiments has also been obtained for the 12- and 14-atom Ni_nAl_m clusters.[39] This indicates that semiempirical potentials can be a valuable tool in determining the structures of alloy clusters and nanoparticles. Because of their computational efficiency, the development of such potentials with a high degree of accuracy for various metals and their combinations is a task of first-rate importance.

3.3 Understanding dynamical (thermal) complexity

As one might expect, the dynamical (or thermal) behaviour of nanoalloys is also affected not only by their size, but by their composition as well. Can one correlate the complexity of nanoalloy dynamics with the complexity of the corresponding structural forms? To which extent is the dynamical (or thermal) behaviour of the pure counterparts of a nanoalloy, as a function of their energy (or temperature), mimicked or, alternatively, altered by the nanoalloy? Can a nanoalloy exhibit novel dynamical (thermal) features that are not found in its pure counterparts? These are the types of questions one would want to have answers to in an attempt to formulate a conceptual framework for analysis and characterization of the dynamical (thermal) behaviour of nanoalloys.

Here we address these questions through the paradigm of Ni_nAl_m 13-mers. The results discussed below are obtained *via* constant-energy molecular dynamics simulations (for details see refs. 7 and 32–35). The quantities used in the analysis are: 1) Averaged kinetic energy $\langle E_{kin} \rangle_t$ per atom, where $\langle \ldots \rangle_t$ denotes time-averaging over long trajectories; (2) temperature T,

$$T = \frac{2\langle E_{kin} \rangle_t}{(3N - 6)k},$$ (4)

where N is the number of atoms in the cluster and k is the Boltzmann constant; (3) root-mean-square bond length fluctuation δ,

$$\delta = \frac{2}{N(N-1)} \sum_{i<j} \frac{(\langle r_{ij}^2 \rangle_t - \langle r_{ij} \rangle_t^2)^{1/2}}{\langle r_{ij} \rangle_t},$$ (5)

and (4) specific heat per atom (in units of the Boltzmann constant) c,

$$c = \left[N - N\left(1 - \frac{2}{3N-6}\right) \langle E_{kin} \rangle_t \langle E_{kin}^{-1} \rangle_t \right]^{-1},$$ (6)

all considered as a function of the cluster total energy per atom.

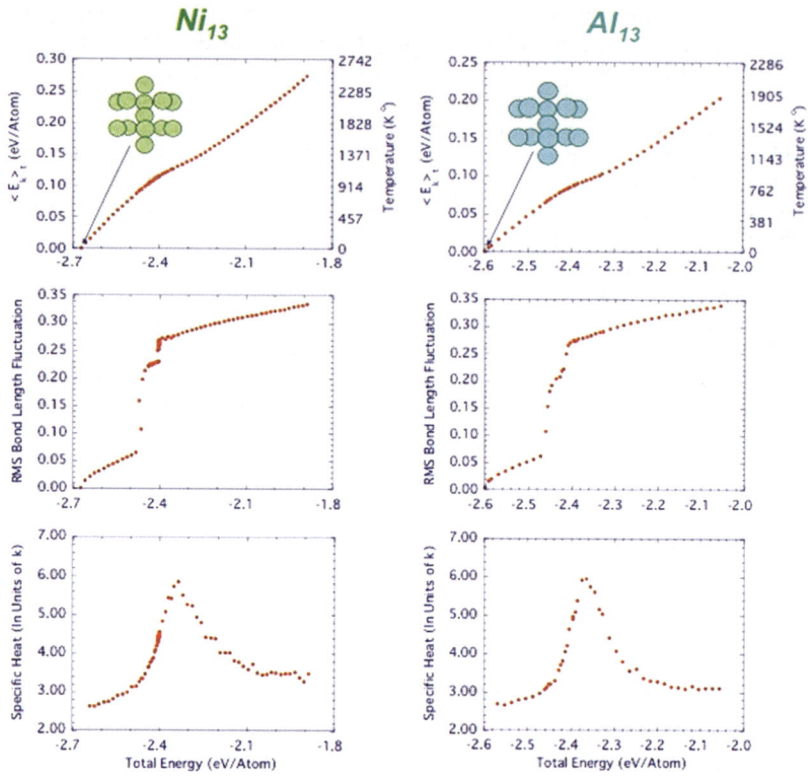

Fig. 25 Caloric curves ($\langle E_{kin} \rangle_t$ and T), root-mean-square bond length fluctuations, and specific heats as a function of the cluster total energy (per atom) for Ni_{13} and Al_{13}.[7,32] See the text for details.

We start by examining the behaviour of the pure Ni_{13} and Al_{13} shown in Fig. 25. Both clusters are icosahedra in their most stable form, and they respond to an increase of their energy in quite a similar fashion. First, they undergo isomerization transitions, as indicated by the change in the slope of their caloric curves and the abrupt changes in their δ graphs. The first abrupt increase in the δ values is associated with surface isomerizations (*i.e.* structural rearrangements that involve only the surface atoms). The second such increase corresponds to the onset of global isomerizations (*i.e.*, structural transformations that involve the central atom as well). Further increase of the energy causes "melting" of the clusters, which is signified by the peaks in the graphs of their specific heats. For further discussion of the melting-like transition in finite systems, a phenomenon that is considerably more peculiar and complex than bulk melting, we refer the reader to refs. 41 and 42 and the citations therein.

To what extent does the dynamical (thermal) behaviour of the pure Ni_{13} and Al_{13} shown in Fig. 25 carry over, or change, when one combines Ni and Al atoms to form Ni_nAl_m 13-mers? The answer to this question is given by the graphs shown in Fig. 26. As clearly displayed by these graphs, with the exception of Ni_2Al_{11} and $NiAl_{12}$, the dynamical (thermal) features of Ni_nAl_m 13-mers largely mimic those of their pure counterparts. The only apparent change is a shift to lower values in energies at which the different transitions listed above become operational as the composition of the 13-mer gets closer to 50/50%. The reason for this shift is the already mentioned lowering of the configurational energies in this composition range (*cf.* Fig. 19 and 23); implicit here is the assumption that the barriers between the different

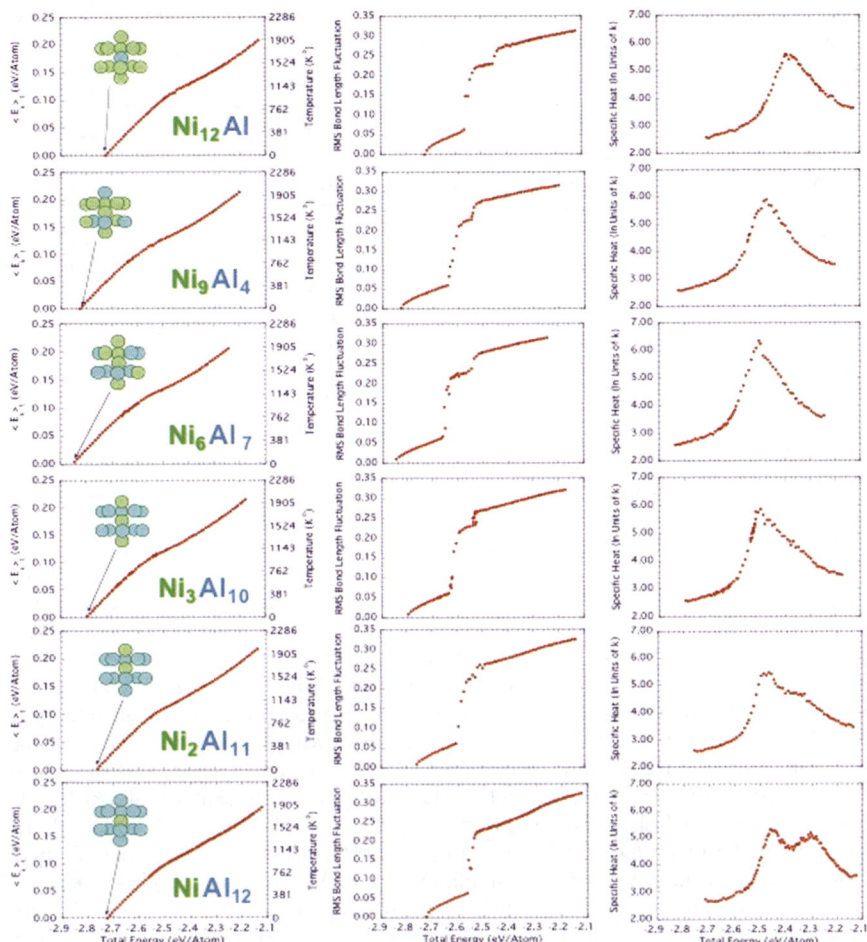

Fig. 26 The same as Fig. 25, but for 13-atom Ni_nAl_m clusters (*cf.* refs. 7, 32–35). For each composition, the shown zero-temperature configuration of the cluster is that of its most stable structure. The Ni atoms are depicted in green, the Al atoms in blue. See the text for details.

equilibrium structures change with the composition in a manner similar to that of the energies of these structures.

Ni_2Al_{11} and $NiAl_{12}$ behave differently. The second abrupt change in the δ graphs becomes blurred for Ni_2Al_{11} and disappears entirely for $NiAl_{12}$. On the other hand, the graphs of the specific heat of these clusters exhibit two peaks (in the case of Ni_2Al_{11}, one of the two is somewhat less developed). A detailed analysis reveals that the first peak in the specific heat of $NiAl_{12}$ signifies the onset of the state of surface melting, in which the 12 surface Al atoms constantly change their positions through diffusive motion, but continue to encapsulate the central Ni atom. As the energy of the cluster is increased further, the graph of the specific heat changes its downward direction to an upward one. This change is caused by the onset of global isomerizations that involve exchanges in the positions of the central (initially Ni) and one of the surface (initially Al) atoms. Further increase of the energy makes this exchange ever more frequent and, eventually, the cluster attains the state of global (or complete) melting, which is signified by the second peak in the graph of its specific heat.

Table 2 Temperatures (in K) of the onset of surface isomerization (T_{si}), global isomerization (T_{gi}), surface melting (T_{sm}), and (global) melting ($T_{(g)m}$).[7] See the text for details

Cluster	T_{si}	T_{gi}	$T_{(g)m}/T_{sm}$
Ni_{13}	810	1050	1190
$Ni_{12}Al$	680	1040	1200
$Ni_{11}Al_2$	775	1020	1215
$Ni_{10}Al_3$	750	1015	1200
Ni_9Al_4	780	1070	1220
Ni_8Al_5	785	1030	1220
Ni_7Al_6	735	1080	1200
Ni_6Al_7	790	1135	1200
Ni_5Al_8	710	1050	1180
Ni_4Al_9	705	1010	1145
Ni_3Al_{10}	680	1000	1085
Ni_2Al_{11}	690	1220	1315/1070
$NiAl_{12}$	680	1180	1365/985
Al_{13}	570	715	850

The peculiarities of the dynamics of $NiAl_{12}$ as a function of energy are rooted in the peculiarities of its spectrum of configurational energies (*cf.* Fig. 18). The icosahedral structure with Al in the center has just a bit higher energy than the most stable homotopes of the next three isomers, which have Ni in the center. However, the barrier for a direct exchange between the central Ni atom and a surface Al atom in the icosahedral isomer is very high. Therefore, the first structural response of $NiAl_{12}$ (initially in its most stable form) to an increase of its energy is a surface isomerization transition to an icosahedron with a defect, where one of the Al atoms moves to the second shell leaving behind a vacancy. In fact, as displayed in Fig. 18, there are three different such isomers with Ni in the center and a single Al atom in the second shell, all characterized by very close energies. Because of the vacancy in the first shell, the barriers that separate the single homotopes of these isomers with Ni in the center are small. As a consequence, at energies needed to transform the most stable form of the cluster into either of these homotopes, all three become accessible and undergo interconversions. Moreover, because these three homotopes are separated by large energy gaps from their counterparts (as well as from the homotopes of other higher energy isomers) with Al in the center (*cf.* Fig. 18), they can absorb significant extra energy with the only consequence being an increase in the rate of interconversions between them. This leads to the state of surface melting. An accompanying result is a larger first increase in the δ value, which then masks the effect of the onset of global isomerizations. As a consequence, the δ graph shows only one abrupt change as a function of energy. The same line of arguments applies to establishing a correlation between the dynamical and structural (*cf.* Fig. 17) peculiarities of Ni_2Al_{11}. The "softening" of the dynamical effects, as compared to the case of $NiAl_{12}$, is a result of the changes in the spectrum of configurational energies caused by replacing the second Al atom by a Ni atom.

Finally, in Table 2 we show the characteristic temperatures of the different stages in the melting-like transition of the pure and mixed 13-atom Ni–Al clusters. The temperature of the onset of surface isomerizations is defined as the one that corresponds to the lowest energy point of the first abrupt change in the δ graphs (Fig. 25 and 26). Except for Ni_2Al_{11} and $NiAl_{12}$, the temperature of the onset of global isomerizations is specified by the lowest energy point of the second abrupt change in the δ graphs. For Ni_2Al_{11} and $NiAl_{12}$, the temperature of the global isomerizations is associated with the energy of the point that separates the two peaks in their specific heat graphs. The melting temperatures of the clusters are defined by the energies of the maxima in the corresponding graphs of the specific heat. For Ni_2Al_{11} and $NiAl_{12}$, the temperature defined by the first maximum corresponds to the onset of surface melting, whereas that defined by the second maximum signifies the onset of global melting.

The data in Table 2 indicate that the temperatures of surface isomerizations in 13-atom Ni–Al clusters lie between the temperatures of surface isomerizations in the pure Ni_{13} and Al_{13}. In general, this is not the case for the temperatures of the other stages. Especially interesting is the pronounced increase in these temperatures for Ni_2Al_{11} and $NiAl_{12}$, not only with respect to the corresponding temperatures for Al_{13} but also those for Ni_{13}. For calibration we note that the melting temperatures of bulk nickel and aluminium are 993 K and 1726 K, respectively.

4   A glance into the future

What is the current status of nanoalloy research and what are its major challenges and issues for the future? Answering these questions is the central goal of this Discussion Meeting.

The field is young and it rewards its practitioners with the discovery of new, often unexpected, and even unique properties and phenomena. But the field also is diverse and complex, which makes navigating it a truly challenging endeavour. The task is to understand, characterize, and eventually take advantage of the novelty and uniqueness of the properties of nanoalloys as defined by their size and composition. The source of the challenge is the inexhaustible number of possible size–composition combinations. Further exacerbating the challenge is the generally very large number of structural forms associated with each combination.

How can one target those nanoalloy sizes, compositions, and structures that will lead to the best properties for specific applications without going through a very large number of trials? The answer to this question can emerge only from a fundamental understanding of the complex interplay between the size and the composition effects, and the extra dependencies that finiteness of size introduces between the different properties. Such an understanding will result in formulation of general rules, principles, and laws that capture different aspects of nanoalloy complexity.

The rate of progress will largely depend on advances and refinements in our tools of inquiry. On the experimental side, chief among the tasks are: achieving a better control over size, composition, and structure at the synthesis stage; exercising more control over the conditions of deposition and/or embedding; and designing new and improved characterization tools. The advances needed in theory and computations include: formulation and implementation of more efficient electronic structure treatments (this aspect is particularly important for metallic systems); construction of more accurate semiempirical many-body potentials for pure metals and alloys for large-scale dynamical and statistical simulations; and formulation of new conceptual frameworks and analysis schemes.

The central issues will largely remain the same: Structural forms, dynamical (thermal) behaviour, electronic properties, optical and magnetic features, and chemical reactivity, all considered as a function of size and composition. However, to an increasing degree, we will have to approach these issues as matters of complexity and study them at different levels of detail relevant to the specific questions posed or applications targeted. The detailed analyses illustrated in section 3 may not only be unfeasible for nanoalloys of larger sizes, but may, in fact, be unnecessary for them. The shortcut to the characteristic properties and features of relevance is the application of appropriately fine- or coarse-grained analysis and characterization tools. But how can we know what is the appropriate level of detail? Or, alternatively, how can we gauge what is the knowledge that is necessary and sufficient for characterization of a nanoalloy, or a class of nanoalloys, within a specific scientific or applied context? Progress in answering these questions will be an important step towards a comprehensive conceptual picture of the field over a broad range of sizes and compositions.

Acknowledgements

I thank E. B. Krissinel for collaboration on Ni–Al clusters. This work was supported by the Office of Basic Energy Sciences, Division of Chemical Sciences, Geosciences, and Biosciences, US Department of Energy under Contract number DE-AC-02-06CH11357.

References

1   L. Aicheson, *A History of Metals*, Interscience, New York, 1960, vol. 2.
2   *The Beginning of the Use of Metals and Alloys*, ed. R. Maddin, MIT Press, Cambridge, Massachusets, 1988.
3   (*a*) J. Jellinek and P. H. Acioli, *J. Phys. Chem. A*, 2002, **106**, 10919; (*b*) J. Jellinek and P. H. Acioli, *J. Phys. Chem. A*, 2003, **107**, 1610.
4   B. von Issendorf and O. Cheshnovsky, *Annu. Rev. Phys. Chem.*, 2005, **56**, 549.
5   See, *e.g.*, (*a*) M. Moskovits, *Annu. Rev. Phys. Chem.*, 1991, **42**, 465; (*b*) M. Moskovits, in *Clusters of Atoms and Molecules*, ed. H. Haberland, Springer, Heidelberg, 1994, vols. 1 and 2; (*c*) M. Moskovits, in *Theory of Atomic and Molecular Clusters with a Glimpse at Experiments*, ed. J. Jellinek, Springer, Heidelberg, 1999; (*d*) M. Moskovits, in *Metal Clusters*, ed. W. Ekardt, Wiley, New York, 1999; and references therein.
6   J. Jellinek and E. B. Krissinel, *Chem. Phys. Lett.*, 1996, **258**, 283.
7   J. Jellinek and E. B. Krissinel, in *Theory of Atomic and Molecular Clusters With a Glimpse at Experiments*, ed. J. Jellinek, Springer, Heidelberg, 1999, p. 277.
8   R. Ferrando, J. Jellinek and R. L. Johnston, *Chem. Rev.*, 2008, DOI: 10.1021/cr040090g.
9   F. Baletto, C. Motett and R. Ferrando, *Phys. Rev. Lett.*, 2003, **90**, 135504.
10 W. J. C. Menezes and M. B. Knickelbein, *Chem. Phys. Lett.*, 1991, **183**, 357.
11 K. Hoshino, T. Naganuma, Y. Yamada, K. Watanabe, A. Nakajima and K. Kaya, *J. Chem. Phys.*, 1992, **97**, 3803.
12 K. Hoshino, K. Watanabe, Y. Konishi, T. Taguwa, A. Nakajima and K. Kaya, *J. Chem. Phys.*, 1994, **231**, 499.
13 K. Hoshino, T. Naganuma, K. Watanabe, Y. Konishi, A. Nakajima and K. Kaya, *Chem. Phys. Lett.*, 1995, **239**, 369.
14 G. M. Koretsky, K. P. Kerns, G. C. Nieman, M. B. Knickelbein and S. J. Riley, *J. Phys. Chem. A*, 1999, **103**, 1997.
15 W. Bouwen, F. Vanhoutte, F. Despa, S. Bouckaert, S. Neukermans, L. T. Kuhn, H. Weidele, P. Lievens and R. E. Silverans, *Chem. Phys. Lett.*, 1999, **314**, 227.
16 E. Janssens, H. Tanaka, S. Neukermans, R. E. Silverans and P. Lievens, *Phys. Rev. B*, 2004, **69**, 085402.
17 www.thebritishmuseum.ac.uk/science.
18 D. J. Barber and I. C. Freestone, *Archaeometry*, 1990, **32**, 33.
19 M. Faraday, *Philos. Trans.*, 1857, **147**, 145.
20 www.rigb.org.
21 W. Benten, N. Nilius, N. Ernst and H.-J. Freund, *Phys. Rev. B*, 2005, **72**, 045403.
22 M. B. Knickelbein, *Phys. Rev. B*, 2004, **70**, 014424.
23 J. Bansmann, S. H. Baker, C. Binns, J. A. Blackman, J.-P. Bucher, J. Dorantes-Dávila, V. Dupuis, L. Favre, D. Kechrakos, A. Kleibert, K.-H. Meiwes-Broer, G. M. Pastor, A. Perez, O. Toulemonde, K. N. Trohidou, J. Tuaillon and Y. Xie, *Surf. Sci. Rep.*, 2005, **56**, 189; and references therein.
24 S. Yin, R. Moro, X. Xu and W. A. de Heer, *Phys. Rev. Lett.*, 2007, **98**, 113401.
25 S. Dennler, J. Morillo and G. M. Pastor, *J. Phys.: Condens. Matter*, 2004, **16**, S2263.
26 J. Guevara, A. M. Llois, F. Aguilera-Granja and J. M. Montejano-Carrizales, *Physica B (Amsterdam)*, 2004, **354**, 302.
27 M. B. Torres, E. M. Fernandez and L. C. Balbas, *Phys. Rev. B*, 2005, **71**, 155412.
28 J. Wang, J. Bai, J. Jellinek and X. C. Zeng, *J. Am. Chem. Soc.*, 2007, **129**, 4110.
29 S. C. Richtsmeier, E. K. Parks, K. Liu, L. G. Pobo and S. J. Riley, *J. Chem. Phys.*, 1985, **82**, 3659.
30 W. F. Hoffman, E. K. Parks, G. C. Nieman, L. G. Pobo and S. J. Riley, *Z. Phys. D*, 1987, **7**, 83.
31 S. Nonose, Y. Sone and K. Kaya, *Z. Phys. D*, 1991, **19**, 357.
32 E. B. Krissinel and J. Jellinek, *Int. J. Quantum Chem.*, 1997, **62**, 185.
33 E. B. Krissinel and J. Jellinek, *Chem. Phys. Lett.*, 1997, **272**, 301.
34 J. Jellinek and E. B. Krissinel, in *Nanostructured Materials: Clusters, Composites, and Thin Films*, ed. V. M. Shalaev and M. Moskovits, ACS Symposium Series, American Chemical Society, Washington DC, 1997, vol. 679, p. 239.

35   J. Jellinek and E. B. Krissinel, in *Novel Materials Design and Properties*, ed. B. K. Rao and S. N. Behera, Nova Science Publishers, Commack, NY, 1998, p. 83.

36   R. P. Gupta, *Phys. Rev. B*, 1981, **23**, 6225.

37   F. Cleri and V. Rosatto, *Phys. Rev. B*, 1993, **48**, 22.

38   M. J. López and J. Jellinek, *J. Chem. Phys.*, 1999, **110**, 8899.

39   E. F. Rexer, J. Jellinek, E. B. Krissinel, E. K. Parks and S. J. Riley, *J. Chem. Phys.*, 2002, **117**, 82.

40   M. Calleja, C. Rey, M. M. G. Alemany and L. J. Gallego, *Phys. Rev. B*, 1999, **60**, 2020.

41   R. S. Berry, T. L. Beck, H. L. Davis and J. Jellinek, in *Evolution of Size Effects in Chemical Dynamics*, ed. I. Prigogine and S. A. Rice, Advances of Chemical Physics, John Wiley, New York, 1998, vol. 70, part 2, p. 75.

42   J. Jellinek and A. Goldberg, *J. Chem. Phys.*, 2000, **113**, 2570.

A study of bimetallic Cu–Ag, Au–Ag and Pd–Ag clusters adsorbed on a double-vacancy-defected MgO(100) terrace†

Giovanni Barcaro and Alessandro Fortunelli*

Received 3rd April 2007, Accepted 21st May 2007
First published as an Advance Article on the web 11th September 2007
DOI: 10.1039/b705105k

Binary M_2Ag_6, M_2Ag_7, M_1Ag_7 and M_1Ag_8 clusters (M = Cu, Au, Pd) adsorbed on an MgO(100) terrace presenting a double vacancy (DV) neutral defect are investigated through a combination of density-functional (DF) biased searches and global optimizations. Alloying allows one to probe the adsorption characteristics of the DV-defected surface. It is found that Au_1Ag_7 and Au_2Ag_6 are core-shell magic clusters, in analogy with the pure silver case. The magic character is reduced for Cu_1Ag_7 and Cu_2Ag_6, because the shortening of Cu–Ag distances is counteracted by a double-frustration effect due to the DV, and is practically absent for Pd_1Ag_7 and Pd_2Ag_6, because the requirements of the metallic bond conduction electron count are in conflict with the metal–surface interaction. However, fluxionality is enhanced for Pd–Ag clusters with respect to the pure silver case, which could influence their catalytic properties.

Introduction

Metal-on-oxide systems have attracted increasing attention in recent years in both science and technology for their interesting properties and their applications in many technological fields.[1–4] The knowledge of the cluster structure is an obvious pre-requisite for a deep understanding of these fascinating properties, and much theoretical effort has been devoted to this subject, primarily focusing on small clusters especially when high accuracy was pursued through the use of sophisticated, first-principles approaches. In this context, the MgO(100) surface has been studied for various reasons, ranging from its widespread use as an inert support, to the simplicity of its theoretical description.[5–8] MgO(100) is in fact an apolar, simple ionic surface without the complications associated with surface reconstruction, etc.[2] Despite this simplicity, the defectivity of the MgO(100) surface cannot be easily ignored when studying metal adsorption on it, as both experimental[9] and theore-tical[10–12] evidence suggests that—e.g.—noble and quasi-noble metals can diffuse rapidly on MgO(100) terraces even at low temperatures. Defects are thus essential as nucleation or trapping centers in the study of the growth process.[13] In this perspective, an important issue concerns the influence of the metal/defected-surface (rather than regular-surface) interaction on the cluster structure, i.e., whether the metal clusters keep their gas-phase structure, or whether the interaction with the

Molecular Modeling Laboratory, Istituto per i Processi Chimico–Fisici (IPCF) del CNR, Via G. Moruzzi 1, Pisa, Italy. E-mail: fortunelli@ipcf.cnr.it; Fax: +39 050 315 2442; Tel: +39 050 3152447

† The HTML version of this article has been enhanced with colour images.

defected site is strong enough to induce structural transitions.[8,14–19] One of the most studied local defects on the regular MgO(100) terrace is the oxygen vacancy (F_s-center), even though, probably due to the kinetics of the defect diffusion,[20] F_s-centers are mostly found on kinks or steps rather than on regular terraces.[21,22] Another surface defect that has attracted considerable interest in recent years is the double-vacancy (DV),[8,16,23–31] whose relaxed structure is shown in Fig. 1. This is a neutral defect and corresponds to the removal of an MgO dimer in the (100) direction of the regular oxide terrace. Its surface concentration is estimated to be appreciable when the MgO(100) surface is prepared *via* UHV cleaving.[27] At variance with the F_s defect, the removal of an MgO dimer induces an appreciable modification in the structure of the oxide slab with an increase in the dimensions of the cavity and a corresponding stabilization energy of about 2.90 eV[8] (see Fig. 1). It is therefore of interest to study the structure of small metal clusters adsorbed on a DV-defected MgO(100) terrace and the behavior of the binding energy as a function of the metal cluster size, to investigate, for example, whether there exist clusters exhibiting a particular stability with respect to neighboring nuclearities (surface magic clusters[32]) which can thus be preferentially produced *via* a properly devised synthetic approach. Essentially because of the difficulties associated with their experimental and theoretical characterization, surface magic clusters (especially on an oxide substrate) are a much less investigated field than gas-phase magic clusters, even though they present obvious advantages with respect to the latter in terms of technological applications. Binary clusters are promising in the search of surface magic clusters, because of the opportunities they offer in terms of structural degrees of freedom and modulation of the electronic, magnetic and mechanical properties of the clusters.

In the present article, we consider a DV-defected MgO(100) terrace, and study the adsorption on it of selected Cu–Ag, Au–Ag and Pd–Ag binary nanoclusters, with a number of metal atoms around eight. As pointed out above, the double vacancy is a likely candidate as a nucleation center for the growth of metal clusters on the MgO(100) surface. Eight is a magic number for the spherical jellium model,[33] and it is expected that clusters of single-electron metals exhibit a particular stability at this size.

In general, a metal cluster is defined to possess a magic character when it is characterized by a high stability from a structural and/or electronic point of view. Structural stability is achieved when the number of atoms is such that a high-symmetry structural motif (icosahedron, decahedron, truncated octahedron, *etc.*) is completed. When the electrons of the cluster complete a valence shell, and a large energy gap (of the order of 1–2 eV) exists between the HOMO and LUMO levels, an electronic stability is also achieved. A magic cluster is usually associated with a "spike" in the binding energy curve as a function of the number of metal atoms, for

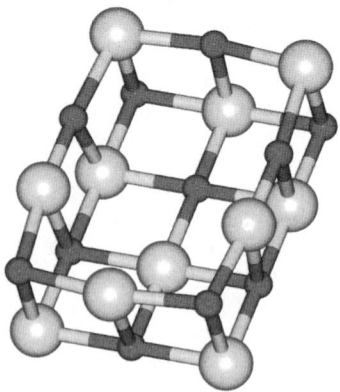

Fig. 1 Relaxed geometry of the MgO(100) surface around a double-vacancy (DV) defect.

example when its incremental formation energy is ~1 eV larger than that of the (N + 1) cluster.

In this work, we will consider clusters of the type M_1Ag_N (N = 7, 8) and M_2Ag_N (N = 6, 7) with M = Cu, Au, Pd (for Pd, also Pd_1Ag_9 clusters will be considered). From a comparison with pure Ag_N clusters at similar nuclearities,[34] the effect of alloying will be investigated on the structural, electronic and energetic properties of these clusters, trying to draw general building principles about the metal/defected-surface interaction and using metals with different electronic and structural characteristics as chemical probes. The choice of M_1Ag_N and M_2Ag_N systems is motivated by the possibility of growing these clusters *via* sequential deposition MBE experiments as: (i) the formation of coinage metal trimers on the double vacancy is energetically not very favored: the difference between the total binding energy of trimers and dimers is in fact only 0.82 eV and 0.74 eV for Cu and Au, respectively;[8] (ii) the energy barriers for the diffusion of Cu and Au dimers on the regular MgO(100) surface are appreciably higher than those of single atoms.[12] One can thus think of saturating the double vacancy on the MgO terraces using an appropriate dosing and deposition flux either with Cu or Au atoms (by exploiting the differential diffusion coefficients of single atoms with respect to dimers) or with Cu or Au dimers (by exploiting the small trimerization energy). Alloying with Pd will then allow one to probe the metal–surface interaction with a very different metal. Another possible synthetic route of these systems passes through the deposition of size-selected gas-phase binary clusters.[35]

Computational approach

Biased searches were performed to locate the global minima of Cu–Ag, Au–Ag and Pd–Ag bimetallic clusters. The starting structures were chosen as the lowest-energy structures obtained *via* DF-BH (Density-Functional Basin-Hopping)[36] searches on pure Ag clusters adsorbed around a DV-defected MgO(100) terrace[34] replacing the silver atoms above one or both the empty sites of the vacancy with M atoms (M = Cu, Au, Pd). In the case of the M_2Ag_N clusters (N = 6, 7) the eight structures considered are displayed in Fig. 2. The corresponding structures for the clusters M_1Ag_N (N = 7, 8) can be derived from those of Fig. 2 by replacing the M atom above the empty oxygen site with a silver atom. From a comparison between Fig. 1 and Fig. 2, it can be noted that the oxide support "closes up" around an adsorbed metal cluster.[8]

Configurations (A)–(D) represent eight-atom clusters. Structure (A) corresponds to the global minimum of the Ag_8 cluster both in the gas phase (where it is characterized by a D_{2d} symmetry[37,38]) and when adsorbed on the DV. This motif can be described as a three-layer structure: in the first layer, we find the metal atom filling the empty Mg site of the vacancy (positioned at almost the same height as the nearby oxide atoms); in the second layer, we find 4 atoms, almost at the vertices of a square: one of them is positioned atop the empty oxygen site of the vacancy, whereas the other three are almost on top of the three oxygen atoms at the border of the cavity. In the third layer, we find the last three atoms of the cluster, positioned in fcc stacking with respect to the atoms of the second layer. Eight corresponds to a magic number of the spherical jellium model[33] and indeed Ag_8 exhibits a HOMO–LUMO gap of about 1.7 eV. In structure (B), the first two metal layers are the same as structure (A) and the difference is in the position of one metal atom of the third layer. For Ag_8, this structure is 0.46 eV higher in energy than structure (A) and its HOMO–LUMO gap is about 1.0 eV. Structure (C) is a bit different: we can describe it in terms of a central triangular bipyramid with three silver atoms grown tetrahedrally on the three triangular faces of the upper pyramid. For Ag_8, this structure is 0.61 eV above the ground-state and its HOMO–LUMO gap is about 1.2 eV. Finally, structure (D) is substantially different from the other three: it can be described as a deformed incomplete tetrahedron of 10 atoms (missing two basal atoms), tilted with respect to the surface and with two silver atoms interacting with two oxygen atoms at the opposite side of the cavity (near the empty O site). For Ag_8, this structure is 0.79 eV above the ground-state and its HOMO–LUMO gap is about 0.65 eV.

The structures of the nine-atom clusters are also shown in Fig. 2: structures (E)–(G) are obtained by adding one silver atom to structure (A), while structure (H) is characterized by a pentagonal bipyramid motif with the lower vertex above the oxygen empty site of the vacancy. In the case of Ag_9, these four structures have half-integer spin and very similar total binding energy (in an interval of 0.2 eV), with structure (H) being the putative global minimum.

In the case of Cu–Ag and Au–Ag clusters, alloying does not introduce significant structural rearrangements: the corresponding optimized configurations are only slightly modified by a minor structural relaxation. Also the energy ordering among these structures does not change with respect to the pure Ag case (at least in the case of the eight-atom metal clusters). We thus assumed that the structures in Fig. 2

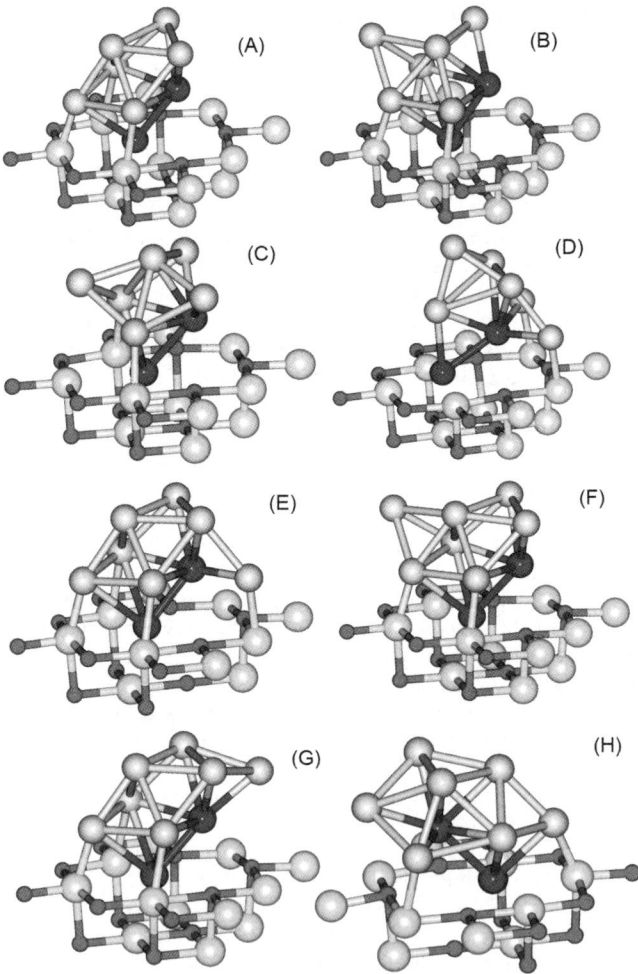

Fig. 2 Schematic drawing of the structures of M_2Ag_6 (A)–(D) and M_2Ag_7 (E)–(H) considered in the present work (M = Au, Cu). In the metal clusters, lighter metal atoms represent Ag, darker metal atoms represent M; in the oxide support, lighter atoms represent O and darker atoms Mg.

represent the lowest-energy isomers also for these bimetallic clusters, and confirmed this *via* selected DF-BH runs, which did not produce any novel low-energy configurations. In the Pd–Ag case, instead, we found that a significant structural rearrangement occurs upon alloying and in this case we performed more accurate DF-BH runs, as described below.

The DFT local calculations were carried out by employing a cluster approach for the description of the MgO support. We chose an $(Mg_{12}O_{12})$ cluster of C_{2v} symmetry embedded in an array of $+2.0$ a.u. point charges and repulsive pseudopotentials on the positive point charges in direct contact with the cluster, see ref. 37 for more details. The atoms of the oxide cluster and the point charges were located at the lattice positions of the MgO rock-salt bulk structure at the experimental lattice constant of 4.208 Å. To account for the lattice relaxation around the vacancy, the coordinates of the three oxygens and of the three magnesium ions at the border of the cavity were relaxed as well as the coordinates of the atoms of the metal cluster, leaving the other atoms frozen.

The DF-BH algorithm is articulated in the following steps: (i) an initial random configuration is chosen, a local geometry optimization is performed, and the final energy (the fitness parameter) is registered as E_1; (ii) starting from the relaxed configuration, the atoms of the metal cluster are randomly displaced, a new local geometry optimization is performed, and the final energy is registered as E_2; (iii) a random number *rndm* between 0 and 1 is generated and the movement of step (ii) is accepted only if $\exp[-(E_2 - E_1)/k_B T] > rndm$ (Metropolis criterion); (iv) steps (ii) and (iii)—the Monte Carlo steps—are repeated a given number of times. Depending on the $k_B T$ parameter (which in this work is set to 1.0 eV), some high-energy configurations are accepted and the search is able to explore different structural motifs of the metal cluster. In the DF-BH approach, the determination of energies and forces is achieved by using a first-principles DF method: this makes the method computationally very demanding.

All the calculations were carried out with the DF module of the NWChem package[39] using the PW91 xc-functional[40] in the spin-unrestricted formalism. The geometry optimizations were performed by using Gaussian-type-orbital basis sets of double-ζ quality,[41] whereas the final energies used to evaluate the binding energies of the metal clusters were calculated *via* a single-point calculation on the relaxed geometry using a triple-ζ plus polarization basis set.[41] Effective-core-potentials were used for all the transition metals involved (Ag, Cu, Au and Pd).[42] Charge density fitting Gaussian-type-orbital basis sets were used to compute the Coulomb potential.[43] The calculations used a Gaussian-smearing technique[44] (with a smearing parameter of 0.005 a.u.) for the fractional occupation of the one-electron energy levels.

Results and discussion

The results of the calculations are shown in Table 1 where the following quantities are reported: (i) the total binding energy (in eV) of the metal cluster, calculated by subtracting the total energy of the system (cluster + surface) from the sum of the energy of the relaxed isolated defected surface and of the isolated atoms that constitute the metal cluster; (ii) the spin of the system; (iii) the HOMO–LUMO gap at the Fermi level: when applying a smearing technique, if the gap is of the same order of magnitude as the smearing parameter, levels above the Fermi energy are also fractionally occupied, but in Table 1 we do not consider this "fake" occupation and report the energy difference between the nominal HOMO and LUMO orbitals; (iv) the incremental formation energy due to the addition of a silver atom: this quantity is calculated by subtracting the total binding energy of the global minimum at nuclearity N from the total binding energy of the global minimum at nuclearity $(N - 1)$: $\Delta E(N) = E(N - 1) - E(N)$, and is a useful indicator of the cluster magicity: the higher $\Delta E(N)$ and the lower $\Delta E(N + 1)$, the more stable is the N-atom cluster.

Table 1 Results for the bimetallic clusters considered in the present work. The total binding energy E_{bnd}, the electronic spin, the HOMO—LUMO gap and the incremental formation $\Delta E(N)$ energy are reported for the various binary alloys in the configurations (A–L) shown in Fig. 1 and 2

	Struct.	E_{bnd}/eV	Spin	Gap/eV	$\Delta E(N) = E(N-1) - E(N)$/eV
Au$_2$Ag$_6$	A	19.22	0	1.61	—
	B	18.75	0	1.11	—
	C	18.52	0	1.29	—
	D	18.57	0	0.54	—
Au$_2$Ag$_7$	E	20.85	1/2	0.63/1.05	1.63
	F	20.58	1/2	0.61/1.05	—
	G	20.80	1/2	0.88/0.98	—
	H	20.79	1/2	0.92/0.75	—
Au$_1$Ag$_7$	A	18.57	0	1.70	—
	B	18.22	0	1.18	—
	C	17.86	0	1.43	—
	D	17.69	0	0.55	—
Au$_1$Ag$_8$	E	20.28	1/2	0.63/1.19	1.71
	F	19.96	1/2	0.57/1.20	—
	G	20.15	1/2	0.91/1.04	—
	H	20.18	1/2	1.00/0.74	—
Cu$_2$Ag$_6$	A	20.25	0	1.31	—
	B	19.66	0	0.86	—
	C	19.99	0	0.78	—
	D	19.61	0	0.45	—
Cu$_2$Ag$_7$	E	21.85	1/2	0.61/0.73	—
	F	21.75	1/2	0.80/0.47	—
	G	21.91	1/2	1.00/0.47	—
	H	21.71	1/2	0.79/0.38	1.66
Cu$_1$Ag$_7$	A	19.64	0	1.23	—
	B	19.18	0	0.90	—
	C	19.23	0	0.84	—
	D	19.23	0	0.71	—
Cu$_1$Ag$_8$	E	21.24	1/2	0.66/0.66	1.60
	F	21.06	1/2	0.63/0.57	—
	G	21.23	1/2	0.97/0.42	—
	H	21.07	1/2	0.88/0.32	—
Pd$_2$Ag$_6$	A → I	21.39	0	0.76	—
	B → B'	21.02	0	1.02	—
	C	21.07	0	0.89	—
	D	21.17	0	1.04	—
Pd$_2$Ag$_7$	E	23.13	1/2	0.52/0.70	1.74
	F	22.76	1/2	0.42/0.76	—
	G	22.92	1/2	0.29/0.56	—
	H	22.77	1/2	0.20/0.54	—
	J	23.07	1/2	0.32/1.01	—
Pd$_1$Ag$_7$	A → I	19.06	1/2	0.45/0.88	—
	B → B'	18.92	1/2	0.47/1.26	—
	C	19.09	1/2	0.71/1.17	—
	D	19.24	1/2	0.57/1.53	—
	K	19.49	1/2	0.77/0.95	—
Pd$_1$Ag$_8$	E	21.54	0	0.71	2.05
	F	21.29	0	0.52	—
	G	21.19	0	0.12	—
	H	21.21	1	0.26/0.22	—
	J	20.69	1	0.32/0.32	—
Pd$_1$Ag$_9$	L	23.34	1/2	1.09/0.70	1.80
	L$_1$	23.09	1/2	0.57/0.09	—
	L$_2$	23.10	1/2	0.69/0.42	—

Au–Ag system

Among the three binary systems considered here, the Au–Ag bimetallic system is the simplest to discuss because of the similarity with the case of pure silver clusters. This is due to the fact that silver and gold atoms have the same electronic configuration ($d^{10}s^1$) and also similar dimensions (their bulk lattice constants are almost the same). The difference between the two metals is in the interaction with the defected surface: 2.34 eV for a single Au atom and 1.95 eV for a single Ag atom.[8,31] In the case of the dimer adsorption, the difference in the binding energy between the two metals increases due to the larger binding energy of the gold dimer: 5.21 eV for Au_2 *versus* 4.23 eV for Ag_2.[8,31] This implies that one (in the case of Au_1Ag_N systems) or two (in the case of Au_2Ag_N systems) gold atoms of the bimetallic cluster will be directly adsorbed on the DV, whereas the silver atoms will form a shell covering these Au atoms. This structural ordering is also favored by the lower surface energy of silver with respect to gold. We have confirmed that this arrangement corresponds to the lower energy ordering by performing calculations on structures where Ag and Au atoms were swapped, always finding a decrease in binding energy. Alloying with gold does not change the energetic ordering of the considered configurations with respect to pure silver clusters, and also the HOMO–LUMO gaps are very similar (especially in the case of the Au_1Ag_N system). Test DF-BH calculations (around 10–15 Monte Carlo steps with $k_BT = 1$ eV for each nuclearity and composition) confirmed that these are the lowest-energy structures of Au–Ag. We can thus conclude that we have singled out two magic bimetallic clusters adsorbed on the defected surface: Au_2Ag_6 and Au_1Ag_7. From Table 1, we see that the ground-states are separated from higher-energy configurations by at least 0.5 eV. The two HOMO–LUMO gaps are of the order of 1.6–1.7 eV and the incremental formation energies of Au_1Ag_8 and Au_2Ag_7 are of the same order of magnitude, and much smaller than that of Au_1Ag_7 and Au_2Ag_6 (which we estimate to be around 2.4–2.5 eV). Alloying with gold thus does not introduce major changes in the energetics of the system with respect to the pure Ag case, except for the fact that we now have core-shell binary clusters.

Cu–Ag system

Also in the case of the Cu–Ag bimetallic system, the larger interaction of the copper atom and dimer with the DV with respect to silver atoms determines the formation of a core-shell structure with a copper core, as confirmed by selected calculations in which Cu and Ag atoms were swapped. Test DF-BH calculations (around 10–15 Monte Carlo steps with $k_BT = 1$ eV for each nuclearity and composition) also confirmed that structures (A–B) are the putative lowest-energy isomers of Cu–Ag. However, with respect to the gold atom, the copper atom is characterized by a reduced size, and replacing one or two Ag atoms with Cu atoms implies a relaxation consisting in a reduction of the silver–copper distances. An analysis of Table 1 shows that the Cu–Ag case is similar to the Au–Ag case, albeit with a general decrease in the magic character of the Cu_1Ag_7 and Cu_2Ag_6 putative global minima. This is due to the fact that the shrinking of Cu–Ag distances in the mixed clusters is counteracted by the interaction with the surface, characterized by a "double frustration" analogous to that experienced by pure and bimetallic clusters interacting with a neutral F_s-center of the MgO(100) surface.[45] Double frustration means that the metal-on-oxide growth around local defects is frustrated not only horizontally by the mismatch of the lattice constant of the metal and of the oxide, but also vertically by the remarkable difference in the equilibrium heights of the atoms directly adsorbed on the vacancy and those adsorbed on the sites around the defect. This double frustration plays a destabilizing role for all the clusters considered in this work (and also for pure silver clusters) but its effect is particularly dramatic in the case of Cu–Ag clusters because of the reduced dimension of the copper atoms and the corresponding shrinking of the Cu–Ag distances. The consequences of this destabilization can be mainly appreciated in the reduction of the HOMO–LUMO gaps at the Fermi level with respect to pure silver clusters and in the fact that the ground-states of the Cu_2Ag_6 and Cu_1Ag_7 systems are separated from the higher-energy configurations by only 0.2–0.3 eV, which corresponds to an increased fluxionality of the clusters. This is at variance with the gas-phase situation, in which the core-shell Cu–Ag clusters are stabilized by a reduced dimension of the internal core.[47] The fluxional character observed in these supported clusters may be important for catalytic activity.[14]

Pd–Ag system

With respect to the previous cases, where structural relaxation did not play a fundamental role, we found that alloying with Pd significantly changed the structural motifs. We thus decided to perform a global minimum search by performing DF-BH calculations for this system. In particular, 3 DF-BH runs, each one composed of 10–15 Monte Carlo steps with $k_BT = 1$ eV were performed for each nuclearity and composition. Starting with the case of Pd_2Ag_6, we observe that the lowest energy structure is based on a Pd atom filling the empty Mg site of the vacancy and a second Pd atom lying outside the vacancy above the empty O site of the vacancy (as usual), but now with the 6 Ag atoms forming a pentagonal bipyramid on top of the latter atom—see structure (I) of Fig. 3. Structures (A–B), when alloyed with Pd_2, undergo a remarkable rearrangement towards the pentagonal bipyramid motif. In particular structure (A) rearranges to (I), while (B) rearranges to structure (B′). Structure (I), which completes the bipyramid, results in the global minimum; a DF-BH search was unable to find any structure with a lower energy than this one. Structure (D), the incomplete tetrahedron, is the most competitive motif not based on the pentagonal bipyramid, being only 0.22 eV above the ground-state. The reason for this behavior is of electronic origin and can be understood by analyzing the nine-atom clusters: as shown in the structure (M) of Fig. 3 (where the electron spin density for Pd_2Ag_7 is plotted), the electron spin is localized on a d electron of the Pd atom inside the DV cage. This Pd atom is thus promoted to a $4d^9 5s^1$ configuration and the 5s electron takes part in the metallic bond, whereas the Pd atom outside the defect keeps its $4d^{10}$ gas-phase configuration. This electronic situation is forced by the requirements of the metallic bond—which tends to complete an 8-electron shell—but is disfavored by the metal–surface interaction which would tend to quench the spin of the Pd atoms interacting with the surface.[48] It is interesting to note that, in a way, this situation now resembles that found for the single oxygen vacancy: after one of the Pd atoms has filled the Mg site of the DV, the oxygen site is left available for the interaction with metal clusters, in a similar way to what happens for the single oxygen vacancy.[45] The major difference with respect to the single oxygen vacancy case is that the electron count is different here, because the Pd atom inside the defect contributes with only one electron, whereas in the neutral F_s case the defect contributes with two electrons. This implies a mismatch between structural and electronic shell closure which, together with the destabilization of high-spin configuration for the atoms interacting with the surface, hinders the formation of binary Pd–Ag magic clusters adsorbed on the DV, at variance with the F_s-defected surface.[46] The requirements of the metallic bond are thus in conflict with the optimal metal–surface interaction. For the Pd_2Ag_6 cluster, which is a spin-restricted system, both Pd atoms are now approximately in a $4d^{10}$ electronic configuration, which however implies an electronic count of 6 (a non-magic number) and a weakening of the metallic bond. In passing, we observe that predictions about the spin properties of these clusters could be experimentally verified by EPR measurements.[22]

From a structural point of view, in the case of Pd_2Ag_7 clusters we observe a close competition between structures (E) and (J). All the clusters at this composition are open shells with one unpaired electron, localized, as already discussed and shown in structure (M) of Fig. 3, on a d electron of the lowest-lying Pd atom. The incremental

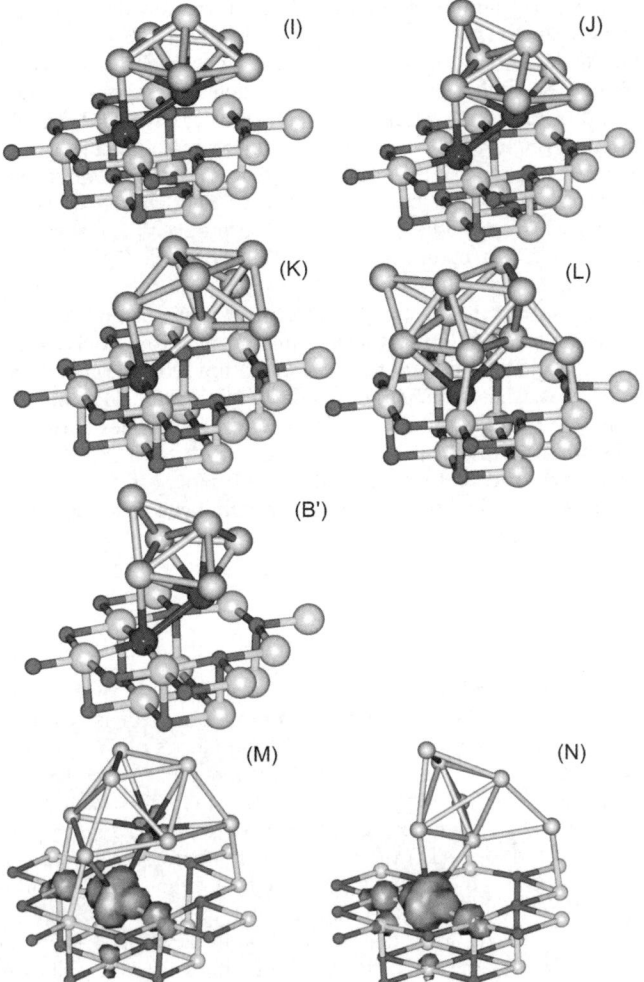

Fig. 3 Schematic drawing of Pd_2Ag_6 (I), Pd_2Ag_7 (J), Pd_1Ag_7 (K) and Pd_1Ag_8 (L) structures. (M) and (N) show the electron spin density plots in the case of structure (E) of Pd_2Ag_7 and structure (D) of Pd_1Ag_7. In the metal clusters, lighter metal atoms represent Ag, darker metal atoms represent M; in the oxide support, lighter atoms represent O and darker atoms Mg.

formation energy of the Pd_2Ag_7 system is higher than that of the other (Au–Ag, Cu–Ag) bimetallic systems. Furthermore, the values of the HOMO–LUMO gap for Pd_2Ag_6 are smaller than in the case of the pure silver clusters and of the Cu–Ag and Au–Ag systems. The Pd_2Ag_6 cluster thus is not magic.

The analysis drawn in the case of the Pd_2Ag_N system holds also in the case of the Pd_1Ag_N system. As it can be seen from structure (N) of Fig. 3, in fact, also for Pd_1Ag_7 the Pd atom inside the vacancy is characterized by an unpaired d electron. For this system, a DF-BH search located a structure lower in energy with respect to structures (E)–(H): in the case of a pure silver coverage of the Pd atom in the cage, in fact, the pentagonal bipyramid becomes unstable and the structure evolves towards a different shape. The new putative global minimum is represented by structure (K). The conduction electron counts at this nuclearity amounts to 8 (7 s electrons coming from the Ag atoms plus one coming from the Pd atom), where the unpaired d electron on the Pd atom determines an half-integer spin value, but the destabilization associated with the unfavorable high-spin metal–surface interaction does not allow a shell-closure effect. By adding a further silver atom, for the Pd_1Ag_8 system the lowest-energy structure results in the (E) one and further DF-BH searches were not able to locate a lower energy minimum. The large value of the incremental formation energy of Pd_1Ag_8 shows that the Pd_1Ag_7 system does not exhibit a magic character. This fact is confirmed by the much lower values of the HOMO–LUMO gap with respect to the pure silver clusters and the Cu–Ag and Au–Ag systems. In order to check whether the Pd_1Ag_8 system itself was a magic structure, we also investigated the Pd_1Ag_9 system, through a biased search conducted by locally optimizing the lowest-energy structures found in the case of the pure Ag_{10} cluster adsorbed on a DV. The lowest-energy structure is the (L) one shown in Fig. 3: this structure is obtained by taking structure (E) of Fig. 2 and adding a further silver atom grown tetrahedrally on a side of the cluster. The electron count for this structure is 10, but, also in this case, the presence of the unpaired d electron on Pd determines the formation of an open-shell configuration and an unfavorable metal–surface interaction. Despite this, the incremental formation energy of Pd_1Ag_9 is not small (1.80 eV), which, together with its low $\Delta E(N)$ (2.05 eV), implies that neither Pd_1Ag_8 is a magic cluster.

To conclude, we observe that the reduced magicity of the Pd–Ag (and also Cu–Ag) DV-supported clusters translate in an enhanced fluxional character. This can be important for catalytic activity.[14]

Conclusions

The structures of binary M_2Ag_6, M_2Ag_7, M_1Ag_7 and M_1Ag_8 clusters (M = Cu, Au, Pd) adsorbed on an MgO(100) terrace presenting a double vacancy (DV) neutral defect have been investigated through a combination of density-functional (DF) biased searches and global optimizations. Alloying allows one to probe the adsorption characteristics of the DV-defected surface. It is found that in the case of Cu–Ag and Au–Ag clusters, alloying does not introduce significant structural rearrangements: the corresponding optimized configurations are only slightly modified by a minor structural relaxation. Also the energy ordering among these structures does not change with respect to the pure Ag case. We found that Au_1Ag_7 and Au_2Ag_6 are core-shell magic clusters. The magic character is reduced for Cu_1Ag_7 and Cu_2Ag_6, because the shortening of Cu–Ag distances is counteracted by a double-frustration effect due to the DV. For the Pd–Ag system, alloying with Pd significantly changes the structural properties of the system, creating a pronounced basin of attraction towards the pentagonal bipyramid motif. Because of the peculiar electronic behavior of the Pd atom(s), the Pd–Ag systems present a mismatch between structural and electronic shell closure which, together with the conflict between the requirements of the metallic bond and the metal–surface interaction, hinders the formation of binary Pd–Ag magic clusters adsorbed on the DV. As a consequence, nuclearities presenting a closed shell configuration are not particularly favored with respect to nuclearities with an half-integer value of the spin, and, in principle, this behavior could be verified through EPR measurements. Finally, we found an enhanced fluxionality in the Cu–Ag and Pd–Ag systems with respect to pure Ag and mixed Au–Ag clusters: this could have an influence on the catalytic properties of these systems.

Acknowledgements

We acknowledge financial support from the Italian CNR for the program "(Supra-) Self-Assemblies of Transition Metal Nanoclusters" within the framework of the ESF EUROCORES SONS, and from the European Community Sixth Framework Project for the STREP Project "Growth and Supra-Organization of Transition and Noble Metal Nanoclusters" (contract no. NMP-CT-2004-001594). Calculations were performed at Cineca within an agreement with Italian CNR-INFM.

References

1   H. J. Freund, *Surf. Sci.*, 2002, **500**, 271.
2   C. R. Henry, *Surf. Sci. Rep.*, 1998, **31**, 235.
3   G. J. Hutchings and M. Haruta, *Appl. Catal., A*, 2005, **291**, 2.
4   C. T. Campbell, *Surf. Sci. Rep.*, 1997, **27**, 1.
5   I. Y. Yudanov, G. Pacchioni, K. Neyman and N. Rösch, *J. Phys. Chem. B*, 1997, **101**, 2786.
6   A. V. Matveev, K. Neyman K, I. Yudanov and N. Rösch, *Surf. Sci.*, 1999, **426**, 123.
7   N. Lopez, F. Illas, N. Rösch and G. Pacchioni, *J. Chem. Phys.*, 1999, **110**, 4873.
8   G. Barcaro and A. Fortunelli, *J. Chem. Theor. Comput.*, 2005, **1**, 972.
9   M. Sterrer, T. Risse, U. Martinez Pozzoni, L. Giordano, M. Heyde, H. P. Rust, G. Pacchioni and H. J. Freund, *Phys. Rev. Lett.*, 2007, **98**, 096107.
10 L. Xu, G. Henkelman, C. T. Campbell and H. Jonsson, *Phys. Rev. Lett.*, 2005, **95**, 246103.
11 G. Barcaro, A. Fortunelli, F. Nita and R. Ferrando, *Phys. Rev. Lett.*, 2005, **95**, 246103.
12 G. Barcaro and A. Fortunelli, *New J. Phys.*, 2007, **9**, 22.
13 G. Haas, A. Menck, H. Brune, J. V. Barth, J. A. Venables and K. Kern, *Phys. Rev. B*, 2000, **61**, 11105.
14 H. Häkkinen, S. Abbet, A. Sanchez, U. Heiz and U. Landman, *Angew. Chem., Int. Ed.*, 2003, **42**, 1297.
15 A. Sanchez, S. Abbet, U. Heiz, W. D. Schneider, H. Häkkinen, R. N. Barnett and U. Landman, *J. Phys. Chem. A*, 1999, **103**, 9573.
16 L. Giordano and G. Pacchioni, *Surf. Sci.*, 2005, **575**, 197.
17 L. M. Molina and B. Hammer, *J. Chem. Phys.*, 2005, **123**, 161104.
18 H. Moseler, H. Häkkinen and U. Landman, *Phys. Rev. Lett.*, 2002, **89**, 176103.
19 A. Bogicevic and D. R. Jennison, *Surf. Sci.*, 2002, **515**, L481.
20 J. Carrasco, N. Lopez, F. Illas and H. J. Freund, *J. Chem. Phys.*, 2006, **125**, 074711.
21 M. Chiesa, M. C. Paganini, E. Giamello, D. M. Murphy, C. Di Valentin and G. Pacchioni, *Acc. Chem. Res.*, 2006, **39**, 861.
22 M. Sterrer, E. Fischbach, T. Risse and H. J. Freund, *Phys. Rev. Lett.*, 2005, **94**, 186101.
23 J. H. Lunsford and J. P. Jayne, *J. Phys. Chem.*, 1966, **70**, 3464.
24 L. Ojamaë and C. Pisani, *J. Chem. Phys.*, 1998, **109**, 10984.
25 A. Bogicevic and D. R. Jennison, *Surf. Sci.*, 1999, **437**, L741.
26 D. Ricci, G. Pacchioni, P. V. Sushko and A. L. Shluger, *J. Chem. Phys.*, 2002, **117**, 2844.
27 C. Barth and C. R. Henry, *Phys. Rev. Lett.*, 2003, **91**, 196102.
28 G. Pacchioni, *ChemPhysChem*, 2003, **4**, 1041.
29 L. Giordano, C. Di Valentin, J. Goniakowski and G. Pacchioni, *Phys. Rev. Lett.*, 2004, **92**, 096105.
30 B. Ealet, J. Goniakowski and F. Finocchi, *Phys. Rev. B*, 2004, **69**, 195413.
31 A. Del Vitto, G. Pacchioni, F. Delbecq and P. Sautet, *J. Phys. Chem. B*, 2005, **109**, 8040.
32 Y. L. Wang and M. Y. Lai, *J. Phys.: Condens. Matter*, 2001, **13**, R589.
33 M. Brack, *Rev. Mod. Phys.*, 1993, **65**, 677.
34 G. Barcaro and A. Fortunelli, *Phys. Rev. B*, submitted.
35 K. Judai, S. Abbet, A. S. Wörz, U. Heiz and C. R. Henry, *J. Am. Chem. Soc.*, 2002, **126**, 2732.
36 E. Aprà, R. Ferrando and A. Fortunelli, *Phys. Rev. B*, 2006, **73**, 205414.
37 G. Barcaro, E. Aprà and A. Fortunelli, *Chem.–Eur. J.*, 2007, **13**, 6408.
38 E. M. Fernàndez, J. M. Soler, I. L. Garzón and L. C. Balbàs, *Phys. Rev. B*, 2004, **70**, 165403.
39 R. A. Kendall, E. Aprà, D. E. Bernholdt, E. J. Bylaska, M. Dupuis, G. I. Fann, R. J. Harrison, J. Ju, J. A. Nichols, J. Nieplocha, T. P. Straatsma, T. L. Windus and A. T. Wong, *Comput. Phys. Commun.*, 2000, **128**, 260.
40 J. P. Perdew, J. A. Chevary, S. H. Vosko, K. A. Jackson, M. R. Pederson, D. J. Singh and C. Fiolhais, *Phys. Rev. B*, 1992, **46**, 6671.
41 A. Schaefer, C. Huber and R. Ahlrichs, *J. Chem. Phys.*, 1994, **100**, 5289.
42 D. Andrae, U. Haeussermann, M. Dolg, H. Stoll and H. Preuss, *Theor. Chim. Acta*, 1990, **77**, 123.
43 F. Weigend, M. Haser, H. Patzel and R. Ahlrichs, *Chem. Phys. Lett.*, 1998, **294**, 143.
44 C. Elsässer, M. Fhänle, C. T. Chan and K. M. Ho, *Phys. Rev. B*, 1994, **49**, 13975.
45 G. Barcaro and A. Fortunelli, *J. Phys. Chem. B*, 2006, **110**, 21021.
46 G. Barcaro and A. Fortunelli, *J. Phys. Chem. C*, 2007, **111**, 11384.
47 G. Rossi, A. Rapallo, C. Mottet, A. Fortunelli, F. Baletto and R. Ferrando, *Phys. Rev. Lett.*, 2004, **93**, 105503.
48 A. Markovits, J. C. Paniagua, N. Lopez, C. Minot and F. Illas, *Phys. Rev. B*, 2003, **67**, 115417.

Global optimisation and growth simulation of AuCu clusters

T. J. Toai, G. Rossi and R. Ferrando*

Received 23rd May 2007, Accepted 21st June 2007
First published as an Advance Article on the web 14th September 2007
DOI: 10.1039/b707813g

Global optimisation techniques and Molecular Dynamics simulations are employed to investigate the structure and the chemical order of AuCu clusters, with composition $Au_{0.75}Cu_{0.25}$, $Au_{0.5}Cu_{0.5}$ and $Au_{0.25}Cu_{0.75}$. Global optimisations by Parallel Excitable Walkers algorithm have located the global minimum configurations of clusters at size $N = 100$, 160 and 200 atoms. Stable clusters do not exhibit any alloy-like ordering, and combine the tendency to surface segregation of the gold atoms with the icosahedral structural motif. As unique exception to the icosahedral trend, an almost perfect decahedron is located at size $N = 100$, exhibiting copper atoms in its outer shell. As regards to the dynamics, the paper deals with the growth of AuCu clusters, both depositing gold and copper atoms onto an heterogeneous seed, and depositing copper atoms onto an homogeneous gold seed. In agreement with the global optimisation results, the former growth process leads to the formation of Ih clusters, whose surface is enriched in gold atoms. The latter deposition process turns into the formation of decahedral clusters with a reversed $Cu_{shell}Au_{core}$ chemical ordering.

1. Introduction

The AuCu system has three bulk ordered alloys, $Au_{0.5}Cu_{0.5}$ (fcc, $L1_0$) $Au_{0.25}Cu_{0.75}$ and $Au_{0.75}Cu_{0.25}$ (fcc, $L1_2$). The alloy stoichiometries are quite well experimentally reproduced in clusters, both when the particles are synthesized by simultaneous reduction methods[1,2] and when they are produced by laser vaporization sources like that used by Pauwels et al.[3] In these experiments, optical, electron diffraction and HREM observations have been used to obtain information about the structural characteristics of the AuCu clusters dispersed in the colloidal solution, deposited on a carbon-coated grid and then dried[1] or deposited on MgO and amorphous carbon.[3] Pal and coworkers suggest a $Au_{shell}Cu_{core}$ arrangement, but focus essentially on the influence of composition on the structural distribution of the particles in the size-range 2–10 nm. On the other hand, the qualitative picture coming from the measurements by Pauwels is that the interaction with MgO forces the system to adopt a solid solution with fcc ordering, while several alternative packings are possible on the other, less interactive substrate. These findings confirm that AuCu is indeed quite a ductile system. This characteristic, together with the interest raised by gold nanoparticles in a variety of research fields, make AuCu an interesting object of investigation. In this paper we address the comparison between the AuCu structural configurations coming from global optimisation and growth simulations. The size

Dipartimento di Fisica, Università di Genova, Via Dodecaneso 33, 16136 Genova, Italy.
E-mail: ferrando@fisica.unige.it

range considered is $30 < N < 300$. As entropic and kinetic effects can influence the growth pattern of clusters,[4,5] it is expected that the structural distributions coming from growth could not coincide with those coming from global optimisations. In particular, we will focus on the fitness of core-shell chemical ordering as varying size and composition, and on the competition between decahedral and icosahedral structural motifs.

2. Model, method and simulation plan

2.1 Potential function

The potential used to model the atomic interactions was proposed by Cleri and Rosato.[6] It is a semi-empirical potential derived within the tight-binding second-moment approximation. Cluster potential energy E is given by the sum over all the atoms of their bonding and repulsive energy:

$$E = \sum_j (E_j^b + E_j^r) \tag{1}$$

Where the bonding term E_j^b is expressed as:

$$E_j^b = -\sqrt{\sum_i \xi_{sw}^2 e^{\left[-2q_{sw}\left(\frac{r_{ji}}{r_{sw}^0}-1\right)\right]}} \tag{2}$$

The repulsive Born–Mayer term E_j^r is:

$$E_j^r = \sum_i A_{sw} e^{\left[-p_{sw}\left(\frac{r_{ji}}{r_{sw}^0}-1\right)\right]} \tag{3}$$

r_{ij} is the distance between the atoms at sites $i.e.$ j; s = A, B is the chemical species of the atom j, while w = A, B is the species of the atom i. r_{sw}^0 is the nearest-neighbours distance. If $s = w$, it coincides with the pure metal nn distance. If the interaction is heteroatomic, $i.e.$ $s \neq w$, r_{sw} is expressed as:

$$r_{AB}^0 = r_{BA}^0 = \frac{r_{AA}^0 + r_{BB}^0}{2} \tag{4}$$

The parameters A_{sw}, p_{sw}, q_{sw}, ξ_{sw} are fitted to several bulk properties of the $Au_{0.25}Cu_{0.75}$ cubic alloy. The potential is able to predict with accuracy the occurrence of the structural order–disorder transition in the alloy. A cut-off analytical function is used from the Au fifth-neighbours distance on.

2.2 Global optimisation method

Small AuCu clusters ($N < 30$) have been studied by means of a genetic[7,8] global optimization approach by Lordeiro et al.[9] This size limit was extended up to 45 atoms by Rapallo et al.[10] Here a Parallel Excitable Walkers (PEW)[11] algorithm is used as an efficient tool to deal with the global optimisation of heterogeneous clusters of larger sizes. The algorithm performs parallel Basin Hopping searches.[12,13] Each walker explores a different region of a geometric order parameter space, and the information about the low-energy clusters belonging to such different geometrical basins is retained during the optimisation. In the code-setup used to optimise the AuCu clusters, minima have been classified according to the percentage of local fivefold symmetries in the cluster. Clusters have been optimised at size $N = 100$, $N = 160$ and $N = 200$, considering all the three alloy stoichiometries.

Our optimisation procedure consists of two parts. First, for each size and composition we run 8 PEW simulations of 5×10^5 steps. In these simulations, the *shake* move is employed. In the shake move, each atom of the cluster is displaced

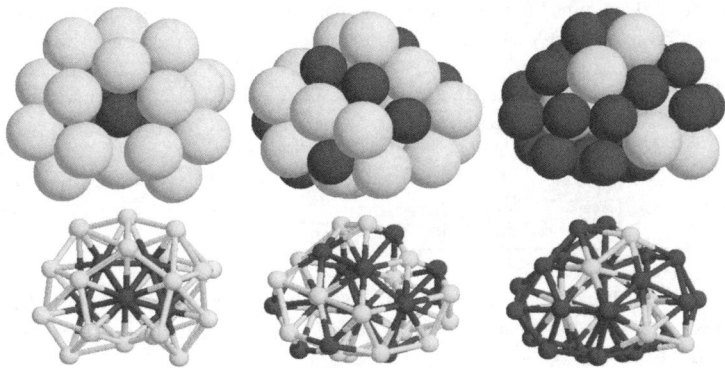

Fig. 1 The 34-atom seeds of the growth simulations, which correspond to global minima located by a basin hopping procedure for the compositions $Au_{26}Cu_8$, $Au_{17}Cu_{17}$ and Au_8Cu_{26}. Dark and light atoms represent, respectively copper and gold atoms.

randomly within a sphere of radius 1.3 Å, centered around its present position. We have verified that this kind of move is the most efficient in exploring different geometric structures. We classify the results of these simulations according to the percentage of (5,5,5) Common Neighbours Analysis signatures. In this way, we are able to build up a database of structures belonging to different geometrical motifs. For the best structures of each motif, we run Basin Hopping simulations in which only the *exchange* moves are employed. In such a kind of move, two atoms of different species are swapped. These exchange runs are made at low temperature in order to optimize chemical ordering in the clusters.

2.3 Growth simulation method

Growth simulations are performed by Molecular Dynamics simulations. Newton equations are implemented by a Velocity Verlet algorithm with a time step $\tau = 7 \times 10^{-15}$ s. Temperature is controlled by means of an Andersen Thermostat, and atoms are deposited one by one with a delay of $\tau_{dep} = 7 \times 10^{-9}$ s.

Two sets of growth simulations have been performed. The simulations of the first set reproduce the growth of clusters with the three alloy stoichiometries. Simulations start from a 34-atom seed whose composition is Au_8Cu_{26}, $Au_{17}Cu_{17}$, $Au_{26}Cu_8$, respectively (see Fig. 1). When an atom has to be deposited, a random number is extracted in order to give the atom its chemical species, according to the desired composition. Growth simulations take place at two temperatures, $T = 400$ K and $T = 500$ K. For each composition and temperature, three independent simulations are performed. Simulations stop at size $N = 200$ atoms. The second set of simulations starts from an homogeneous icosahedral gold core of 147 atoms. Copper atoms are then deposited onto the seed, up to size $N = 310$ atoms. This allows us to make some comparisons between the structure and the chemical order of clusters with $Au_{0.5}Cu_{0.5}$ composition which are grown by two different deposition processes.

3. Results

3.1 Global optimisation results

3.1.1 A Homogeneous clusters. The global optimisation of homogeneous Au and Cu clusters at size $N = 100$, $N = 160$ and $N = 200$ has been performed first. The results are summarized in Table 1. At the smallest size, gold and copper have quite a different behaviour, gold preferring to adopt an fcc ordering and copper favouring the icosahedral configurations. It is worth noting that homogeneous Cu minima based on the icosahedral motif can have both a Mackay (Ih-M) and an anti-Mackay

Table 1 Global minima for homogeneous Au and Cu clusters, and heterogeneous $Au_{0.25}Cu_{0.75}$, $Au_{0.5}Cu_{0.5}$, $Au_{0.75}Cu_{0.25}$

Composition	$N = 100$	$N = 160$	$N = 200$
Au	Fcc	Fcc	Dh
Cu	Ih-M	Ih-aM	Dh
$Au_{0.25}Cu_{0.75}$	Dh	Ih-aM	Ih-M
$Au_{0.5}Cu_{0.5}$	Ih-M	Ih-aM	Ih-M
$Au_{0.75}Cu_{0.25}$	double Ih	Ih-aM	Ih-M

(Ih-aM) external layer. At size $N = 200$, the decahedral motif is favoured by both the metals, even if in the case of copper there is a strict competition between the Dh and the Ih-M motif.

3.1.1 B $Au_{0.75}Cu_{0.25}$. As a general characteristic of the mixed composition clusters, one can notice that the icosahedral motif is preferred to the decahedral one. Darby *et al.*[14] demonstrated that the introduction of a single copper impurity into a homogeneous Au_{55} cluster can shift its global minimum structure from the amorphous to the icosahedral basin. According to the global optimisations, at size $N = 100$, $N = 160$ and $N = 200$ the gold-rich clusters adopt different Ih-based structures. The one-hundred atom cluster is a double icosahedron, containing two interpenetrating Ih_{55} and a small anti-Mackay patch (see Fig. 2). Both the Ih_{55} have a copper atom in their central site, then there is a mixed gold and copper layer constituting the surface of both the Ih_{13}, and finally the surface is composed by gold atoms only. This can thus be considered a three-shell configuration. At size $N = 160$, the global minimum cluster is a Ih_{147} partially covered by an anti-Mackay shell. The cluster has again a gold surface. Finally, at size $N = 200$ the global optimisation procedure has located a Mackay icosahedron (Ih_{147} plus a Mackay patch). It is worth noting that this structure is in close competition with the icosahedron covered by the anti-Mackay patch, differing by less than 0.12 eV. The $L1_2$ alloy ordering in the bulk is thus not reproduced at these small sizes. This finding is somehow a validation of the model, since Yasuda and Mori[15] found by TEM observations that solid solution becomes more stable than the ordered $L1_2$ low-temperature phase even at room temperature when the cluster diameter is reduced below approximately 5 nm. The location of stable $Au_{shell}Cu_{core}$ particles is coherent also with other previous findings of Ascencio *et al.*[2]

3.1.1 C $Au_{0.5}Cu_{0.5}$. All the three sizes analysed present an icosahedral global minimum. The perfect Ih_{55} and Ih_{147} are covered by an external Mackay shell at size $N = 100$ and $N = 200$, while at the intermediate size $N = 160$ the anti-Mackay

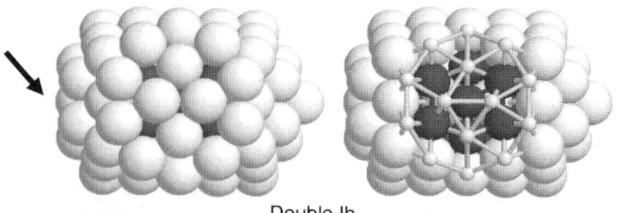

Double Ih_{55}

Fig. 2 The double icosahedron which is the global minimum structure for $Au_{0.75}Cu_{0.25}$ at size $N = 100$ (dark atoms represent copper, light atoms represent gold). One of the Ih_{55} misses a vertex atom, as indicated by the arrow on the left. The double Ih is partially covered by an anti-Mackay patch, indicated by the small atoms on the right.

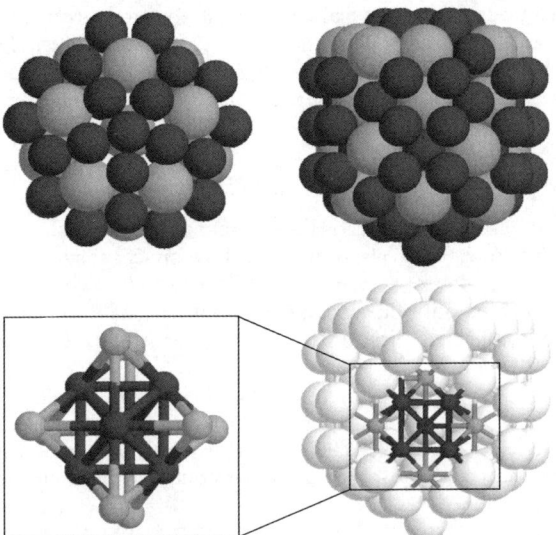

Fig. 3 The Dh $Au_{25}Cu_{75}$ global minimum cluster. In the first row, a top and side view of the structure. Dark and light atoms represent, respectively copper and gold. In the second row, the right snapshot highlights the position of a fragment of a unit cell of the $L1_2$ alloy within the decahedral arrangement.

patch is preferred to the Mackay one. Even at this intermediate composition there is a tendency to form core-shell structures.

3.1.1 D $Au_{0.25}Cu_{0.75}$. The picture is somehow different at the copper-rich composition. At size $N = 100$, the global minimum structure is indeed a decahedron, namely the perfect Marks Dh_{101} missing a vertex atom. This Dh accommodates gold and copper atoms so as to reproduce the chemical order of the $L1_2$ bulk alloy (see Fig. 3). In such a structure, the core-shell order is reversed, and the copper atoms are positioned on the external (100) facets of the cluster (the radial distribution of the cluster is depicted in Fig. 4). The exception of a Dh global minimum structure is due to the combination of a geometrical and a chemical driving force. The former is the

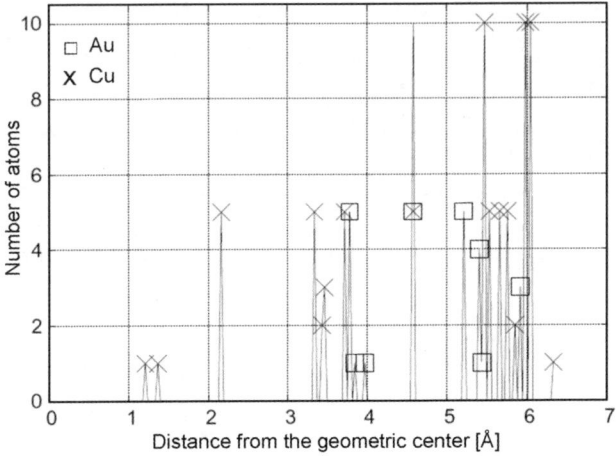

Fig. 4 The radial distribution of atoms in the global minimum of the cluster $Au_{25}Cu_{75}$. Cu atoms occupy the external (100) facets.

presence of a magic size for the decahedral motif, and the latter is the possibility to accommodate the $L1_2$ chemical order into each tetrahedron of the Dh. At size $N = 160$ the anti-Mackay patch covers the Ih_{147}, in the same way as for the other stoichiometries. The Mackay stacking is the lowest energy one at size $N = 200$.

3.2 Growth results

The results of the growth simulations are summarized in Table 2. All the clusters grown from an heterogeneous 34-atom seed follow an icosahedral pattern, so that at size $N = 200$ their structure is made of a Ih_{147} core, plus an external, incomplete icosahedral patch. This outer layer can exhibit either a Mackay or an anti-Mackay pattern (see Fig. 5). The smallest amount of copper in the cluster ($Cu_{0.25}Au_{0.75}$) is

Table 2 Structure and chemical order of the 200-atom clusters resulting from the growth simulations upon a 34-atom seed. From the left: temperature of the growth simulations, final structure of the cluster at size $N = 200$, number of Cu atoms in the cluster, percentage of Au atoms on the surface, percentages of homogeneous and heterogeneous nearest-neighbours bonds

$Au_{0.75}Cu_{0.25}$	Structure	Cu atoms	% of Au on surf.	% Au–Au	%Cu–Cu	%Au–Cu
$T = 400$ K						
1	Ih_{147} + M patch	53	92	49	10	41
2	Ih_{147} + M patch	46	95	54	9	37
3	Ih_{147} + aM patch	60	90	44	15	41
$T = 500$ K						
1	distorted Ih	49	92	52	8	40
2	Ih_{55} + M patch	56	98	45	13	42
3	distorted pIh	51	96	51	38	11

$Au_{0.25}Cu_{0.75}$	Structure	Cu atoms	% of Au on surf.	% Au–Au	%Cu–Cu	%Au–Cu
$T = 400$ K						
1	Ih_{147} + aM patch	157	21	4	59	37
2	Ih_{147} + M patch	141	32	7	50	43
3	Ih_{147} + M patch	140	32	8	49	43
$T = 500$ K						
1	Ih_{147} + M patch	145	33	6	54	40
2	Ih_{147} + M patch	144	36	6	54	40
3	Ih_{147} + M patch	151	28	5	59	36

$Au_{0.5}Cu_{0.5}$	Structure	Cu atoms	% of Au on surf.	% Au–Au	%Cu–Cu	%Au–Cu
$T = 400$ K						
1	Ih_{147} + M patch	106	59	19	30	51
2	Ih_{147} + aM patch	99	60	22	26	52
3	Ih_{147} + aM patch	94	66	24	26	50
$T = 500$ K						
1	Ih_{147} + aM patch	99	66	21	29	50
2	Ih_{147} + M patch	92	75	25	29	46
3	Ih_{147} + M patch	109	58	19	34	47

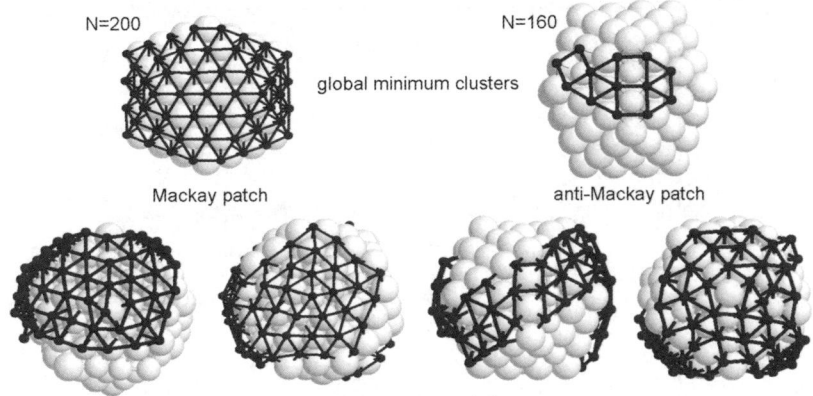

Fig. 5 Icosahedral clusters with Mackay (left) and anti-Mackay (right) patches. The atoms of the patches are dark and smaller, regardless of their species. In the top row, the Mackay and anti-Mackay global minimum clusters at size $N = 200$ and $N = 160$ for the $Au_{0.75}Cu_{0.25}$ composition. In the bottom row, on the left, two examples of clusters grown with a Mackay patch upon the Ih_{47} core. On the right, two examples of anti-Mackay patches.

thus sufficient to boost the formation of icosahedral structures during the growth process. In such icosahedral clusters, regardless of the composition, some segregation of gold to the surface of the cluster can be observed. As reported in Table 2, the percentage of Au atoms on the surface almost always exceeds the gold concentration in the whole cluster. Nevertheless, clusters are not perfectly core–shell. The Ih_{13} core of the clusters is never composed by Cu atoms only. Also in $Cu_{0.75}Au_{0.25}$ clusters, the central Ih_{13} has at least one gold atom in its 12-atom shell. The chemical order of the clusters is thus characterized by a copper-rich core, a mixed intermediate shell, and a gold-rich external shell. This shows that AuCu clusters can assume a three-shell arrangement. This arrangement is however different from the one obtained in simulations of AgCu and AgNi growth (where there was no intrashell intermixing) but it has a striking resemblance to the three-shell pattern experimentally observed in AuPd.[16]

A different scenario arises when the growth of the cluster is simulated starting from an homogeneous Au core and depositing Cu atoms upon it. It has been proved that in other heterogeneous metallic systems, the deposition of A atoms onto a seed of B atoms can determine the formation of core-shell[17] or three-shell onion-like clusters.[18] We thus analysed the growth of an Au icosahedral core of 147 atoms up to 310 atoms ($Au_{147}Cu_{163}$). The choice of an icosahedral seed, despite the fact that homogeneous gold does not exhibit an icosahedral arrangement in this size range, has two main reasons. First of all, the icosahedral motif should become the most favorable as soon as a few copper atoms approach the cluster, coherently with the results of global optimisations. Moreover, the icosahedral structure could favour the effects[19,20] of fast-alloying, encouraging the small copper atoms to incorporate to the core of the cluster and help to release some volume strain.

At the end of the 6 growth simulations, both at $T = 400$ K and $T = 500$ K, five clusters have a decahedral morphology, and one is ordered according to an fcc structure with stacking fault. In order to locate the size at which the Ih → Dh transition takes place, one can perform a local minimisation of the snapshots collected during growth, and estimate the stability of each configuration by the Δ index, defined as:

$$\Delta_N = \frac{E - 147E_{coh}^{Au} - NE_{coh}^{Cu}}{N^{2/3}} \tag{5}$$

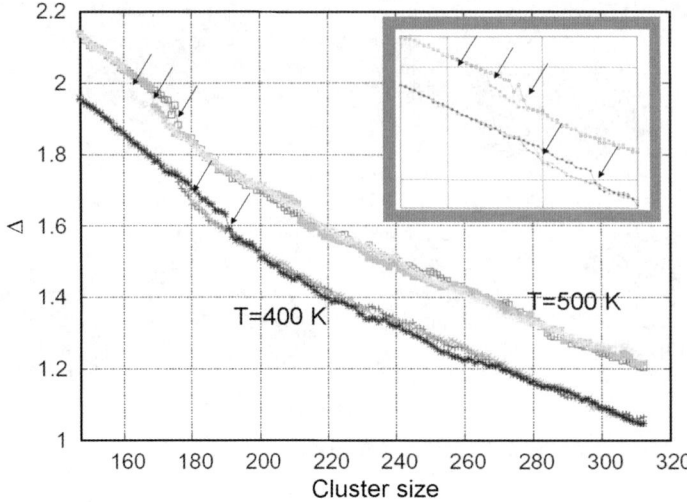

Fig. 6 Δ values referred to the snapshots collected through the simulation of the deposition of Cu atoms on a Au$_{147}$ icosahedral core.

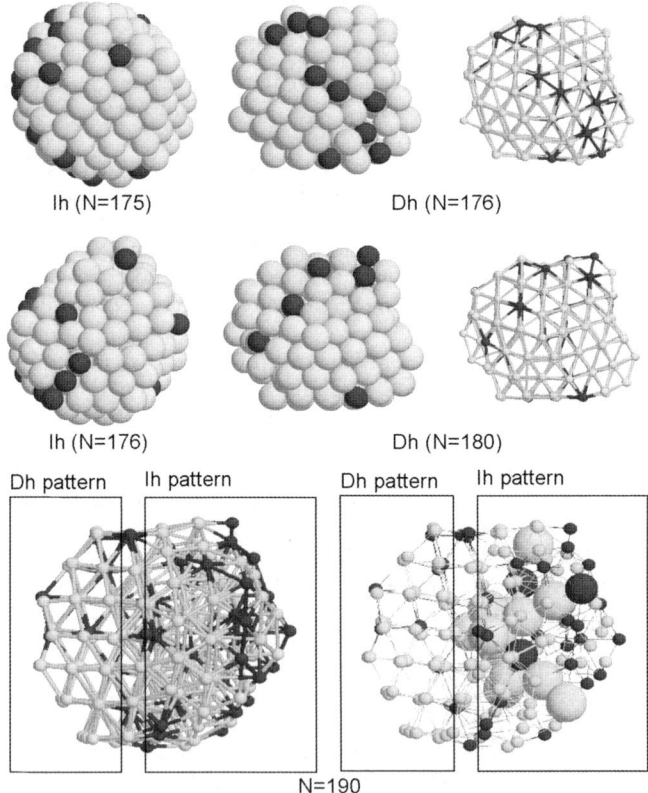

Ih (N=175)   Dh (N=176)

Ih (N=176)   Dh (N=180)

Dh pattern   Ih pattern   Dh pattern   Ih pattern

N=190

Fig. 7 Snapshots from the simulation of the growth of Cu atoms on an Au$_{147}$ icosahedral core. Simulations have been led at $T = 400$ K. In the first and second row, the transition from the icosahedral motif (on the left) to the decahedral one (right) in two independent growth simulations. On the bottom, a snapshot from a growth simulation showing a cluster which is transforming from Ih to Dh. The right snapshot shows in space fill representations the atoms belonging to the fivefold axis of the incomplete icosahedron.

Low Δ values refer to stable structures. The denominator is introduced to compensate the scaling of Δ with the size of the cluster, while the drift observed in Fig. 6 is due to the increase of the more cohesive atoms, Cu, in the cluster. Drops in the Δ curves refer to the structural transition from the icosahedral to the decahedral motif. Please note that, in this size range, the icosahedral-like skeletal structure has the lower energy, as predicted by global optimisation. The stability of the icosahedral motif is nevertheless strictly related to the chemical order of the cluster, and it is well accomplished only by the partial core–shell ordering previously described. Our simulation proves that, in the timescale of the simulated growth upon a Au_{147} core, copper atoms have neither the energy nor the time needed to incorporate the gold surface, and the decahedral arrangement becomes preferred to the icosahedral one. In Fig. 7 some snapshots coming from the simulations at $T = 400$ K are shown. It is worth noting that the existence of $Au_{core}Cu_{shell}$ decahedra has been previously postulated by Ascencio *et al.*,[2] who recognized that during the synthesis of bimetallic colloidal AuCu particles different configurations can coexist in the sample, and not in all the cases do the clusters exhibit their lowest energy structure and chemical order. Molecular Dynamics simulation of the heating of such decahedral particles proved that the chemical order of the particle is reversed before melting, so as to achieve the more stable $Au_{shell}Cu_{core}$ configuration.

4. Conclusions

In this paper both global optimisation techniques and Molecular Dynamics simulations are used to study the structure and the chemical order of AuCu nanoclusters. The sizes considered extended up to 200 atoms for the global optimisation study, and up to 300 atoms for the clusters resulting from growth simulations.

The picture arising from the search for the lowest-energy configurations is characterized by the prevalence of the icosahedral motif, that is typical of homogeneous copper clusters and is instead quite unusual for homogeneous gold clusters at equilibrium.[21] Icosahedral structures are characterized by the tendency of gold to segregate at the surface of the cluster. At the compositions $Au_{0.5}Cu_{0.5}$ and $Au_{0.25}Cu_{0.75}$ the segregation is not complete, and clusters with some gold atoms in the core are found as global minimum configurations.

The structural and chemical analysis of the clusters resulting from the growth simulations of AuCu clusters, by deposition onto an heterogeneous 34-atom seed, confirms the tendency to build up icosahedral clusters. It is likely to observe both anti-Mackay and Mackay patches over cluster surfaces, and such patches are usually enriched in gold atoms. Following the suggestion coming from some previous experimental studies,[22] the deposition of Cu atoms onto a gold core of 147 atoms has been simulated. Despite the fact that a small amount of copper should lead to the formation of icosahedral and partially core–shell structures, at the growth temperature of 400 and 500 K no incorporation of the copper atoms into the core has been observed. The cluster indeed transforms from Ih to Dh before getting to the size $N = 200$. This again confirms the output of the global optimisations, as the Dh structure located at size $N = 100$ and composition $Au_{0.25}Cu_{0.75}$ exhibits Cu atoms in its outer sites. In the decahedral structure, copper atoms can find a better surface arrangement than in the icosahedral structure, where surface bonds are quite stretched. Moreover, in the decahedral geometries there is no driving force causing Cu atoms to reach the cluster center fast, at variance with the case of icosahedral clusters.

References

1   U. Pal, J. F. Sanchez Ramirez, H. B. Liu, A. Medina and J. A. Ascencio, *Appl. Phys. A*, 2004, **79**, 79.
2   J. A. Ascencio, H. B. Liu, U. Pal, A. Medina and Z. L. Wang, *Microsc. Res. Tech.*, 2006, **69**, 522.

3   B. Pauwels, G. Van Tendeloo, E. Zhurkin, M. Hou, G. Verschoren, L. Theil Khun, W. Bouwen and P. Lievens, *Phys. Rev. B*, 2001, **63**, 165406.
4   F. Baletto, C. Mottet and R. Ferrando, *Phys. Rev. Lett.*, 2000, **84**, 5544.
5   F. Baletto, C. Mottet and R. Ferrando, *Phys. Rev. B*, 2001, **63**, 155408.
6   F. Cleri and V. Rosato, *Phys. Rev. B*, 1993, **48**, 22.
7   R. L Johnston, *Dalton Trans.*, 2003, **22**, 4193.
8   N. T. Wilson and R. L. Johnston, *J. Mater. Chem.*, 2002, **12**, 2913.
9   R. A. Lordeiro, F. F. Guimarães, J. C. Belchior and R. L. Johnston, *Int. J. Quantum Chem.*, 2003, **95**, 112.
10 A. Rapallo, G. Rossi, R. Ferrando, A. Fortunelli, B. C. Curley, L. D. Lloyd, G. M. Tarbuck and R. L. Johnston, *J. Chem. Phys.*, 2005, **122**, 194308.
11 G. Rossi and R. Ferrando, *Chem. Phys. Lett.*, 2006, **423**, 17.
12 D. J. Wales and J. P. K. Doye, *J. Phys. Chem.*, 1997, **28**, 5111.
13 D. J. Wales, *Energy Landscapes*, Cambridge University Press, Cambridge, 2003.
14 S. Darby, T. V. Mortimer Jones, R. L. Johnston and C. Roberts, *J. Chem. Phys.*, 2002, **116**, 1536.
15 H. Yasuda and H. Mori, *Z. Phys. D*, 1996, **37**, 181.
16 D. Fevrer, A. Torres-Castro, X. Gao, S. Sepulveda-Guzman, U. Ortiz-Mendez and M. J. Yacaman, *Nano Lett.*, in press.
17 F. Baletto, C. Mottet and R. Ferrando, *Phys. Rev. B*, 2002, **66**, 155420.
18 F. Baletto, C. Mottet and R. Ferrando, *Phys. Rev. Lett.*, 2003, **90**, 135504.
19 H. Yasuda, H. Mori, M. Komatsu, K. Takeda and H. Fujita, *J. Electron Microsc.*, 1992, **41**, 267.
20 H. Mori, H. Yasuda and T. Kamino, *Philos. Mag. Lett.*, 1994, **69**, 279.
21 K. Koga and K. Sugawara, *Surf. Sci.*, 2003, **529**, 23.
22 H. Yasuda and H. Mori, *Z. Phys. D*, 1994, **31**, 131.

PAPER | www.rsc.org/faraday_d | Faraday Discussions

Mechanical properties of bimetallic crystalline and nanostructured nanowires†

Marc Hou,*[a] Oksana Melikhova[ab] and Stoyan Pisov[ac]

Received 29th March 2007, Accepted 13th June 2007
First published as an Advance Article on the web 3rd October 2007
DOI: 10.1039/b704706a

Nanowires are basic components of interconnects at the nanoscale level in electronic as well as in electromechanical devices. Presently, there is a fast growing interest in their synthesis as well as in their mechanical testing. Focused ion beams now allow machining pillars with diameters as small as a few tens of nanometres and nanoindenter systems allow measuring strains at the atomic scale and compressive stresses up to the 10 GPa range. Such pillars typically contain less than millions of atoms, which makes their modelling and the modelling of their mechanical properties at the atomic scale realistic. A few Molecular Dynamics studies are presently available, discussing deformation mechanisms in thin narrow crystalline nanowires, but the literature about nanoalloy wires and nanostructured wires, as they can be synthesized from clusters, is almost non-existent. In the latter, the dislocation activity may be inhibited, leading to specific mechanical properties. By means of large scale computations, we use Ni_3Al to discuss the mechanical properties of crystalline and nanostructured nanowires. We also compare wires to their bulk counterparts. Both isothermal and isoenergetic—whereby mechanical work converts into heat in the system-deformation mechanisms are considered. The comparison between pair correlation functions, stress distributions, configuration analysis and strain stress relations capture most of the stress-induced evolution mechanisms of nanowires with different diameters and structures, including elastic properties, dislocation activity, grain rotation and boundary motion, local melting, superplasticity and fracture. A structural transition which may be martensitic is predicted for the first time at the nanoscale level, suggesting possible shape memory properties of nanoalloy nanowires.

1. Introduction

Size scale effects in materials have gained a tremendous interest since observation and characterization techniques down to the atomic scale allowed studying nano-structures systematically. As mechanical properties are concerned, the breakdown of the Hall–Petch relation[1,2] in nanostructured metals gave impetus to the search for

[a] Physique des Solides Irradiés et des Nanostructures CP234, Université Libre de Bruxelles, Bd du Triomphe, B-1050, Bruxelles, Belgium. E-mail: mhou@ulb.ac.be; Fax: +32 2 6505227; Tel: +32 2 6505735
[b] Department of Low Temperature Physics, Faculty of Mathematics and Physics, Charles University, V Holesovickach 2, CZ-180 00, Prague 8, Czech Republic
[c] Faculty of Physics, University of Sofia, 5 James Bourchier str., 1164, Sofia, Bulgaria

† The HTML version of this article has been enhanced with colour images.

alternatives to dislocation driven plasticity when the grain sizes range below 10 nm.[3] More recently, the advent of efficient cluster sources,[4] of mass selected and focused low energy cluster beams[5] make it possible to design cluster assembled materials with specific and controlled properties. Moreover, thanks to the possibility of mixing virtually all elements in clusters, whatever they are miscible or not, the way is open toward the design of cluster assembled materials formed by mixtures that do not exist in conventional bulk materials. Still more recently, the technological interest in nanodevices motivated advances in identifying changes of properties induced by the nanostructure geometry. There is presently a significant enhancement of the effort in measuring the mechanical properties at the nanoscale in systems, like wires, where surfaces and interfaces may play an important role. This concerns the mechanisms of crack propagation[6,7] as well as, more generally, the mechanical response to an external stress.[8-10] In this respect, the possibility was demonstrated to cut a nanowire of controlled diameter (down to ~ 100 nm) from a material with a Focused Ion Beam (FIB),[9,10] of which the mechanical properties can be studied with a conventional nanoindenter. In parallel, atomic scale modeling studies were initiated of deformation mechanisms in nanostructured systems with large open volumes[11] and of thin nanowires (<10 nm diameter).[12-14] There still remains a gap however between experimentally tested and modeled wires as the size range is concerned as well as the time limitations of Molecular Dynamics (MD) techniques employed which needs to be bridged. Moreover, to our knowledge, although experimental studies of alloy nanowires are presently available, MD studies of nanowires made by more than one element are very sparse.[15]

Bridging the gap seems nowadays possible thanks to refinements of the FIB technique allowing to design thinner wires[16] and to the use of large scale computations with parallel versions of atomic scale modeling codes and dedicated algorithms.

This paper focuses on the interplay between mechanical deformation, geometrical and structural changes of nanowires made of two different metallic elements. The Ni$_3$Al alloy is selected as a case study. Molecular dynamics is used, ran parallel on 12 to 27 processors, for modeling wires with diameters in the 10 nm range and with a simulation strategy which aims at circumventing the problem of restricted timescale. The wires used are described in Section II together with the MD technique, the Section III is devoted to mechanical uniaxial deformations in the elastic regime, the Section IV discusses deformation in the vicinity of the yield strength, Section V in the plastic regime and the whole deformation picture is summarized in Section VI.

2. The model

Although the experimental testing of nanowires with an indenter presently mainly concerns compressive tests, the mechanical responses to both compressive and tensile stresses are considered in what follows. The wires themselves are modeled as realistically as possible, using modeling conditions aiming at reproducing real experimental ones as much as possible. The modeling is at the atomic scale, however, because of the large number of atoms involved (ten thousand to one hundred thousand in what follows), classical mechanics is used. In this section, we briefly describe the Molecular Dynamics (MD) schemes used, the model nanowires and their mechanical testing method. A discussion of the testing conditions follows.

2.1. Molecular dynamics potentials

The Ni–Al system is a good case study using classical mechanical schemes because semi-empirical potentials describe its structural, mechanical and thermodynamic properties rather well. A number of studies report about Ni–Al properties using the second moment approximation of the tight binding model and, as in our previous work, we use the parameterization in ref. 17 of the functionals suggested in ref. 18. We also parameterize an EAM potential[19] on the basis of the same mechanical properties, using the functionals in ref. 20. For convenience, the former will be termed "TB potential" and the latter "EAM potential" in what follows. Using both, the sensitivity to the stress–strain relationship and of the deformation mechanisms on the model potential will be emphasized.

2.2. Model nanowires

In the present report, phenomena involving interfaces between a wire and a substrate or an indenter tip are not discussed and self-standing infinite wires are modeled by applying periodic boundary conditions in the axial direction. The role of heterogeneous interfaces will be discussed in a later report.

Nanowires may be synthesized by several routes and here we focus on those obtained by cutting a matrix with a FIB[9,16] and on those that could be obtained by means of aerodynamically focused cluster beams.[21] Such wires are self-standing on a substrate and their mechanical response to an axial compressive stress can be measured with a conventional indenter device.[9] Recently, tensile tests on gold nanowires were also reported.[7]

The matrices out of which nanowires are cut can be of any composition and structure. We consider here the Ni_3Al $L1_2$ bi-metallic system. Crystalline Ni_3Al nanowires are modeled as cylinders with a [001] axis. Two diameters are used, of 7 nm and 18 nm, respectively. For convenience, the latter will be termed "thick wire" in what follows. The size in the axial direction before deformation of the former is 7 nm and that of the latter 6 nm.

Ni_3Al cluster assembled materials were obtained by accumulating clusters on a substrate with supersonic velocities, mimicking a Low Energy Cluster Beam Deposition (LECBD) experiment. The model matrices used are those studied in ref. 21, whose bulk mechanical properties are described in ref. 11, 22 and 23. Cylinders are cut in the simulated Ni_3Al cluster assembled boxes in order to model a nanowire. The cluster assemblies, display open volumes of the same size (2–5 nm) as the clusters forming the material. Since this size is also comparable to the wire diameter considered here, in order to increase the contact area between clusters, the wire was compacted by 20% before mechanical testing, using the compression procedure described below. For convenience, this wire will be denoted "n-wire" in what follows.

2.3. Mechanical testing

Changing the length of the box in the [001] direction, the stress tensor on each atom and the macroscopic stress tensor are calculated as:

$$\sigma_i^{\alpha\beta} = \frac{1}{\Omega}\left(m_i v_i^\alpha v_i^\beta + \sum_{j\neq i}^N F_{ij}^\alpha r_{ij}^\beta\right) \text{ and } \sigma^{\alpha\beta} = \sum_{i=1}^N \sigma_i^{\alpha\beta} \tag{1}$$

where N is the number of atoms, v is the atomic velocity, F_{ij} is the force component of particle j on particle i, r_{ij} is the distance between particles i and j, α and β refer to the vector components. The axial stress, σ_{zz} and the pressure, $P = -\frac{1}{3}tr\|\sigma\|$, are naturally deduced.

As it comes from eqn (1), estimating the stress requires estimating the wire volume, Ω. Its definition is however ambiguous because the spatial atomic distributions in deformed and in nanostructured wires are heterogeneous and profound surface roughness may take place. Therefore, in order to use a definition which is consistent with all cases studied, we take Ω as the volume of the parallelepiped bounding the wire.

The Young modulus is estimated by linear regression as

$$E = \frac{4}{\pi} \left[\frac{d\sigma_{zz}}{d\varepsilon_{zz}} \right]_{\varepsilon_{zz}=0} \quad (2)$$

where σ_{zz} and ε_{zz} are respectively the stress and the strain components in the axial direction. The factor $4/\pi$ corrects eqn (1) for the cylindrical shape of the wire, which remains about unchanged in the case of small elastic deformations. It still remains approximate in the case of the n-wire where empty volumes and surface roughness are inherent, even when it is not deformed.

The Poisson ratio was also estimated by linear regression analysis of the radial strain ε_{rr} versus the axial strain ε_{zz} as

$$\nu = \left[\frac{d\varepsilon_{rr}}{d\varepsilon_{zz}} \right]_{\varepsilon_{zz}=0} \quad (3)$$

The radial strain ε_{rr} is defined by $\varepsilon_{rr} = \Delta R/R_0$, where $\Delta R = R - R_0$ is the difference between the mean distance of atoms from the wire axis after and before deformation.

In the atomic scale modeling, one has to face two problems involving the MD strategy. One is the conversion of mechanical energy into heat and the second is the short timescales inherent to MD (typically in the nanosecond range) as compared to those involved in real experimental tests (typically in the range of seconds). As the conversion of work into heat is concerned, we repeat all simulations with two strategies. One is to couple the system to a thermostat with a characteristic energy exchange time which is short (a few picoseconds) with regard to the characteristic time involved in deformation (a few nanoseconds). The model used therefore accounts for electron–phonon coupling in the same way as in ref. 25 and considering the electronic system as a reservoir at constant temperature. The electron–phonon coupling time at 300 K is taken as 2 ps in Ni_3Al.[22] The second strategy ignores electron–phonon coupling and the system is considered isolated thermally, thus allowing for the temperature to increase with strain. In the case of large deformations, the subsequent temperature rise may be significant (above 1000 K), although not sufficient to induce the melting of the wire. In this way, two extreme conditions are met in between which real experiments should necessarily occur.

The second problem has already been faced in similar studies.[12,13,28] The problem is that only strain rates several orders of magnitude larger than in real experiments can be accessed by MD. Such unrealistic strain rates have not only the consequence of too fast heating—which can be overcome with a suitable thermostat—but also to include strong uncontrolled dynamic events like shock waves. The method used here to circumvent this problem is similar, in its principle, to that employed in ref. 24 which consists in deforming the sample stepwise rather than continuously and constraining it to thermal and mechanical equilibrium between each step. In the limit of small steps, this corresponds to an equilibrium evolution which is closer to experiment.

As standard in this work, strain was modified by steps of either 1 or 5% and the system was brought to thermal equilibrium, canonically or microcanonically, before proceeding with the next step. The time interval between two deformation steps was 50 ps. It was checked that no dynamic event introduced a significant bias by inserting alternate quenching and thermalization phases in between strain steps in order to insure that neither different stress–strain relationship nor different deformation mechanisms were induced. The same procedure was used for compressive and tensile testing.

3. Elastic regime

In this section, we discuss the elastic deformation of nanowires, with a main focus on the Ni_3Al single crystal at constant temperature. No substantial difference is found

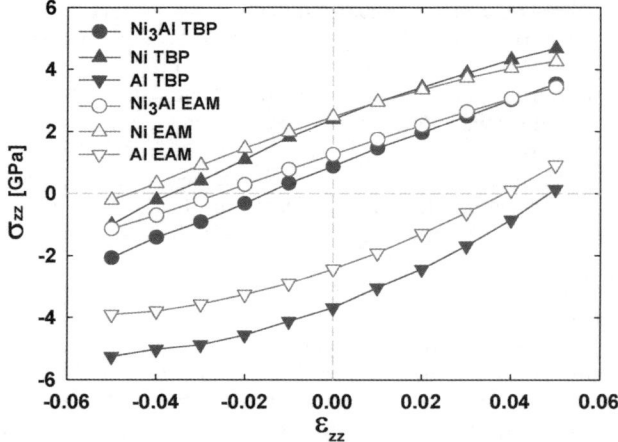

Fig. 1 Strain–stress relation in the case of small elastic deformations of the Ni_3Al crystalline wire with the EAM (empty symbols) and the TB (filled symbols) potentials. Circles represent the results for the total stress, σ_{zz}, top down triangles represent the stress on the Al subsystem and bottom down triangles those for the Ni subsystem.

at constant energy. Small bulk elastic deformations obey Hooke's law and the first question is to know if it is obeyed by nanowires as well. In order to explore this question, different elongations in the axial direction, ε_{zz}, were considered around equilibrium ($\varepsilon_{zz} = 0$) and the corresponding stress, σ_{zz}, was measured. Strains of no more then 5% are considered. In the case of Ni_3Al, the distinction between stress on the Al and the Ni subsystems is made.

Fig. 1 provides the relationship between stress and strain in this range as obtained with both potentials for the $L1_2$ crystalline nanowire. A good linearity is found for both cases of tensile and compressive stresses. Estimates of the Young modulus, E, and of the Poisson ratio, ν, are given in Table 1. The Young modulus obtained with the EAM potential is significantly smaller than with the TB potential while Poisson ratios are equal. Both estimates of E are smaller than estimated with the TB potential in an infinite single crystal in $\langle 100 \rangle$ directions,[23,24] also shown in Table 1 for comparison.

Repeating the deformations with a wire having a larger diameter has the consequence of a reduced surface to volume ratio and an estimate of a Young modulus closer to the bulk value is anticipated, as is indeed found in Table 1. This contrasts with experimental bending experiments indicating that the Young modulus may not depend significantly on the wire diameter in the case of gold wires with diameters above 40 nm.[26] Since the only difference between the wire and the bulk crystal is the occurrence of a surface, it is tempting to interpret the lower E-value as a

Table 1 Young modulus (E) and Poisson ratios (ν) of different wires (superscript "W") and bulk samples (superscript "B"). When necessary, "n-" denotes that the wire is nanostructured

Potential	Sample	Temperature/K	E/GPa	ν
TB	Ni_3Al^W	300	71.5(6)	0.39(1)
		600	59.3(6)	0.42(1)
	Thick Ni_3Al^W	300	88(1)	0.37(7)
	n-Ni_3Al^W		33(2)	0.40(1)
	$Ni_3Al^B_{\langle 100 \rangle}$ [22,23]		100	0.39
	n-Ni_3Al^B		59.6	0.35(10)
EAM	Ni_3Al^W	300	46(2)	0.39(1)
	n-Ni_3Al^W		25(1)	0.41(1)

surface effect. The evidence for a sizeable contribution of surface to stress is also seen in Fig. 1 for $\varepsilon_{zz} = 0$. In this case the total configuration energy of the wire is checked to be minimal, however, the overall stress in the axial direction is non zero (close to 1 GPa) and tensile, which is consistent with surface relaxation. Mean Poisson ratios are in reasonable agreement. Poisson ratios were found close to linearly decreasing functions of the strain in all samples where an elastic behaviour is identified. Expectedly, the effect of temperature is to lower the Young modulus while no significant effect of temperature is predicted on the Poisson ratio. The same characteristics were found with the TB potential in ref. 24 in the case of a bulk crystal.

Finding different E-values in Ni_3Al nanowires with the EAM and the TB potentials is *a priori* surprising since both potentials are fitted at equilibrium on the bulk elastic constants, more precisely, the TB potential on C_{11}, C_{12} and C_{44}, and the EAM potential on the bulk modulus. The reason why EAM and TB potentials predict different E may have its origin in the detail of the elastic deformation, which is far from obvious. The first point to note, as seen in Fig. 1, is that the Ni and the Al subsystems behave differently under stress. Hooke's law is not well verified for the subsystems considered separately but it results from a balance between sublattices responses. It is worth already mentioning here that in case of large elastic deformations (typically $|\varepsilon_{zz}| > 0.05$) the EAM and TB potentials predict increasingly different stresses with increasing strain in the Ni and in the Al subsystems. This will be discussed in more detail in the section devoted to the yield strength regime. The complexity of stress distributions in the elastic regime is illustrated by Fig. 2 showing the distribution of partial pressures on the Ni and the Al subsystems at different ε_{zz} values, as obtained with both potentials. In the absence of deformation, Ni atoms are underpressurized and most Al atoms overpressurized. This was already found in ref. 11 both for an infinite solid and spherical nanoclusters. In the absence of strain ($\varepsilon_{zz} = 0$), the pressure distributions display secondary structures seen in Fig. 2. As Al is concerned, they are induced by the surface tension. Because of its curvature, the cylinder surface displays all crystallographic orientations normal to the [001] z-axis, resulting in different values of the tension and in the occurrence of more then one secondary structure in Fig. 2. These structures are different when calculated with different potentials, as well as their dependencies on ε_{zz}. Expectedly, in both cases, the main peak, which was identified as corresponding to perfectly coordinated atoms ($Z = 12$), shifts toward lower pressures as ε_{zz} increases. However, the dependency of partial pressures is stronger with the TB potential, consistently with a faster increase of σ_{zz} and with a higher Young modulus. While Al partial pressure distributions essentially evolve with strain by peak shifts, new structures emerge in the partial pressure distributions of Ni at large strains, on top of those associated with surface tension. The reason for this is that ε_{zz} is normal to {001} planes stacked as alternate sequences of pure Ni and mixed Ni–Al planes in which, for the same ε_{zz}, the shear stress is different and, consequently, the partial pressures too.

When the diameter of the wire is increased, the surface to volume ratio decreases and the signature of the surface in the structure of pressure distributions decreases too.

Increasing the temperature has the effect to broaden the structures but it does not modify the dependency on strain of the overall features of the pressure distributions.

When, rather than single crystalline, the Ni_3Al is nanostructured, deformations are never fully elastic and the range in which the elastic behaviour is significant is much reduced as compared to the crystalline wire. In the example of the n-Ni_3Al wire described in Section II, this elasto-plastic behaviour is approximately limited to the strain range ($-0.05 < \varepsilon_{zz} < 0.02$) when estimated with both potentials. The Young modulus is also lower, as was found in the absence of surface effects, and the Poisson ratio close to the experimental value of an ordinary polycrystalline material.[27] Because of the occurrence of internal surfaces and grain boundaries, the structures in the pressure distributions are all smeared out, at least as far as the Ni subsystem is concerned. The distributions are shown in Fig. 2e and 2f to be quite broad, in the

This journal is © The Royal Society of Chemistry 2008

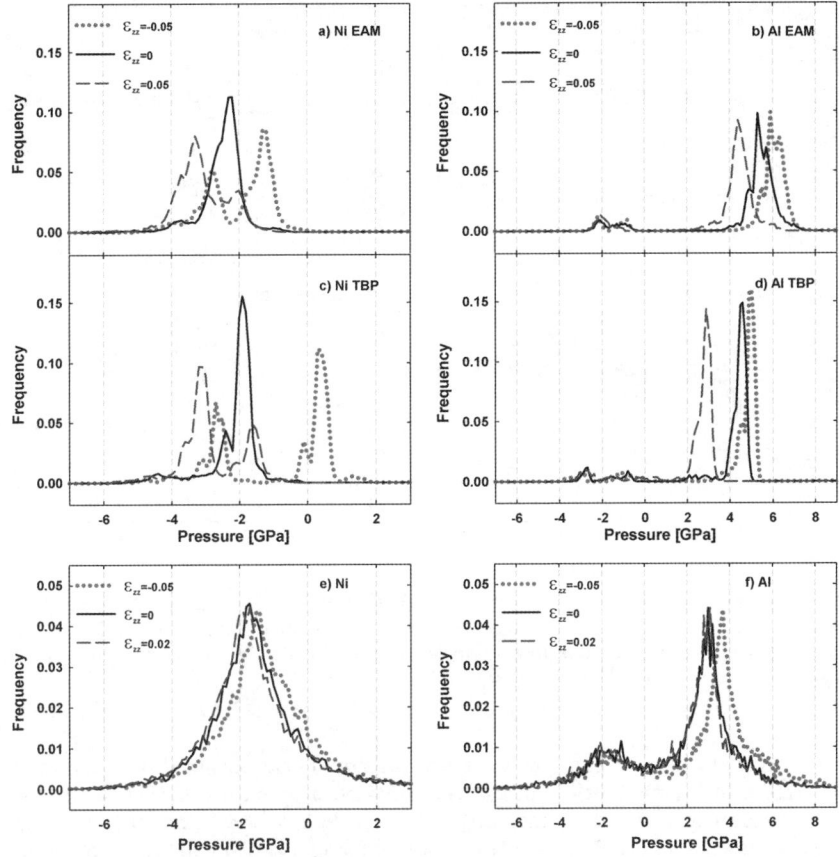

Fig. 2 Pressure distributions on the Ni and the Al subsystems predicted with the EAM (a, b) and the TB potentials (c, d) for $\varepsilon_{zz} = -0.05$, $\varepsilon_{zz} = 0.0$ and $\varepsilon_{zz} = 0.05$. Pressure distributions for the n-Ni$_3$Al wire obtained with the TB potential are given in (e) and (f) for $\varepsilon_{zz} = -0.05$, $\varepsilon_{zz} = 0.0$ and $\varepsilon_{zz} = 0.02$. The bin size is 0.1 GPa.

range between 1 and 2 GPa. This width depends on the potential. As far as Al is concerned, pressure distributions are bimodal with a main peak at positive pressure and a peak at negative pressure associated to surface atoms.

To summarize, Fig. 2 allows distinguishing the four contributions to the stress–strain curves in Fig. 1, namely, the contribution of the Al subsystem, of the Ni subsystem, of surface Ni and of surface Al atoms. The results in Fig. 1 and thus in the Young modulus are a consequence of a balance between these contributions, which depends on the model potential. To our knowledge, no parameterization of EAM and TB potentials is available in the literature, which accounts for inhomogeneous elastic deformations like those evidenced in the present case.

Comparing the results in ref. 23 with the present ones shows that, both in the case of the single crystal and of the n-wire, the effect of the n-wire surface is to decrease the Young modulus by almost a factor of 2 as compared with the bulk. The surface allows for limited plastic deformations in the n-wire, even at the lowest strains.

4. The yield strength regime

In the case of the L1$_2$ crystalline wires, Hooke's law was found to result from a balance between the mechanical response of the Ni and of the Al subsystems when the deformations are small. The stress–strain relations are depicted in Fig. 3 over a

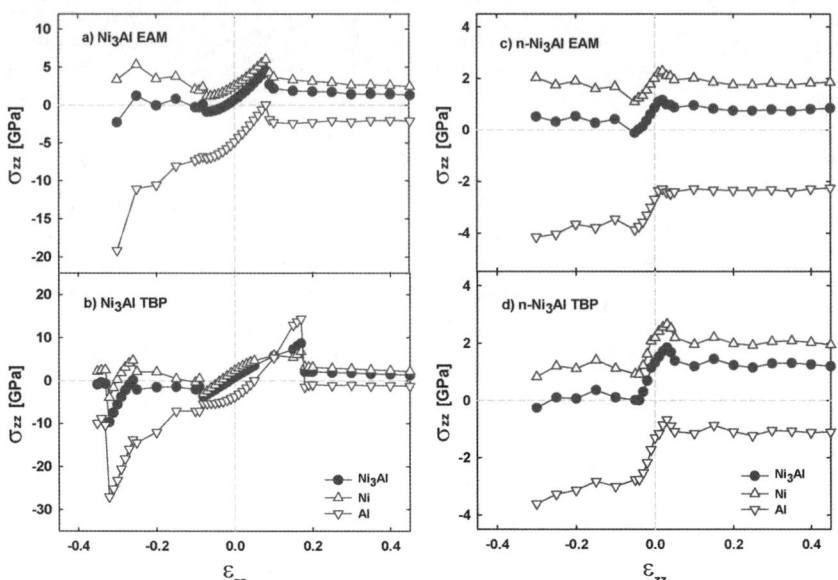

Fig. 3 Stress–strain relations. (a) and (b): calculated for the Ni_3Al wire with 7 nm diameter, (c) and (d): calculated for the n-Ni_3Al wire (a), (c) EAM potential, (b,d) TB potential. Circles: relation for the whole wire, top down triangles: relation for the Al subsystem, bottom down triangles: relation for the Ni subsystem.

large range of deformations. The yield strength could be determined by tracking the occurrence of the first nucleus of a permanent defect by means of 3D visualization techniques. It corresponds to the sudden drop of the stress, σ_{zz}, seen in Fig. 3 when ε_{zz} is increased above zero. In the case of compression, mainly with the TB potential, the discontinuity of σ_{zz} is not clearly occurring at the yield strength and may not be captured in a real experiment. It is worth mentioning here that, since the nucleation of permanent defects is a stochastic process, the simulation time might not be sufficient to capture the first one as the magnitude of the strain is increased and the estimated yield strength must be considered as an upper limit. Considering the whole range of elastic deformations ($-0.09 < \varepsilon_{zz} < 0.18$ with the TB potential and $-0.08 < \varepsilon_{zz} < 0.09$ with the EAM potential) it is seen in Fig. 3 that the offset from Hooke's law is generally small, except in the case of compressive stress with the EAM potential where anharmonicity is sizeable.

A close examination of each subsystem when the elastic deformation gets close to the yield strength reveals significant lattice distortions. These distortions show up in the pair correlation functions (pcf) depicted in Fig. 4. In both the compressive and the tensile regimes, the first peak in the Al pcf is split in this range, which is the signature of the wire structure to become tetragonal. The first neighbor peak in the Ni pair correlation function does not split, neither in the case of compression, nor in the case of extension. As a result of the global stress, pure Ni(001) planes thus undergo distortions preserving the distances between first neighbor pairs. The consequence is particularly large with the TB potential, in which case, the Al sublattice, under compressive stress at equilibrium, is predicted even more tensile than the Ni subsystem at large elastic extensions (Fig. 3).

As far as the n-wire is concerned, the range of elasto-plastic deformation is more limited ($-0.06 < \varepsilon_{zz} < 0.04$ with the TB potential and $-0.06 < \varepsilon_{zz} < 0.03$ with the EAM potential) and, given this small strain amplitude, lattice distortions in the vicinity of the yield strength are not observed similar to those in the crystalline wire.

We now consider the transition across the yield strength.

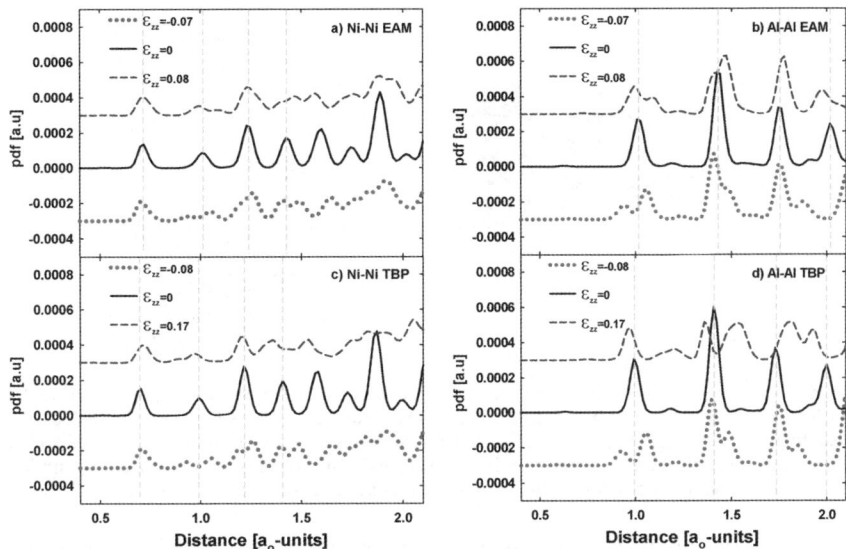

Fig. 4 Pair correlation functions before deformation (middle curve) and at yield strength for extension (upper curve) and compression (bottom curve). (a) $g(r)$ for Ni pairs for EAM (upper figure) and TBP (lower figure) (b) $g(r)$ for Al pairs for EAM (upper figure) and TBP (lower figure).

The strain–stress curves in Fig. 3 show that the elastic/plastic transition is smooth in n-wires and may be sharp in crystalline ones, indicating that the mechanisms at work are different. Beyond the transition, the stress stored elastically during deformation is released to a large extent.

If one takes the stress at $\sigma_{zz}(\varepsilon_{zz} = 0)$ as a reference, Fig. 3 shows that the magnitude of stress under tensile and compressive deformations in the plastic regime are of the order of 1 GPa and do not depend much on the magnitude of the deformation. As will be discussed in the next section, dependencies at the largest deformations shown correlate with necking and structural changes. This may be an indication that the crystalline wire becomes nanostructured in the plastic regime giving rise to grain gliding. In the case of grain gliding, the residual stress relates to friction. The situation is however more complex and is now analyzed in detail.

The nucleation of plastic deformation in the crystalline $L1_2$ wire displays some similarity with bulk materials, as far as dislocations are involved.

Permanent deformations were found to nucleate at the surface. Fig. 5 is a typical example of a configuration where a kink is generated by accommodating a $1/2[001]$ dislocation with its Burger's vector oriented parallel to the external stress. Atoms sitting at the surface in the elastic regime are now seen to form a (001) platelet in Fig. 5a. Such a dislocation is sessile.

For this reason, it only evolves with stress by increasing its Burger's vector and a (001) crack propagates subsequently through the wire section. A similar crack might have been observed experimentally in a tensile test of a gold nanowire[7] although the planar fracture was indexed as {111}. Alternatively, another dislocation system nucleates in the disordered core of the $1/2[001]$. $1/6[112]$ type dislocations were identified to nucleate this way, that are glissile and give rise to {111} stacking faults in the $L1_2$ structure. Both situations are represented in Fig. 5. Such {111} mechanisms contribute to release the stress stored elastically in the lattice. A network of planar {111} faults is formed in this way.

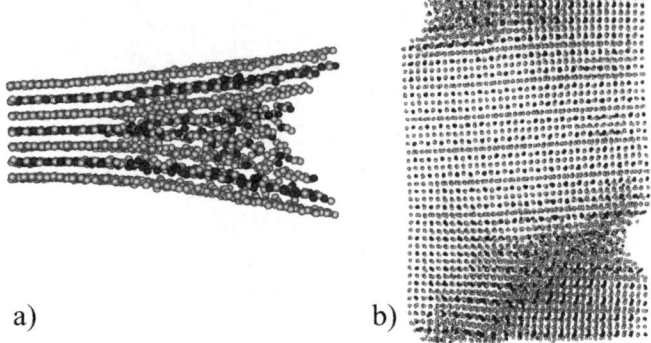

Fig. 5 Nucleation of a plastic defect. (a) Atomistic detail of a 1/2[001] dislocation nucleated at the Ni$_3$Al wire surface and generating a kink. The strain $\varepsilon_{zz} = 0.18$ (b) a glissile dislocation, close to 1/6[112] nucleated in the core of the 1/2[001] sessile dislocation on the right of 5(b). It propagates through the wire section, inducing a second kink. Because of the periodic boundary conditions in the [001] direction, this second kink is seen at the top of the wire.

One example obtained in the case of compression is depicted in Fig. 6. In the case of compression, such planar faults are clearly dominating the nanostructural evolution.

Such a dense network of faults as seen in Fig. 6 is the precursor of a martensitic structural transition described and discussed in the next section.

5. The plastic regime

We start the discussion with the case of a tensile stress.

A typical evolution of a nanowire morphology is depicted in Fig. 7 at different large tensile strains. The case of a Ni$_3$Al initially crystalline nanowire with a diameter of 7 nm is shown. The EAM potential is used. The same deformation steps as in macroscopic wires are found, namely an elongation with reduction of the wire diameter followed by necking and rupture. These steps were found under extension in all the nanowires considered, with both potentials, at constant temperature or energy, as it was also reported in the previous literature.[12,13,28] Their occurrence is thus not dependent on the nanostructure. As discussed in Section 4, depending upon the plastic deformation nucleation mechanism, the ductile regime may be vanishingly limited (generation of a (001) crack), in which case, the necking threshold merges with the yield stress.

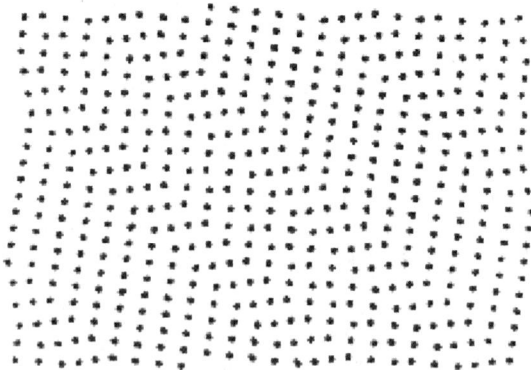

Fig. 6 Typical network of planar {111} faults as obtained at a compressive strain of $\varepsilon_{zz} = -0.2$ in the Ni$_3$Al wire. For the sake of clarity, only the Al subsystem is shown.

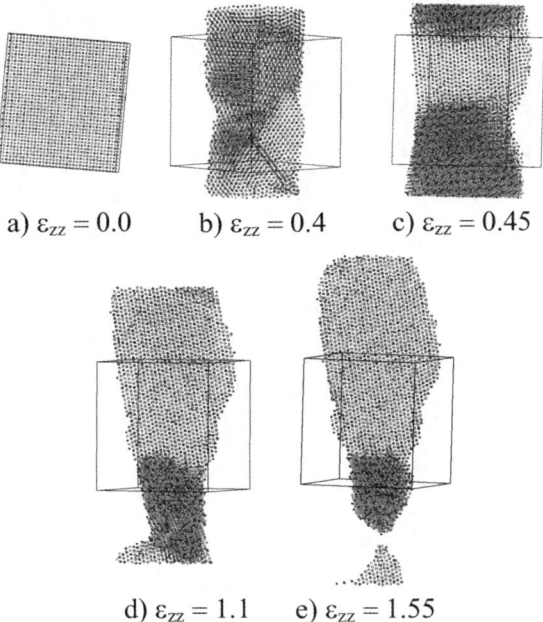

a) $\varepsilon_{zz} = 0.0$ b) $\varepsilon_{zz} = 0.4$ c) $\varepsilon_{zz} = 0.45$

d) $\varepsilon_{zz} = 1.1$ e) $\varepsilon_{zz} = 1.55$

Fig. 7 Snapshots of atomic configurations of the $L1_2$ nanowire at different tensile strains obtained with the EAM potential at 300 K. The bounding box of the undeformed model wire is shown in each frame.

When such a crack does not occur, the snapshot at $\varepsilon_{zz} = 0.4$ in Fig. 7 shows the consequence of dislocation nucleation at a surface kink, as predicted with the EAM potential. Dislocations propagated through the whole wire section, subdividing it into a few nanograins with different orientations. Grain rotation is observed subsequent to the occurrence of local torques during deformation. The same behaviour is observed with the TB potential, however, in this case, the evolutions at larger tensile strain differ. With the EAM, a partial recrystallization is observed as soon as the strain gets larger than $\varepsilon_{zz} = 0.4$, as depicted in Fig. 7. Increasing strain induces grain growth and rupture is achieved in the neck at one of the remaining grain boundaries at $\varepsilon_{zz} = 1.55$. Less ductility is found with the TB potential (rupture arises at $\varepsilon = 0.85$). Beyond the yield strength, it also fragments into nanograins, mainly by {111} slips through the whole wire section and {111} facets develop at the surface. These facets are apparent in the snapshots of the thick wire deformation shown in Fig. 9 and discussed later in this section.

We first turn to compressive stress in the $L1_2$ wire and next to deformations in the *n*-wire. In the case of compressive strains, similarly to the case of extension, dislocations nucleate at the surface when the yield strength is reached, leading to a dense network of planar or twinning faults (Fig. 6) and a structural transformation is observed when the strain becomes more negative. At $\varepsilon_{zz} = -0.3$, the wire becomes suddenly crystalline again, eventually including one twin boundary. This behaviour is characteristic of a martensitic transformation, suggesting that Ni_3Al nanowires may have the properties of a shape memory alloy. Further compression induces a further partition into nanograins and was not carried out further than $\varepsilon_{zz} = -0.4$.

The martensitic transformation observed is characterized by an increase of the magnitude of stress which is sizeable in Fig. 3 for $-0.25 > \varepsilon_{zz} > -0.3$, mainly taken over by the Al subsystem. When the magnitude of strain is smaller, stress is released by a dislocation mechanism whereby dislocations cross the whole wire section without pinning and contribute to the surface topography similar to that discussed below in the case of the thick $L1_2$ wire (see Fig. 8).

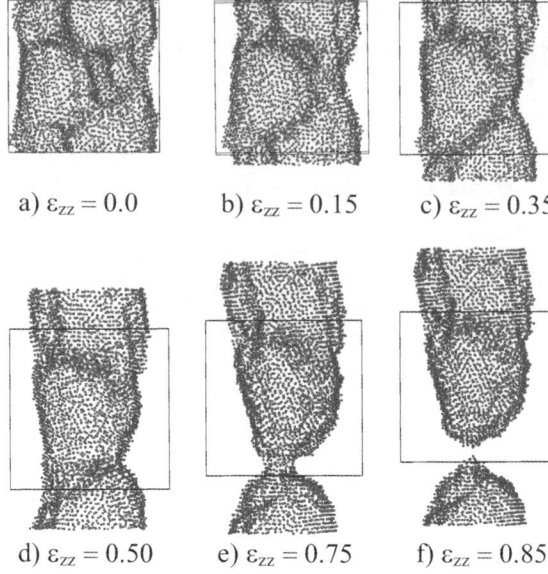

a) $\varepsilon_{zz} = 0.0$ b) $\varepsilon_{zz} = 0.15$ c) $\varepsilon_{zz} = 0.35$

d) $\varepsilon_{zz} = 0.50$ e) $\varepsilon_{zz} = 0.75$ f) $\varepsilon_{zz} = 0.85$

Fig. 8 Same as Fig. 7 in the case of the *n*-wire. In this figure however, only miscoordinated atoms are displayed, mapping the surface and the interfaces.

When the wire is nanostructured, no such slip system is available across the whole wire section. Therefore, the necking mechanism is different, as shown in Fig. 8. As a consequence, no other surface roughening than the formation of a smooth neck is observed. The open volumes do not enlarge and no crack propagation is observed. As shown in Fig. 8, two steps take place in the plastic deformation. The first one consists in the formation of a neck which grows until the wire section at the neck becomes equal to a nanograin diameter. This is seen in Fig. 8 by comparing the snapshots 8(a) to 8(d). A chain of nanograins is formed in this way, as shown in Fig. 8(d). Necking and rupture then occur by grain sliding at one of the grain boundaries.

The ductile mechanisms described so far were found with both potentials, at different temperatures. Constant energy calculations showed no different behaviour. It involves grains with size equal to the wire diameter. This is not possible however when the wire is too thick or, as is the case in the *n*-wire, when the nanostructure inhibits the propagation of dislocations and the subsequent structural transformation.

When the L1$_2$ wire is thick, using the TB potential, necking nucleates according to the dislocation mechanism depicted for the thinner wire. The deformation is depicted in Fig. 9. As in the case of the thinner nanowire deformation described above, {111} facets characterize the surface roughness. Owing to the large surface area, more than one dislocation source are at work simultaneously with the consequence of the partition of the wire into nanograins. Only few have boundaries crossing the whole wire section. Some voids are also found to form at multiple grain boundary junctions. The fragmentation into nanograins inhibits further dislocation activity and further strain becomes dominated by grain sliding, enhancing necking and leading to rupture. In Fig. 9, this grain sliding makes the difference between the snapshots at $\varepsilon_{zz} = 2.6$ and $\varepsilon_{zz} = 3.0$. No crack is found to initiate at voids, which size, once formed, decreases with increasing strain. It should be noticed that such a deformation sequence allows for a particularly large range of ductility. With the present wire, the yield strength occurs just like for the thinner wire at $\varepsilon_{zz} = 0.17$ but rupture only occurs at $\varepsilon_{zz} = 3.4$.

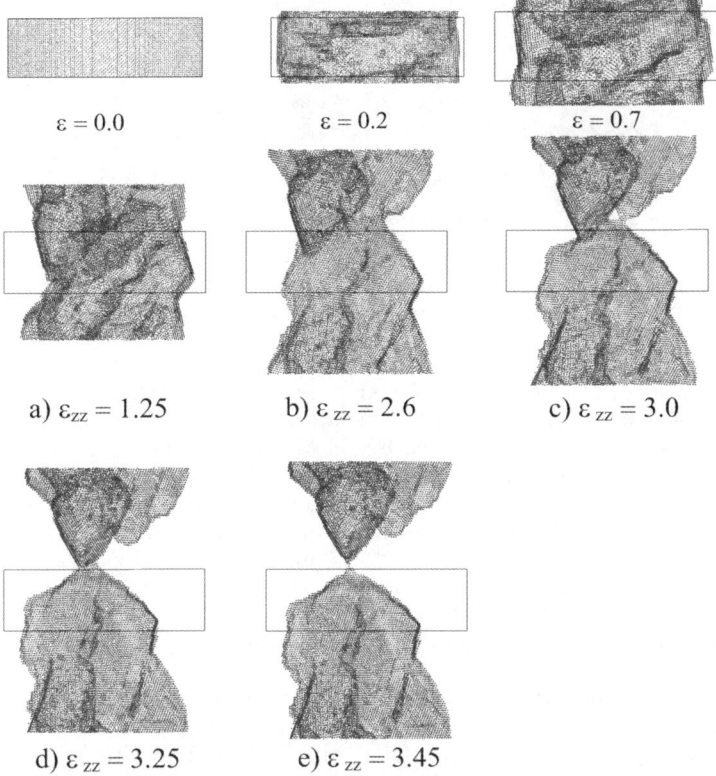

$\varepsilon = 0.0$ $\varepsilon = 0.2$ $\varepsilon = 0.7$

a) $\varepsilon_{zz} = 1.25$ b) $\varepsilon_{zz} = 2.6$ c) $\varepsilon_{zz} = 3.0$

d) $\varepsilon_{zz} = 3.25$ e) $\varepsilon_{zz} = 3.45$

Fig. 9 Same as Fig. 8 in the case of the thick wire, but using the TB potential.

The following picture emerges from the four cases depicted at this point. At the scale of the whole wire, the evolution of the deformation under tensile stress is the same as for ordinary polycrystal. It is characterized by narrowing, necking and rupture. At the atomic scale, the EAM and TB potentials predict different evolutions. In the monocrystalline wire, both predict as a first step the nucleation of dislocations at the surface with the possible consequence of portioning the wire into nanograins by way of glissile dislocations. Alternatively, a (001) crack propagates by increasing the Burger's vector of a sessile 1/2[001] dislocation. In the former case, the next steps depend on the potential. The EAM predicts a partial recrystallization into grains with sizes equal to the diameter of the wire and then, rupture takes place at one of the grain boundaries. The TB predicts the grain sliding and the rupture to be the consequence of sliding.

When the wire is nanostructured before deformation, no grain is formed with size equal to the wire diameter and the first step is a partial necking which stops when the wire diameter reduces to the grain size. Grain sliding and to some extent re-crystallization then occurs, depending on the potential. When the wire diameter is large enough, several surface dislocation sources are at work simultaneously, fragmenting the wire into grains smaller than its diameter. Its further evolution involves no substantial dislocation activity and it is similar to that of an initially nanostructured nanowire.

We now focus on the structural aspects associated with the above mentioned transformation observed for tensile deformation with the EAM potential and compressive deformation with both. Pair correlation functions associated with the wire configurations at $\varepsilon_{zz} = 0$, $\varepsilon_{zz} = 1.55$ in Fig. 7 (EAM potential) and at

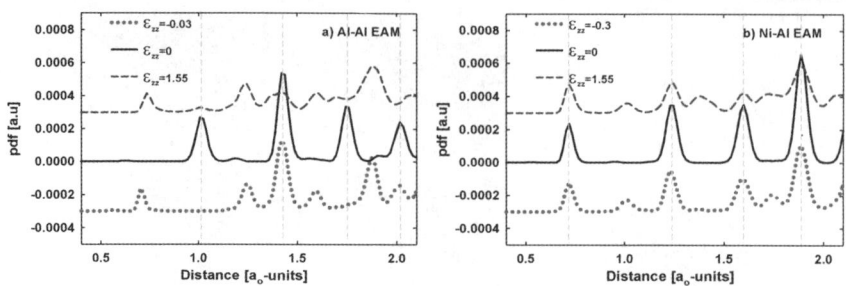

Fig. 10 Pair correlation functions as estimated for strains $\varepsilon_{zz} = -0.3$, $\varepsilon_{zz} = 0$ and $\varepsilon_{zz} = 1.5$ in the initially monocrystalline wire. (a) pcfs associated with Al–Al pairs, (b) pcfs associated with Ni–Al pairs.

$\varepsilon_{zz} = -0.3$ are displayed in Fig. 10 for Al–Al and Al–Ni pairs. The positions of the four first neighbour peaks in the $L1_2$ structure are shown by vertical dotted lines.

The $L1_2$ structure is characterized by Al–Al first neighbour pairs at one cubic lattice cell edge distance and by the lack of Al–Ni pairs separated by this distance. The situation is clearly different at $\varepsilon_{zz} = 1.55$ and $\varepsilon_{zz} = 0.3$, demonstrating that the stress induced recrystallization takes place with another structure. Moreover, the peak positions for $\varepsilon_{zz} = 1.55$ and $\varepsilon_{zz} = -0.3$ are the same, showing that tensile and compressive deformations lead to the same structural change. As attested by the pair correlation functions, this structure is cubic. A block of atoms with this structure and two adjacent cube cells sampled in the wire strained at $\varepsilon_{zz} = -0.3$ are depicted in Fig. 11.

When compressing the wire, the first permanent deformation which occurs beyond the yield strength is a mosaic of planar faults throughout the wire (Fig. 6). Simultaneously, the cross-section of the wire evolves from circular to elliptical (not shown in a figure) which is the signature of the structural modification. The stress induced by further compression, to $\varepsilon_{zz} = -0.3$, is released (Fig. 3(a) and 3(b)), the planar faults disappear and the lattice becomes perfectly ordered into the structure depicted in Fig. 11. Strain beyond $\varepsilon_{zz} = -0.3$ induces a next series of planar faults and slips into the new structure, that were not analysed in detail.

Although the 0 K configuration energy of the structure depicted in Fig. 11 is higher then in $L1_2$ wire, it corresponds to a local minimum in the energy map of the system. This is attested by the fact that the structure remains beyond rupture at $\varepsilon_{zz} = 1.55$. It was also checked by increasing the temperature until melting, which induced no transition back to $L1_2$. Periodic boundary conditions in z were released in the wire compressed at $\varepsilon_{zz} = -0.3$ and the wire segment allowed to evolve freely.

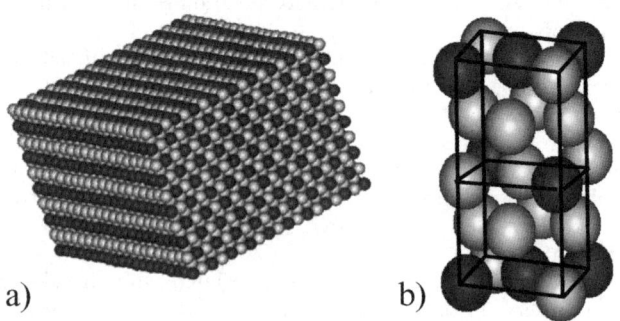

Fig. 11 Structure obtained by compression with both potentials and by tension with the EAM potential. (a): a block of atoms in the compressed wire is shown; (b): two adjacent cubic unit cells sampled in the same wire.

With the EAM potential, the structure was maintained as well. With the TB potential however, heating induces the transition back to the L1$_2$ structure suggesting a possible shape memory effect. It should be noticed that since compression induces the modifications of the wire symmetry from circular to elliptical cross section, such a transformation is not expected in an infinite single crystal. It is thus predicted to be specific to nanowires, and perhaps to nanoclusters and nanocluster assemblies.

6. Summary

Elastic and plastic deformation of Ni$_3$Al nanowires were examined by means of a dedicated Molecular Dynamics scheme whereby deformation is induced stepwise, leading the system along sequences of thermal equilibrium states with external boundary constrains. Monocrystalline and nanostructured nanowires were considered.

In the elastic regime, Hooke's law could be verified. The linear stress–strain relationship was found to be the consequence of balancing non-linear responses of the Ni and the Al subsystems and the elastic deformations were found inhomogeneous. Young moduli are lower than for bulk Ni$_3$Al but increase with the wire diameter.

The detail of the nucleation of plastic deformation mechanisms could be identified. Nucleation was always found to take place at the wire surface. In the monocrystalline wire, surface atoms are fed into the lattice, locally, forming a 1/2[001] edge dislocation. This one generates either a crack leading to rupture or a series of glissile dislocations generating deformation. At this point, in the case of tensile stress, predictions with the TB and the EAM potentials are different. The fragmentation into nanograins and their sliding until rupture is predicted with the TB while a dense system of parallel planar {111} faults is generated with the EAM potential. In the case of compression, this latter network is predicted by both. Further stress induces a martensitic deformation leading to a metastable structure. This structural transition may be reversible and may be useful in shape memory applications. It is specific to nanoscale systems with surfaces.

Finally, it should be commented that, even though rate effects are inhibited by the MD scheme employed, processes occurring in the long term are still not included in the model. For instance, the nucleation of dislocations is stochastic and, with short term simulations like those presented here, infrequent mechanisms may be overlooked. More serious may be the question of atomic diffusion and its role in nucleation and in grain sliding. Such problems should be addressed with further work. Nevertheless, qualitatively, the mechanisms and the structural evolutions predicted here are not expected to be inconsistent with phenomena occurring on larger timescales. Comparison with experiment should be very useful at this point.

Acknowledgements

S. P. is thankful to the Science Policy Office of the Federal government of Belgium for a grant in the frame of the IAP 5-1 project "Quantum size effects in nanostructured materials". O. M. is thankful for a grant of the Université Libre de Bruxelles and of the Fonds National de la Recherche Scientifique of Belgium, under agreement 2.4520.03F.

References

1   E. D. Hall, *Proc. Phys. Soc. London*, 1951, **B64**, 747.
2   N. J. Petch, *J. Iron Steel Inst.*, 1953, **174**, 251.
3   Gleiter, *Adv. Mater.*, 2000, **48**, 1.
4   T. G. Dietz, M. A. Duncan, D. E. Powers and R. E. Smalley, *J. Chem. Phys.*, 1981, **74**, 6511.

5 P. Mélinon, V. Paillard, V. Dupuis, A. Perez, P. Jensen, A. Hoareau, J. P. Perez, J. Tuaillon, M. Broyer, J. L. Vialle, M. Pellarin, B. Baguenard and J. Lerme, *Int. J. Mod. Phys.*, 1995, **B139**, 339.
6 Y. Takahashi, H. Hirakata and T. Kitamura, *Proc. Int. Symposium on Electronic Materials and Packaging*, 2005, **55**.
7 D. Lee, X. Wei, M. Zhao, X. Chen, S. C. Jun, J. Hone, J. W. Kysar, Modeling and Simul, *Mater. Sci. Eng.*, 2007, **15**, S181.
8 J. R. Greer and W. D. Nix, *Appl. Phys.*, 2005, **A80**, 1625.
9 M. D. Uchic, D. M. Dimiduk, J. N. Florando and W. D. Nix, *Science*, 2004, **305**, 986.
10 M. D. Uchic and D. M. Dimiduk, *Mater. Sci. Eng.*, 2005, **400–401**, 268.
11 E. E. Zhurkin, G. Hautier and M. Hou, *Phys. Rev. B*, 2006, **73**, 094108.
12 P. S. Branicio and J.-P. Rino, *Phys. Rev. B*, 2000, **62**, 16950.
13 U. Landman, R. N. Barnett and W. D. Luedtke, *Z. Phys. D*, 1997, **40**, 282.
14 A. Kushima, Y. Umeno and T. Kitamura, *Model. Simul. Mater. Sci.*, 2006, **14**, 1031.
15 K. R. S. S. Subramanian, R. B. Venkat and J. Babu, *J. Phys. Chem. C*, 2007, **111**, 2430.
16 T. Kitamura, in *Abstracts of the Workshop on Modeling of Extended Defects and Phase Transformations at Material Interfaces*, ed. A. Kiejna, University of Wroclaw, Wroclaw, Poland, 2006.
17 F. Gao, D. Bacon and G. Ackland, *Philos. Mag. A*, 1993, **67**, 275.
18 G. J. Ackland and V. Vitek, *Phys. Rev. B*, 1990, **41**, 10324.
19 S. M. Foiles, M. I. Baskes and M. S. Daw, *Phys. Rev. B*, 1986, **33**, 7983.
20 R. A. Johnson, *Phys. Rev. B*, 1990, **41**, 9717.
21 P. Milani, P. Piseri, E. Barborini, A. Podestà and C. Lenardi, *J. Vac. Sci. Technol., A*, 2001, **A19**, 2025.
22 M. Hou, V. Kharlamov and E. Zhurkin, *Phys. Rev. B*, 2002, **66**, 195408.
23 E. E. Zhurkin, G. Hautier and M. Hou, *J. Alloys Compd.*, 2007, **434–435**, 559.
24 E. E. Zhurkin, T. Van Hoof and M. Hou, *Phys. Rev. B*, 2007, **75**, 224102.
25 M. W. Finnis, P. Agnew and A. J. E. Foremann, *Phys. Rev. B*, 1991, **44**, 567.
26 B. Wu, A. Heidelberg and J. J. Boland, *Nat. Mater.*, 2005, **4**, 525.
27 H. Yasuda, T. Takasugi and M. Koiwa, *Acta Metall. Mater.*, 1992, **40**, 381.
28 S. J. A. Koh, H. P. Lee, C. Lu and Q. H. Cheng, *Phys. Rev. B*, 2005, **72**, 085414.

Solid-solution precursor to melting in onion-ring Pd–Pt nanoclusters: a case of second-order-like phase change?†

Florent Calvo

Received 22nd February 2007, Accepted 9th March 2007
First published as an Advance Article on the web 11th September 2007
DOI: 10.1039/b702732j

The thermodynamical behaviour of icosahedral, multilayer Pd–Pt clusters is addressed using a combination of simulation tools, mainly parallel tempering Monte Carlo. A preferential swapping trial move is introduced to increase the chance of successfully exchanging Pd and Pt atoms in the cluster. The 2-, 3- and 4-shell, Pd-rich clusters have been studied. We generally find that the clusters melt at a temperature significantly below the bulk melting point at the same corresponding composition. More interestingly, for the smaller clusters melting is initiated by a solid–solution intermediate phase in which the overall icosahedral frame remains, but the Pd and Pt atoms can swap sites. The transition to this solid–solution phase is seen to have a continuous, second-order like character, which is interpreted from the similarity between the present system with the ferromagnetic Ising model on the 3D cubic lattice. As the cluster grows, the onion-ring structure becomes thermodynamically unstable. The 4-layer cluster already exhibits a solid–solution in its core at temperatures as low as 100 K. The bulk behaviour is thus recovered at very small scales.

1. Introduction

As in bulk compounds, metallic clusters exhibit significant changes in their physical or chemical properties upon mixing: size, in addition to composition, can induce very rich diagrams. Due to their expected technological applications in the fields of electronics, magnetism, optics, and catalysis, a lot of work has been devoted to the theoretical study of noble metal clusters at the levels of first-principles[1–6] or empirical model[6–30] calculations.

While Pd and Pt are naturally good catalysts, their selectivity and activity can be further improved by alloying them.[31,32] In particular, Pd–Pt bimetallic compounds are very efficient for the hydrogenation of aromatics,[33] hydrogen-mediated ring opening reactions,[34] and methane combustion,[35] to name but a few. With their high surface/volume ratio, Pd–Pt bimetallic clusters are expected to show a further enhancement in their catalytic properties. EXAFS studies by Toshima, Harada and coworkers[36,37] and by Kolb and coworkers[38] have emphasized the geometry and composition dependence of the catalytic activity. Fiermans and coworkers[39] and Kolb *et al.*[40] found evidence in their measurements that the surface of Pd–Pt clusters are rich in palladium, while the core is rich in platinum. These observations

---

LASIM, Université Claude Bernard Lyon I, Lyon, France. E-mail: fcalvo@lasim.univ-lyon1.fr

† The HTML version of this article has been enhanced with colour images.

---

confirmed previous experimental findings by Toshima and coworkers[40] who determined the UV-Vis absorption spectra and transmission electron micrographs of these clusters, and by Rousset and coworkers[12] who found that Pd atoms evaporate preferentially. More recently, Bazin and coworkers[41] found further evidence for this core/surface segregation in Pd–Pt nanoalloys deposited on alumina, based on a combination of X-ray absorption spectroscopy, transmission electron microscopy, and volumetric H_2O_2 titration.

Nanoalloy Pd–Pt clusters have also received attention from the theoretical point of view. Most studies we are aware of attempt to characterize the extent of segregation,[13] possibly at finite temperature,[29] or to locate the most stable structures using unbiased global optimisation.[6,19,23,28,30] All these studies concluded that Pd atoms are predominantly located on the surface of the clusters, at vertices or at edges, while Pt atoms occupy the core, in agreement with available experiments. The Pd and Pt metals have similar atomic radii close to 1.38 Å, comparable cohesion and surface energies,[27] and their electronegativity difference is negligible. Hence the structure of their clusters are not expected to be strongly disturbed by alloying them. As was shown recently by Rossi and coworkers in the case of the 38-atom cluster,[28] only for specific compositions does the favoured cluster structure deviate from the notorious octahedral shape. In contrast, the more spherical icosahedral clusters may exhibit particularly stable shell structures consisting of alternating Pd and Pt layers, the first (core) and last (surface) layers being made of Pt and Pd atoms, respectively.[29] Such structures were predicted theoretically[17,21,22,24–26] based on energetic, entropic, or even dynamical arguments. Experimental evidence for these onion-ring structures has been provided by several groups.[42–46]

In this paper we address the issue of thermal stability of these onion-ring icosahedral Pd–Pt bimetallic clusters. The semi-grand-canonical theoretical investigation reported by Cheng et al.[29] suggests that the onion-ring structure enhances the stability of Pd–Pt nanoalloys, even at moderate temperatures. Even though this method does not keep the composition of the clusters fixed, its discrete character[47] may be suitable for sampling the set of structures based on the same geometry, but differing by permutations of unlike atoms, the so-called "homotops".[14] However, in order to determine the melting point, the isomers corresponding to the liquid state must be sampled as well, including also their numerous homotops. Homotops are usually separated by high energy barriers, hence their sampling is more convenient using non-local identity swapping moves in configuration space. The advantages of such a Monte Carlo (MC) approach over continuous molecular dynamics were recognized long ago in the simulation community.[48] When used prior to local optimisation, identity swapping is very powerful to quickly explore the various homotops that belong to a specified discrete structure. However, not relaxing the geometry leads to most particle exchange moves being rejected, the bigger particles having difficulty fitting into the space occupied by the smaller particles. Such problems are especially important at low temperatures, even for the similar elements Pd and Pt. Configurational bias[49,50] can significantly improve the chances of successful particle swap moves, at the expense of generating many trials. Here we propose a new method for increasing swapping efficiency, inspired by the preferential sampling biasing scheme of Owicki and Scheraga.[51] Combined with parallel tempering Monte Carlo simulations, this approach allows us to compute the caloric curve of bimetallic onion-ring clusters containing up to 309 atoms (4 shells). In order to interpret the caloric curves and to further characterize the melting behaviour of these Pd–Pt nanoalloys, we have calculated several other quantities such as a mixing order parameter or the radial distribution functions for each metal. We have also studied the inherent structures for these clusters, in order to identify the important geometries underlying all homotops. As will be seen below, the distributions of inherent structures clearly illustrate how the Pd–Pt clusters evolve toward the stable bulk solid phase, a solid solution.[52]

This journal is © The Royal Society of Chemistry 2008

This article is organized as follows. In the next section, we briefly describe the potential used to model the clusters energy surface and the sampling algorithm, including the original preferential swapping scheme. The results are presented and discussed in Section 3, with an emphasis on the role of homotops at finite temperature. We finally summarize and conclude in Section 4, suggesting possible extensions of the present work.

2. Methods

2.1 Potential

The cluster sizes studied in the present work are relatively large ($55 \leq n \leq 923$) with respect to the present capabilities of electronic structure methods, for which only single point or local optimisation studies[53] are currently feasible in this range. Here, following several previous investigations on Pd–Pt nanoalloys[6,13,19,23,28–30] we use a semi-empirical many-body Gupta potential to represent metallic bonding. This potential is based on the second moment approximation to the electronic density of states in tight-binding theory, and contains a pairwise repulsive term, as well as a many-body cohesion, all involving exponential forms:

$$V(\boldsymbol{R}) = \sum_{i=1}^{n} \left\{ \sum_{j \neq i} \varepsilon_{ij} \exp(-p_{ij}\rho_{ij}) - \left[\sum_{j \neq i} \zeta_{ij}^2 \exp(-2q_{ij}\rho_{ij}) \right]^{1/2} \right\}, \tag{1}$$

$$\rho_{ij} = \frac{r_{ij}}{r_{ij}^{(0)}} - 1. \tag{2}$$

In the above equations, $\boldsymbol{R} = \{\boldsymbol{r}_i\}$ is the atomic configuration of the cluster, r_{ij} is the distance between atoms i and j, p_{ij}, q_{ij}, ε_{ij}, ζ_{ij}, and $r_{ij}^{(0)}$ are parameters fitted to reproduce experimental values of the lattice parameters, elastic constants and cohesive energies of the bulk metals. Following previous efforts by the Johnston group,[19,23] we have taken the parameters for interactions between alike atoms from Cleri and Rosato,[54] and the parameters for interactions between unlike atoms from Massen et al.[19]

Using the Gupta potential with these parameters, we have investigated the three icosahedral Pd–Pt clusters containing between 2 and 4 shells, that is having 55, 147, or 309 atoms, respectively. The absolute binding energies of these nanoalloys are −4.0649 eV per atom for $Pd_{43}Pt_{12}$, −4.3193 eV per atom for $Pd_{104}Pt_{43}$ and −4.4628 eV per atom for $Pd_{205}Pt_{104}$. We also studied the 6-shell cluster $Pd_{567}Pt_{356}$ cluster, with cohesion energy −4.6185 eV per atom.

As recently shown by Paz-Borbón and coworkers,[6] the present Gupta potential may have a limited accuracy for small Pd–Pt clusters, more stable isomers being occasionally found at the density-functional theory level. However, we are here mostly interested in the moderate to large sizes regime, for which the many-body potential is mainly devoted.

2.2 Preferential swapping

The thermal behaviour of the Pd–Pt bimetallic clusters was addressed using parallel tempering Monte Carlo simulations,[55] improved with the recent all-pair exchange strategy,[56] in the canonical ensemble. In order to sample the homotops corresponding to a given isomer, we have added identity swapping moves between pairs of Pd and Pt atoms. The basic swapping move consists of randomly selecting one atom of each type, and exchanging their identity, the move being accepted using the standard Metropolis probability at the current temperature T. While such moves efficiently accelerate sampling of configuration space in binary systems,[47,49,50,57,58] their acceptance rate becomes increasingly low at high densities or low temperatures due to the difficulty of inserting the large particle at the location of the small particle. Faller and de Pablo[49] and Flenner and Szamel[50] employed a configurational bias technique to reduce this problem, by generating several (typically 50) trial moves to enhance the chances of getting one of them effectively accepted.

Here we use a different approach, based on the preferential sampling scheme originally introduced by Owicki and Scheraga[51] to improve convergence in the simulation of solutes. The idea is to select the Pd–Pt pairs which have greater chances of being swapped, and to correct for this bias afterwards in the acceptance rate. Since there are fewer Pt atoms than Pd atoms in the clusters, we first randomly select one Pt atom, labelled i, with equal probability among all such atoms. Instead of selecting the Pd atom j randomly, we allocate a weight w_j^{old} to all Pd atoms, based on their distance to atom i, such that

$$w_j^{old} = \frac{1}{r_{ij}^{\nu}}.$$
(3)

The exponent ν entering eqn (3) above is an arbitrary parameter chosen as $\nu = 2$. The weights are normalised using their sum W_{old} as

$$W_{old} = \sum_j w_j^{old}.$$
(4)

One specific atom j is then randomly selected based on its normalised probability w_j^{old}/W_{old}. The corresponding unnormalised weights w_j^{new} of all Pd atoms after the identity swapping are then calculated using a similar formula. In order to restore the detailed balance, the trial move $i \rightleftharpoons j$ is accepted with the unbiased Metropolis rule:[51]

$$acc\,(i \rightleftharpoons j) = \min\left\{1, \frac{W_{old}}{W_{new}}\exp\left[-\frac{V_{new} - V_{old}}{k_B T}\right]\right\},$$
(5)

with V_{old} and V_{new} being the potential energies of configurations before and after the swapping move, respectively.

Our Monte Carlo simulations therefore consist of three types of move, namely random displacements of a random atom, replica exchange, and identity swapping. In practice, one replica exchange move is attempted once every 10 Monte Carlo sweeps and during each step of the MC sweep identity exchange is attempted with 10% probability. Concerning the simulation details, between 32 and 40 replicas were used with an arithmetic progression in temperature in the range 50–1500 K, with extra replicas being added at low temperatures and in the 500–1000 K range where the melting point was expected to be found. 10^7 MC sweeps were performed for each replica. For all simulations, all replicas were initiated in the onion-ring structure, which is the global minimum at least for the smallest, 2-shell cluster. For larger clusters, we have not attempted to locate the actual global minimum using dedicated optimization methods such as basin-hopping or genetic algorithms. However, parallel tempering Monte Carlo, which also act as a global minimization method, was not able to find structures more stable than the onion-ring isomer for the 3- and 4-shell clusters.

2.3 Order parameters

The Monte Carlo scheme described in the previous section was used to calculate the constant volume heat capacity $C_v(T)$ of the Pd–Pt nanoalloys. For this purpose the potential energy distributions obtained at all replicas were processed into the microcanonical density of states using histogram reweighting. The canonical quantities were then calculated from suitable Laplace transformation of the density of states. We also computed the radial distribution functions $f(R_{pd})$ and $f(R_{pt})$ with respect to the centre of mass. The similarity between homotops, or the extent of mixing within a given isomer, has been quantified using a mixing index γ defined as the thermal average of a function based on the coordination numbers $\mathcal{N}$ associated with unlike and alike pairs:

$$\gamma = \left\langle \frac{\mathcal{N}_{PdPd} + \mathcal{N}_{PtPt} - \mathcal{N}_{PdPt}}{\mathcal{N}_{PdPd} + \mathcal{N}_{PtPt} + \mathcal{N}_{PdPt}} \right\rangle, \qquad (6)$$

with $\mathcal{N}_{AB}$ being the number of bonds between atoms A and B in the cluster, a bond being formed if the distance between them lies below 3 Å. The above definition is somewhat similar to other expressions based on binding energies,[14,58] however it is much more straightforward and has a simple interpretation. For a perfectly segregated, infinite and stoichiometric system, $\mathcal{N}_{AB}$ is negligible with respect to both $\mathcal{N}_{AA}$ and $\mathcal{N}_{BB}$, hence $\gamma = 1$. For a perfectly mixed structure (as in the rock salt crystal), $\mathcal{N}_{AA} = \mathcal{N}_{BB} = 0$ and $\gamma = -1$. As the stoichiometries differ from 50%, these two limiting values are no longer reached. For a layered cluster such as the onion-ring Pd–Pt nanoalloys studied in the present work, there are significantly more Pd atoms than Pt atoms and the mixing index γ lies in the range 0.15–0.3; higher values being found for the smaller cluster. As they are heated up beyond the melting point, mixing is expected to be more complete, therefore γ should decrease at high temperatures.

3. Results and discussion

3.1 Heat capacities

The variations of the canonical capacities of the 2-, 3- and 4-shell Pd–Pt clusters are represented in Fig. 1 *versus* canonical temperature. In addition to the onion-ring bimetallic clusters, we also simulated the homogeneous clusters having the same size. While the palladium clusters exhibit multishell icosahedral ground state structures, platinum clusters have significantly deformed lowest-energy minima, such as the "amorphous" Pt_{55} previously reported by Massen and coworkers.[19] For Pt_{147} and Pt_{309}, we did not attempt to locate precisely the global minimum geometries, however after preliminary parallel tempering simulations, deformed icosahedra in which a vertex atom opens and enters the neighbouring fivefold ring (also referred to as a 'rosette' deformation by Aprà and coworkers[59]) were found at energies lower than those of the perfect icosahedra.

Common observations can be made for each of the three clusters. All heat capacity curves exhibit one main peak indicative of the solid-like–liquid-like transition. Pure Pd clusters melt at a temperature generally lower than pure Pt clusters; the melting point of bimetallic compounds lying in between these limits. This is in agreement with the known phase diagram in the bulk,[62] and will be further discussed below. The area below the peak, which is a direct estimate of the latent heat of fusion, is always higher for the nanoalloy than for pure clusters. This suggests a special energetic stability of the solid-like state, related here to the magic character of the onion-ring pattern.[29] The latent heat of the larger pure Pd and Pt clusters are comparable. However, due to its amorphous character, Pt_{55} melts without involving a high latent heat with respect to the icosahedral Pd_{55}.

Looking now at the variations of the curves with increasing size, the expected behaviour of a first-order-like transition is recovered: the melting peak becomes narrower and higher, evolving toward the bulk δ singularity.[60] The center of the melting peak smoothly increases with size, which is again the expected trend—as in most materials.[60] Interestingly the heat capacity of Pd_{147} displays an important shoulder on the low-temperature side, in the 650–750 K range. A smaller shoulder is also noticeable for Pt_{147}. By performing periodic quenching of the cluster configurations taken along the MC trajectories, the inherent structures of the cluster were gathered and could be identified as surface reconstructed. These isomers differ from

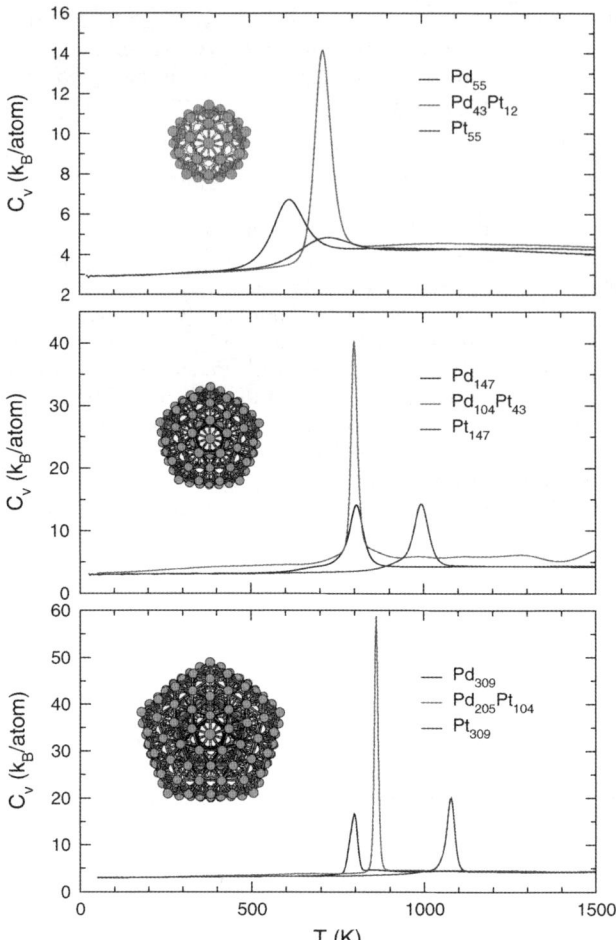

Fig. 1 Calculated canonical heat capacities of icosahedral 2-shell (upper panel), 3-shell (middle panel) and 4-shell (lower panel) Pd_nPt_m clusters. For each size, the three curves corresponding to pure Pd ($m = 0$), pure Pt ($n = 0$), or onion-ring, Pd-rich clusters are represented.

the Mackay icosahedral global minimum from the location of the outer layer, which adopts a more disordered and polytetrahedral anti-Mackay character. Such a surface reconstruction premelting behaviour has also been found in the 309-atom Lennard-Jones cluster by Noya and Doye.[61] In the bimetallic 3- and 4-shell clusters, the heat capacity also has a very flat premelting shoulder extending over several hundreds of Kelvins. In the larger $Pd_{205}Pt_{104}$ this shoulder is also present, albeit rather hidden by the very high melting peak.

3.2 Comparison with the bulk phase diagram

The various melting points of the pure and bimetallic clusters extracted from our simulation results have been represented altogether as a function of the platinum fraction in Fig. 2. The experimental solidus–liquidus line modelled by Okamoto[62] has been superimposed on this graph for the bulk Pd–Pt alloy. We have tested the present Gupta potential against these data by performing additional parallel tempering Monte Carlo simulations for bulk Pd–Pt compounds, using 108 particles distributed on a fcc lattice, with periodic boundary conditions in the minimum image

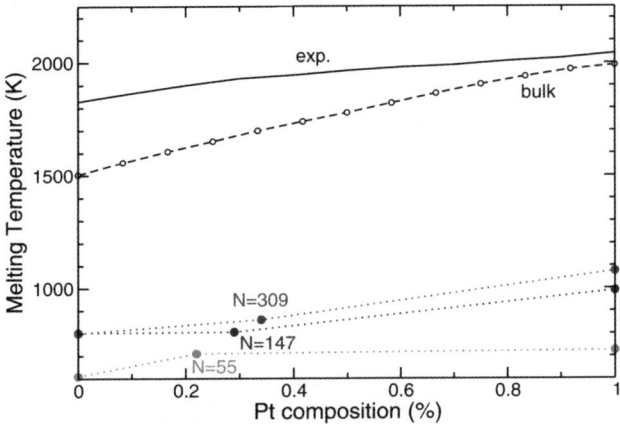

Fig. 2 Melting temperature of a 108-atom sample of bulk Pd–Pt for various platinum fractions, obtained by Monte Carlo simulations (circles) and compared with the experimental solidus–liquidus line.[62] The melting temperatures of the 2-, 3- and 4-shell icosahedral clusters are also shown as triangles, squares, and diamonds, respectively.

convention. Pt and Pd atoms were initially randomly located on the lattice sites, and the preferential swapping scheme was also used to enhance equilibration. For a better comparison with experiment, all these simulations of bulk compounds were performed at constant zero pressure, with volume moves attempted once every single MC sweep.

As can be seen from Fig. 2, the present Gupta potential underestimates the bulk melting point of palladium by a significant amount (about 18%), but rather correctly predicts platinum to melt near 2000 K (2.5% error). Even though the error decreases proportionally to the Pt fraction, the experimental trend is recovered. Of course, due to the small number of particles used to mimic the bulk compound, we cannot exclude finite-size effects in our simulations, and probably the melting points obtained with larger samples will be closer to experiment.

The melting temperatures of the onion-ring nanoalloys lie well below the bulk melting point, even predicted from the Gupta potential. Pd-rich onion-ring structures have a single Pd/Pt composition, which only depends on the number of layers, and would reach 50% as many layers are added. While the data are scarce for each size, the melting point is seen to generally increase with composition, in a less regular fashion in clusters than in the bulk, but with a comparable influence of composition on its magnitude. In particular, we do not find evidence for strong size effects such as a non-monotonic composition dependence on the melting temperature, as could be observed, *e.g.* in rare-gas binary clusters.[58,63]

3.3 Mixing

The phase diagram sketched in Fig. 2, which separates the liquid from the solid, actually involves a solid solution in the bulk.[52] We now turn to characterizing the atomic structure of bimetallic clusters below their melting point. A first useful quantity is the thermally averaged mixing order parameter defined in eqn (6). The variations of γ *versus* temperature are represented in Fig. 3 for the 2-, 3- and 4-shell onion-ring clusters.

The three curves $\gamma(T)$ exhibit similar, generally decreasing variations, with a strong drop in the melting region, followed by a plateau. To better interpret these data, we have performed Monte Carlo simulations in the solid-like region, keeping the onion-ring structure by disabling both exchange moves between replicas and identity swapping moves between unlike atoms. The corresponding variations of γ, also

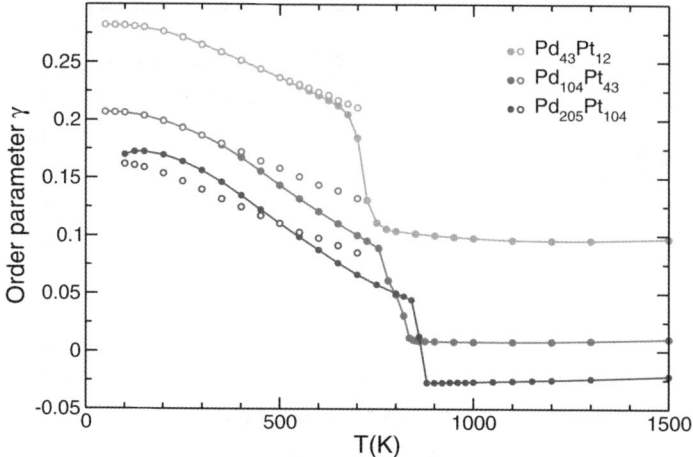

Fig. 3 Mixing order parameter γ [see eqn (6)] of the 2-, 3- and 4-shell onion-ring icosahedral clusters, obtained from Monte Carlo simulations. The open symbols were obtained from simulations restricted to sample the thermal fluctuations in the onion-ring geometry only.

represented in Fig. 3, quantify the role of thermal atomic fluctuations around the equilibrium geometry on the variations of the mixing order parameter. For the smaller cluster, these fluctuations lead to an important part of the variations in γ at moderate temperature, and swapping moves are rarely performed up to about 600 K. For the 3-shell $Pd_{104}Pt_{43}$ cluster the deviations are marked above 400 K and, for $Pd_{205}Pt_{104}$, the curves do not match even at 100 K. These differences show that some Pd and Pt atoms switch locations at thermal equilibrium, below the cluster melting point. For the largest cluster, the most stable structure (with the lowest free energy) is no longer the onion-ring geometry at temperatures as low as 100 K.

Further insight into the mixing process can be obtained from the radial distributions of atoms with respect to the centre of mass. While icosahedral clusters are not perfectly spherical, the radial distributions of atoms within a shell do not overlap with each other until the fifth and sixth shells. These distributions have been represented in Fig. 4 for the 4-shell cluster, and in Fig. 5 for the 6-shell cluster, at 300 and 600 K.

At 300 K, the first icosahedral shell of $Pd_{205}Pt_{104}$ contains as many atoms of each metal, as seen from the equal peaks at 2.7 Å. The second shell contains a majority of Pd atoms and a few Pt atoms. The other remaining layers are no longer pure and also affected, albeit to a lesser extent. At 600 K, the peaks thermally broaden, and unlike atoms swap more frequently: the relative proportions of Pd atoms in the 2nd and 4th shells and of Pt atoms in the 3rd shell all increase. In the larger, 6-shell cluster, mixing is more important at 300 K and significantly affects the four innermost layers. At 600 K, the layered structure seems lost on the outer shells, possibly due to surface melting or surface reconstruction. From the very similar radial densities of the two metals, the onion-ring structure is fully lost except on the outer parts, where the higher fraction of palladium leads to a surface richer in this metal.

Based on the previous figures, bimetallic Pd–Pt clusters behave essentially as a solid solution below the melting point. However, the solid solution does not fill the entire cluster, but proceeds from the core to the surface as the temperature increases. This core melting process, which was suggested before[64] and found in a simulation with a specially tailored pair potential,[65] has not yet been reported in any realistic cluster, to the best of our knowledge. The more frequent exchanges between Pd and Pt atoms in the cluster interior can be caused by the strain experienced by icosahedra, which is larger in the core than on the surface. The higher strain in the icosahedral geometry explains why the cluster changes to less compact,

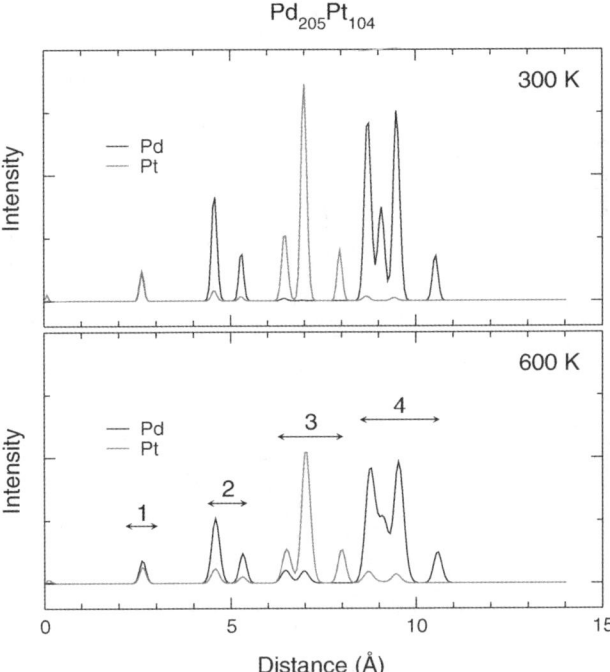

Fig. 4 Radial atomic density of Pd and Pt atoms in the 4-shell onion-ring cluster, at 300 K (upper panel) and 600 K (lower panel). The horizontal arrows indicate the extent of the successive shells in the onion-ring icosahedral structure.

decahedral or cubic geometries at large sizes.[66] As the number of icosahedral layers increases, the central shells are compressed and the interatomic distances shorten. The distance between Pd and Pt atoms from neighbouring layers decreases also, some strain is then relieved by mixing unlike atoms within the same layer. Since the exchanges are favoured by a higher density, they do not occur often on the less dense surface, which further amplifies the core-melting effect.

Neither in the heat capacity nor in the mixing index γ did we find a strong change that would provide evidence for a phase change rounded by size effects associated with the appearance of the solid solution from the well-defined onion ring structure. The transition to the solid solution is continuous over hundreds of kelvins below the melting point. A simple analogy can be made with the 3D Ising model on the ferromagnetic fcc lattice, where Pd and Pt atoms are replaced by spins up or down. The constant Pd/Pt composition is then analogous to a constant magnetization, with the Pd–Pt identity swapping move corresponding to a double spin flip ($\uparrow \downarrow$ into $\downarrow \uparrow$). The order–disorder transition on this lattice model is second-order and continuous in the thermodynamical limit, which is consistent with our observation of a continuous transition in the present case of Pd–Pt nanoalloys. The transition toward the solid solution phase may then be a unique case of a second-order-like phase change in finite clusters.[67] Such a conjecture is supported by the absence of any latent heat, only a small increase without a peak being seen in the heat capacity.

3.4 Inherent structures

We now seek to interpret the above results in terms of the energy landscape of the clusters, paying particular attention to the homotops of the icosahedral structure. The instantaneous configurations periodically taken from the Monte Carlo trajectories were locally optimised using a conjugate gradient algorithm. Each

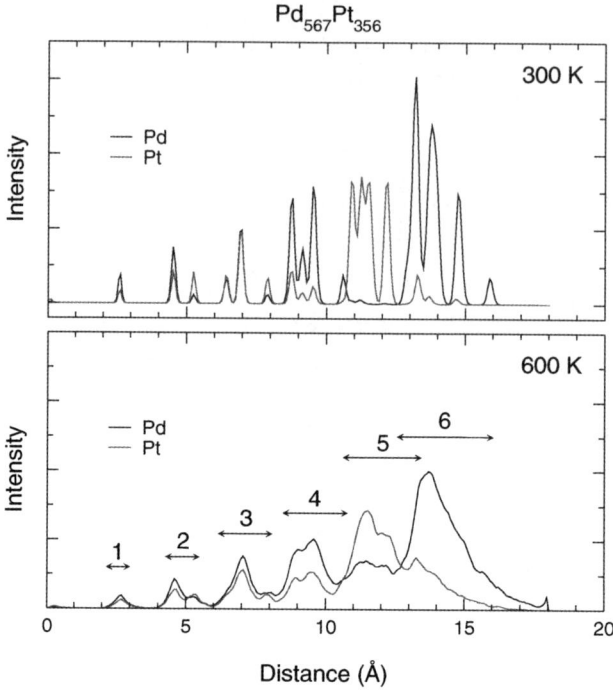

Pd$_{567}$Pt$_{356}$

300 K

— Pd
— Pt

Intensity

600 K

— Pd
— Pt

Intensity

1   2   3   4   5   6

0   5   10   15   20

Distance (Å)

Fig. 5 Radial atomic density of Pd and Pt atoms in the 6-shell onion-ring cluster, at 300 K (upper panel) and 600 K (lower panel). The horizontal arrows indicate the extent of the successive shells in the onion-ring icosahedral structure.

configuration was also quenched in its homogeneous form, assuming all atoms are of the Pd type. The isomers obtained with this second quench form the basis of the homotops in bimetallic clusters. In particular, we are interested here in the homotops that belong to the multilayer icosahedron. 20 000 configurations were optimised for each size, resulting in 10 863 (2-shell cluster), 12 783 (3-shell cluster) and 15 345 (5-shell cluster) different minima. Optimising these configurations assuming an homogeneous Pd cluster lead to 52, 3380, and 4739 different homotops of the icosahedral isomer for the 2-, 3- and 4-shell clusters, respectively. The densities of inherent structures are represented in Fig. 6 for these clusters. On this figure, the samples of isomers that are homotops to the Mackay icosahedra have been high-lighted. As expected, these homotops are found on the low energy boundary of the isomer spectra.

In the smaller cluster Pd$_{43}$Pt$_{12}$, all icosahedral homotops are the first excited isomers, separated from the non-icosahedral, liquid-like structures by more than 0.1 eV per atom. Isomers based on the Mackay icosahedron are a small subset of the existing inherent structures. In the intermediate size cluster Pd$_{104}$Pt$_{43}$, the density of inherent structures is mainly bimodal, and an important peak containing mainly homotops of the multilayer icosahedron is found with a negligible gap from the onion-ring global minimum. In the largest cluster Pd$_{205}$Pt$_{104}$, a high and narrow peak at low energies reveals that there are many homotops of the icosahedron that are very close in energy. The inherent structure distribution is also bimodal at this size.

These figures provide several useful pieces of information that can be directly related to the other quantities discussed in the previous sections. The bimodal distribution of isomers is related to the bimodality of the instantaneous potential energy distribution (not plotted here) found when performing the histogram

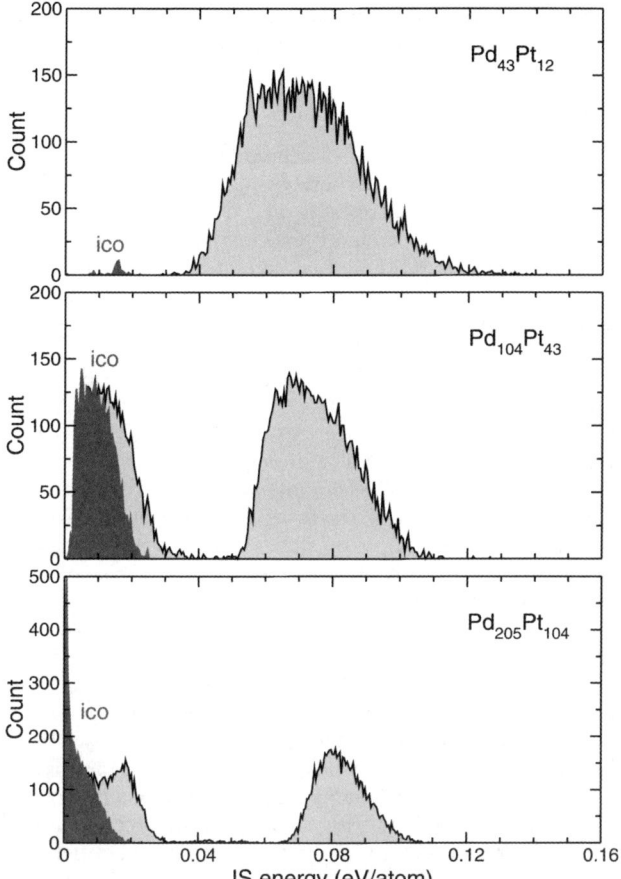

Fig. 6 Density of inherent structures obtained from periodic quenching along the Monte Carlo trajectories for the 2-shell (upper panel), 3-shell (middle panel), and 4-shell (lower panel) Pd-rich, onion-ring clusters. The subset of isomers corresponding to homotops of the Mackay icosahedron is highlighted in dark shaded histograms. The reference energy is the onion-ring multilayer structure.

reweighting analysis, which further shows that melting is a first-order transition rounded by size effects.[60] The two peaks in the distributions naturally distinguish the low-energy solid-like isomers from the high-energy liquid-like structures. The large number of solid-like isomers is mainly due to the presence of homotops, but not exclusively. Numerous isomers based on icosahedra with surface defects are also very close in energy and constitute the high energy part of the solid-like peak in the inherent structures distribution. These defects can be local (holes or floaters), or more extended as in the anti-Mackay reconstructed surfaces observed in simulation by Noya and Doye.[61] The gap between the two peaks in the distributions increases with the cluster size, in agreement with the increase in the melting temperature.

The distributions of icosahedral homotops and defective icosahedra significantly overlap for the larger clusters, and from the evolution of the distributions shown in Fig. 6 it seems more and more easy to excite the onion-ring structure to one other homotop. The loss of the onion-ring structure in the 4-shell cluster at 100 K is reflected on the vanishingly small excitation energy. In larger clusters the equilibrium state will be the superposition of many homotops, even at nearly zero temperature, and the bulk solid–solution behaviour is recovered.

4. Conclusion

The energetic and thermal stability of bimetallic compounds is greatly affected by their composition. In the present paper, we have investigated the influence of temperature on the equilibrium properties of Pd–Pt nanoalloys with icosahedral onion-ring structures. This geometry has been shown to be particularly stable for Pd-rich clusters,[29] which seems to be confirmed here on the especially high peaks in the heat capacity curves, suggesting high latent heats of melting. However, a contrasting picture emerged from looking at the atomic details of the isomers involved below the melting point. The Monte Carlo simulations carried out in the present work were improved using a preferential swapping scheme that increases the chances that identity swapping moves between unlike atoms are accepted. This improvement allowed the various homotops of a given structure to be sampled more efficiently than they would otherwise be with conventional techniques.

Our simulation results indicate that onion-ring Pd–Pt nanoalloys essentially melt at a temperature which increases with the Pt fraction, in a similar way as for the bulk alloy. The present parameterization of the potential slightly underestimates the melting point in the bulk, especially for pure palladium and this underestimation will probably affect the predictions of the potential for clusters. Rescaling the potential could for instance be anticipated as a correcting attempt. Prior to melting, the Pd–Pt clusters undergo a preliminary melting phase in which the Pd and Pt atoms can occupy various locations among the Mackay icosahedral lattice. Considering the homotops that are most likely sampled, a solid–solution behaviour was found to proceed from the core to the surface as the temperature increases. The solid–solution state takes place here as a precursor to melting, and for large clusters the onion-ring structure itself is no longer the free energy minimum at low temperatures. From the absence of latent heat, the analogy with the constant-magnetization ferromagnetic 3D Ising model, and the very continuous variations in all order parameters, we conjectured that this continuous transition may be more akin to a second-order phase change, rounded by size effects, rather than the more usual first-order like changes.

Although the present work was limited to the specially stable onion-ring structures, we believe that most results will hold at other compositions that do not allow for perfect alternating layers. In particular, we expect that the solid–solution precursor will be present also in clusters with a nonicosahedral global minimum. In very large nanoalloys, the close-packed cubic geometry should become favoured over the strained icosahedra,[66] and the relative locations of Pd and Pt atoms in the global minimum may differ from the onion-ring structure. However, the present results suggest that one should not worry too much about finding the global minimum of large Pd–Pt bimetallic clusters, because even a low temperature will result in the solid–solution being the free energy minimum.

Beyond this study, it would be worth trying to characterize the transition toward the solid–solution state using more quantitative tools, to address the issue of the order of this transition. Trying to build a general size–temperature phase diagram[66] would also be interesting, since the solid–solution precursor is highly size-specific. Of course, the effects of composition would have to be dealt with properly, and a simple coarse-grained approach could be to select several fractions and to interpolate between the obtained results. The preferential swapping scheme introduced in this paper, which was a key to efficiently sampling the multiple homotops of the starting structure, could be applied straightforwardly to other heterogenenous systems in which convergence of the simulation is a fundamental issue. For instance, the well-documented binary Lennard-Jones glass forming system[49,50] provides a natural testing ground for the method, on which we shall work in the near future.

Acknowledgements

The author wishes to thank Dr R. L. Johnston for very useful discussions on Pd–Pt bimetallic clusters.

References

1   M. Calleja, C. Rey, M. M. G. Alemany, L. J. Gallego, P. Ordejón, D. Sánchez-Portal, E. Artacho and J. M. Soler, *Phys. Rev. B*, 1999, **60**, 2020.
2   A. S. Chacko, M. Deshpande and D. G. Kanhere, *Phys. Rev. B*, 2001, **64**, 155409.
3   A. S. Bromley, G. Sankar, C. R. A. Catlow, T. Maschmeyer, B. F. G. Johnson and J. M. Thomas, *Chem. Phys. Lett.*, 2001, **340**, 524.
4   A. B. K. Rao, S. Ramos de Debiaggi and P. Jena, *Phys. Rev. B*, 2001, **64**, 024418.
5   V. Bonačić-Koutecký, J. Burda, R. Mitrić, M. Ge, G. Zampella and P. Fantucci, *J. Chem. Phys.*, 2002, **117**, 3120.
6   L. O. Paz-Borbón, R. L. Johnston, G. Barcaro and A. Fortunelli, *J. Phys. Chem. C*, 2007, **111**, 2936.
7   A. P. Ballone, W. Andreoni, R. Car and M. Parrinello, *Europhys. Lett.*, 1989, **8**, 73.
8   I. L. Garzon, X. P. Long, R. Kawai and J. H. Weare, *Chem. Phys. Lett.*, 1989, **158**, 525.
9   A. M. Schoeb, T. J. Raeker, L. Q. Yang, T. Wu, T. S. King and A. E. DePristo, *Surf. Sci.*, 1992, **278**, L125.
10 G. E. López and D. L. Freeman, *J. Chem. Phys.*, 1993, **98**, 1428.
11 A. L. Zhu and A. E. DePristo, *J. Chem. Phys.*, 1995, **102**, 5342.
12 J. L. Rousset, A. M. Cadrot, F. J. Cadete Santos Aires, A. Renouprez, P. Mélinon, A. Perez, M. Pellarin, J. L. Vialle and M. Broyer, *J. Chem. Phys.*, 1995, **102**, 8574.
13 J. L. Rousset, B. C. Khanra, A. M. Cadrot, F. J. Cadete Santos Aires, A. Renouprez and M. Pellarin, *Surf. Sci.*, 1996, **342–354**, 583.
14 (a) J. Jellinek and E. B. Krissinel, *Chem. Phys. Lett.*, 1996, **258**, 283; (b) E. B. Krissinel and J. Jellinek, *Chem. Phys. Lett.*, 1997, **272**, 301.
15 M. J. López, P. A. Marcos and J. A. Alonso, *J. Chem. Phys.*, 1996, **104**, 1056.
16 M. C. Vicéns and G. E. López, *Phys. Rev. A*, 2000, **62**, 033203.
17 F. Baletto, C. Mottet and R. Ferrando, *Phys. Rev. B*, 2002, **66**, 155420.
18 S. Darby, T. V. Mortimer-Jones, R. L. Johnston and C. Roberts, *J. Chem. Phys.*, 2002, **116**, 1536.
19 C. Massen, T. V. Mortimer-Jones and R. L. Johnston, *J. Chem. Soc., Dalton Trans.*, 2002, **23**, 4375.
20 M. S. Bailey, N. T. Wilson, C. Roberts and R. L. Johnston, *Eur. Phys. J. D*, 2003, **25**, 41.
21 F. Baletto, C. Mottet and R. Ferrando, *Phys. Rev. Lett.*, 2003, **90**, 135504.
22 G. Rossi, A. Rapallo, C. Mottet, A. Fortunelly, F. Baletto and R. Ferrando, *Phys. Rev. Lett.*, 2004, **93**, 105503.
23 L. D. Lloyd, R. L. Johnston, S. Salhi and N. T. Wilson, *J. Mater. Chem.*, 2004, **14**, 1691.
24 T. V. Hoof and M. Hou, *Phys. Rev. B*, 2005, **72**, 115434.
25 A. M. M. Marcelo, A. D. Sergio and P. M. L. Ezequiel, *J. Chem. Phys.*, 2005, **123**, 184505.
26 (a) G. F. Wang, M. A. van Hove, P. N. Ross and M. I. Baskes, *J. Phys. Chem. B*, 2005, **109**, 11683; (b) G. F. Wang, M. A. van Hove, P. N. Ross and M. I. Baskes, *J. Chem. Phys.*, 2005, **122**, 024706.
27 A. Rapallo, G. Rossi, R. Ferrando, A. Fortunelli, B. C. Curley, L. D. Lloyd, G. M. Tarbuck and R. L. Johnston, *J. Chem. Phys.*, 2005, **122**, 194308.
28 G. Rossi, R. Ferrando, A. Rapallo, A. Fortunelli, B. C. Curley, L. D. Lloyd and R. L. Johnston, *J. Chem. Phys.*, 2005, **122**, 194309.
29 D. Cheng, W. Wang and S. Huang, *J. Phys. Chem. B*, 2006, **110**, 16193.
30 D. Cheng, S. Huang and W. Wang, *Chem. Phys.*, 2006, **330**, 423.
31 J. H. Sinfelt, *Bimetallic Catalysts: Discoveries, Concepts, and Applications*, Wiley, New York, 1983.
32 N. Toshima and T. Yonezawa, *New J. Chem.*, 1998, **11**, 1179.
33 B. Pawelec, R. Mariscal, R. M. Navarro, S. van Bokhorst, S. Rojas and J. L. G. Fierro, *Appl. Catal., A*, 2002, **225**, 223.
34 N. Györffy, L. Tóth, M. Bartók, J. Ocskó, U. Wild, R. Schlögl, D. Teschner and Z. Paál, *J. Mol. Catal. A: Chem.*, 2005, **238**, 102.
35 K. Persson, A. Ersson, A. M. Carrera, J. Jayasuriya, R. Fakhrai, T. Fransson and S. Järås, *Catal. Today*, 2005, **100**, 479.
36 N. Toshima, M. Harada, T. Yonezawa, K. Kushihashi and K. Asakura, *J. Phys. Chem.*, 1991, **95**, 7448.
37 M. Harada, K. Asakura, Y. Ueki and N. Toshima, *J. Phys. Chem.*, 1992, **96**, 9730.

38   U. Kolb, S. A. Quaiser, M. Winter and M. T. Reetz, *Chem. Mater.*, 1996, **8**, 1889.
39   L. Fiermans, R. D. Gryse, G. D. Doncker, P. A. Jacobs and A. J. Martens, *J. Catal.*, 2000, **193**, 108.
40   N. Toshima, T. Yonezawa and K. Kushihashi, *J. Chem. Soc., Faraday Trans.*, 1993, **89**, 2537.
41   D. Bazin, D. Guillaume, C. Pichon, D. Uzio and S. Lopez, *Oil Gas Sci. Technol.*, 2005, **60**, 801.
42   S. Mandal, P. R. Selvakannan, R. Pasricha and M. Sastry, *J. Am. Chem. Soc.*, 2003, **125**, 8440.
43   R. Harpeness and A. Gedanken, *Langmuir*, 2004, **20**, 3431.
44   N. Toshima, M. Kanemaru, Y. Shiraishi and Y. Koga, *J. Phys. Chem. B*, 2005, **109**, 16326.
45   J. He, I. Ichinose, T. Kunitake, A. Nakao, Y. Shiraishi and N. Toshima, *J. Am. Chem. Soc.*, 2003, **125**, 11034.
46   B. J. Hwang, L. S. Sarma, J. M. Chen, C. H. Chen, S. C. Shih, G. R. Wang, D. G. Liu, J. F. Lee and M. T. Tang, *J. Am. Chem. Soc.*, 2005, **127**, 11140.
47   D. H. Robertson, F. B. Brown and I. M. Navon, *J. Chem. Phys.*, 1989, **90**, 3221.
48   R. G. Donnely and T. S. King, *Surf. Sci.*, 1978, **74**, 89.
49   R. Faller and J. J. de Pablo, *J. Chem. Phys.*, 2003, **199**, 4405.
50   E. Flenner and G. Szamel, *Phys. Rev. E*, 2006, **73**, 061505.
51   J. C. Owicki and H. A. Scheraga, *Chem. Phys. Lett.*, 1977, **47**, 600.
52   R. Hultgren, P. D. Deai, D. T. Hawkins, M. Gleiser and K. K. Kelley, *Values of the Thermodynamic Properties of Binary Alloys*, American Society for Metals, Berkeley, 1981.
53   P. Nava, M. Sierka and R. Ahlrichs, *Phys. Chem. Chem. Phys.*, 2003, **5**, 3372.
54   F. Cleri and V. Rosato, *Phys. Rev. B*, 1993, **48**, 22.
55   (a) R. H. Swendsen and J.-S. Wang, *Phys. Rev. Lett.*, 1986, **57**, 2607; (b) G. J. Geyer, in *Computing Science and Statistics: Proceedings of the 23rd Symposium on the Interface*, ed. E. K. Keramidas, Interface Foundation, Fairfax Station, 1991, p. 156; (c) K. Hukushima and K. Nemoto, *J. Phys. Soc. Jpn.*, 1994, **65**, 1604.
56   (a) F. Calvo, *J. Chem. Phys.*, 2005, **123**, 124106; (b) P. Brenner, C. R. Sweet, D. VonHandorf and J. A. Izaguirre, *J. Chem. Phys.*, 2007, **126**, 074103.
57   T. S. Grigera and G. Parisi, *Phys. Rev. E*, 2001, **63**, 045102.
58   F. Calvo and E. Yurtsever, *Phys. Rev. B*, 2004, **70**, 045423.
59   R. Aprà, F. Baletto, R. Ferrando and A. Fortunelli, *Phys. Rev. Lett.*, 2004, **93**, 065502.
60   P. Labastie and R. L. Whetten, *Phys. Rev. Lett.*, 1990, **65**, 1567.
61   E. G. Noya and J. P. K. Doye, *J. Chem. Phys.*, 2006, **124**, 104503.
62   H. Okamoto, *J. Phase Equilibria*, 1991, **12**, 617.
63   D. D. Frantz, *J. Chem. Phys.*, 1997, **107**, 1992.
64   R. S. Berry, in *Large Clusters of Atoms and Molecules*, ed. T. P. Martin, 1996, Kluwer, Dordrecht, p. 281.
65   D. V. Anghel and M. Manninen, *Eur. Phys. J. D*, 1999, **9**, 437.
66   J. P. K. Doye and F. Calvo, *Phys. Rev. Lett.*, 2001, **86**, 3570.
67   A. Proykova and R. S. Berry, *Z. Phys. D: At. Mol. Clusters*, 1997, **40**, 215.

Atomistic computer simulations on the generation of bimetallic nanoparticles

M. M. Mariscal,* N. A. Oldani, S. A. Dassie and E. P. M. Leiva

Received 23rd April 2007, Accepted 11th May 2007
First published as an Advance Article on the web 19th September 2007
DOI: 10.1039/b706149h

Computer simulations on the generation of bimetallic nanoparticles are presented in this work. Two different generation mechanisms are simulated: (a) cluster–cluster collision by means of atom dynamics simulations; and (b) nanoparticle growth from a previous seed through grand canonical Monte Carlo (gcMC) calculations. When two metal nanoparticles collide, different structures are found: core/shell, alloyed and three-shell (A–B–A). On the other hand, the growth mechanism at different chemical potentials by means of gcMC reveals the same results as atom dynamics collisions do.

1. Introduction

Nanoparticles are currently the subject of intense investigation. Research is in progress from a variety of viewpoints, ranging from basic research-oriented studies to the most sophisticated technological applications.[1–3] Such tiny objects, characterized by having a diameter of a few nanometers (1–10 nm), are leading to a new class of materials with properties in the limit between bulk materials and atoms.

Since the days of *Michael Faraday*, who first scientifically elucidated the preparative method for aqueous dispersions of metal nanoparticles,[4] metal nanoparticles have often been produced by chemical reduction of the corresponding metal salts in solution in the presence of suitable stabilizers.[5] Today, metal nanoparticles can be prepared by physical and chemical methods. Showing a parallelism with top–down and bottom–up nanostructuring procedures, these methods start from two different and somehow opposite approaches. Physical methods, which often involve vapor deposition, consist in subdividing bulk precursors to nanoparticles. Chemical procedures, on the other hand, start from reduction of metal ions to metal atoms, followed by controlled aggregation of atoms.

Special attention has been drawn lately to the study of bimetallic nanoparticles. Such nanoparticles, composed of two different metallic elements, are of greater interest than monometallic ones, from both the scientific and technological viewpoints, due to their potential to improve the catalytic properties of metal particles.[6,7] This is so because bimetallization can improve catalytic properties of the original single-metal catalysts and create new properties, which may not be achieved by monometallic catalysts. These effects, caused by the addition of a different metal component, can often be explained in terms of an ensemble and/or a ligand effect in catalysis.

One of the most active areas of research at present is the growth of new materials in the nanometre scale. Special attention is being devoted to the study of bimetallic nanoparticles, which are of great interest due to their numerous potential

Unidad de Matemática y Física, INFIQC—Facultad de Ciencias Químicas, Universidad Nacional de, Córdoba, Argentina. E-mail: marmariscal@fcq.unc.edu.ar

applications in several fields, starting with catalysis towards medical applications.[8] Bimetallic nanoparticles are interesting because the spatial distribution of atoms changes the chemical and physical properties of the nanoparticle.[9,10]

Another burgeoning area of research is currently the fabrication of nanoparticles with different materials to produce core–shell structures, since such a coating allows modification and tailoring of the physical and chemical properties of core materials depending on the synthetic conditions. Furthermore, core–shell nanoparticles are expected to have unique properties that are not originally present in either core or shell materials.[11–13]

2. Bimetallic nanoparticles generation

2.1 Atom dynamics simulations

Very recently, we have proposed a new approach to generate bimetallic nanoparticles through the collision of two single metal nanoparticles. The model system has been explored by means of atom dynamics simulations AD (numerical resolution of the Newton's equation of motion) using the many body *embedded atom method* (EAM)[14] potential to compute the interactions among the various atomic components of the system. In the present work, we have extended our study and performed a series of simulations on the collision of metal nanoparticles having different magic structures: an icosahedron (Ih) of 147 atoms and a truncated octahedron (TO) of 201 atoms, and different chemical nature, say X and Y. A schematic illustration of the simulation model is represented in Fig. 1, where the centres of mass of both metal clusters (X_{147} + Y_{147}) were aligned in the z direction at an initial distance Δz_0, bigger than the cut-off radius used in the EAM potential. The same initial velocity $\vec{v}_0$, but with opposite sign, was applied to each cluster to bring them into collision.

The starting geometries of all metal clusters were constructed in order to represent a truncated octahedron (TO) or an icosahedron (Ih) of different sizes. Before collision, all clusters were equilibrated for a period of 30 ps at 300 K, and later cooled down to 0 K in order for the initial structures to relax.

After impact, the large cluster formed was equilibrated in a constant-total-energy mode until the potential energy of the system approached a steady state.

In an arbitrary system of coordinates, the initial momentum of a system consisting of two separate clusters with different masses and the same velocities but opposite signs is not cancelled out. Thus, the final nanoparticle has a nonzero linear momentum due to momentum conservation in the constant-energy AD simulation. However, if the velocities are referred to the coordinates of the centre of mass (*c.m.*, C reference system[15]), the linear momentum is cancelled out and the dynamic description of the system becomes simple. Similarly, for an arbitrary choice of the system of coordinates, the kinetic energy of the system E_k can be written as

$$E_k = E_k^{c.m.} + \frac{1}{2} M v_{c.m.}^2 \qquad (1)$$

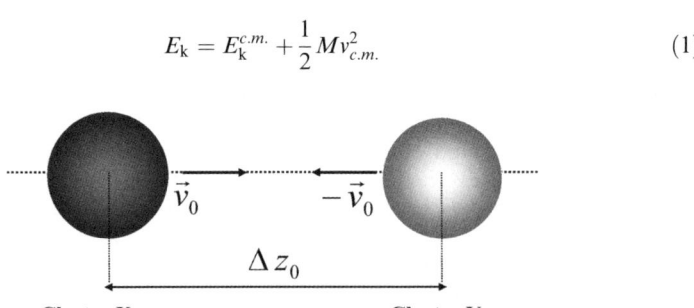

Cluster X_N Cluster Y_N

Fig. 1 Model system for the atom dynamics simulation. Two metal clusters are placed at a distance Δz_0 from each other, with a given initial velocity v_0 of the same magnitude but opposite direction.

Fig. 2 Snapshots taken during the evolution of two Ih_{147} clusters upon collision: (a) initial stage, (b) diffusion stage, and (c) final configuration of the $Co_{core}-Ag_{shell}$ nanoparticle. (black spheres: Co and gray spheres: Ag).

where $E_k^{c.m.}$ is the kinetic energy of the system referred to the c.m., M denotes the total mass, and $v_{c.m.}$ is the velocity of the c.m. We report in all cases $E_k^{c.m.}$, as this is the part of the kinetic energy that may vary due to changes in the potential energy of the system.

2.1.1 Core/shell nanoparticles. Simulations of candidate systems to form core–shell structures were performed within the atom dynamics approach described above. According to this, both metal clusters were placed initially at a distance of 16 Å from each other, with opposite velocities $\vec{v}_0$ and $-\vec{v}_0$, respectively; and after a few picoseconds they collided. Fig. 2 shows the atomic configuration during the time evolution of these clusters. The configuration of the $Ih-Co_{147}$ and $Ih-Ag_{147}$ relaxed clusters at the initial stage is shown in Fig. 2(a). Frame (b) shows an intermediate stage where diffusion of Ag atoms on the Co cluster surface takes place, presenting the wrapping-like behaviour observed for the Pt/Au system.[16] The core–shell structure formed after 0.2 ns can be observed in frame (c). Different simulations were performed to analyze the influence that the magnitude of the initial velocity $\vec{v}_0$ has on the collision, but the same class of core–shell structures were observed in the range of $\vec{v}_0 = 1$–500 m s^{-1}. Similar results (core–shell structures) were observed for the systems: Ni/Au, Cu/Au, and Pt/Au.[16]

In order to gain information on the composition of the nanoparticles around the centre of mass of the nanocluster, we define a radial distribution function as follows:

$$g_r(r) = \frac{1}{m}\sum_{k=1}^{n}\delta\left(k\,dr - r_i\right) \qquad (2)$$

where m represents a normalization factor, k is the number of increments in length (dr) in the distance r measured from the c.m. of the cluster, and r_i is the distance from the ith-atom to the c.m. of the nanoparticles.

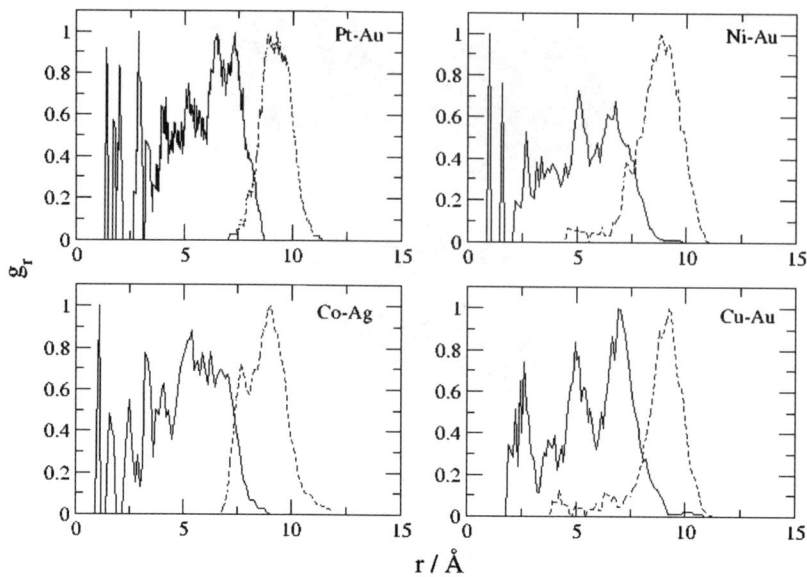

Fig. 3 Radial distribution functions g_r, for different core–shell systems. Full line: g_r of the core atoms (Pt, Ni, Co, Cu). Dotted lines: g_r of the atoms that form the shell (Au and Ag).

This function gives information on the chemical composition of the nanoparticles as a function of the distance from the *c.m.* Fig. 3 shows the radial distribution functions for four selected systems which were observed to yield core–shell structures.

2.1.2 Alloyed nanoparticles.

As a second type of nanoparticles formation by the cluster–cluster collision method we have chosen in this study the Pd–Au and Pd–Cu systems, first due to their important catalytic applications,[17,18] and in the second place, because of the miscibility of the two metals. Since it is well known that palladium and gold form bulk alloys,[19] a certain degree of alloying is expected during the cluster–cluster collision for a wide range of $\vec{v}_0$.

Collision of Pd with Cu clusters produces a kind of structure different from those shown above, that we will call alloyed nanoparticles. This involves a full mixing of atoms, leading to a solid solution. In our previous work the same behaviour was observed for Pd–Au.[16]

For this system we have performed atom dynamics simulations with the "magic" number clusters Ih_{147}, Ih_{309}, and TO_{201}. The final result is in all cases a nanoalloy, where the atoms are placed at random positions. The time evolution of the atomic coordinates is illustrated in Fig. 4. The snapshot in frame (a) shows the initial stage of the relaxed clusters Pd–Ih_{309} and Cu–Ih_{309} right before collision, (b) presents an intermediate stage where the interdiffusion of Cu and Pd atoms can be observed in the interfacial region, and frame (c) shows the system configuration after 2 ns, where the formation of the nanoalloy is evident. To quantify the mixing between Pd and Cu, the radial distribution function $g_r(r)$ is plotted in frame (d).

It is well known from previous experimental work that bimetallic nanoparticles present optical properties[20] different from those of monometallic ones, and also that those properties depend on the composition of the alloy. In the case of Pd–Cu system, strained overlayers are absent; however, the full mixing leads to a new lattice parameter, which changes the nearest-neighbour distance of the atoms in the nanoalloys as compared with that of the pure metal nanoparticles.

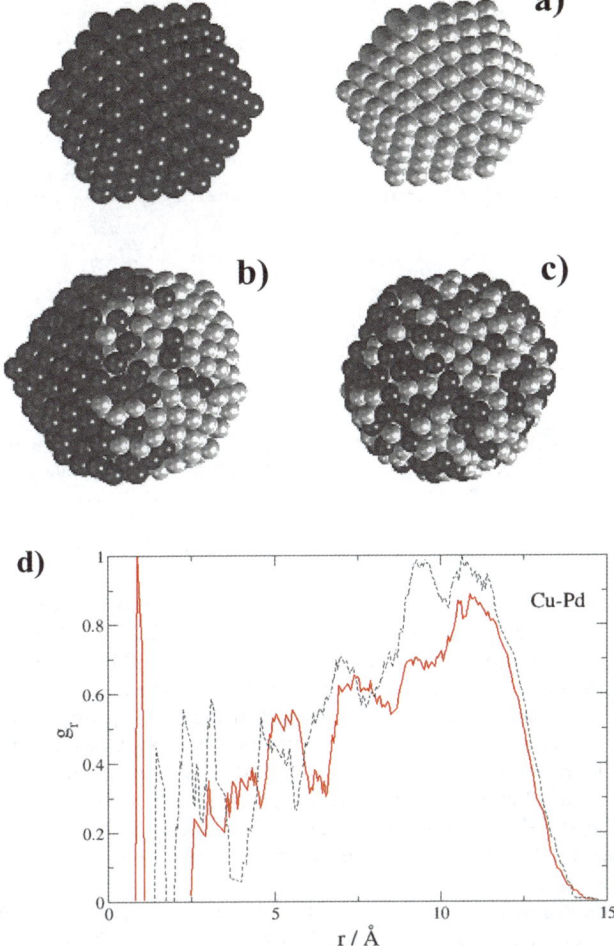

Fig. 4 Snapshots taken during the evolution of two Ih_{309} clusters upon collision (a) initial stage, (b) diffusion stage, and (c) final configuration of the CuPd alloyed nanoparticle. (d) The radial distribution function shows the mixing of Cu and Pd inside the cluster ($r < 7.5$ Å), (Cu: dotted line and Pd: full line).

For quantitative purposes, it is useful to define the pair distribution function, $g_p(r)$ as follows:

$$g_p(r) = \frac{V}{N^2} \left\langle \sum_i^N \sum_{j\neq i}^N \delta\left[r - r_{ij}\right] \right\rangle \qquad (3)$$

where N is the total number of atoms, V is the volume, r_{ij} denotes the distance between atoms i and j, and the brackets represent a time average. The $g_p(r)$ function gives the probability of finding an atom of any type at a distance r, and it allows characterization of the lattice structure during the generation of the nanoalloy. Fig. 5 shows the pair distribution function before and after collision of $Pd–Ih_{309}$ with $Cu–Ih_{309}$. The most probable nearest neighbours distances (NN) before collision are 2.55 Å and 2.75 Å for Cu and Pd, respectively. These values are marked with arrows on the upper panel. After collision and formation of the nanoalloy, the distribution of NN distances around the Cu and Pd atoms show a broadening. Besides the expected contributions of the pure elements at 2.55 Å and 2.75 Å, a new maximum is evident

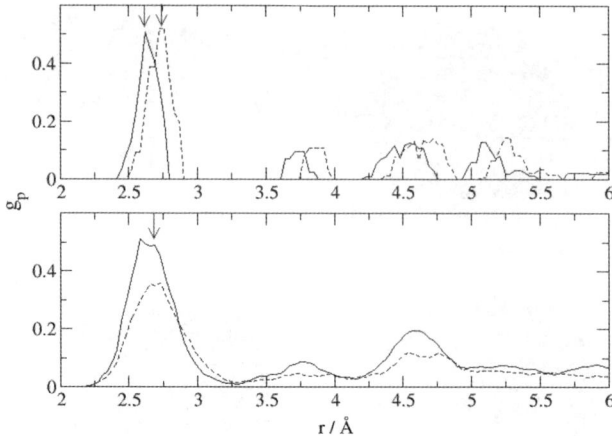

Fig. 5 (Upper panel) Pair distribution function (g_p) for Cu–Ih$_{309}$ (solid line) and Pd–Ih$_{309}$ (dotted line) before collision and (lower panel) after formation of the alloyed nanoparticle. Note how the first nearest neighbors are shifted after collision.

at 2.65 Å (marked with an arrow on the lower panel) indicating the formation of a new structure with a different lattice parameter.

It must be noticed that the surface energies of both metals are very similar[14] and that the alloy heat of formation of an impurity of Pd in Cu, and *vice versa*, is negative, indicating a favorable mixing.[14] Further insight into the remarkably different behaviour of this system in relation to the core/shell ones, where atoms of a given type are segregated to the outer part of the cluster, can be gained from an analysis of the energy change obtained by bringing a type-A atom from the surface of a cluster made of type-B atoms to the centre of that cluster.[16] These energy changes are reported in Table 1. It is clear that the penetration of atoms which were found to be in the shell is strongly disfavored on energetic grounds. On the other hand, it is clear why core-type atoms are better inside the nanoparticle from an energetic viewpoint.

Both, the inclusion of a Pd atom in a Cu cluster and that of a Cu atom in a Pd cluster yield negative results, indicating a favorable mixing of the system on energetic grounds. The same is valid for the Au/Pd system.

Table 1 Energy change ΔE obtained by bringing a type-B atom from the surface of a cluster made of type-A atoms to the centre of that cluster. (Negative values indicate that exchange is favorable)

B/A	ΔE/eV
Co/Ag	−0.34
Ag/Co	+1.02
Pt/Au	−0.10
Au/Pt	+0.76
Ni/Au	−0.22
Au/Ni	+0.85
Cu/Au	−0.44
Au/Cu	+0.70
Cu/Pd	−0.43
Pd/Cu	−0.32
Pd/Au	−0.36
Au/Pd	−0.19

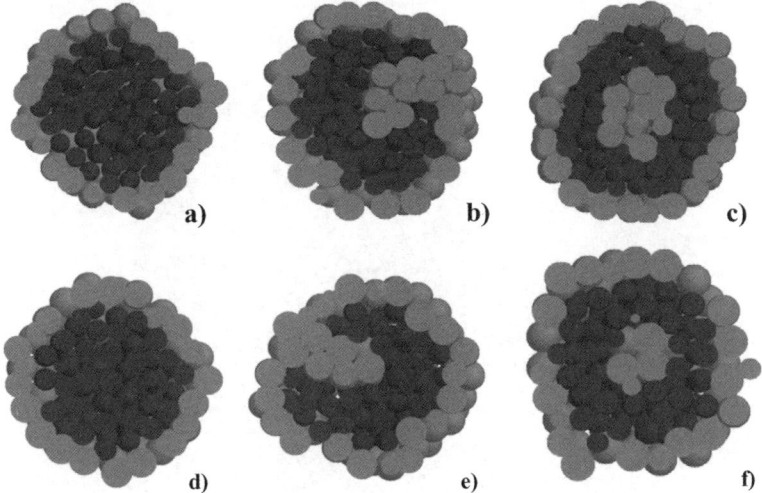

Fig. 6 Formation of three shell nanoparticles. (a–c) TO Ag–Cu–Ag, and (d–f) Ih Ag–Cu–Ag (Cu: black; Ag: gray).

2.1.3 Three-shell onion-like nanoparticles.
Three-shell onion-like bimetallic nanoclusters (A–B–A structure) in which the A atoms are enriched in the outer and inner atomic shells, while the B atoms are enriched in the second atomic shell have attracted theoretical interest in recent years.[16,21–23] This category of structure is also generated by our collision method.

In order to define some patterns in the formation process we have performed simulations with the same initial conditions, that is, with Ih_{147} and TO_{201} clusters, as were used in the above-mentioned systems.

After collision of Cu and Ag TO_{201} clusters, the formation of a core/shell structure was observed at *ca.* 5.3 ns as shown in Fig. 6(a). As the simulation proceeded, few Ag atoms from the outermost layers started to diffuse towards the centre of the cluster. A cross section of the cluster at that point (6 ns) is shown in Fig. 6(b). Finally, after 7.2 ns a well defined three-shell layer nanoparticle (A–B–A structure) is formed (see Fig. 6(c). In the case of colliding Ih structures, essentially the same results are observed. However, a minor amount of Ag atoms diffuse to the centre to form a very small core (see frames d–f in Fig. 6).

The characteristics of these three-shell layers can be quantified by looking at the three-dimensional radial distribution function $g_r(r; t)$ of the silver and copper atoms. Fig. 7 shows $g_r(r; t)$ for Ag and Cu during the time evolution after collision of two TO_{201} clusters. Note that at the beginning Ag is absent at the core of the cluster (between 2–6 Å); however, as the dynamics of the system progress (*ca.* 5.8 ns) a small amount of Ag appears in the centre of the cluster (spikes evident in the figure). The $g_r(r; t)$ function of the Cu atoms on the TO cluster shows the concomitant expulsion of Cu atoms from the centre at the same time.

To monitor the randomization of the positions of N atoms initially occupying an fcc lattice, a translational order parameter λ can be introduced, where λ is defined as:

$$\lambda = \frac{1}{3}\sum_3 \lambda_\alpha \quad \alpha = x, y, z \tag{4}$$

where λ_α is given by:

$$\lambda_\alpha = \frac{1}{N}\sum_i^N \cos\left(\frac{4\pi\alpha_i}{a}\right)$$

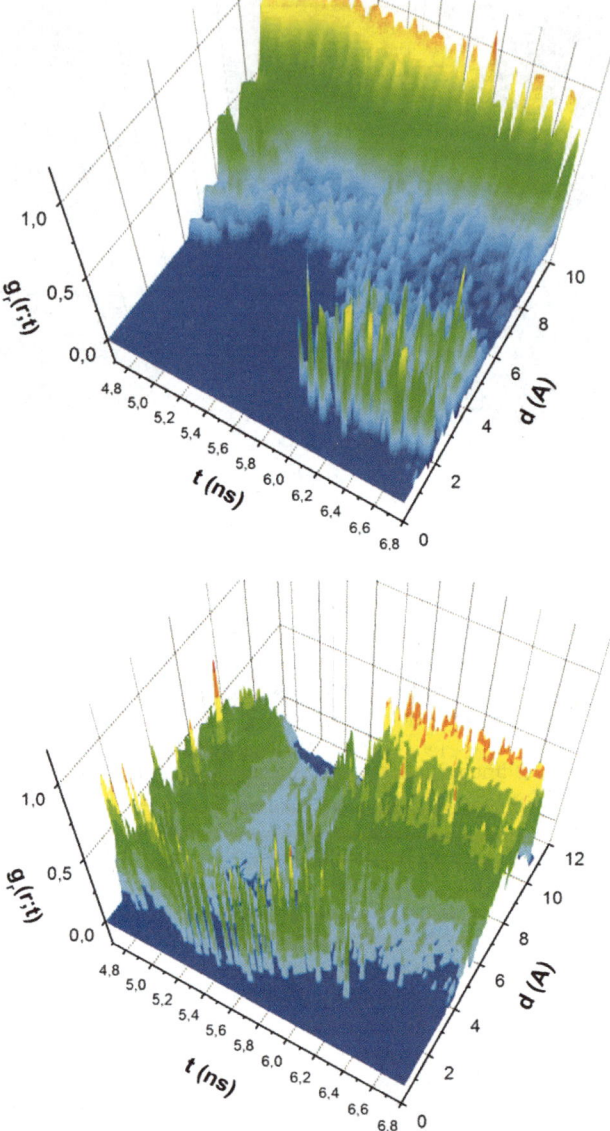

Fig. 7 Three dimensional radial distribution function for Ag (top) and Cu (bottom) on the three-shell TO nanoparticle, respectively. The colours denote the increasing value of the radial distribution function in the blue to red sequence.

a is the length of one of the fcc unit cells. Initially, when all atoms occupy fcc lattice sites, $\lambda = 1$ because the positional components α_i are all integer multiples of $\frac{1}{2}a$. When the lattice is completely disordered, λ fluctuates around zero because then the atoms are distributed randomly about the original lattice sites.

The order parameter functions for TO and Ih clusters are shown in Fig. 8. As can be seen, the Ag atoms lose their initial structured arrangement more strongly than Cu, and this can be explained in terms of the interdiffusion process which later on produces the three-shell structure.

As in the growth procedure used by Baletto *et al.*[21] we have found a three-shell onion-like structure after collision of two single metal clusters. Although the

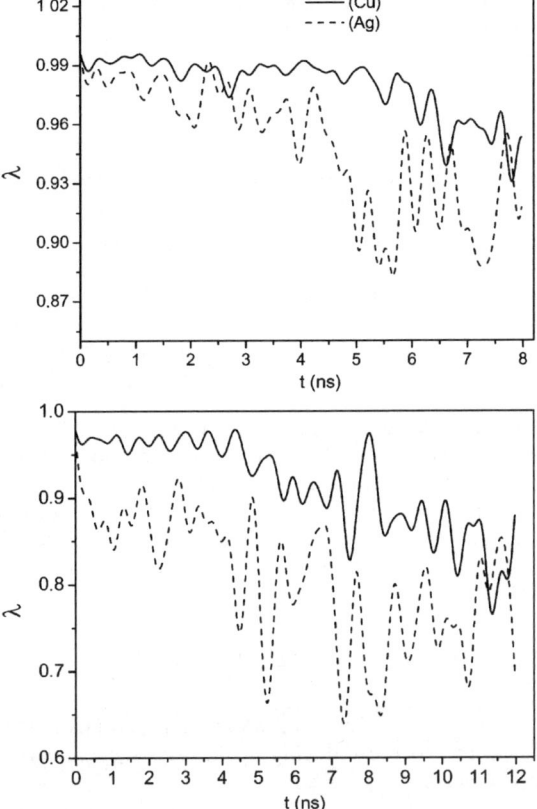

Fig. 8 Order parameter function λ as a function of time for TO_{201} Ag–Cu–Ag cluster (upper panel) and $Ih1_{47}$ (lower panel).

diffusion mechanism is essentially the same, the structures are slightly different. In the simulations performed by Baletto[21] the temperature was controlled at 400–600 K, and at 400 K a perfect onion-like structure was found to form in an fcc TO cluster. However, core–shell structures were obtained when increasing the temperature and Ih structures were used as core templates. In the present case, the temperature is variable, and therefore we reach values of this quantity that depend on the initial dynamic conditions at impact. In our simulations, when the temperature of the cluster reached *ca.* 700–800 K a core/shell nanostructure was obtained, in agreement with the findings of Baletto *et al.*[21]

These simulation results suggest that it is possible, even when the simulation model (representing also a different experiment) is completely different, to obtain three-shell (A–B–A) structures for the system Ag–Cu. In the present simulations, this occurs over a wide range of temperatures (initial velocity of impact) and of initial configurations.

2.2 Grand canonical Monte Carlo simulations of core–shell structures

The most frequent growth mechanism for metal nanoparticles in solution is the growth of precursor nuclei under given oversaturation conditions. Metal deposition from solution is usually accomplished by adding to a solution containing the metal ions to be deposited a reducing agent Red that oxidizes itself to the species Ox. This reaction allows further formation of nuclei and their subsequent growth *via* the redox reaction. When clusters of a single metal M are generated, we have for a one-electron Red/Ox system the half-reactions:

$$\text{Red} \rightarrow \text{Ox} + e^-$$

$$\text{M}^{z+} + ze^- \rightarrow \text{M (deposit)}$$

According to this reaction scheme, the generation of a cluster made of N atoms can be written as the chemical reaction:

$$zN\text{Red} + N\text{M}^{z+} \rightarrow zN\text{Ox} + N\text{M (cluster)} \tag{5}$$

where the corresponding free energy change is given by:

$$\Delta G_{\text{cl},N} = \Delta G^0 + RT \ln \frac{a_{\text{Ox}}^{zN} a_{\text{M}_{\text{cl}}}^{N}}{a_{\text{Red}}^{zN} a_{\text{M}^{z+}}^{N}} \tag{6}$$

where ΔG^0 denotes the standard free energy change, a_{Red} is the activity of the reducing agent, and a_{Ox} is the activity of the corresponding oxidized species, $a_{\text{M}^{+y}}$ is the activity of the metal ion solution being reduced and $a_{\text{M}_{\text{cl}}}$ is the activity of the metal atoms in the metallic particle formed. We will conduct the following discussion only in thermodynamic terms, so that our conclusions will be valid for those systems where close-to-equilibrium conditions are fulfilled. Eqn (6) can be written in terms of standard potential differences:

$$\Delta G_{\text{cl},N} = -NzF\Delta E^0 + RT \ln \frac{a_{\text{Ox}}^{zN} a_{\text{M}_{\text{cl}}}^{N}}{a_{\text{Red}}^{zN} a_{\text{M}^{z+}}^{N}} \tag{7}$$

where ΔE^0 is given by the difference of standard potentials corresponding to the metal, say $E^0_{\text{M}^{+y},\text{M}}$, and to the reducing agent, say $E^0_{\text{ox,Red}}$. Substitution of these quantities in eqn (7) yields:

$$\Delta G_{\text{cl},N} = NzF \left[E^0_{\text{Ox,Red}} - \frac{RT}{F} \ln \frac{a_{\text{Red}}}{a_{\text{Ox}}} - \left(E^0_{\text{M}^{z+},\text{M}} + \frac{RT}{zF} \ln a_{\text{M}^{z+}} \right) \right] + NRT \ln a_{\text{M}_{\text{cl}}} \tag{8}$$

This equation can be interpreted as follows. The term in brackets has the meaning of an overpotential that the reducing agent generates with respect to the bulk deposition potential of the metal M at the ion activity $a_{\text{M}^{+y}}$. The second term in the r.h.s. of eqn (8) can be written as $N(\mu_{\text{M,cl}} - \mu_{\text{M,bulk}}) = \Phi(N)$, where $\mu_{\text{M,cl}}$ and $\mu_{\text{M,bulk}}$ are the chemical potentials of the atoms in the metal cluster and in the bulk, respectively. With these considerations we arrive at the more familiar equation:[24]

$$\Delta G_{\text{cl},N} = -NzF\eta + \Phi(N) \tag{9}$$

which is used in textbooks to analyze nucleation and growth phenomena. The physical meaning of $\Phi(N)$ is an excess of free energy, required to generate a cluster of N atoms from the bulk metal. Fig. 9a illustrates qualitatively the typical curves of Gibbs energy excess vs. the number of particles that are usually employed to analyze the growth of a cluster made of a single metal M. The monotonic growing curve corresponds to the behaviour $\Phi(N)$ ($\eta = 0$), and the increase in the free energy is due to the larger size of the area of the growing cluster with N. The shape of the curve is due to the energetic cost of generating this area. When an overpotential η is applied, the curve exhibits a maximum, as expected from eqn (9). This critical point corresponds to the critical cluster size N^*. All clusters with more atoms than N^* will grow at η, and all clusters with $N < N^*$ will dissolve. Clusters with $N = N^*$ should remain, but this is an unstable point.

We will first consider qualitatively the growth process of a core–shell structure, made of a fixed number of atoms of metal S, on which a second metal M is

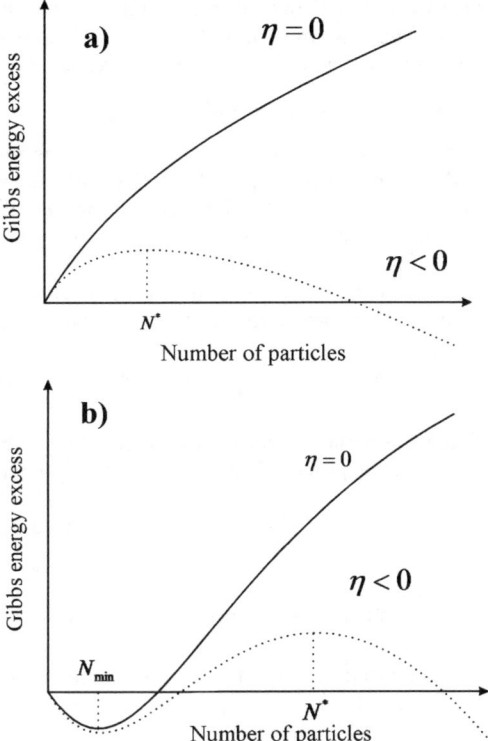

Fig. 9 Qualitative scheme of excess Gibbs energy as a function of the number of atoms for the formation of a cluster. (a) Cluster made of a pure metal M. (b) Core shells of a metal M deposited on a substrate S. Full line corresponds to zero overpotential. Dotted line corresponds to a negative overpotential.

deposited. If the interaction of M with S is strong enough, it can outweigh the area effect, and the $\Phi(N)$ *vs.* N curve will look like that in Fig. 9b. For small N values, the excess function $\Phi(N)$ will be a decreasing one, defining a minimum in the free energy curves. At non-zero overpotentials, the problem of the instability of the critical clusters persists. However, in the case of core–shell structures, clusters with $N < N^*$ will dissolve until $N = N_{min}$, corresponding to the core–shell structure. On the other hand, clusters $N < N_{min}$, will also be unstable, but now they will be forced to grow until they reach the state where $N = N_{min}$.

Now we will deal with simulation of the growth of the core–shell structures mentioned above. The following analysis presupposes that the deposition rate is relatively small, so that it can be assumed that metal deposition occurs at a constant chemical potential, μ. The natural ensemble to simulate this process is the grand-canonical one, where the parameters fixed during a simulation are the volume of the simulation cell V, the temperature T, and the chemical potential μ of the atoms being deposited. Thus we have performed grand-canonical Monte Carlo (gcMC) to simulate the growth of bimetallic nanoparticles.

The generalities of this technique are very well described in textbooks such as Allen & Tildesley,[25] so here we address just a few points specific to our simulation.

The embedded atom potentials were again used to compute the interaction between various particles. The Metropolis Monte Carlo (MMC) algorithm[25] was used to sample the configurations that minimize the free energy of the system. We note that the present approach does not consider the real time evolution of the system.

The sample scheme in our gcMC procedure, includes three types of trial moves:

• Motion of a particle. This is attempted within a cube which is small in relation to the size of the system, and the new configuration is accepted with the probability:

$$W_{j \to i} = \min(1, \exp - (\Delta U_{ij}/k_B T)) \tag{10}$$

where ΔU_{ij} is the potential energy change associated with the motion of the atom calculated by the EAM potential. Both types of metal atoms were subject to this type of motion which mimics the vibration of atoms around their equilibrium position as well as self-diffusion. This type of move is important to allow for the relaxation of the atoms, driven by the interatomic forces.

• Particle insertion. An attempt is made to insert a particle at a random position on the surface of the cluster. The new configuration is accepted with the probability:

$$W_{N \to N+1} = \min\left(1, \frac{V_{acc}}{\Lambda^3(N+1)} \exp(\mu - \Delta U_{N+1,N}/k_B T)\right) \tag{11}$$

where V_{acc} is the volume where the particles are created, Λ is the thermal De Broglie wavelength, and $\Delta U_{N+1,N}$ is the potential energy change associated with the creation of a particle.

• Particle removal. A particle is chosen at random and a removal attempt is accepted with the probability:

$$W_{N \to N-1} = \min\left(1, \frac{\Lambda^3 N}{V_{acc}} \exp(-\mu - \Delta U_{N-1,N}/k_B T)\right) \tag{12}$$

where $\Delta U_{N-1,N}$ is the potential energy change associated with the removal of a particle.

2.2.1 Core/shell nanoparticles.
Growth simulations by means of the gcMC technique were performed for the Pt–Au system. An initial or precursor core is needed to start the simulation. In our case, an fcc TO_{201}–Pt cluster was used, which was previously relaxed, performing 1×10^4 MC steps in the grand canonical ensemble until the potential energy of the system reached a steady state. With this configuration a series of simulations were performed at several chemical potentials of the Au atoms, μ_{Au}. In Fig. 10 the final configurations after 1×10^5 MC steps for different μ_{Au} are shown. Analysis of the atomic configurations shows that the growth mechanism can be understood in terms of the energetics of the system. In all cases Au atoms start to deposit on the surface of the Pt core. At potentials more negative than the bulk binding energy of Au (−3.93 eV), the new gold overlayer nucleates initially at the {100} planes of the TO Pt_{201} cluster. This is probably due to the fact that the Pt–Au interaction is stronger than the Au–Au interaction: adsorption on (100) facets maximizes Pt–Au bonds. As the chemical potential becomes more positive, an overlayer of Au covers the full surface sites leading to a perfect core/shell nanoparticle, as found in the atom dynamics simulations shown above. This

Fig. 10 Simulation of the growth of a bimetallic Pt-core/Au-shell nanoparticle by means of grand canonical Monte Carlo simulations at chemical potentials different from those of the depositing metal μ. From left to right, μ was −4.20, −3.95, and −3.85 eV, respectively. Note that at low chemical potentials the Au overlayer starts to grow at the (100) facets.

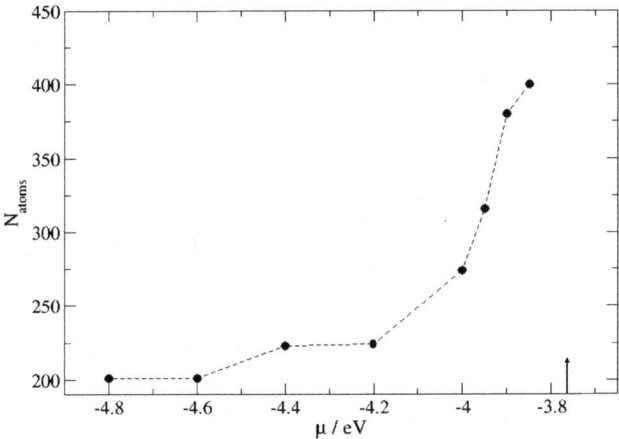

Fig. 11 Number of Au atoms deposited on the surface of an fcc-TO Pt_{201} core at different chemical potentials μ. The arrows mark the bulk deposition potential for Au.

leads to a growth pattern which is somewhat different from the one obtained by Baletto et al.[26] growing pure gold on TO Au_{201} by molecular dynamics at a constant temperature. These authors have considered metal deposition on a cluster under a flux of adatoms, performing a detailed analysis of adatom diffusion on the surface of the cluster. In the present case, we would rather simulate metal deposition taking place via a chemical reaction under chemical potential control by a gcMC technique. Thus, time is absent in the present considerations and our simulations rather describe a number of states of the systems in the neighbourhood of equilibrium at a given μ. If the Au chemical potential is further increased, a bulky Au deposit starts to form on the cluster.

The above-mentioned features are also reflected in a plot of the number of Au atoms in the cluster as a function of μ_{Au} (Fig. 11). As stated above, it can be noticed that the Au overlayer starts to grow at potentials more negative than the bulk binding energy of Au, allowing for the existence of a core/shell where only a monolayer of Au is covering the Pt core. Two steps are evident in the plot, corresponding to the sequential decoration of the (100) and (111) facets of the cluster. This Pt particle decoration by a monolayer of Au is the counterpart of the electrochemical phenomenon known as *underpotential deposition* (UPD). This is the name given to a phenomenon were a metal is deposited on a foreign substrate at electrode potentials which are more positive than the corresponding bulk deposition potential. While the driving force is provided in that case by the electrochemical potential of the electrons in the electrode, in the present situation the driving force is given by the chemical potential applied.

The present results can also be understood considering the binding energy of a Au monolayer on Pt(100) (-4.238 eV atom^{-1}) and on Pt(111) (-4.142 eV atom^{-1}).[27] These results also point out that Au deposition on Pt(100) facets is the most energetically favorable process, followed by Au deposition on Au(111), which is in turn more favorable than the formation of bulk Au.

The pair distribution functions of the Au atoms at different chemical potentials are shown in Fig. 12. At low coverages the Au layer is strained, because of the misfit between Pt/Au, that is, the Au atoms try to follow the fcc lattice of the Pt core. As the number of Au atoms on the surface of the cluster increases, the most probable Au–Au nearest neighbours distance shifts to slightly larger values, approaching that of bulk fcc Pt (see vertical dotted lines in Fig. 12). In the case of the Pt core, neither compression nor expansion of the nearest neighbour distance of the fcc lattice is observed when it is covered by the Au shell.

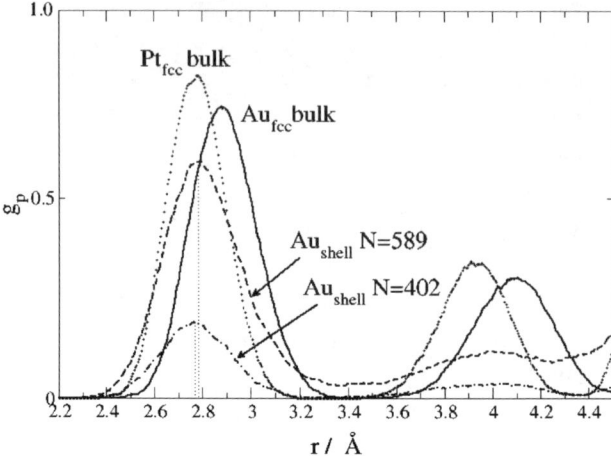

Fig. 12 Pair distribution function for the Au overlayer at different coverages. First and second nearest neighbors only are shown. The pair distribution function of a Pt and Au fcc bulk metals are shown for comparison.

The mean square atomic displacement of the atoms can be used to analyze their vibrational mobility around their equilibrium positions. It can be measured directly from the atomic coordinates from the simulation as follows:

$$\sigma^2 = \frac{1}{N}\sum_i^N (\langle r_i^2 \rangle - \langle r_i \rangle^2)$$

(13)

where r_i is the current position of the i-atom. The calculated σ^2 for Pt atoms in a TO–Pt$_{201}$ and Pt-core/Au-shell clusters are compared in Fig. 13. At low temperatures the root-mean-square bond-length fluctuation is small. As temperature increases, higher values of σ^2 are observed. Larger atomic displacements of the Pt atoms in the Pt/Au core–shell structure are observed, as compared with the pure Pt$_{201}$ core, indicating that the Pt core loses its hardness after deposition of Au.

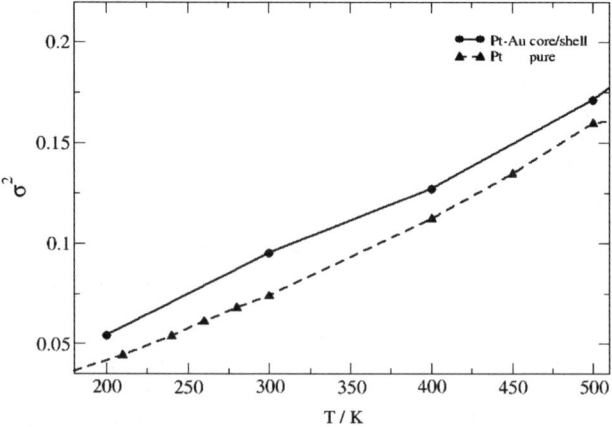

Fig. 13 Mean square atomic displacement σ^2 as a function of temperature for a pure Pt core cluster (dotted line), and for a Pt/Au core–shell nanoparticle (solid line).

 This journal is © The Royal Society of Chemistry 2008

3. Conclusions

Two computer simulation techniques have been used to generate bimetallic nano-particles with different structures and chemical compositions. On the one hand, collisions of two metal nanoparticles of different chemical nature have been studied by means of atom dynamics simulations. The final structures obtained after collision are in good agreement with previous experimental and theoretical results using other methods.

Bimetallic nanoclusters of *core–shell* Pt_{core}–Au_{shell}, Ni_{core}–Au_{shell}, Cu_{core}–Au_{shell} and Co_{core}–Ag_{shell}, *alloyed* Pd–Au and Cu–Pd, and *three-shell onion-like* (A–B–A) Cu–Ag structures were found after the collision of two clusters with a given initial velocity. While the nature of the emerging nanostructure (core–shell, alloy, and three-shell) is determined by the chemical nature of the colliding clusters, the initial kinetic energy is an important control parameter. As a second type of simulation, we have performed grand canonical Monte Carlo calculations to study the growth mechanisms at constant chemical potential. The results obtained for the Pt–Au system are in good agreement with the first method. Pt_{core}–Au_{shell} structures were observed in which the Au overlayer is formed at underpotentials. Structural parameter functions reveal that the Au shell is strained because of the misfit between Pt–Au.

The different structures found with both simulation techniques are understood on energetic grounds.

Acknowledgements

The authors acknowledge the financial support from Consejo Nacional de Investi-gaciones Científicas y Técnicas (CONICET), Agencia Córdoba Ciencia, Secyt UNC, Program BID 1728/OC-AR PICT No. 06-12485 y PICT No. 06-15115 and language assistance by Karina Plasencia.

References

1   Y. Huang, X. Duan and C. M. Lieber, *Science*, 2001, **291**, 630.
2   H. Fan, Y. Lu, A. Stump, S. T. Reed, T. Baer, R. Schunk, V. Perez-Luna, G. P. Lopez and C. J. Brinker, *Nature*, 2000, **405**, 56.
3   J. J. Green, E. Chiu, E. S. Leshchiner, J. Shi, R. Langer and D. G. Anderson, *Nano Lett.*, 2007, **7**, 874.
4   M. Faraday, *Philos. Trans. R. Soc. London*, 1857, **147**, 145.
5   M. Kerker, *J. Colloid Interface Sci.*, 1986, **112**, 302.
6   J. H. Sinfelt, *J. Catal.*, 1973, **29**, 308.
7   N. Toshima, *J. Macromol. Sci., Pure Appl. Chem.*, 1990, **27A**, 1225.
8   J. H. Sinfelt, in *Bimetallic Catalysts Exxon Monograph*, John Wiley, New York, 1983.
9   T. Shibata, B. A. Bunker, Z. Zhang, D. Meisel, C. F. Vardeman II and J. D. Gezelter, *J. Am. Chem. Soc.*, 2002, **124**, 11989.
10 Z. Y. Li, J. Yuan, Y. Chen, R. E. Palmer and J. P. Wilcoxon, *Appl. Phys. Lett.*, 2005, **87**, 243103.
11 J. He, I. Ichinose, T. Kunitake, A. Nako, Y. Shirashi and N. Toshima, *J. Am. Chem. Soc.*, 2003, **125**, 11034.
12 P. V. Kamat, *J. Phys. Chem. B*, 2002, **106**, 7729.
13 S. Sun, C. B. Murray, D. Weller, L. Folks and A. Moser, *Science*, 2000, **287**, 1989.
14 (*a*) S. M. Foiles, M. I. Baskes and M. S. Daw, *Phys. Rev. B*, 1986, **33**, 7983–7991; (*b*) H. N. G. Wadley, X. W. Zhou, R. A. Johnson and M. Neurock, *Prog. Mater. Sci.*, 2001, **46**, 329.
15 M. Alonso and E. J. Finn, *Fundamental University Physics*, Addison-Wesley, Massachu-setts, 1967, vol. 1.
16 M. M. Mariscal, S. A. Dassie and E. P. M. Leiva, *J. Chem. Phys.*, 2005, **123**, 184505.
17 J. H. Sinfelt, Y. L. Lam, J. A. Cusamano and A. E. Barnett, *J. Catal.*, 1976, **42**, 227.
18 W. M. H. Sachtler and R. A. van Santen, *Adv. Catal.*, 1997, **26**, 69.
19 R. Hultgren, P. D. Desai, D. T. Hawkins, M. Gleiser and K. K. Kelley, *Selected Values of the Thermodynamics Properties of Binary Alloys*, American Society for Metals, Metals Park, OH, 1973, vol. 1.

20 N. Toshima and T. Yonezawa, *New J. Chem.*, 1998, 1179.
21 F. Baletto, C. Mottet and R. Ferrando, *Phys. Rev. Lett.*, 2003, **90**, 135504.
22 M. C. Fromen, J. Morillo, M. J. Casanove and P. Lecante, *Europhys. Lett.*, 2006, **73**, 885.
23 D. Cheng, S. Huang and W. Wang, *Phys. Rev. B*, 2006, **74**, 064117.
24 E. Budevski, G. Staikov and W. J. Loenz, *Electrochemical Phase Formation and Growth, An Introduction to the Initial Stages of Metal Deposition*, VCH, Weinheim, 1996.
25 M. P. Allen and D. J. Tildesley, *Computer Simulation of Liquids*, Clarendon Press, Oxford, 1987.
26 F. Baletto, C. Mottet and R. Ferrando, *Surf. Sci.*, 2000, **446**, 31.
27 O. A. Oviedo, E. P. M. Leiva and M. I. Rojas, *Electrochim. Acta*, 2006, **51**, 3526.

Dynamical equilibrium in nanoalloys

F. Lequien,[a] J. Creuze,*[ab] F. Berthier[ab] and B. Legrand[c]

Received 10th April 2007, Accepted 2nd May 2007
First published as an Advance Article on the web 25th September 2007
DOI: 10.1039/b705281b

Using Monte Carlo simulations on a lattice–gas model, we study the
segregation isotherm of a cluster made of thousands of atoms for a system
that tends to phase separate, e.g., Cu–Ag. We show that the Ag segregation
involves the vertices first, then the edges and finally the (111) and (100)
facets. In these facets, the segregation starts on the outer shells, leading to a
heterogeneous chemical composition. When the nominal Ag concentration
(or the chemical potential difference $\Delta\mu$ between Ag and Cu), is increased a
dynamical equilibrium replaces the progressive evolution of the segregation
towards the core of the facets: the whole facet oscillates between one
pseudo Ag-pure state and another one corresponding to a rather Cu-pure
core surrounded by Ag-enriched outer shells. A remarkable consequence is
that very different concentrations can be observed for facets of equivalent
orientation. This dynamical equilibrium occurs in a $\Delta\mu$ range that is very
close to the critical value $\Delta\mu_c$ associated with the first-order phase transition
of the Fowler–Guggenheim type that affects the surfaces of semi-infinite
alloys. These results, which have been obtained in the grand-canonical
ensemble, can also be derived in the canonical ensemble due to a sufficient
number of facets that behave with each other as a reservoir.

I. Introduction

The use of bimetallic clusters in fields as various as catalysis and magnetism requires
control of their surface properties (chemical composition, structure).[1,2] To specify
the finite-size effect on the surface segregation, the comparison of the segregation
isotherms in semi-infinite alloys and the outermost shells of clusters is an interesting
tool. A recent study has thus shown that the anisotropy of the superficial segregation
between the (111) and (100) facets of the cuboctahedron could strongly differ from
the segregation predicted between the (111) and (100) surfaces of semi-infinite
alloys.[3] More precisely, computations relying on a Ising model fitted for the Cu(Ag)
system revealed that while the usual anisotropy of semi-infinite surfaces, corresponding to a more important enrichment of the more open surface (here the (100) surface
versus the (111) one) is established for clusters larger than about 12 431 particles, for
smaller clusters the (111) facets are more Ag-enriched than the (100) facets.[3] If the
anisotropy of the superficial segregation varies with the cluster size, what happens to
the superficial phase transitions that occur in phase-separating systems like Cu–Ag?
In semi-infinite alloys, the surfaces of such systems display segregation isotherms of
the Fowler–Guggenheim type.[4,5] They are characterized by the existence of a critical

[a] Université Paris Sud, LEMHE/ICMMO, UMR 8182, Bât. 410, F91405, Orsay Cedex,
France
[b] CNRS, LEMHE/ICMMO, UMR 8182, Bât. 410, F91405, Orsay Cedex, France
[c] SRMP-DMN, CEA Saclay, F91191, Gif/Yvette Cedex, France

temperature below which the segregation isotherm undergoes a first-order phase transition that corresponds to a jump of the superficial concentration for a critical bulk concentration or equivalently for a critical difference of chemical potentials between both metals.[5] At room temperature, in the Cu(Ag) system, the infinite surface shifts from an almost Cu-pure state to an almost Ag-pure state for a very low Ag bulk concentration (less than 0.1%).[6–8] This work aims at studying the evolution of such a phase transition when one switches from surfaces of semi-infinite alloys to cluster facets. We therefore consider the same lattice–gas model with pair interactions as mentioned here above, and specify in Section II how the main parameters have been fitted to the Cu–Ag system, and its extension to the cuboctahedron case. In Section III, we present the segregation isotherms obtained by Monte Carlo simulations at 300 K, and analyze in detail both the mean concentrations of the (100) facets and the instantaneous concentrations of each facet. As a result, we emphasize the existence of a dynamical equilibrium that corresponds to the extension of the first-order phase transition of infinite surfaces to facets. We finally discuss the questions arising, such as the possibility to observe this two-state equilibrium in an experimental way.

II. Model

To benefit from the previous study of the segregation isotherms in a mean-field framework,[3] we considered the same energetic model. It is an Ising model governed by mainly two parameters in the superficial segregation context: $\tau = (V_{AA} - V_{BB})/2$ and $V = (V_{AA} + V_{BB} - 2V_{AB})/2$, where V_{AA}, V_{BB} and V_{AB} are the nearest-neighbour pair interactions between A–A, B–B and A–B atoms. The main features of the Cu–Ag system are reproduced by ascribing $\tau = (V_{AgAg} - V_{CuCu})/2 = 46$ meV and $V = -30$ meV.[3] τ is positive due to the lower cohesion energy of Ag with regard to Cu and leads to Ag segregation on the less coordinated sites.[5] The negative value of V imposes the tendency of the Cu–Ag alloy to phase separate and leads to the existence of critical temperatures for both the bulk and the surfaces.[5] In the mean-field approximation, for a monolayer model of the (111) and (100) surfaces, these critical temperatures are given by:

$$k_B T_c^{bulk} = -6V, k_B T_c^{(111)} = -3V, k_B T_c^{(100)} = -2V,$$

i.e., $T_c^{bulk} = 2088$ K, $T_c^{(111)} = 1044$ K and $T_c^{(100)} = 696$ K. Recall that the mean field approximation overestimates the critical temperatures by a crystallographic ratio:[9] 1.22 for the bulk, 1.64 and 1.76 for the (111) and (100) surfaces, respectively.[10] The exact critical temperatures (obtained by Monte Carlo simulations for instance) are then equal to 1711 K for the bulk, 637 K and 395 K for the (111) and (100) surfaces, respectively. Note that this rigid lattice approach can be extended to partially take into account atomic relaxations by considering effective values for the energetic parameters, as demonstrated in the case of grain boundaries and reconstructed surfaces in the Cu–Ag system.[11,12]

We study the 3871-atoms cuboctahedron depicted in Fig. 1 with this energetic model. Adding to 12 vertices of coordination 5 and 24 edges of coordination 7, the cuboctahedral cluster owns 6 square (100) facets (coordination 8) and 8 triangular facets (111) of coordination 9. For the cluster considered here and said to be of order 10, the edges are made of 11 atoms including the vertices. Note that this size is well beyond the critical size that separates the stability domains of the icosahedron from structures issued from the FCC lattice (cuboctahedron and especially the truncated octahedron), even if decahedral structures are highly competitive up to 10^4 atoms for Cu and Ag clusters.[13,14]

To determine the distribution of Cu and Ag atoms onto the different sites of the cluster, we performed Monte Carlo simulations in the grand-canonical ensemble,[15,16] where the nominal concentration of the $Ag_c Cu_{1-c}$ cluster is determined by the chemical potential difference $\Delta\mu = \mu_{Ag} - \mu_{Cu}$. Similar to the studies of massive

This journal is © The Royal Society of Chemistry 2008

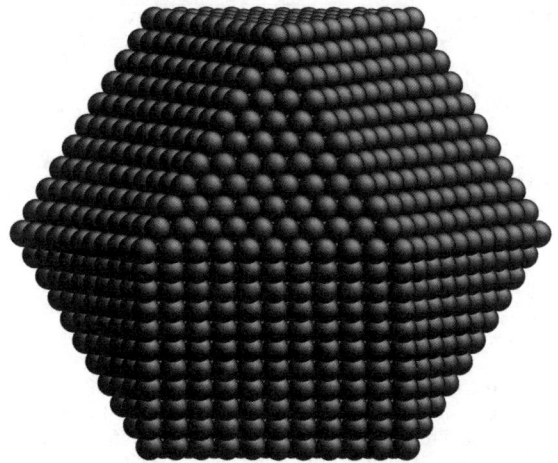

Fig. 1 3871 atom cuboctahedron. Vertices, edges, (100) square facets and (111) triangular facets are clearly visible.

alloys, the choice of the grand-canonical ensemble instead of the canonical one (where the nominal concentration is imposed by the relative proportion of the Ag and Cu atoms initially introduced) allows a better description of the phase transition characteristics.[9] We shall discuss later comparison with experiments, which are closer to a canonical-ensemble study. A standard Metropolis algorithm is used[16,17] and the averages are evaluated on 2×10^5 MC macro-steps, a similar number of steps being used to reach equilibrium. A MC macro-step corresponds to N_{at} propositions of chemical switch, N_{at} being the total number of atoms of the cluster. This allows us to obtain the mean concentrations on the vertices, the edges and the (111) and (100) facets. These concentrations can be compared with those derived from the mean-field framework, where one defines four classes p of sites (vertices, edges, (111) and (100) facets) and mean coordination numbers Z_{pq} between the sites of classes p and q.[3] The concentration assumed to be homogeneous on each site of a class p is then given by:

$$\frac{c_p}{1 - c_p} = \exp\left(-\frac{\Delta H_p^{\text{perm}} - \Delta\mu}{k_B T}\right) \qquad (1)$$

ΔH_p^{perm} is the variation of enthalpy during the permutation of a Cu atom into an Ag atom on a site of the class p:

$$\Delta H_p^{\text{perm}} = Z_p(\tau - V) + 2V \sum_q Z_{pq} c_q \qquad (2)$$

and Z_p is the coordination number of a site of the class p.[3] This kind of model allows one to follow the evolution of a phase transition with the cluster size but neglects the possibility to have heterogeneous concentrations inside a class, for instance between the edges and the core of the facets or between two facets with the same orientation. We shall see that this limitation is very drastic under certain conditions.

III. Results

III.1 Segregation isotherms

We show on Fig. 2 the segregation isotherms for the sites in the vertices, edges and (100) and (111) facets obtained at 300 K by Monte Carlo simulations and compare them with those obtained in the mean-field approximation. Note that for the values

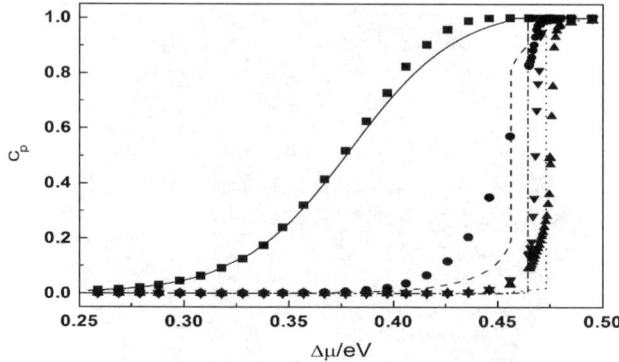

Fig. 2 Segregation isotherms obtained by Monte Carlo simulations (symbols) and within a mean-field approximation (lines). The concentrations for the different sites are given as a function of the chemical potential difference $\Delta\mu$. Vertices (squares/solid line), edges (circles/dashed line), (100) facets (up triangle/dotted line) and (111) facet (down triangles/dashed-dotted line).

of $\Delta\mu$ we consider, only the surface is enriched in Ag, the underlying layers and the core of the cluster remaining almost Cu-pure.[3] Due to the lowest coordination number of the vertices, the related isotherm is shifted towards the lowest values of $\Delta\mu$. Increasing $\Delta\mu$ leads to an Ag enrichment on the edges, then on the (111) facets and finally on the (100) facets. Note that the relative position of the isotherms for the two types of facets does not follow the order of increasing coordination numbers. The mean-field approach shows that this is due to the coupling with the edges, the classical hierarchy being recovered for sizes larger than 12 431 atoms.[3] If Monte Carlo simulations and Mean-Field approach are in a good agreement about the relative positions of the isotherms, they differ on the nature of these isotherms. The mean-field model predicts the existence of a first-order phase transition for the edges and facets isotherms, whereas Monte Carlo simulations lead to continuous isotherms for any kind of site. Nevertheless, if the facet isotherms are continuous, they display very stiff variations for a $\Delta\mu$ range close to the critical values $\Delta\mu_c$ derived from the mean-field framework. We verified that these isotherms are perfectly reversible, the same curves being obtained by progressively increasing or decreasing $\Delta\mu$.

III.2 Facet inhomogeneity

The relative positions of the edge and facet isotherms indicate that the edges are Ag-enriched before the segregation phenomenon reaches the facets. One can then wonder whether the sites of the facet that are close to the edges are more enriched than the center of the facets, *i.e.*, the concentration profile of a facet decreases from the edges towards the core. To answer this question we show on Fig. 3a the equilibrium concentration for each site of the cluster surface at $\Delta\mu = 0.476$ eV. At this value of $\Delta\mu$, vertices edges and (111) facets are Ag-pure, while the mean concentration of the (100) facet equals 0.68 (Fig. 2). Fig. 3a reveals an important inhomogeneity inside the (100) facets. By analyzing the facet as a sequence of concentric square shells (the center atom, then shells made of, respectively 8, 16, 24 and 32 atoms), one observes that the two outermost shells (32 and 24-atom shells) are Ag-enriched, while the centermost shells remain rather Cu-pure. A deeper analysis shows that the outer concentric shells are also inhomogeneous, the segregation being stronger on the vertices than on the middle of each edge.

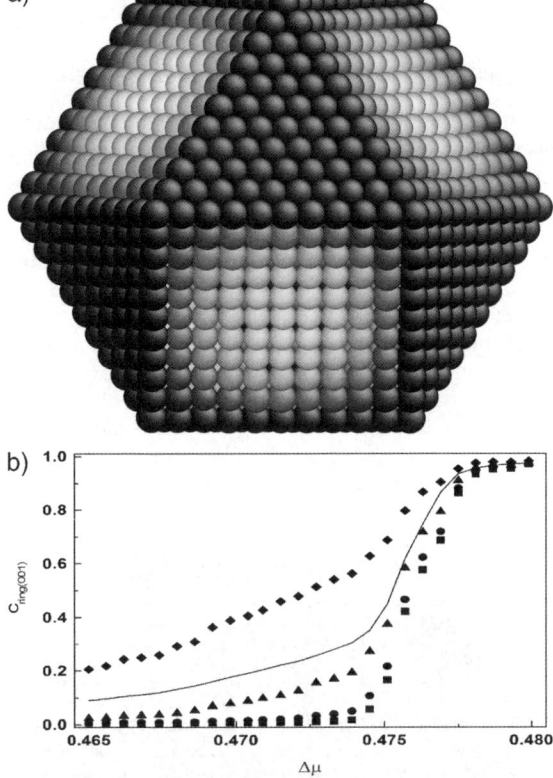

a)

b)

Fig. 3 (a) Equilibrium concentration for each atomic site obtained by MC simulations at 300 K and $\Delta\mu = 0.476$ eV. The grey levels vary between white ($c_p = 0$, Cu-pure site) and black ($c_p = 1$, Ag-pure site); (b) Segregation isotherms at 300 K for the various concentric shells of the (100) facets. Central shell (9 atoms): squares, second shell (16 atoms): circles, third shell (24 atoms): triangles and fourth shell (32 atoms): diamonds. The isotherm for the facet itself is given by the solid line.

Fig. 3b depicts the variation of the concentration of the concentric square shells of the (100) facets as a function of $\Delta\mu$. We gathered the central atom with the innermost 8-atom shell and we display the concentration of these 9 central sites, the one of the 3 following shells, and the mean concentration of the facet (already depicted on Fig. 2 but here rescaled in $\Delta\mu$). This figure shows 3 regimes for the segregation isotherm of the (100) facets. In the first regime (for $\Delta\mu \leq 0.474$ eV), the concentration of the two outer shells increases regularly, while the two center shells remain almost Cu-pure. This yields the heterogeneity of the facet displayed on Fig. 3a. In the second regime ($0.474 \leq \Delta\mu \leq 0.477$ eV), the concentrations of all the shells increase very rapidly and simultaneously. This last point allows us to exclude a mechanism of progressive Ag-enrichment of the facet from the outer shells that would be analogous to the one observed in regime 1. This regime is responsible for the stiff increase of the isotherm associated with the mean concentration of the (100) facets that we shall detail in the next paragraph. Finally, for $\Delta\mu \geq 0.477$ eV, the isotherms reach the saturation, and the (100) facets are homogeneously Ag-pure.

The behaviours of the (111) facets and the (100) facets are similar: the three regimes can be observed, with a progressive Ag-enrichment from the outer shell in regime 1, a stiff and simultaneous increase of the isotherms in regime 2 and the saturation in regime 3. Note that in the three regimes there is no symmetry break between the facets of the same kind: the concentration remains the same for all the equivalent sites from a facet to the other as depicted on Fig. 3a.

III.3 Dynamical equilibrium

We detail here the nature of regime 2 for the (100) facets that occur in a very narrow range of $\Delta\mu$. This range is very close to the critical value $\Delta\mu_c$ of the first-order phase transition that occurs in the isotherms derived from the mean-field approximation (Fig. 2). Fig. 4 displays two instantaneous configurations of the cluster surface for $\Delta\mu = 0.476$ eV. In the same instantaneous configuration the (100) facets have very different concentrations: some are almost Ag-pure (like regime 3), while the configurations of the others are very similar to those observed in regime 1 (Ag-enriched outer shells and almost Cu-pure inner shells). Moreover, from one instantaneous configuration to the other, the concentration of a same facet can evolve strongly from an almost Ag-pure configuration to a regime 1-type config-uration. Thus, each facet undergoes a dynamical equilibrium between two states: in the first one the inner shells are almost Cu-pure, while these shells are almost Ag-pure in the second one. This dynamical equilibrium is illustrated in Fig. 5a, which shows the fluctuations of the concentration of the innermost crown (built by the 9

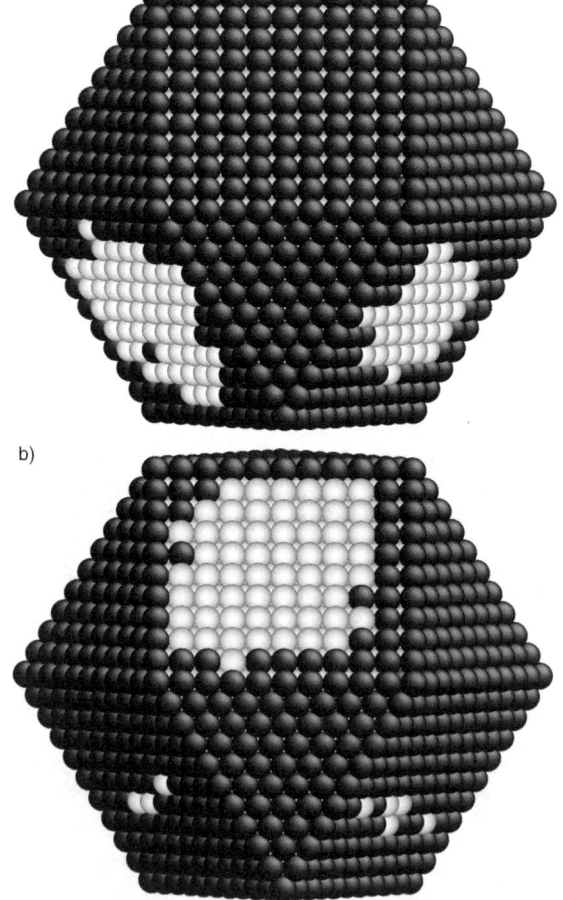

Fig. 4 Two snapshots obtained by Monte Carlo simulations at 300 K for $\Delta\mu = 0.476$ eV (regime 2). Cu atoms are in white and Ag atoms are in black.

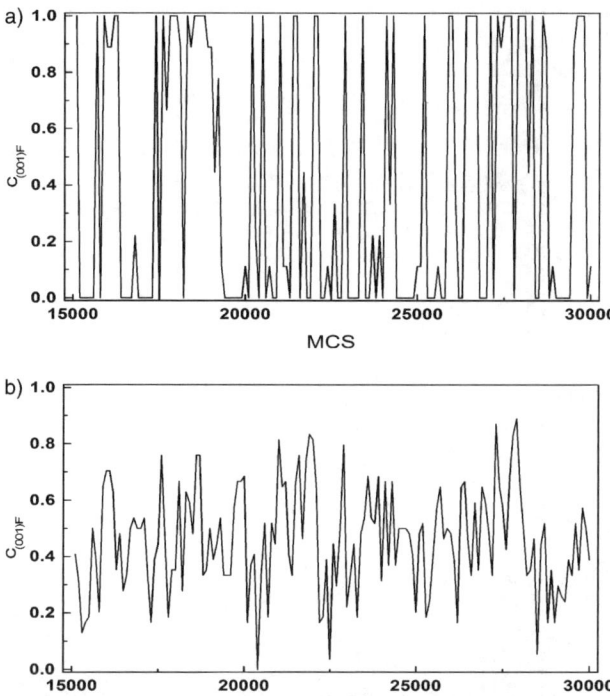

Fig. 5 Instantaneous concentration for the central shell (9 sites) of a given (100) facet at 300 K for $\Delta\mu = 0.476$ eV (regime 2) as a function of the number of MC steps (a). Instantaneous concentration for the central shells of all the 6 (100) facets (54 atoms) as a function of the number of MC steps in the same conditions as in (a), see Section IV.1.

core atoms) for a given (100) facet. This bi-stability between two states that characterizes the regime 2 is associated with the bimodality of the configurational density of states (CDOS) shown in Fig. 6. These histograms $n(c)$ are defined as the number of times the instantaneous concentration of the innermost shell of a given facet is between c and $c + dc$, the densities being normalized to 1. In regimes 1 (Fig. 6a) and 3 (Fig. 6c), these densities are monomodal whereas they become bimodal in regime 2 (Fig. 6b). The sharp rise of the isotherm in regime 2 corresponds to an increasing occupancy of the almost pure Ag state ($c \approx 1$) and a decreasing occupancy of the almost pure Cu state ($c \approx 0$). Note that the concentrations associated to these two states ($c \approx 0.01$ and $c \approx 0.97$) remain almost constant in the whole regime 2, as this is the case for the solubility limits in an infinite phase-separated system.

IV. Discussion

IV.1 Grand-canonical and canonical ensembles for clusters

The fact that the number of facets is large in a cluster (six (100) facets and eight (111) facets for a cuboctahedron) has some important consequences on the dynamical equilibrium. In particular, whereas the instantaneous concentration of a given facet shows very large fluctuations during the dynamical equilibrium (Fig. 5a), the instantaneous concentration for all the facets of equivalent orientation displays fluctuations of smaller amplitude (Fig. 5b). This comes from the fact that the 6 (100) facets represent a sufficient number to realize a reasonable sampling of the CDOS at each Monte Carlo step. Moreover, this indicates that a simulation in the canonical

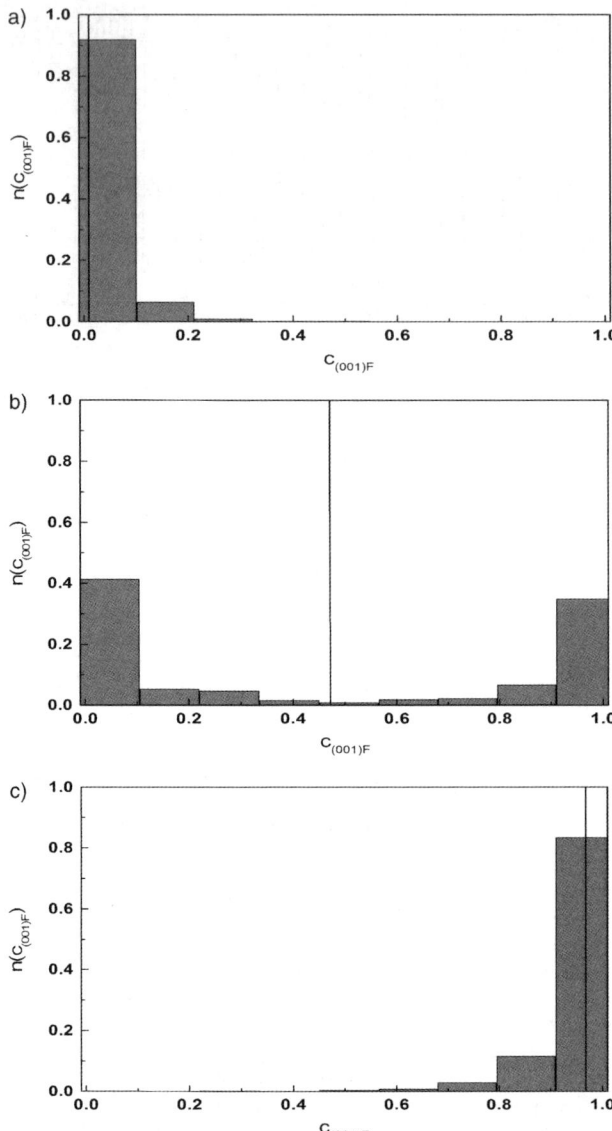

Fig. 6 Configurational density of states (CDOS) for the central shell (9 sites) of a given (100) facet obtained by MC simulations at 300 K. CDOS representative of (a) regime 1 ($\Delta\mu$ = 0.473 eV), (b) regime 2 ($\Delta\mu$ = 0.476 eV) and (c) regime 3 ($\Delta\mu$ = 0.48 eV). The vertical line shows the equilibrium concentration of this central shell.

ensemble is also able to reveal the existence of this dynamical equilibrium. For a given nominal concentration, the instantaneous concentration of each facet can fluctuate between two states, the facets behaving with each other as a reservoir.

Such an effect has been put forward as evidence in the simpler case of linear chains.[18] Monte Carlo simulations have been realized in the following three ensembles:

– the Grand-Canonical ensemble (GC) for a linear chain of 11 sites (corresponding to the edge length of the cluster considered in this study), in which $\Delta\mu$ is fixed;

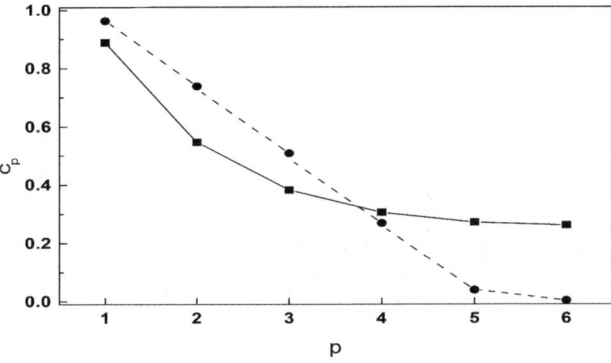

Fig. 7 Concentration profiles for a linear chain of 11 atoms at $T = 100$ K and $c_{nom} = 5/11$ (or the corresponding value of $\Delta\mu$ in the grand-canonical ensemble). Solid line: GC and C_{Multi} ensembles, dotted lines: C_{Mono} ensemble. The profile being symmetric relatively to $p = 6$, only half of the profile is shown.

– the canonical ensemble for the same chain, hereafter referred to as "Canonical mono-object" and noted C_{Mono}, in which the nominal concentration c_{nom} is fixed;
– the canonical ensemble for 24 chains of 11 sites, all these chains being in coexistence to reproduce the 24 edges of the cluster. This ensemble will be referred to as "Canonical multi-object" and noted C_{Multi}, in which the nominal concentration of the 24 chains is fixed.

Whereas the concentration profiles in the grand-canonical and "canonical multi-object" ensembles are similar and show the usual positive curvature, the profiles in the "canonical mono-object" ensemble are characterized by a linear decrease up to the core concentration (Fig. 7). The analysis of the profiles in the Grand-canonical ensemble reveals that the chain concentration presents fluctuations of large amplitude around the mean value determined by $\Delta\mu$. The CDOS related to this GC profile is shown in Fig. 8a. The CDOS is bimodal and presents a broad peak in the low concentration side and a narrow one for the high concentrations. The profiles corresponding to each nominal concentration are very similar to those obtained in the C_{Mono} ensemble; in particular they exhibit a linear attenuation (Fig. 8b). Note that averaging these linear profiles along the weights given by the CDOS leads to the convex profile shown in Fig. 7.

The C_{Multi} ensemble leads to the same results as the grand-canonical one, the CDOS on each chain being similar to the one obtained in the GC ensemble. This means that each chain behaves with each other as a reservoir, establishing the equivalence between the grand-canonical ensemble and the "canonical multi-object" ensemble. From this point of view, essential properties of the cluster are not only their finite size but also the presence of a large number of equivalent objects such as the edge or the facets of different orientations.

IV.2 Wetting and dynamical equilibrium

The present Monte Carlo simulations have shown unexpected behavior for the enrichment of the less cohesive element in the cluster facets. In the first regime, the increase of the nominal concentration (or the chemical potential difference between both constituents) leads to segregation occurring mainly on the vertices and then on the edges and in the outermost layers of the facets. It would have been conceivable that this process leads to a wetting phenomenon similar to the behavior observed for the semi-infinite Cu–Ag alloys. In this last case, the planes parallel to the surface are *successively* affected by a change from an almost Cu-pure state towards an almost Ag-pure state.[19–21] The concentric shells which form a facet would have been able to adopt similar *successive* changes. However, this process is prevented by the

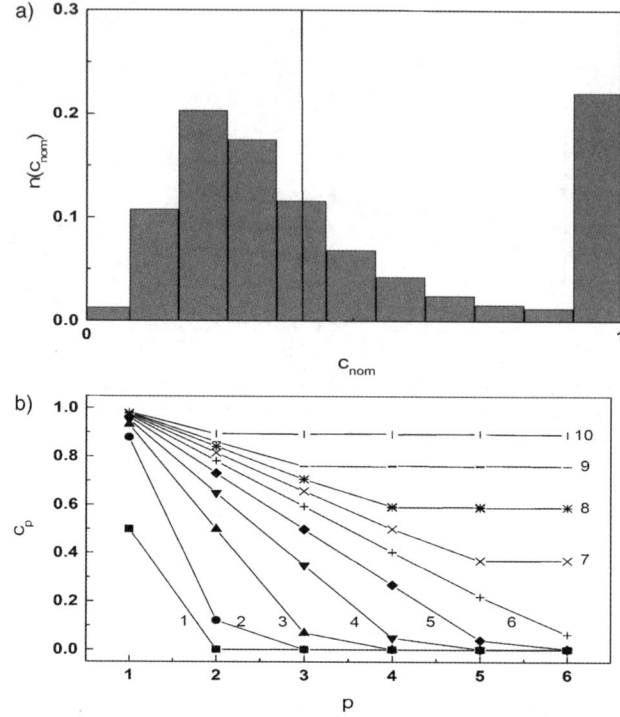

Fig. 8 CDOS for a linear chain of 11 sites obtained by MC simulations in the grand-canonical ensemble at 100 K for a value of $\Delta\mu$ corresponding to $c_{nom} = 5/11$ (a); Concentration profiles for the different states defining the CDOS, the nominal concentration associated with the curve labelled i is equal to $i/11$. The profile being symmetric relatively to $p = 6$, only half of the profile is shown (b).

occurrence of a dynamical equilibrium affecting *simultaneously* all the concentric shells of the facet, in analogy with the Fowler–Guggenheim phase transition for the infinite surface.[4,5] Preliminary studies on the linear chain allows us to discuss the key factors that drive the competition between wetting and dynamical equilibrium, or between wetting and phase transition when using a mean-field formalism. For a

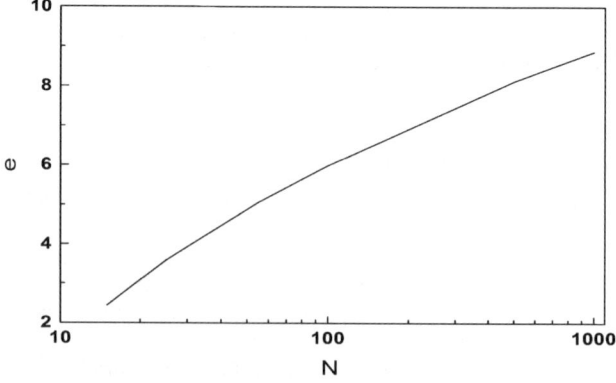

Fig. 9 Maximum thickness of the wetting layer e_{max} (expressed in site number) as a function of N (in logarithmic scale) for a linear chain of N sites at 300 K for $\tau = 46$ meV and $V = -30$ meV.

finite linear chain of N sites, the thickness e of the Ag-rich wetting layer formed at each end of the chain increases logarithmically as a function of $(\Delta_{\mu_c} - \Delta\mu)$, where $\Delta\mu_c^{\infty}$ is the critical value for the infinite chain. More precisely,

$$e \equiv -\xi \ln(\Delta\mu_c^{\infty} - \Delta\mu), \tag{3}$$

where ξ is a characteristic length driving the damping of the concentration profile.[18] Wetting stops for a critical chemical potential difference $\Delta\mu_c^N$ smaller than $\Delta\mu_c^{\infty}$. This comes from the fact that when the Ag-pure state occurs, the two interfaces between the Ag wetting layers and the Cu core disappear. This leads to a free energy gain stabilizing the Ag-pure state but which is negligible for an infinite chain. It can

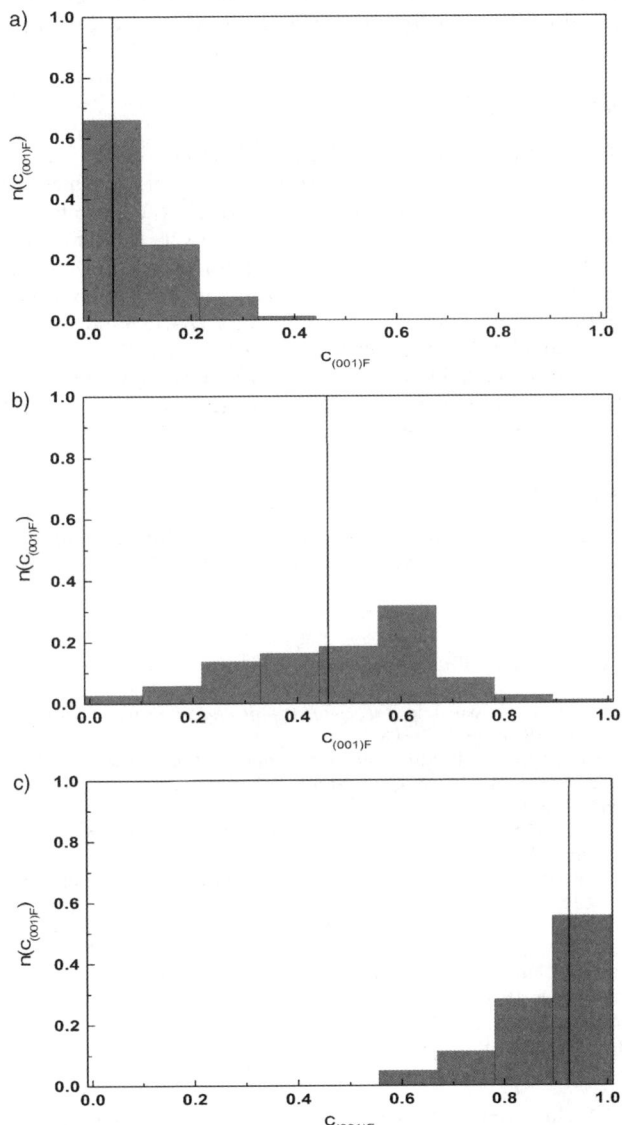

Fig. 10 CDOS for the central shell (9 sites) of a given (100) facet obtained by MC simulations at 1100 K: $\Delta\mu = 0.31$ eV (a), $\Delta\mu = 0.465$ eV (b) and $\Delta\mu = 0.53$ eV (c). The vertical line shows the equilibrium concentration of this central shell.

be shown that $(\Delta\mu_c^N - \Delta\mu_c^\infty)$ varies as $1/N$.[18] This allows us to express the maximum thickness of the wetting layer for a finite chain of N sites:

$$e_{max} \equiv -\xi \ln(\Delta\mu_c^\infty - \Delta\mu_c^N) \equiv \xi \ln N. \tag{4}$$

This relation is illustrated in Fig. 9 for the energetic parameters described in Section II and can be used to describe the behavior of the edges for the cuboctahedron. This figure clearly indicates that the finite character of the linear chain leads to a maximum thickness of the wetting layer which remains very small in comparison to the length of the chain. This is consistent with the behavior observed for the facet. Due to the free energy gain arising from the elimination of the interface between the Ag-rich outermost shell and the Cu-rich innermost shell, the dynamical equilibrium stops the growth of the wetting shell.[18] Note that this behavior described here for the (100) facets is also observed for the (111) facets, that also present a dynamical coexistence.

V. Conclusion

The observation of the dynamical equilibrium described in this study represents a real experimental challenge. First, this requires a technique which allows one to determine the concentration on a given facet and not the concentration averaged over all the facets of equivalent orientation. Moreover, the temperature must be sufficiently high to activate the diffusion required for the dynamical equilibrium but not too high to remain below a critical temperature separating the regime of bimodal CDOS (dynamical equilibrium) and monomodal CDOS (standard equilibrium) as the one obtained at 1100 K and shown in Fig. 10.

This study will be extended to other cluster structures, in particular to icosahedral and decahedral structures in which interatomic distortions are susceptible to modify strongly segregation phenomenon.[22-24]

Acknowledgements

We are very grateful to Isabelle Braems for her help during all this work and we thank Robert Tétot, Christine Mottet and Guy Tréglia for helpful discussions.

References

1   J. H. Sinfelt, *Bimetallic Catalysts: Discoveries, Concepts and Applications*, Wiley, New York, 1983.
2   J. Jellinek and E. B. Krissinel, in *Theory of Atomic and Molecular Clusters*, ed. J. Jellinek, Springer, Berlin, 1999, pp. 277–308.
3   F. Lequien, J. Creuze, F. Berthier and B. Legrand, *J. Chem. Phys.*, 2006, **125**, 094707.
4   R. H. Fowler and E. H. Guggenheim, *Statistical Thermodynamics*, Cambridge University Press, Cambridge, 1960.
5   G. Tréglia, B. Legrand, F. Ducastelle, A. Saúl, C. Gallis, I. Meunier, C. Mottet and A. Senhaji, *Comput. Mater. Sci.*, 1999, **15**, 196.
6   R. Tétot, F. Berthier, J. Creuze, I. Meunier, G. Tréglia and B. Legrand, *Phys. Rev. Lett.*, 2003, **91**, 176103.
7   I. Braems, F. Berthier, J. Creuze, R. Tétot and B. Legrand, *Phys. Rev. B*, 2006, **74**, 113406.
8   F. Berthier, J. Creuze, R. Tétot and B. Legrand, *Phys. Rev. B*, 2002, **65**, 195413.
9   F. Ducastelle, in *Order and Phase Stability in Alloys*, North-Holland, 1991.
10 M. E. Fisher, *Rep. Prog. Phys.*, 1967, **30**, 615.
11 F. Berthier, B. Legrand and G. Tréglia, *Acta Mater.*, 1999, **47**, 2705.
12 J. Creuze, F. Berthier, R. Tétot and B. Legrand, *Phys. Rev. B*, 2000, **62**, 2813.
13 F. Baletto, R. Ferrando, A. Fortunelli, F. Montalenti and C. Mottet, *J. Chem. Phys.*, 2002, **116**, 3856.
14 C. Mottet, J. Goniakowski, F. Baletto, R. Ferrando and G. Tréglia, *Phase Transitions*, 2004, **77**, 101.
15 S. M. Foiles, *Phys. Rev. B*, 1985, **32**, 7685.
16 K. Binder, *The Monte Carlo Method in Condensed Matter Physics*, Springer Verlag, Berlin, 1995.

17 N. Metropolis, A. W. Metropolis, M. N. Rosenbluth, A. H. Teller and E. Teller, *J. Chem. Phys.*, 1953, **21**, 1087.
18 B. Legrand, F. Lequien, J. Creuze and F. Berthier, Document Technique DMN/SRMP/NT/2006–36, 2006, vol. 137, pp. 131.
19 A. Saúl, B. Legrand and G. Tréglia, *Phys. Rev. B*, 1994, **50**, 1912.
20 A. Saúl, B. Legrand and G. Tréglia, *Surf. Sci.*, 1995, **331–333**, 805.
21 J. Creuze, F. Berthier, R. Tétot and B. Legrand, *Surf. Sci Lett.*, 2001, **491**, L651.
22 G. Rossi, A. Rapallo, C. Mottet, A. Fortunelli, F. Baletto and R. Ferrando, *Phys. Rev. Lett.*, 2004, **93**, 105503.
23 A. Rapallo, G. Rossi, R. Ferrando, A. Fortunelli, B. C. Curley, L. D. Lloyd, G. M. Tarbuck and R. L. Johnston, *J. Chem. Phys.*, 2005, **122**, 194308.
24 V. Moreno, J. Creuze, F. Berthier, C. Mottet, G. Tréglia and B. Legrand, *Surf. Sci.*, 2006, **600**, 5011.

General Discussion

Professor Sir Thomas opened the discussion of Professor Jellinek's paper: I agree with much of what you say. But there are a few points that I would like to air which, I believe, will help to eliminate some confusion when we talk about nanometals and nanoalloys.

First, when you have only 13 atoms, as you have described, it is essential to talk about "clusters" not "nanoalloys". These two terms tend to be used synonymously: they should not be. A 13 atom bimetallic entity has energy levels, whereas a so-called nanoalloy of 100 nm diameter contains around 10^8 atoms, so that we must talk about energy bands. The notion of a LUMO in a cluster is quite distinct from the Fermi level in a 10^8 atom entity.

Second, I agree also that emphasis has to be placed on structural determinations of these clusters, whether they float in a vacuum or are anchored to a solid support. Why have there not been many attempts (to my knowledge) in using gas-phase electron diffraction (of the kind first carried out by Pauling and 70 years later used in a time-resolved fashion) to determine the precise structure of say $He_{12}Ni$ or Ni_6H_7? It may well transpire, from the determined intra-molecular (or intra cluster) bond lengths that there is a well-defined structure with chemical bonds formed by spin pairing. (I am not sure that a Gupta potential, good as it is, will yield precise "molecular" structures for Al_nNi_m ($m + n = 13$)).

Lastly, although I did not mention it, I feel we have to be careful when we read papers dealing with nanoalloys or nanometals that are shrouded with an organic (usually polymeric) surfactant, to keep them suspended in a fluid. It is always better to work with ligand-free nanoclusters wherever possible.

Professor Jellinek responded: The defining hallmark of nanosystems is that their properties change with the system size. Whereas the origins of the "nano" part of the term "nanosystems" can be traced to the nanometer (one billionth of a meter), the characteristic size of what is considered to be a nanosystem evolved from hundreds to tens, to just a few nanometers, and even to the subnanometer scale. In my view, the most adequate operational definition of nanoscience is as the science of systems and phenomena whose properties change with the size. It includes small-size nanosystems (clusters), medium-size nanosystems (up to a few tens of nanometers) and large-size nanosystems (upper tens and hundreds of nanometers). Of course, the properties in the different size ranges are different and, what is even more important, they change with the size of the systems differently. In the small-size range, most properties exhibit non-monotonic variations with the size (often dubbed as emergent phenomena). In the larger-size ranges, these changes become monotonic and can be characterised by scaling laws. In some sense, the small-size range is more interesting as it presents a particularly strong size-dependence of properties and an unprecedented variety and variability of features. The goal of nanoscience, including that of nanoalloys, is to understand the size-evolution of properties of systems at all size ranges and the transition from emergent to scaling behaviour. Bimetallic clusters are superb "laboratories" for addressing the complex issues of the role of size, structure, composition, temperature, and other control parameters on the electronic, optical, magnetic, and chemical properties of nanoalloys at a very detailed level.

As to the scarcity of electron diffraction experiments, to my knowledge the difficulty stems from the usually insufficient density of the samples. Joel Parks from the Rowland Institute at Harvard is the champion of such experiments at present.

The main goal of our explorations of Ni/Al clusters with the many-body Gupta potential was to use them as a paradigm for addressing fundamental questions such as: (1) What are the different possible structural forms of bimetallic clusters?;

(2) How do they change with the cluster size and composition? (3) Can one introduce a rational classification within the immensely large manifolds of the possible structural forms of nanoalloys? (4) What defines the propensity of the component elements in a nanoalloy to mix or to segregate? (5) What are the appropriate or even optimal indicators/measures of the degree of mixing or segregation? (6) How do the dynamical properties of nanoalloys change with their size and composition?; (7) What underlies the size- and composition-specific peculiarities of nanoalloy dynamics? The answers to these questions are quite robust and do not depend on the fine details of the potentials, provided they are globally adequate (*e.g.*, of many-body character). The Gupta potential for Ni/Al systems turned out to be adequate not only for providing answers to the mentioned questions, but also for defining the correct structural forms, including those of Ni/Al 13-mers, as verified by the subsequent experiments.[1]

1   E. F. Rexer, J. Jellinek, E. B. Krissinel, E. K. Parks, and S. J. Riley, *J. Chem. Phys.*, 2002, **117**, 82.

Professor Fortunelli asked: In the disentanglement of structural and electronic *etc.* properties, do you not have to take into account the structural changes induced by the change in composition?

Professor Jellinek replied: The disentanglement is only an apparent one as the semiempirical potentials we are using are fitted to measured data, which, of course, incorporate the electronic and other effects. The structural changes due to the change of composition, both in the equilibrium interatomic distances and, possibly, in the atomic packing, are fully incorporated and accounted for by complete reoptimization of the structures for each new composition. In fact, we have found that some isomeric forms (types of atomic packing) that are supported by some compositions may not survive as stable equilibrium conformations for others.

Professor El-Shall asked: You mentioned that N_2 is adsorbed on Ni but not on Al surfaces. This behavior occurs in condensed phases for extended surfaces, but in 13 atom clusters the situation could be different since N_2 could bind to the cluster by typical van der Waals' forces. Are the binding energies of N_2 to the small Al and Ni clusters sufficiently different to reflect the condensed phase behavior? Is the 13 atom cluster a good model for the extended surface?

Professor Jellinek responded: The binding arrangements and energies of atoms and molecules on surfaces of clusters/nanoparticles are, as a rule, very similar to those on the corresponding extended surfaces. That is the basis of our argument. In that sense, the surface of an icosahedral 13-mer is not very different from the (111) surface of a fcc lattice. This view allowed us to fully explain all the details of the measured N_2 adsorption data on Ni/Al clusters of various sizes and all possible compositions in terms of the structural forms of these clusters we predicted computationally.[1–4]

1   E. F. Rexer, J. Jellinek, E. B. Krissinel, E. K. Parks, and S. J. Riley, *J. Chem. Phys.*, 2002, **117**, 82.
2   J. Jellinek and E. B. Krissinel, *Chem. Phys. Lett.*, 1996, **258**, 283.
3   E. B. Krissinel and J. Jellinek, *Int. J. Quantum Chem.*, 1997, **62**, 185.
4   J. Jellinek and E. B. Krissinel, in *Theory of Atomic and Molecular Clusters with a Glimpse at Experiments*, ed. J. Jellinek, Springer, Heidelberg, 1999, pp. 277–308.

Professor Ferrando asked: What is the further evidence supporting the classification into homotopic classes, mixing energy *etc.*, besides the example of AlNi for size 13?

Professor Jellinek replied: The definitions of homotops and homotopic classes as well as those of mixing energy and mixing coefficient are general and do not depend either on the size or the composition of a nanoalloy or any other two-component system. Partition into homotopic classes and subclasses (defined, *e.g.*, by the type of the central or the most coordinated atom(s)) allows for hierarchical classification and understanding of the energy ordering of the different structural forms within the immense manifolds of possible nanoalloy, or more generally two-component, conformations. Further evidence for 12- and 14-atom Ni/Al clusters is given in ref. 1.

1   E. F. Rexer, J. Jellinek, E. B. Krissinel, E. K. Parks and S. J. Riley, *J. Chem. Phys.*, 2002, **117**, 82.

Professor Bond communicated: Professor Jellinek reminds us that calculated structures represent the situation at zero Kelvin where thermal motions are essentially frozen out, and he goes on to say that for higher temperature states the motion of surface atoms can be modelled. Now most conceivable applications will be conducted at ambient or higher temperature, where the surface will be disordered. How then can it be a realistic objective for future work to design clusters *ab initio* for specific uses? It seems to me that attention has to be focussed on the electronic properties of single atoms while in motion, *i.e.* while not occupying specific lattice sites, rather than on geometric factors associated with specific groupings of surface atoms derived from static models.

Professor Jellinek communicated in reply: Although the melting temperature of clusters and nanoparticles is, as a rule, suppressed as compared to the bulk, many of them, especially those of metals and alloys, have well-defined structures at ambient and even elevated temperatures. The thermal motion of the atoms is executed around these well-defined structures, and one thus can talk about the structure–reactivity correlation or lack of it. At higher temperatures, the particles undergo structural (isomerization) transition with a possible stage of coexistence of different isomeric forms. At still higher temperatures the particles melt. A stage of surface melting, or partial melting, may precede the stage of complete melting. As this happens, the structural aspects become, of course, less relevant.

Dr Baletto opened the discussion of Professor Fortunelli's paper: How were the calculations of M_xAg_y (M = Au, Pd) ($x = 1, 2$; $y = 8, 9$) on the (3×3) MgO surface performed?

Professor Fortunelli responded: The calculations have been performed using both a basis set of Gaussian-type-orbitals and one of plane waves, with a unit cell of (3×3) but also (4×4). All these calculations give qualitatively identical and quantitatively similar results. This has been already noted in the literature, and is partly due to compensation effects and error cancellations.

Dr Calvo commented: Your discussion of the magic character of the mixed (M_nAg_p) clusters is based on the binding energy of the silver atoms, as well as the HOMO–LUMO gap. Shouldn't it rely on the binding energy of the other M atom as well?

Professor Fortunelli responded: The binding energy reported in the table in our article is a total binding energy, which thus includes the interaction of the whole cluster (both silver and copper or gold or palladium atoms) with the surface. The differential binding energy $\Delta E(N)$ is also reported and is useful because it gives one an estimate of the binding energy of the added silver atom, and thus of the fragmentation energy of the cluster, which is both an important indicator of its stability and provides necessary information about partial detrapping leading to Ostwald ripening (thus providing suggestions about kinetic formation effects).

Dr Calvo remarked: The fluxionality you refer to mainly involves the silver 'cage', the Cu or Pd atoms remaining in the vacancy. How could this affect the catalytic properties?

Professor Fortunelli responded: Catalytic properties are known to be surface-driven. Silver exhibits some catalytic activity, whence the importance of the structure of the silver 'cage'. Furthermore, we have found[1] that Ag_8 has a much lower interaction energy with the oxygen molecule than the neighboring sizes, due to the compact character of its 'magic' structure and also to the effects that electronic shell closure has on its chemical properties. We think that this line of investigation (subtle effects induced by electronic and structural shell closure) is worthwhile pursuing in the search for more *selective* metal catalysts, and we plan to continue along this line in the future.

1   G. Barcaro and A. Fortunelli, *Phys. Rev. B*, 2007, **76**, 165412.

Dr Grönbeck asked: Could you comment on the changes in cluster/surface interaction as a function of cluster size and composition? To understand the reason for the enhanced stability of Ag_8 it could be good to separate cluster/surface and internal cluster interaction energies.

Professor Fortunelli answered: I agree that it would have been useful to decompose the total binding energy into adhesion and metal binding components. We have often reported and discussed this decomposition in previous publications on analogous systems. In the present case, we limited ourselves to discussing this type of information in the text, without providing explicit figures. Anyway, it results that Ag_8 has a good adhesion to the surface, but that its enhanced stability is mainly due to the electronic shell closure effect.

Professor Johnston asked: I gather that the double vacancy is a neutral defect. Is there much charge transfer? I presume a charged defect would lead to charge transfer to or from the metal cluster which would show up as a change in the magic number?

Professor Fortunelli replied: The double vacancy is a neutral defect. Experimentally, when MgO(100) surfaces are prepared by UHV cleavage, a thermal annealing is usually performed which gets rid of most charges on the surface. However, charged defects and thus charged clusters growing on them can certainly be considered, in which case the electron count would differ, thus entailing differences in the sizes at which electronic shell closure is achieved. In the neutral double vacancy, the charge transfer from the surface to the adsorbed clusters is limited, and of the same order as that on the regular surface. It has been estimated in our calculations, but not discussed explicitly.

Professor Fortunelli commented: There have been some questions about the order of binding energies among the various cluster sizes and compositions. In general, there is no simple rule which is always valid for predicting this. Much depends on the relative strength of the metal–surface *vs* metal–metal interaction, and this is a subtle effect because Cu, Au, and Pd have binding energies in the bulk that are not too dissimilar, but Pd usually interacts with the MgO surface somewhat more strongly than Cu or Au. However, it can be observed that Cu–Ag clusters have a binding energy larger than the corresponding (same composition) Au–Ag clusters. This is also true for Pd–Ag clusters with respect to both Au–Ag and Cu–Ag clusters, but with the exception of Pd_1Ag_7 in which a strong structural rearrangement takes place (as described in the paper). It can also be observed that M_2Ag_N clusters have a binding energy larger than the corresponding (same total number of atoms) $M_1Ag_{(N+1)}$ clusters: this is essentially because Pd, Cu and Au all have an interaction with the surface larger than Ag (as proven by the fact that they always occupy *internal* positions in the cluster), and also because as a norm alloying is effective in stabilizing these clusters.

Professor Lievens remarked: 8-atom noble metal clusters indeed do correspond to electronic shell closures, provided they have a 3-dimensional shape. For 2-dimensional structures, 6 is likely to be a shell closure. Is there any evidence for clusters with a 2-dimensional shape to bind on double vacancy defects?

Professor Fortunelli responded: There is no evidence of planar six-atom clusters adsorbed on the double vacancy (DV) for silver, nor—we think (but we have not checked this in detail)—for the binary M–Ag (M = Cu, Au, Pd) systems that we have considered in this work. This is because when a six-atom silver cluster is adsorbed on the DV, it changes its shape to a compact (non-planar) configuration. This is connected with the general fact that electronic shell closure is not very strong in silver, due to electronic wave function interference at the given bond distances. However, Au_6 has a planar structure both in the gas phase and when adsorbed on the DV, and it possesses a somewhat enhanced stability, so that it might be considered as a sort of surface magic cluster: this is still unpublished work, the corresponding manuscript is in preparation. In general, we may expect to find more such examples when considering clusters not of noble metals but of transition metals.

Professor Sir Thomas said: I wonder why you took the double vacancy at the MgO surface. In my reading of the catalysis field, I encounter much more frequently experiments that involve single atoms (*e.g.* Pd) bound to an F-centre (the colour centre formed when you have an electron trapped at an oxygen vacancy). There is evidence of charge-transfer involving the F-centre and the metal cluster.

Professor Fortunelli responded: The type of defect on the surface depends on the method of preparation.

When MgO(100) single crystal surfaces are prepared by UHV cleavage, there are experimental indications that the most common defects are double vacancies, as shown in ref. 27 of our article. So, our model is thought to be fully realistic.

We have studied the growth of metal clusters adsorbed on an Fs-center of an MgO(100) terrace as well, and found that binary Pd–Ag clusters can give rise to magic behavior, as described in the poster by Giovanni Barcaro. However, recent experiments (see ref. 20–22 in our article) indicate that Fs-centers are not located on terraces, but at edges or corners of terraces, which casts some doubts on previous literature on the subject, or at least calls for a more realistic modeling of the defective surface.

Mr Neumann communicated: I would like to refer to your given HOMO–LUMO energy gap of Ag_8 clusters on MgO surfaces (1.7 eV). Is it possible to generate free charge carriers (*e.g.* electrons and defect electrons (p-holes)) in the cluster after illumination with light? Did you include optically excited states in your calculations? Do you know something about the stability of excited clusters and the lifetime of photogenerated charges?

Professor Fortunelli communicated in reply: Your questions are extremely interesting. We have performed some preliminary calculations of the optical response of deposited Ag clusters, including absorption, excited states, *etc.* However, we are just at the beginning, and I would like to cross-check the calculations before I commit myself to definite answers. Due to the presence of the support, these calculations are more complicated than for gas phase clusters. I can only say that calculations are in progress that should answer your questions, and that I will let you know as soon as I have absolutely sound results available.

Professor Hou opened the discussion of Dr Rossi's paper: One of the interesting outcomes of your work is the possible competition between gold segregation and $L1_2$ phase formation as suggested by your Fig. 3 in the case of Cu_3Au decahedral clusters. But why don't you find any phase formation in $CuAu_3$ which would be favoured by gold segregation? Could the reason be that, as happens with several semi-empirical potentials parameterized on the properties of an A_3B $L1_2$ alloy, the phase stability of the AB_3 alloy is not well predicted?

Dr Rossi replied: We agree that the use of the RGL potential—as proposed by Cleri and Rosato—predicts the behaviour of Au_1Cu_3 better than the behaviour of Au_3Cu_1. As regards to the global minimisation, one can thus be more confident about the global minima obtained for the Au_1Cu_3-like compositions. Concerning cluster growth, the situation is further complicated by the fact that the relative concentrations of Au and Cu can change during the process. In this case, the choice and use of a single parametrization is in our opinion the only reasonable approach at present. Concerning the fact that we do not observe the $L1_2$ ordering in the global minimum clusters of Au_3Cu_1 composition, we can state that the best clusters are I_h because the Cu concentration allows a good relaxation of the volume strain, while the decahedral minima are much higher in energy.

Professor Johnston commented: From calculations that we have performed in my group we can confirm that the Gupta potential for Cu–Au performs worse for bulk $CuAu_3$ than Cu_3Au.

Professor Ferrando remarked: The Gupta potential does not include charge transfer. This probably means that the tendency to intermixing is underestimated, and as a consequence, the tendency of gold to surface segregation is overestimated.[1]

1   M. Zhang and R. Fournier, *J. Mol. Struct. (THEOCHEM)*, 2006, **762**, 69.

Professor Fortunelli commented: What is the origin of the fact that the intermediate shell in Au–Cu clusters is not ordered? (from both growth simulations and global optimisation).

Dr Rossi replied: The chemical order of the intermediate shell in Au–Cu clusters is structure-dependent. Most of the global minimum clusters obtained at the sizes and compositions analysed belong to the I_h structural family. Two driving forces are competing to set the chemical order of the icosahedral clusters. The first is a geometrical driving force: the smaller atoms tend to occupy the volume site of the cluster, the larger atoms tend to be placed on the surface. The second force has a chemical nature, and leads to maximise the number of heterogeneous bonds. The competition between such forces leads to the stabilisation of I_h structures with a Cu-enriched core, a mixed intermediate shell and a Au-rich surface. Moreover, the ordered phase can not be stabilised inside each tetrahedron composing the I_h, because of frustration effects.[1]

1   G. Rossi, R. Ferrando and C. Mottet, *Faraday Discuss.*, 2008, **138**, DOI: 10.1039/b705415g

Dr Baletto commented: Please provide information on:

(1) The transformation from icosahedral to decahedral clusters as size increases observed in your simulations.

(2) Diffusion of gold and copper atoms on the cluster surface.

(3) A possible explanation of surface segregation in terms of diffusion/incorporation.

Dr Rossi responded:

(1) The opposite solid–solid transition, from D_h to I_h, has been previously proved to be induced by the growth of stacking-fault islands on the (111) facets of the D_h.[1] The transformation of the icosahedral Au-rich cluster into a decahedral cluster which we observe during the growth does not involve the melting of the structure, as far as can be appreciated looking at the snapshots collected during the simulation. At present, we can state that the transition involves the progressive destruction of the five-fold symmetry axes through the volume of the cluster.

(2) and (3). The time scale of the MD simulations are not suitable to observe any incorporation of the Cu atoms into the Au core. Concerning surface diffusion, we have not yet quantified the diffusion barriers associated with the displacement of Cu or Au atoms through the different surface facets.

1   F. Baletto and R. Ferrando, *Rev. Mod. Phys.*, 2005, **77**, 371.

Dr Calvo said: What would you get in your growth simulations if you had started from a perfect truncated octahedron or a decahedron rather than an anti-Mackay icosahedron?

Dr Rossi answered:

In order to reply to the question, we ran four growth simulations depositing Cu atoms onto a magic Au 201-atom truncated octahedron. Two simulations were performed at $T = 500$ K, two at $T = 600$ K.

(a) $T = 500$ K

At this temperature no liquid clusters are observed during the growth process. At size $N = 300$ the cluster has still a fcc structure, but the truncation of the starting Au seed disappears or is limited to small 2×2 (100) facets. 85% of the Cu atoms occupy surface sites, while the remaining atoms can be found in subsurface sites. These are predicted to be the favorable sites for the smaller atoms in fcc structures also for other size-mismatched heterogeneous systems (such as AgPd, AgCu, AgNi).[1]

(b) $T = 600$ K

At this temperature the 201-atom cluster is liquid. It undergoes a liquid–solid transition at size $N = 300$, when the cluster solidifies into a I_h structure. A I_h to D_h transition can be located at about $N = 325$ atoms. Going on with the growth up to $N = 350$, one eventually obtains a D_h cluster which is slightly Au-enriched in the core. Such a cluster is shown in Fig. 1.

It is worth noting that a I_h to D_h transition just below the melting point of Au clusters (size > 3 nm) has been experimentally observed by Koga.[2]

As a general remark, one can point out that other structural changes could be associated with the incorporation of copper in the center of the cluster. Nevertheless, the time scale of this event can not be reproduced by MD growth simulations at present.

1   F. Baletto, C. Mottet and R. Ferrando, *Phys. Rev. Lett.*, 2003, **90**, 135504.
2   K. Koga, T. Ikeshoji and K. Sugawara, *Phys. Rev. Lett.*, 2004, **92**, 115507.

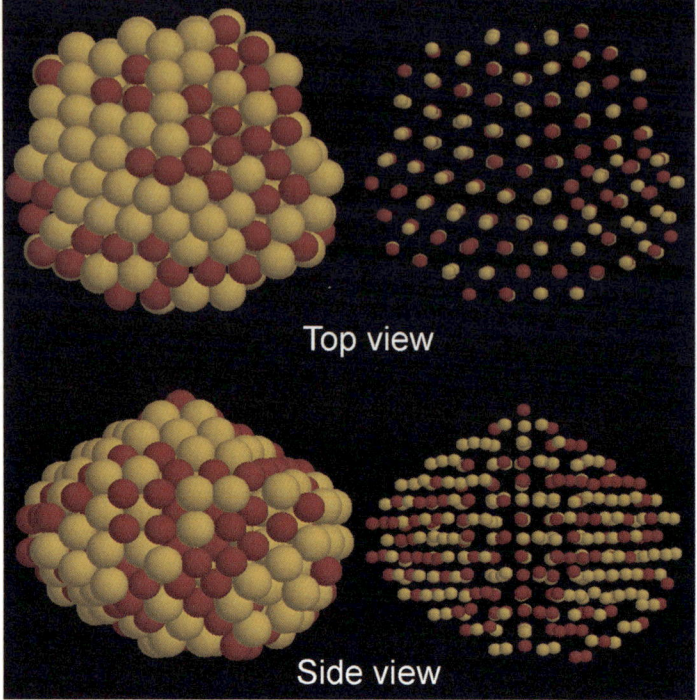

Top view

Side view

Fig. 1

Professor Mattei asked:

(1) The presented simulations on Au–Cu nanoclusters with a number of atoms up to 200 support the segregation of Au at the surface. Is the extent of such segregation dependent on the number of atoms for a given alloy composition, *i.e.*, is there any trend on increasing the cluster size (also beyond the 200 atoms limit)?

(2) Is the Au surface segregation present also for alloy clusters grown at temperatures higher than 500 K?

Dr Rossi answered:

(1) In our opinion, some segregation effects should persist also for larger clusters, but this segregation effect is essentially limited to the most external layer (single atom thickness) of the cluster, which can be Au-rich.

(2) Yes, the effect can still be observed in clusters grown at $T = 600$ K. In a cluster with 350 atoms (about 2 nm diameter), for example, and composition Cu 58% and Au 42%, the surface composition is reversed to Au 55%, Cu 45%.

Professor Sir Thomas opened the discussion of Professor Hou's paper: In the edge dislocation shown on the left in your Fig. 5, presumably its sessile nature arises from a Cottrell cloud that can be accommodated in the expanded region of the lattice at the core of the dislocation. So far as your glissile dislocation is concerned, I wonder whether it is its screw component that is responsible for its glissile nature because there is no expansion of the lattice at the core of such a dislocation.

Professor Hou replied: The burgers vectors of the dislocations shown are given in the Figure title. The glide plane of the $\frac{1}{2}[001]$ dislocation (Fig. 5a) is (001). (001) is perpendicular to the applied stress, which favours its sessile character. The glide plane of 1/6[112] is (11−1). It is parallel to the burgers vector and this dislocation—

which is also edge—may thus glide. The change in volume, as we already found in cluster assembled Ni_3Al simulations (ref. 11 in our paper) is easily accommodated by surface relaxation. This feature is specific to the nanoscale.

Beyond this typical example, the dislocations we observed could not all be identified as pure edge although they generally end up in planar {111} faults. We failed indexing burgers vectors on the fly because the dislocation core areas are too large as compared to the wire diameter in order to allow identifying a burgers circuit unambiguously.

Professor Johnston said: Do your calculations indicate enrichment of the strained region by one of the component metals?

Professor Hou responded: I answer this question together with that of Professor Ferrando (see below), which is similar.

Our simulations are indeed too short to predict segregation in strained regions. This problem was addressed in our previous work using the same potential and Metropolis Monte Carlo simulations. We found that:

(1) A limited segregation of Al is possible at the surface of isolated clusters (no more than 10%).[1] This segregation state is not significantly altered by assembling the clusters into a material and does not affect its mechanical properties significantly (see ref. 11 and 23 in our paper).

(2) Given these results, I do not think that segregation at the induced wire nanostructure interfaces would substantially affect comparison with experiment.

1   E. E. Zhurkin and M. Hou, *J. Phys. C*, 2000, **12**, 6735.

Professor Ferrando asked: During your simulations, the time scale is too short to equilibrate chemical ordering.

Could this be a real problem in comparing with experiments, since experiments take place on a much longer time scale?

Professor Hou replied: See my previous reply to the question of Professor Johnston.

Professor Ricolleau asked: Does the inequivalence between the young modulus of the wires and of the bulk come from a size effect or the anisotropy of the wires?

Professor Hou responded: We compare the Young modulus of the wires— estimated in the axial [001] direction—to the Young modulus estimated in a bulk single crystal in the same direction as well (ref. 23 of our paper). The difference is thus a size effect.

Dr Calvo asked: The cylindrical to elliptical transition you find is reversed when heating up. Is it also reversible when you pull back on your sample to where you initially started?

Professor Hou answered: We did not try, and we thank you for the suggestion. We find it reversed when heating up only after compression, with the TB potential. This means that the energy barrier encountered is small. In this case, it should be reversible when pulling back.

Professor Bond communicated: It is worth noting that many of the binary combinations of metals that are the subject of theoretical and experimental study in this Discussion would exhibit phase separation in the bulk state. This is quite clear for the Pt–Au system, where the mutual solubilities at both ends of the composition range are quite small.[1] We noted with some surprise when writing *Catalysis by Gold*[2] that a number of authors claimed to be able to make homogeneous PtAu particles of compositions that lay well within the miscibility gap, a surprise that is shared by Sermon *et al.*[3] Reports of lattice parameter determination have however confirmed that this is possible.[1] I am not aware of any experimental measurements of the size at which a bimetallic particle transforms into a biphasic one. There has been much thought given in the literature, particularly in respect of gold, to the size at which metallic properties start, *i.e.* at which a band structure is formed. Similar considerations might apply to bimetallic particles (nanoalloys), *i.e.* phase separation may start when there is a sufficient number of both kinds of atoms present for bands to exist, after which the conditions for mutual solubility no longer apply, and this probably corresponds to the onset of metallic conduction. This change has been discussed in a very qualitative way in the case of the Pt–Au system.[1] Just as the transition of monometallic particles to the metallic state is not abrupt, and indeed depends on the method used to assess it, so the transition of bimetallic particles to a biphasic structure may be ambiguous, depending for example on the presence of stabilising ligands or chemisorbed molecules. I see comparatively little use of XRD to assess the homogeneity of small particles, and the size limitation for this condition deserves greater attention.

One can hardly expect the surface atoms in a small particle or cluster to exhibit precisely the same chemical properties shown in chemisorption and catalysis as those of atoms on the surface of large particles. Even on such particles, isolated atoms of one type, not interacting with others of the same kind, must show atomic behaviour, so that this kind of 'electronic factor' by which catalytic activity varies with composition should exist alongside the more usually noted 'geometric factor'. It is perhaps one of the limitations of DFT, at least for an old-fashioned chemist, that the nature of the orbitals available on surface atoms for bonding ligands or for chemisorbing molecules is not explicitly revealed in the process, so the chance to discuss the chemistry of surfaces is somewhat limited.

1   G. C. Bond, *Platinum Met. Rev.*, 2007, **51**, 63–68.
2   Geoffrey C. Bond, Catherine Louis and David T. Thompson, *Catalysis by Gold*, Catalytic Science Series Vol. 6, Imperial College Press, London, 2006.
3   Ken A. Grant, Kelei M. Keryou and Paul A. Sermon, *Faraday Discuss.*, 2008, **138**, DOI: 10.1039/b708810h

Professor Jellinek responded: It is quite conceivable that the energy-driven preference for mixing or segregation of the component elements in nanoalloys may change with the size and/or the percentile composition of the particles. However, even when energetically not the most preferred one, the mixed or the segregated (*e.g.*, core–shell) phase can be generated experimentally and in simulations as metastable states by defining appropriate assembly conditions. The question about the possible correlation between the size (or amount of component elements) driven transition to metallicity and the preference for segregation is an open and complex one in part because of the complex nature of the size-induced transition to metallicity (*cf.*, *e.g.* ref. 1–3; and citations therein).

The chemical reactivity of nanoalloys is a function of their size and composition and, in general, may not be reducible to the attributes of an "isolated" surface atom. Incidentally, DFT does a pretty good job in providing a qualitative picture of the orbitals and the charge density distributions.

1   P. H. Acioli and J. Jellinek, *Phys. Rev. Lett.*, 2002, **89**, 213402.
2   J. Jellinek and P. H. Acioli, *J. Phys. Chem. A*, 2002, **106**, 10919.
3   J. Jellinek and P. H. Acioli, *J. Phys. Chem. A*, 2003, **107**, 1670.

Professor El-Shall opened the discussion of Dr Calvo's paper: The Lindemann index, which measures the average root-mean-square fluctuation in intermolecular separations of atoms in the cluster, has been used to study the melting and coexistence phenomena in atomic clusters. Is this index applicable to the melting of Pd–Pt clusters? Can it predict the same melting behavior you found using the canonical heat capacities? What is the signature of the second-order transition in this case?

Dr Calvo answered: The Lindemann index is useful when performing single trajectory molecular dynamics or Monte Carlo simulations, but is not applicable to multiple-trajectories parallel tempering simulations. In the present case, its variations with increasing temperature would show a sharp rise near the melting point, but would not be able to detect the alloying transitions, because the rate constants associated with the identity swap processes are much too low with respect to the time scale of molecular dynamics. Hence it would more or less show the same conclusions regarding melting, but would hardly provide any useful information about the preliminary alloying transition.

Professor Johnston said: Does the melting start from the gold core (in the Au–Ag cluster mentioned) or the silver shell? Note that the cohesive energy of silver is much lower than gold.

Dr Calvo responded: You are right, and this is an important point which shows some contrasting behaviour with Pd–Pt clusters.

Inspecting the radial distributions functions for Au and Ag at increasing temperature clearly shows that temperature-induced alloying is initiated on the surface and progresses toward the core. This could have some important consequences for catalytic applications.

Professor Sir Thomas asked: Your work, and especially one of the slides you showed during the Discussion, prompts me to recall the elegant work of a Swiss physicist (whose name I have forgotten) that was reported during the Faraday Discussion (on nanoparticles) that I organised in the Royal Institution in 1991 to coincide with the 200th anniversary of Faraday's birth. By elegant electron diffraction and electron microscopy this Swiss physicist (I think in the late 1970s) showed how the melting point of gold varied as a function of particle size. As I recall, the melting point of a 5 nm particle of Au was as low as about 400 K.

Dr Calvo replied: What you are referring to here are the nice results obtained by Buffat and Borel[1] on supported gold nanoclusters having a few nanometers in diameter. These authors found one of the most striking pieces of evidence that the depression in the melting point scales as the inverse of the cluster radius.

1   Ph. Buffat and J.-P. Borel, *Phys. Rev. A*, 1976, **13**, 2287.

Professor Bond commented: It is possible to take Professor Sir John Meurig Thomas's remark on the particle size dependence of melting temperature a little further. In the case of gold, and perhaps generally, surface mobility starts well below the melting temperature, and the surfaces of metallic particles of 1–2 nm in size are likely to be semi-molten at ambient temperature. What significance can therefore be attached to the results of computational procedures purporting to assign definite structures to such small structures? Such models can have only limited usefulness in understanding chemisorption and catalysis.

Dr Calvo answered: The issue you raise is indeed a real concern in computational approaches to such problems. The lower thermal stability of smaller nanoparticles may not be of real consequence here, due to their still high melting points, much higher than room temperature. However, as you correctly point out, finite temperatures can stabilize various defective structures, not only with molten surfaces but also, in the present case of Pd–Pt clusters, molten cores (in a statistical sense). The preferential structures found at room temperature could thus differ from the unique lowest-energy structure, however locating the most stable free-energy conformations is a difficult task, for which I believe that Monte Carlo methods are appropriate. Even though the global minimum structure itself is not necessarily relevant for small bimetallic clusters, such simulations provide a detailed statistical description at the level of atomistic detail, hence a more realistic model for possible applications.

Professor Jellinek commented: Theory and simulations predict and describe the high mobility of atoms on surfaces of particles as a consequence of surface melting induced by heating (*cf. e.g.* ref. 1 and 2).

1   Z. B. Güvenç and J. Jellinek, *Z. Phys. D*, 1993, **26**, 304.
2   E. B. Krissinel and J. Jellinek, *Chem. Phys. Lett.*, 1997, **272**, 301.

Dr Baletto stated: The melting of binary clusters may take place through a sequence of transitions including alloying transitions, surface melting, *etc.*

Dr Calvo responded: There are probably no general rules, but yes, the various premelting phenomena known in homogeneous metal clusters should combine with alloying transitions into a richer sequence. In alloys of size-matched elements, melting could involve a structural transition and, on top of it, an alloying transition among the various homotops. However, the situation should be different for size-mismatched elements, as the stable structures and related transitions are much more likely to be composition- and size-specific. Some qualitatively new phenomena could then appear.

Professor Johnston asked: Perhaps what we should address here is which effects are due to the fact that we have a two component nanoparticle, rather than size effects which are observed for pure metal clusters. Can we see new finite size effects for bimetallic clusters?

Dr Calvo answered: This question will have different answers for different nanoalloys. Size-matched elements may lead to extra order/disorder transition having no counterpart in pure clusters. Size-mismatched elements, on the contrary, could produce finite size effects of their own, with stronger size- and composition-dependencies.

Dr Baletto asked: Is it possible to define order parameters to describe the melting process including transition through different homotops/isomers?

Dr Calvo replied: Characterizing properly the finite size phase changes may indeed be difficult, and only few quantities have a well defined interpretation valid up to the bulk limit. The heat capacity is one of them, and is generally appropriate for locating the melting point, even though some more complicated behaviours have been reported.[1] The questions of premelting and alloying transitions are harder to address in general. Purely dynamical indices (Lindemann parameter, diffusion constant...) are of limited interest due to the very short times of molecular dynamics simulations.

Radial distributions are also not practical for relatively large clusters. For alloying transitions, dedicated order parameters should be carefully calculated. A good picture of the various transitions often emerges from the energy landscape, obtained by performing regular quenches along a reasonably ergodic simulation.

1   F. Calvo and F. Spiegelman, *J. Chem. Phys.*, 2004, **120**, 9864.

Professor Johnston stated: Another question we can ask is what is the necessary level of detail we need as a function of cluster size? For small clusters we can talk about global minima and individual atom motions—perhaps we merely need to define classes of clusters? Related to this, we probably need to introduce different order parameters for different size regimes.

Professor Ferrando opened the discussion of Professor Leiva's paper: You have shown in your slides that you also obtain three-shell structures in the collision of Ag and Cu clusters. Three-shell AgCu structures are metastable.[1] Have you tried different collision temperatures and velocities to determine what is the range in which three-shell structures form?

1 F. Baletto, C. Mottet and R. Ferrando, *Phys. Rev. Lett.* 2003, **90**, 135504.

Professor Leiva responded: We have tried collision velocities of 40, 100, 300 and 400 m s^{-1} with these clusters. In the first and second cases we obtained three shell structures, while in the remaining ones we got core–shell structures. A more detailed study will be undertaken in the future.

Professor Jellinek commented: In addition to head-on collisions, one can also consider non-head-on (*i.e.*, non-zero impact parameter) collisions. They will result in some part of the collision energy being channeled into the overall rotation of the resulting bimetallic particles. It is of interest to explore the effects of the collision impact parameter and the resulting rotation on the geometric and elemental (homotopic) structures of the collision-generated nanoalloys.

Dr Calvo commented: How do you get "final configurations" without removing the extra kinetic energy? That would imply that the simulation leads to single, well defined isomers that do not change in time—a stationary state. Otherwise, different structures could well be found later if you can wait long enough.

Professor Leiva responded: There was no energy removal in our simulations. The use of the word "final" was probably misleading. With this I meant the final part of the simulation, that typically lasted of the order of ten nanoseconds. At these times the kinetic energy (alternatively the potential energy) remains stable in the nanosecond range and different systems presented different behavior (atom mixing *vs.* core–shell structures). The long time behavior (microsecond and more) is still an open question and would of course be interesting to analyze.

Professor Hou remarked: I am surprised by the evolution from core–shell to onion Ag–Cu cluster configuration as depicted in Fig. 6 of your paper. The Metropolis Monte Carlo simulations in the papers of Lequien *et. al.*[1] and of Dzhurakhalov and Hou[2] predict the core–shell structure to be thermodynamically stable. If this is correct, why does your system evolve from stable to metastable?

1 F. Lequien, *Faraday Discuss.*, 2008, **138**, DOI: 10.1039/b705281b
2 Abdiravuf A. Dzhurakhalov and Marc Hou, *Phys. Rev. B*, 2005, **76**, 045429.

Professor Leiva responded: A comparison between the present and the former results is not straightforward. First of all, the simulations belong in principle to different ensembles and simulations in different ensembles may lead to different conclusions in the case of nanosystems. Second, the sampling of the configuration space is different in the three cases. Third, we are not sure how close to equilibrium our configurations were at the final stage of the simulation. Fourth, the results may depend on the sizes of the clusters involved in the collision. A detailed study changing the cluster sizes and collision conditions is necessary to give a definite answer to this question.

Professor Hou opened the discussion of Ms Lequien's paper: Very interestingly, you show that, when the edges are isolated from the rest of the cluster, each edge may serve as a reservoir for the others and the expected bimodal configurational density of states is correctly predicted in the canonical ensemble. I wonder if this can be applied to a physical cluster since, for geometrical reasons, the direct exchange of particles between two edges is impossible.

Ms Lequien replied: First of all, a general argument: reaching chemical equilibrium for a cluster (as for any system) requires that a diffusion mechanism is operating on all sites. Therefore, it is not because edges do not have common sites that they cannot be in mutual equilibrium. It is necessary to differentiate the two following facts:

The edge equilibrium isotherms are distinctly separated from other isotherms (facets and core). This justifies the study of the (dynamical) equilibrium of a network of chains in mutual equilibrium by discarding the other sites of the cluster; the diffusion mechanism, which makes it possible to reach the equilibrium. This can be due to superficial diffusion, bulk diffusion or specific mechanisms taking place only in clusters. Whatever the pertinent diffusion mechanism is, this does not affect the prediction of the (dynamical) equilibrium.

Professor Ricolleau remarked: In the regime where all the shells are equivalent, the $\Delta\mu$ range is quite narrow. Which is the equivalent flux of incident atoms if we try to find the regime experimentally?

Ms Lequien replied: It should first be remembered that this is an equilibrium study that does not involve the growing notion underlying "the flux of incident atoms". In experiments, following the results of our simulations supposes that the total number of incident atoms deposited with respect to the chosen experimental condition, or the nominal concentration at the end of the deposition procedure, can be determined. Considering for instance the (001) facets of the 3871-atom cuboctahedron, the dynamical equilibrium appears when the average concentration lies between 0.3 and 0.9, which corresponds to a variation of the total number of silver atoms from about 500 silver atoms (pure copper (001) facets) to about 1000 silver atoms (pure silver (001) facets), that is to say for a nominal concentration ranging from 13% up to 25% (for the (111) facet the limits are 200 and 500 Ag atoms so between 5% and 13%). This 25% concentration corresponds to the core shell structure.

These nominal concentrations depend on the cluster size and shape. For a given shape, the following formula gives a rough estimate of the interval for the facets dynamical equilibrium:

$$[(N_{\text{vertice}} + N_{\text{edge}})/N_{\text{total}}, N_{\text{shell}}/N_{\text{total}}]$$

where N_k is the number of sites of type k: k = vertice, edge, shell or total.

Dr Calvo addressed Ms Lequien and Dr Mottet: To talk about "dynamical" equilibrium, especially if you have experimental justification in mind, suggests that you could access some aspects of the actual kinetics. Maybe you could estimate the energy pathways leading from two such states in coexistence by looking at the history of the MC simulation. The energy barrier (once converted to a temperature) is of primary importance for the kinetics.

Ms Lequien replied: The term "dynamical equilibrium" is used to differentiate an equilibrium corresponding to a single free energy minimum from the present case with two minima and where the system, in equilibrium, explores these two minima with the same probability when the F(111) average concentration (or the F(100) average concentration) is close to 0.5.

In the absence of knowledge of what the main mechanism of diffusion in bimetallic clusters is, the actual kinetics of the pathway between the two states is still unknown. The present equilibrium simulations in the Grand Canonical ensemble are unable to give some information about this. However, due to the fact that there are several facets with equivalent crystallographic orientation in the cuboctahedron, the observation of equivalent facets in different chemical states within the same cluster is possible due to the dynamical equilibrium even if the flip dynamics of every facet from one state towards the other one cannot be predicted in the absence of a realistic diffusion model.

Dr Mottet said: First a remark following the discussion raised by M. Hou and F. Calvo about the physical meaning of the so-called "dynamical equilibrium" and why they did not estimate the kinetic barrier or describe the pathway to pass from one configuration (illustrated in Fig. 4) to the other. In any case, the results presented by F. Lequien *et al.* are related to a possible dynamic behaviour (even if estimated, this barrier could be very high as compared to the temperature where the segregation process is still effective). The study performed here is purely an equilibrium investigation (using standard Metropolis Monte Carlo simulations) and the result called "dynamical equilibrium", if leading to a misunderstanding because of the antonymic terms of either "dynamics" and "equilibrium", is nevertheless a referenced expression, (see ref. 1 or related subject (ref. 2) or ref. 3) to describe an equilibrium in *phase* space (and not a dynamical effect of matter with possible diffusion inside the cluster). Whereas to imagine the two possible states of the binodal distribution (as illustrated by the CDOS, Fig. 6b) in the same cluster, it could be more appropriate (and physically relevant) to consider a large collection of particles in equivalent conditions where half of them would adopt one configuration and the other half, the other configuration.

Then I had one question: By considering another shape like the Wulff polyhedron where the (111) facets are more extended than the (100) ones, could we expect to get the same kind of dynamical equilibrium than on the (100) facets or does it depend essentially on the orientation of the facets and not on their size?

1   R. S. Berry, T. L. Beck, H. L. Davis and J. Jellinek, *Adv. Chem. Phys.*, 1988, **90**, 75.
2   Pierre Labastie and Robert L. Whetten, *Phys. Rev. Lett.*, 1990, **65**, 1567.
3   David J. Wales and R. Stephen Berry, Phys. Rev. Lett., 1994, **73**, 2875.

Ms Lequien replied: I first thank you for your answer which explains what I mean by "dynamical equilibrium".

Concerning the difficulty in thinking about both states in the same cluster, I do not fully agree. The flips of the different facets of equivalent orientations are not correlated in the individual dynamical equilibrium and there is therefore a similar probability to observe both states in the same cluster as in different clusters.

Concerning your question, there are two kinds of facets in the cuboctahedron: (111)F and (001)F. The dynamical equilibrium appears in both cases. The facet size should not be very important for the existence of a dynamic equilibrium. We can predict that the range (for instance in $\Delta\mu$) where the equilibrium takes place decreases when the facet size increases. The difference of energy between two states being proportional to the number of facet sites N, increasing N implies that for a given value of $\Delta\mu$ close to $\Delta\mu_c$ the most stable state will become preponderant. On the other hand, facets must be large enough (the notion of dynamic equilibrium for a single site is not pertinent!).

Within the rigid-lattice formalism, the dependence of segregation and dynamical equilibrium with the facet orientation is similar with the one obtained for the phase transition for infinite surfaces: the higher Z (coordination number obtained for surfaces, "or facets") is, the higher T_c is.

On the contrary, beyond the rigid-lattice formalism, the possible coupling between chemical and structural change will make the analysis more complex. Recent simulations mentioned above, that take into account atomic relaxations using second-moment tight-binding interatomic potentials, show that the (001) facets have a collective dynamical equilibrium which is associated with a structural transition, while the (111) facets have a behaviour very similar to the one observed with the rigid-lattice model.

Professor Broyer remarked: What kinds of experiments may be done to measure the segregation, core–shell structures or atoms on each facet? More generally, can you compare your calculations with experiments?

Ms Lequien answered: Performing numerical simulations, I am probably not the right person to answer your question. However, note that the core/shell structure obtained by the present simulations compares fairly well with the TEM results (see for example Cyril Langlois et al.[1]). One aim of this presentation is to motivate experimentalists to go beyond the determination of only a superficial concentration (what is already not easy!) and to detail the concentration on the different sites and even on the different facets of equivalent orientation. The direct observation of these different concentrations is probably not possible yet even if we can hope that STM observations with resolution of the chemical nature of the atoms can provide some data about it. Moreover, we can hope to characterize the superficial concentration, or even specific sites concentration, by observing their consequences on some chemical reactions occurring on these sites.

1   C. Langlois, D. Alloyeau, Y. Le Bouar, A. Loiseau, T. Oikawa, C. Mottet and C. Ricolleau, *Faraday Discuss.*, 2008, **138**, DOI: 10.1039/b705912b

Dr Calvo communicated: The potential you use belongs to the Ising class, which may not be too realistic for metals. Have you tried a many-body form, with additive repulsion but where the attractive part is put onto a square root (for instance)? I do not think that the simulation times would increase dramatically.

Ms Lequien communicated in reply: It is true that the Ising model is not adapted to represent the complete energy of metals or metallic alloys. However, using tight-binding calculations taking into account the diagonal disorder allows one to show that the part of the energy which depends on the chemical configuration can be described with an effective Ising model. This has been demonstrated both for the bulk[1,2] and for the surfaces.[3] This justifies the use of an Ising model to study ordering or segregation phenomena when the size mismatch between the elements is small.

However, modelling on a rigid lattice when the atomic radii differ significantly can only be seen as a first approach. We therefore performed new simulations in the Cu–Ag system with a second-moment tight-binding interatomic potential (attractive part with a square root form and additive repulsion). These very recent simulations confirm totally the bi-stability of the (111) facets. Note, however, that these new simulations were performed on the Wulff polyhedron but the results can be extended reasonably to the cuboctahedron.

1   F. Ducastelle and F. Gautier, *J. Phys. F*, 1976, **6**, 2039.
2   F. Ducastelle and G. Tréglia, *J. Phys. F*, 1980, **10**, 2137.
3   G. Tréglia, B. Legrand and F. Ducastelle, *Europhys. Lett.*, 1988, **7**, 575.

Professor Ferrando remarked: You simulate AgCu nanoparticles are on a rigid lattice. Since Cu and Ag atoms present a large size mismatch, effects of local relaxation are important. Can these effects change your picture of dynamical equilibrium in AgCu clusters?

Ms Lequien replied: There is indeed a difference between Cu and Ag atomic radii: $r(Ag) = 0.145$ nm and $r(Cu) = 0.128$ nm. The present work can be seen as a first approach concerning the existence of chemical dynamical equilibrium in bimetallic clusters. In this case, the rigid-lattice formalism is a good beginning. However, to take into account this large size mismatch, some other MC simulations with relaxations have been done for a Wulff polyhedron (see previously). The results show that a dynamical equilibrium takes place for each (111) facet. For the (001) facets, a chemical dynamical equilibrium is also observed but it is mixed with a structural dynamical equilibrium: when the silver content increases in these facets, the (001) facets adopt a pseudo-hexagonal structure (similar to the structure observed for the Ag/Cu(001) deposition). This change provokes some distortions on the whole cluster, that confer a collective character to the dynamic equilibrium of (001) facets, all the (001) facets flipping collectively from one state to the other one, contrary to the individual dynamical equilibrium observed with the rigid-lattice formalism.

As for the infinite alloys, simulations on rigid lattices allow one to explore the main characteristics of the phase diagrams of bimetallic clusters but they must be later supplemented by specific simulations for a given system, for which the possible size mismatch and atomic relaxations must be taken into account.

Professor Jellinek commented: The dynamical behavior of clusters/nanoparticles as a function of energy (microcanonical case) or temperature (canonical case) is complex, but there is nothing mysterious or inextricable about it. Clusters/nano-particles undergo a melting-like transition as their energy (temperature) is increased. The main difference between this transition and the bulk melting, which is a first order phase transition in crystalline materials, is that the former takes place over a finite range of energies (temperatures). The degree of complexity of this transition, in general, and the number and nature of different stages in it, in particular, depend on the particle size and composition. Identification of these stages and characterization of their nature in terms of the dynamical mechanisms involved requires long-time numerical simulations. These can be performed only with computationally efficient potentials for the description of the interatomic interactions. Therefore, development of such potentials for metals and alloys (they have to be of many-body character), which are not only efficient but also accurate (*i.e.*, appropriate for the size range considered), is a task (a challenging one) of paramount importance.

Optical properties and relaxation processes at femtosecond scale of bimetallic clusters

M. Broyer,*[a] E. Cottancin,[a] J. Lermé,[a] M. Pellarin,[a] N. Del Fatti,[a] F. Vallée,[a] J. Burgin,[b] C. Guillon[b] and P. Langot[b]

Received 23rd July 2007, Accepted 25th July 2007
First published as an Advance Article on the web 16th October 2007
DOI: 10.1039/b711282n

The optical properties of Au–Ag and Ni–Ag clusters are measured by linear optical absorption spectroscopy and the time-resolved pump–probe femtosecond technique allowing a study of the influence of alloy or core–shell structure.

1. Introduction

The reduction of size of a material to a nanometric scale leads to drastic modification of its electronic and optical properties. In noble metals, the dielectric confinement leads to the emergence of the giant plasmon surface resonance which renders the nanoparticles of these materials very attractive in the context of optical data storage media, optical logic devices or biosensing. For example the remarkable enhancement of photoabsorption and emission processes occurring in molecules near a metallic nanoparticle is generally interpreted as resulting from a local amplification of the electromagnetic field.[1] The use of mixed clusters of two materials such as gold and silver allows tuning of the surface plasmon resonance in a spectral range suitable for a given experiment.[2,3]

The intrinsic confinement effects and the strong interactions close to the metal surface are also at the origin of the exaltation of the non linear optical response. These properties are mainly ruled out by the electrons and their interactions inside the nanoparticles such as electron–electron and electron–phonon interactions. Time-resolved femtosecond techniques have been shown to be powerful tools for investigating the different electron scattering processes. For pure metal clusters, these processes have been recently investigated,[4,5] yielding for instance, information on the intrinsic electron–electron and electron–phonon coupling. The results obtained in silver and gold nanoparticles demonstrated that electron–lattice energy exchanges strongly increase as the size decreases.

We present in this paper the measurement of the optical response of bimetallic clusters Au–Ag and Ni–Ag prepared by Low Energy Cluster Beam Deposition (LECBD). We investigated both the evolution of the surface plasmon resonance and of the electron–phonon scattering with the cluster composition. The influence of the structure, core–shell or alloy is also discussed.

[a] Université Lyon 1, CNRS, LASIM, UMR 5579, Bât. A. Kastler, 43 Boulevard du 11 novembre 1918, F-69622 Villeurbanne, France
[b] Centre de Physique Moléculaire optique et Hertzienne, CNRS and Université Bordeaux 1, 351 Cours de la Libération, 33405 Talence, France

2. Experimental

Clusters are produced by the Low Energy Cluster Beam Deposition (LECBD technique). A laser vaporization source and a continuous flow of helium are used to produce the cluster beam. Briefly, the second harmonic of a Nd:YAG pulsed laser (532 nm) is focused onto a metallic rod of composition Ni_xAg_{1-x} or Au_xAg_{1-x}. The produced bimetallic plasma is cooled by the He gas and combines into clusters which expand with the inert gas through a nozzle. The cluster beam is then collimated with a skimmer into a high vacuum chamber (10^{-7} Torr) and codeposited with the transparent alumina matrix on various substrates depending on the measurements to be performed. The alumina matrix is evaporated thanks to an electron gun and has an amorphous structure with a porosity of about 45% compared to anodic alumina. As a consequence, the effective dielectric function of our porous alumina is slightly smaller ($\varepsilon_m = 2.7$) than the bulk value ($\varepsilon_m = 3.1$). By varying the helium pressure or by using a He–Ar mixture, one is able to vary the mean size of our sample. Moreover the metal volumic concentration in the sample is kept below 5% to minimize the cluster coalescence and possible optical coupling between neighbouring clusters. The typical nanocomposite samples then consist of a 1 cm × 1 cm square Suprasil substrate of 1 mm thickness with about 200 nm of alumina doped with metal clusters.

The stoichiometry of the obtained clusters is probed through Rutherford Back Scattering (RBS) or by Energy Dispersive X-ray (EDX) experiments. We found that the cluster composition is within 10% the same as the rod composition.[2,6,7] The size distributions are determined from Transmission Electron Microscopy (TEM). Clusters are almost spherical and randomly distributed in the transparent matrix. The size distribution follows a log-normal law with a standard distribution of about 40% of the mean diameter.

The optical absorption spectra of these samples are recorded with a Perkin-Elmer spectrophotometer in the energy range 1.55 to 6.52 eV. The Suprasil substrate is totally transparent in this spectral range but the transmission of the alumina matrix is decreasing from 4.5 eV to the UV domain, due to the granular structure of this porous matrix. The absorption spectra are recorded under Brewster incidence using p-polarized light in order to avoid Fabry–Perot fringes due to multilayers reflections in our samples.

The relaxation processes following the optical excitation of these clusters is then studied at the femtosecond scale. The pump–probe femtosecond experiments are performed with a home-made femtosecond Ti-Sapphire oscillator delivering 20 fs near infrared pulses ($\lambda \approx 860$ nm) at 80 MHz repetition rate. The pulse train is split into two parts. The first one is used to create the nonequilibrium electron distribution with a pump pulse at the fundamental of the laser, $\hbar\omega_{pp} \approx 1.5$ eV. The second part, at the doubled laser frequency, $\hbar\omega_{pr} \approx 3$ eV, is used to probe the induced sample transmission change $\Delta T/T$. The two beams were sent in a standard pump–probe setup with mechanical chopping of the pump beam at 1.5 kHz and lock-in detection of the probe beam transmission change ΔT (defined as the difference between the perturbed and the unperturbed sample transmission). Taking advantage of the high stability and high repetition rate of our setup, very high sensitivity $\Delta T/T$ measurements were performed with a noise level in the 10^{-7} range.

3. Results and discussion

A. Optical spectra

(a) Optical absorption spectra of Au/Ag clusters. The optical extinction spectra of Au/Ag mixed clusters have been measured as a function of size and of the Au/Ag relative concentration. In our experiments the particle size varies typically from 1.5 nm to 7 nm. In this range the diffusion may be neglected and extinction cross section corresponds to the absorption cross section. Fig. 1 shows the absorption (extinction) cross section as a function of energy for nanoparticles of various relative compositions and of about 2 nm in diameter. The spectra are dominated by a single resonance corresponding to the surface plasmon resonance. This indicates the formation of alloy nanoparticles[8,9] because core–shell Au/Ag nanoparticles have two distinct resonance peaks. This is not surprising because gold and silver are miscible metals in any proportion and have very similar Wigner–Seitz radius. The plasmon resonance peak regularly shifts from the silver to the gold value when the gold fraction x increases in the Au_xAg_{1-x} clusters, as observed by El Sayed and collaborators for larger sizes.[8] The measurement of the surface plasmon resonance for very small diameters (2 nm to 4 nm) allows us to study the quantum effects (electronic confinement). Fig. 2 shows for each nanoparticle composition that the resonance peak is blue-shifted when the size increases. The results may be interpreted by a semi-quantum model where the s electrons motion is calculated by Time Dependent Local Density Approximation (TDLDA) and the dynamic screening by the d electrons is taken into account through their contribution ε_d to the bulk dielectric function.[10,11] The two main ingredients responsible for the size evolution of the surface plasmon frequency at very small size are: (1) the electronic spill out

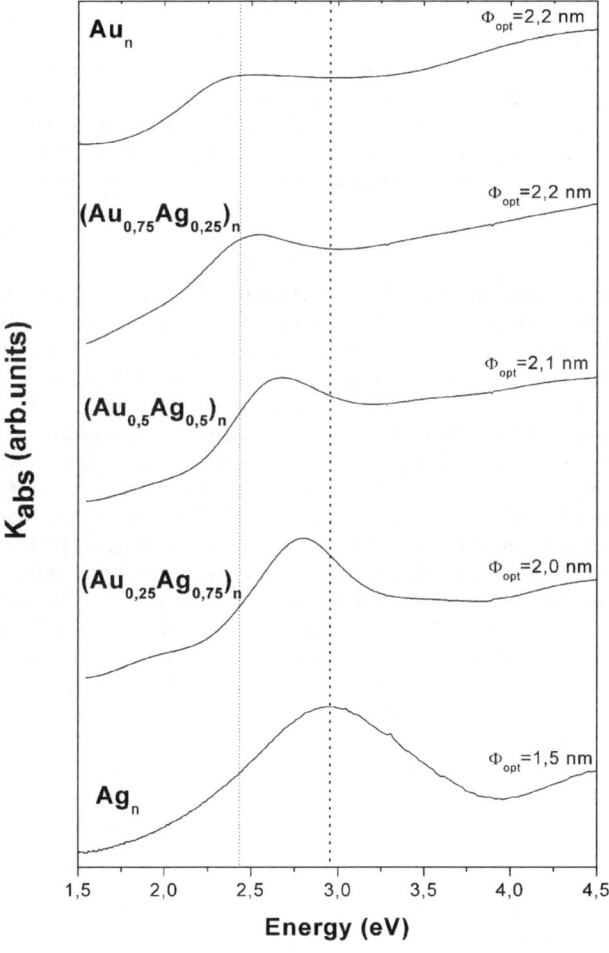

Fig. 1 Optical absorption spectra *versus* energy of alumina embedded Au_xAg_{1-x} clusters as a function of the gold fraction x. All clusters have diameters Φ_{opt} close to 2 nm.

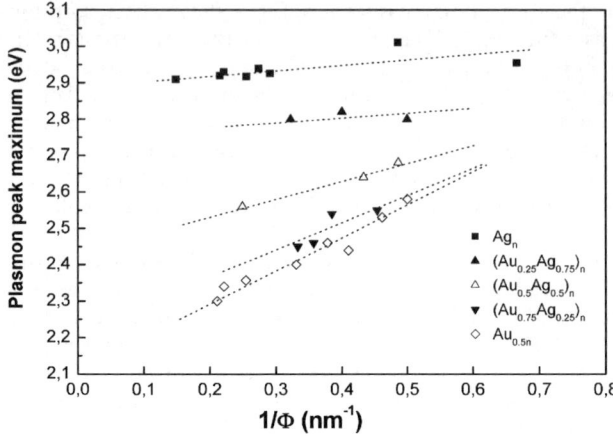

Fig. 2 Size evolution of the peak surface plasmon maximum for Au_xAg_{1-x} embedded in an alumina matrix, as a function of size ($1/\Phi_{opt}$) and composition defined by x. The straight lines are just guides for the eyes.

beyond the cluster radius which is a pure quantum effect, (2) the inner surface shell of vanishing ionic core polarizability due to d electrons. The first effect leads to a red shift and the second one to a blue shift. In gold the blue shift is the most important while in silver two effects are almost compensated. This is due to the fact that the ε_d value (*i.e.* the dynamic screening) is higher in gold than in silver. Detailed calculations for Au_xAg_{1-x} are given in ref. 2 and the experimental results are in good agreement with semi-quantum calculations using the dielectric constants of the alloys.[12]

(b) Optical absorption spectra of Ni/Ag clusters. The cluster structure may be first guessed from the bulk properties of nickel and silver. First the atomic Wigner–Seitz radius of Ag and Ni are noticeably different, respectively 3.02 and 2.6 a.u. and a pronounced lattice mismatch is expected. Second the surface energy of silver is almost two times smaller than that of nickel clusters. Therefore to minimize its energy the mixed cluster will tend to have a majority of silver atoms on the surface. Furthermore, the phase diagram of Ni/Ag shows that silver and nickel are completely immiscible. These bulk properties strongly suggest that mixed Ni/Ag nanoparticles will form a core–shell structure with silver on the surface. This core–shell structure has been confirmed by Low Energy Ion Scattering (LEIS) experiments indicating that the cluster surface is mostly composed of silver atoms.[7]

Fig. 3 shows the evolution of absorption spectra *versus* the nickel proportion for nanoparticles having about the same optical diameter $\Phi_{opt} \approx 2.6$–2.7 nm. ($\Phi_{opt} = (\langle \Phi^3 \rangle)^{1/3}$). A surface plasmon resonance is observed in the same spectral range as the one obtained for pure silver clusters and is only slightly blue shifted but this resonance is more and more broadened and damped when the nickel proportion increases. For pure nickel clusters, no clear resonance peak is observed as predicted by the Mie theory for pure nickel. The experimental results are compared to the one calculated within the Mie Theory in two cases. In the first case (Fig. 3b), the theoretical absorption is obtained for an alloyed homogeneous sphere with a dielectric function taken as the composition weighted averaged of the dielectric function of both materials. In the second case corresponding to core–shell structure, we use the Mie theory extended to more than one interface. In the dielectric function of silver, we take into account the damping factor $\Gamma_R = \Gamma_\infty + A v_F / l_{path}$. v_F is the Fermi velocity of free electrons and A is taken to be equal to 1 in our calculations. l_{path} is the mean free path of the electrons. It is equal to the radius of the cluster for

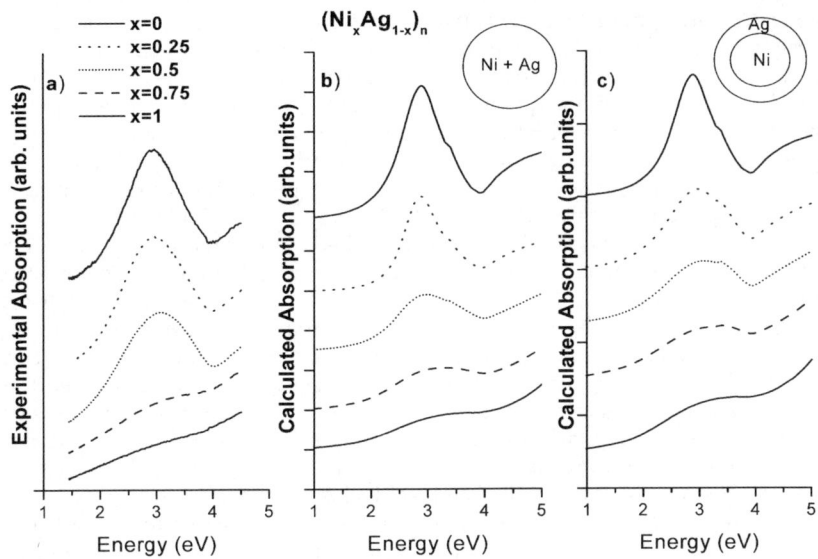

Fig. 3 Evolution of the optical absorption spectra of Ni_xAg_{1-x} as a function of x for nanoparticles with an optical diameter in the range 2.6–2.7 nm. (a) Experiment. (b) Theoretical spectra with the homogenous alloy model. (c) Theoretical spectra when the reduction of the mean free path in the silver shell is taken into account (see text).

pure metal or for alloys. For core–shell clusters, as proposed by Grandqvist and Hunderi,[13] we use $l_{path} = [(R - R_c)(R^2 - R_c^2)]^{1/3}$.

The agreement between experiments (Fig. 3a and 3c) and theory for the core–shell structure is very good. The progressive damping of the surface plasmon resonance is well represented by the calculations. However the experiment may also be considered to be in reasonable agreement with the "alloy" model of Fig. 3b. This means that, for the Ni/Ag system, it is difficult to discriminate between core–shell and alloy structure only from the optical spectra. This is true in particular if the dielectric constant of the "hypothetic" alloy is not known. This contrasts with the Au/Ag system for which the dielectric constant of the Au/Ag alloy is known and where the core–shell structure exhibits two peaks corresponding roughly to the resonance peaks of, respectively silver and gold. In conclusion, for the Ni/Ag system, the whole set of experimental results (optics and LEIS) is in favour of a core–shell structure even if probably the interface is not perfect. Moreover for the smallest clusters studied here, the number of silver atoms is probably not sufficient to form a complete shell and we may have segregation without a real core–shell structure.

B. Electron phonon relaxation processes

(a) Au/Ag clusters. Our results on mixed clusters are an extension of those obtained previously on pure noble metal nanoparticles. It relies on the close connection between the energy distribution of the conduction electrons and the optical response, *i.e.* the metal dielectric function. This can be separated into an intra-conduction band (free electrons) contribution and an interband one, related to transitions from d band to the conduction band. The absorption threshold $\hbar\Omega_{ib}$ for these transitions is in the ultraviolet-visible range, with $\hbar\Omega_{ib} \approx 4$ eV for silver and $\hbar\Omega_{ib} \approx 2.3$ eV for gold. In time-resolved experiments, energy is first selectively injected in the conduction electrons by the pump pulse. The different steps of their relaxation, *i.e.* internal thermalization of the electron gas and its cooling, are followed by monitoring the time-dependent sample transmission change $\Delta T/T$ using a time-delayed probe pulse. It has been shown that, for the probe photon energy well below $\hbar\Omega_{ib}$, the signal amplitude is almost insensitive to the details of the electron distribution and is proportional to the electron gas excess energy $\Delta u_e(t_D)$ (defined as the difference between the electron gas energy at time t_D and the one before perturbation, t_D being the pump–probe delay). For this off-resonant probing condition, the time evolution of the measured signal thus directly reflects the energy losses of the electrons to their environment, permitting its precise analysis.

The time evolution $\Delta T/T$ measured in Au_xAg_{1-x} alloy nanoparticles with pump and probe photon energies $\hbar\omega_{pp} = 1.5$ eV and $\hbar\omega_{pr} = 3$ eV shows a mono-exponential decay after typically $t_D = 500$ fs. This behavior reflects that states close to the threshold of the interband transitions are probed with 3 eV. The measured signal is thus sensitive to the energy redistribution below the Fermi energy and consequently, at the establishment time of the electronic temperature. The measured signal at a shorter time is thus sensitive to the energy redistribution below the Fermi energy and consequently, at the establishment time of the electronic temperature.[5] Above 500 fs, the exponential signal decay is related to the electron energy loss time to the lattice, as it is expected for weak excitation regime studies; and it corresponds to the electron–lattice energy exchange time τ_{e-ph} which may also called, the electron–phonon time. For silver and gold clusters, this electron phonon time decreases when the cluster size decreases. This means that the electron–lattice energy exchanges strongly increase for small sizes ($\Phi \leq 10$ nm). This size behaviour is in qualitative agreement with a model based on an inner surface shell of vanishing ionic core polarization of the d electron which means that the dynamic screening by the d electrons is not efficient in this surface shell. This model is similar to the model used to explain the blue shift of the surface plasmon resonance when the size decreases for gold and silver clusters. The agreement of this model with experimental results for gold or silver clusters is only qualitative and the detailed interpretation is probably more complicated than this simple model.[5,14] The aim of our paper is not to discuss again this interpretation but to examine this effect for mixed clusters.

Fig. 4 shows the measured electron–phonon time τ_{e-ph} for $Au_{0.5}Ag_{0.5}$ clusters. We see in this figure, that τ_{e-ph} for $Au_{0.5}Ag_{0.5}$ decreases with size and have values intermediate between the gold and silver values. As for the pure metal nanoparticles a reduction of the decay time is observed for smaller sizes yielding evidence for a confinement-induced increase of the electron–lattice interaction in small alloy clusters. However, the range of the studied sizes is reduced because the LECBD technique does not permit obtaining nanoparticles of an average diameter above 4–5 nm. For all studied sizes, the measured time τ_{e-ph} corresponds to the average of those obtained in pure gold and silver nanoparticles of the same size. This result is

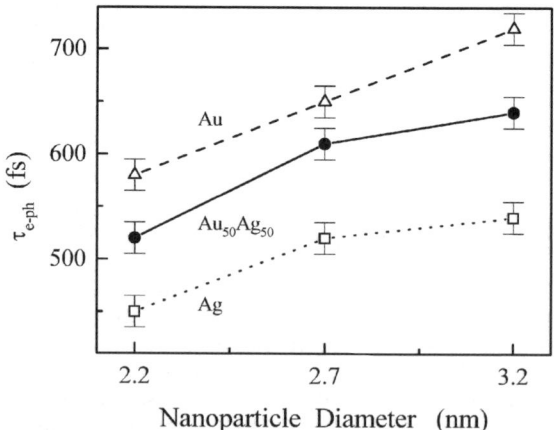

Fig. 4 Measured electron–lattice energy exchange time τ_{e-ph} for gold, $Au_{0.5}Ag_{0.5}$ alloys and silver clusters with mean diameters 2.2, 2.7 and 3.2 nm.

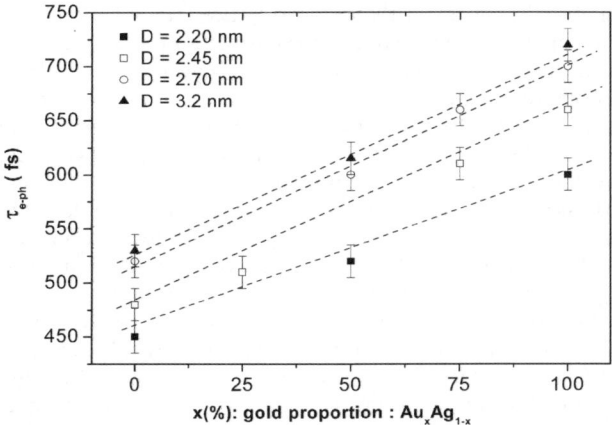

Fig. 5 Measured electron–lattice energy exchange time τ_{e-ph} as a function of x for $Au_{0.5}Ag_{0.5}$ alloy particles in alumina matrix. The straight lines are guides for the eyes. Each straight line is related to the experimental results obtained for a given size. D is the diameter of the clusters.

confirmed by performing a systematic investigation of the electron–lattice energy exchanges in Au_xAg_{1-x} alloy clusters for different stoichiometries in the diameter range 2.2–3.2 nm. The time τ_{e-ph} measured as a function of the gold concentration ($x = 0, 0.25, 0.5, 0.75, 1$) are shown in Fig. 5. In the same way as the linear optical properties, the electron–phonon interactions in an alloy structure reflect simply its stoichiometry. Moreover, they indicate that the disorder introduced by the alloy effect plays a negligible role in the electron–lattice coupling. This result may be qualitatively understood because the dielectric constant of the alloy varies regularly with the gold fraction x.[12]

(b) Ni/Ag clusters. The electron–lattice energy exchanges in Ni/Ag clusters have been studied by similar femtosecond pump–probe experiments with, respectively $\hbar\omega_{pp} = 1.5 \, eV$ and $\hbar\omega_{pr} = 3 \, eV$. Fig. 6 shows the results obtained as a function of x, the Ni fraction in Ni_xAg_{1-x} clusters for two cluster sizes 3.2 and 4.8 nm. The values obtained for Ag and Ni films are also indicated on this figure. We must remark that it was the first measurement of τ_{e-ph} for pure Ni clusters and Ni film. We observe a decrease of τ_{e-ph} for pure Ni clusters as the size decreases. This means that a

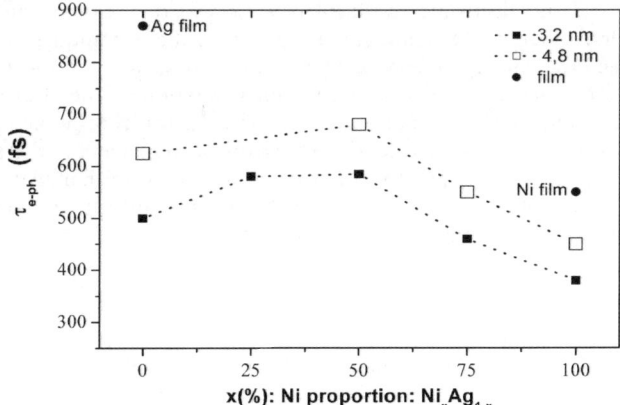

Fig. 6 Measured electron–lattice energy exchange time τ_{e-ph} for Ni_xAg_{1-x} clusters as a function of composition x. The results are shown for two cluster diameters of, respectively 3.2 and 4.8 nm. The dashed lines are just guides for the eyes.

confinement-induced increase of the electron lattice interaction is also observed in pure Nickel nanoparticles. This effect appears therefore quite general, at least for transition metal clusters (τ_{e-ph} has been not directly measured for alkali clusters in a similar weak excitation regime).

For Ni_xAg_{1-x} clusters, the evolution as a function of x is very different from the case of Au_xAg_{1-x}. This is not really surprising because for a core–shell structure there are more interfaces. However the presence of interfaces is rather in favour of an acceleration of the electron–lattice energy exchanges which would rather result in a stronger decrease of τ_{e-ph}. In contrast, in Fig. 6, τ_{e-ph} tends to increase for mixed clusters and to have a maximum for $x = 0.5$. This effect is not easily interpreted. We can only suggest some possible trends. For small silver clusters, we expect that, at very small sizes (from few atoms to 100 atoms), the gap between electronic states increases and will become larger than the thermal energy kT, due to the electronic confinement (quantum effect). This is especially true for silver because the d levels are well separated in energy from the s levels (interband threshold of 4 eV). Therefore, we expect that τ_{e-ph} which decreases with the size in the range from the bulk to 2 nm, will increase for very small sizes below 1–2 nm and will tend finally toward the nanosecond scale for the Ag atom or the Ag_2 molecule. This increase of τ_{e-ph} at very small sizes for silver clusters may be called "molecular regime" and may explain the behaviour of Fig. 6. If the shell of silver is not well defined and fragmented in small silver clusters, this "molecular regime" may influence the global increase of τ_{e-ph}. On Fig. 6, we can remark that the maximum as a function of x is more pronounced for clusters of 3.2 nm than for clusters of 4.8 nm. This is in favour of our interpretation since the core–shell structure is more defined for larger clusters. Nevertheless, this interpretation remains speculative and experiments are in progress to observe this "molecular regime", firstly on pure silver clusters.

4. Conclusion

We have shown that the Low Energy Cluster Beam Deposition technique is a very efficient method to produce clusters of any bimetallic material. By performing the co-deposition of these clusters with a transparent matrix, the optical properties may be easily studied. The linear optical properties may be used to probe the core–shell and alloy structure. The optical signature of this structure can only be clear if the two materials have a clear and different optical response. For instance this optical signature is clear for Au/Ag clusters but much less so for Ni/Ag clusters because the optical absorption spectrum of pure Ni clusters is quite flat and does not show any characteristic features. Moreover we have been able to investigate both the dielectric confinement and the electronic confinement (also called quantum effect) in these mixed clusters. Finally the femtosecond pump–probe techniques allow us to investigate the electron–lattice energy exchanges in these systems and to study the influence of the structure. The quantitative interpretation of the electron–phonon coupling in metallic clusters remains challenging even for pure noble clusters. We have also discussed the possible role of the electronic confinement in these processes. Future experiments will measure the electron–lattice energy exchange time τ_{e-ph} for very small clusters (diameters smaller than 2 nm) and perform the same kind of optical measurements on single nanoparticles.

References

1 M. Moskovits, Rev. Mod. Phys., 1985, 57, 783.
2 E. Cottancin, J. Lermé, M. Gaudry, M. Pellarin, J. L. Vialle, M. Broyer, B. Prével, M. Treilleux and P. Mélinon, Phys. Rev. B., 2000, 62, 5179.
3 M. Gaudry, J. Lermé, E. Cottancin, M. Pellarin, J. Vialle, M. Broyer, B. Prével, M. Treilleux and P. Mélinon, Phys. Rev. B, 2001, 64, 085407.
4 C. Voisin, D. Christofilos, N. Del Fatti, F. Vallée, B. Prével, E. Cottancin, J. Lermé, M. Pellarin and M. Broyer, Physical Review Letters, 2000, 85, 2200.

5   A. Arbouet, C. Voisin, D. Christofilos, N. Del Fatti, F. Vallée, J. Lermé, G. Celep, E. Cottancin, M. Gaudry, M. Pellarin, M. Broyer, M. Maillard, M. P. Pileni and M. Treguer, *Phys. Rev. Lett.*, 2003, **90**, 177401.

6   J. L. Rousset, Cadrot, F. J. C. S. Cadete Santos Aires, A. Renouprez, P. Mélinon, A. Perez, M. Pellarin, J. L. Vialle and M. Broyer, *J. Chem. Phys.*, 1995, **102**, 8574.

7   M. Gaudry, E. Cottancin, M. Pellarin, J. Lermé, L. Arnaud, J.-R. Huntzinger, J.-L. Vialle, M. Broyer, J.-L. Rousset, M. Treilleux and P. Mélinon, *Phys. Rev. B*, 2003, **67**, 155409.

8   S. Link, Z. L. Wang and M. A. El-Sayed, *J. Chem. Phys.*, 1999, **103**, 3529–3533.

9   U. Kreibig and M. Vollmer, *Optical properties of Metal Clusters*, Springer, Berlin, 1995.

10 E. Cottancin, G. Celep, J. Lermé, M. Pellarin, J.-R. Huntzinger, J.-L. Vialle and M. Broyer, *Theor. Chem. Acc.*, 2006, **116**, 514.

11 B. Palpant, B. Prével, J. Lermé, E. Cottancin, M. Pellarin, M. Treilleux, A. Pérez, J. L. Vialle and M. Broyer, *Phys. Rev. B.*, 1998, **53**, 1963.

12 K. Ripken, *Z. Phys.*, 1972, **250**, 228.

13 C. Granqvist and O. Hunderi, *Z. Phys. B: Condens. Matter*, 1978, **30**, 47.

14 J. Lermé, G. Celep, M. Broyer, E. Cottancin, M. Pellarin, A. Arbouet, D. Christophilos, C. Guillon, P. Langot, N. Del Fatti and F. Vallée, *Eur. Phys. J. D*, 2005, **34**, 199.

Stability and dissociation pathways of doped Au_nX^+ clusters (X = Y, Er, Nb)†

Nele Veldeman,[a] Ewald Janssens,[a] Klavs Hansen,[b] Jorg De Haeck,[a] Roger E. Silverans[a] and Peter Lievens*[a]

Received 19th April 2007, Accepted 4th May 2007
First published as an Advance Article on the web 14th September 2007
DOI: 10.1039/b705920e

Size dependent stabilities, fragmentation pathways and dissociation energies of a series of gas phase cationic doped gold clusters, Au_nX^+ ($3 \leq n \leq 20$; X = Y, Er and Nb), and pure Au_n^+ clusters were investigated in photofragmentation experiments. Size dependent stability patterns were obtained and the branching between monomer and dimer evaporation was studied. For bare gold, the competing neutral monomer and dimer evaporation channels were found to be in agreement with earlier studies. For doped clusters, monomer evaporation is the most likely fragmentation channel with the exception of $Au_{18}Y^+$ and $Au_{20}Y^+$ for which gold dimer evaporation is also observed. Relations between the evaporative activation energies and both the experimental abundances and the fragment yield were derived based on unimolecular rate constants. The dissociation energies from this analysis show an odd–even staggering and enhanced stabilities for certain cluster sizes, in agreement with simple electronic shell model predictions.

1. Introduction

Gold nanoclusters have attracted a lot of attention in recent years in particular due to the potential applications as catalytic particles.[1] The electronic properties and size dependent stability of gold clusters have been characterized by photoelectron spectroscopy, photoionization spectroscopy, photofragmentation, and collision-induced dissociation studies.[2–8] Quantum confinement of the gold valence 6s electrons produces stability patterns that to a first approximation can be described in terms of an electronic shell structure.[4,9] Closed electron shells appear if the number of delocalized electrons correspond to the so-called *magic numbers*: 2, 8, 18, 20, 34, 58, 92. These magic sizes were observed in different experimental studies.

Since the demonstration with ion mobility measurements that charged gold clusters are planar up to surprisingly large sizes,[10,11] also the geometric structures of gold clusters attracted a lot of attention.[12–14] A number of unexpected geometries have recently been discovered, such as tetrahedral Au_{20},[15,16] fullerene-like Au_{26},[17]

[a] Laboratorium voor Vaste-Stoffysica en Magnetisme & INPAC—Institute for Nanoscale Physics and Chemistry, K.U.Leuven, B-3001, Leuven, Belgium. E-mail: peter.lievens@fys.kuleuven.be
[b] Department of Physics, Göteborg University, Göteborg SE-412 96, Sweden. E-mail: klavs@physics.gu.se

† Electronic supplementary information (ESI) available: Metastable fragmentation pathways of Au_n^+, Au_nY^+, Au_nEr^+ (n = 3–20), and Au_nNb^+ (n = 3–14). See DOI: 10.1039/b705920e

Au_{32},[18] and Au_{50},[19] and hollow Au_n ($n = 16$–18) cages.[20] Even golden nanotubes have been predicted.[21]

Not only bare gold clusters, but also their transition metal doped[22–31] and carbon (or carbon cluster)[32] doped counterparts have been actively studied. A single transition metal atom can strongly interact with a small gold cluster and, for certain cluster sizes and number of valence electrons, have a strong influence on the electronic shell structure and/or the cluster geometry. Combined photofragmentation mass spectrometry and density functional theory studies on Au_n clusters doped with open 3d shell atoms (Sc, Ti, V, Cr, Mn, Fe, Co, Ni, and Zn), have shown that the planar structure can be stabilized by a dopant atom and that in particular six-electron species are more stable.[22–26] It was also predicted that a central transition metal atom stabilizes a golden cage structure if the number of valence electrons satisfies the 18-electron rule.[27,28] This explains the high abundance obtained in mass spectra for $Au_{16}Sc^+$, $Au_{16}Y^+$, $Au_{15}Ti^+$ (ref. 7 and 22) as well as the unique stability of the icosahedral $Au_{12}W$.[29–31]

The strength of the interaction between a dopant atom and the host cluster is determined by the size of the cluster (geometric stabilization) and the number of dopant valence electrons (electronic shell closure). The combined effect of these parameters is reflected in the total binding energy and the differential binding energy (dissociation energy). The clusters are intrinsically stable and some excess energy, provided by, e.g., photoexcitation, is needed to induce fragmentation. After photo-excitation the internal energy is statistically distributed over the vibrational modes of the system within a time scale shorter than the experimentally relevant time scales in mass spectrometric studies. The result is that the cluster does not instantaneously dissociate, even if the excitation energy is above the dissociation threshold. The delayed fragmentation in these so-called unimolecular decomposition processes is characterized by several parameters intrinsic to the clusters, and by the fact that the clusters are evaporating freely in an evaporative ensemble.[33,34] By applying statistical models, which predict the unimolecular dissociation rate assuming energy partitioning over the various modes, the recorded fragmentation yields after a well defined time can be converted into dissociation energies.

The delayed fragmentation yield has been measured for a multitude of mono-metallic clusters composed of monovalent atoms: Li_n^+ ($n = 4$–42),[35] Na_n^+ ($n = 5$–40),[33] K_n^+ ($n = 5$–200),[36] Cu_n^+ ($n = 2$–17),[37] Cu_n^- ($n = 2$–8),[38] Ag_n^+ ($n = 3$–21),[39] Ag_n^- ($n = 2$–11),[40] Au_n^+ ($n = 3$–23),[2,41–43] and Au_n^- ($n = 2$–15).[44] These studies showed that for small and for odd-sized clusters, monomer and dimer decay are competing channels, while all even-numbered ionic clusters evaporate a neutral monomer. In contrast to the extensive amount of accumulated information on decay pathways of pure clusters, there is little information available on the influence of the composition on the decay pathways of mixed clusters. We are only aware of two recent studies on some size selected $Ag_nCo_m^+$ and Si_nCr clusters.[45,46]

In this work we study the dissociation pathways and size dependent stability of cationic Au_nX^+ (X = Y, Er and Nb, $n = 5$–20) clusters. Statistical models are used to determine the monomer and dimer dissociation energies from both the recorded abundances and the fraction of mass selected clusters that undergo delayed fragmentation in the time window of the experiment. The dopant elements Y, Er, and Nb have been chosen to extend our earlier work, using 3d dopant atoms,[22–24] towards 4d transition metal atoms (Y, Nb) and lanthanide (Er) dopants.

2. Experimental

The clusters are produced in a dual-target dual-laser vaporization source, which is extensively described elsewhere.[47] Two independent pulsed Nd:YAG lasers vaporize the surface of two translating rectangular plate targets. Helium gas is introduced into the source by a pulsed supersonic valve. Subsequently, the mixture of atoms, clusters and inert gas expands into the vacuum, resulting in a cold molecular cluster beam.

The source parameters are optimized to achieve gold clusters containing only one or a few dopant atoms. After passage through a skimmer, the cluster beam enters the extraction region of a reflectron time-of-flight (RTOF) mass spectrometer.

To study neutral species, charged particles are deflected out of the beam and the remaining neutral beam is irradiated with laser light stemming from an ArF excimer laser ($\lambda = 193$ nm). Depending on the laser fluence, roughly two regimes can be differentiated. In the low fluence regime (<0.5 mJpp cm^{-2}) multiphoton absorption is unlikely and clusters will be excited or ionized by absorption of one photon only. In the high fluence regime (>5 mJpp cm^{-2}) the clusters are heated by multiphoton absorption. Cooling occurs through fast sequential evaporation of atoms and larger fragments. The resulting photofragments reveal a size distribution with higher abundances for clusters with enhanced stability. Following acceleration in the extraction zone, the clusters are mass separated in the field free drift region of the RTOF mass spectrometer. During the drift period, metastable parent cluster ions can undergo further evaporation (metastable or delayed fragmentation) with parent and fragment particles proceeding at the center-of-mass velocity.[33] In the reflectron the clusters are decelerated, turned around, and reaccelerated resulting in spatial and temporal separation of fragments with different masses.

A mass gate, located in between the extraction zone and the reflectron, allows for mass selection and offers the possibility to determine the evaporation channels of size selected clusters. The pulsed mass gate consists of a plane of parallel wire segments, formed by two electrically isolated sets of wires, similar to the device developed by Vlasak et al.[48] If the two wire sets are held at the potential of the field free region, the ion beam, travelling orthogonal to the plane of wire segments, is unaffected (gate open). Applying opposite voltages of equal magnitude (250 V) on the set of wires results in an electrical field perpendicular to the ion velocity and deflection of the ions (gate closed). By holding the gate closed until an ion packet of interest is about to reach the selector and then pulse the opening of the gate for the duration of the passage time, mass selective ion transmission is obtained. The resolution of the mass selection is determined by the planarity of the parallel wire system, the rise and fall times of the pulsed field, and the position of the mass gate, which should be located in the time focus of the two-field extraction zone. Pulsed voltage switches (Behlke HTS 21-03) with a short rise and fall time of 15 ns are used and a mass selection resolution of $m/\Delta m = 60$ at $m = 1000$ amu was obtained. The timing is controlled by precise delay generators (DG 535, Stanford Research Systems). For each cluster size the time delay is optimized and the gate is typically opened for 600 ns.

3. Stability patterns and fragmentation channels

The recorded mass spectra for photofragmented Au$_n^+$ and Au$_n$X$^+$ (X = Y, Er, Nb; $n = 1$–30) are shown in Fig. 1. The dotted lines mark the pure Au$_n^+$ clusters, peaks corresponding to singly doped Au$_n$X$^+$ species are connected by a solid line. Metastable fragmentation pathways of Au$_n$X$^+$ for $n = 9$–14 (X = Au, Er, Nb) and $n = 11$–14, 18, 20 (X = Y) are shown in Fig. 2. The evaporation channels over an extended size range ($n = 3$–20) can be found in the ESI.† From the recorded metastable fragmentation, relative monomer and dimer evaporation yields are deduced (see Fig. 3).

A. Au$_n^+$

The photofragmentation spectrum for Au$_n^+$ (Fig. 1a) exhibits a distinct stability pattern with intensity steps at $n = 3, 9, 19$ and 21, corresponding to 2, 8, 18 and 20 delocalized valence s electrons. These are the so-called magic numbers for simple metal clusters and are related to the completion of electron shells for delocalized valence electrons enclosed in a potential well, as described extensively in the literature.[4,9,22] In addition to clear drops in abundance after the magic sizes a clear odd–even staggering is visible.

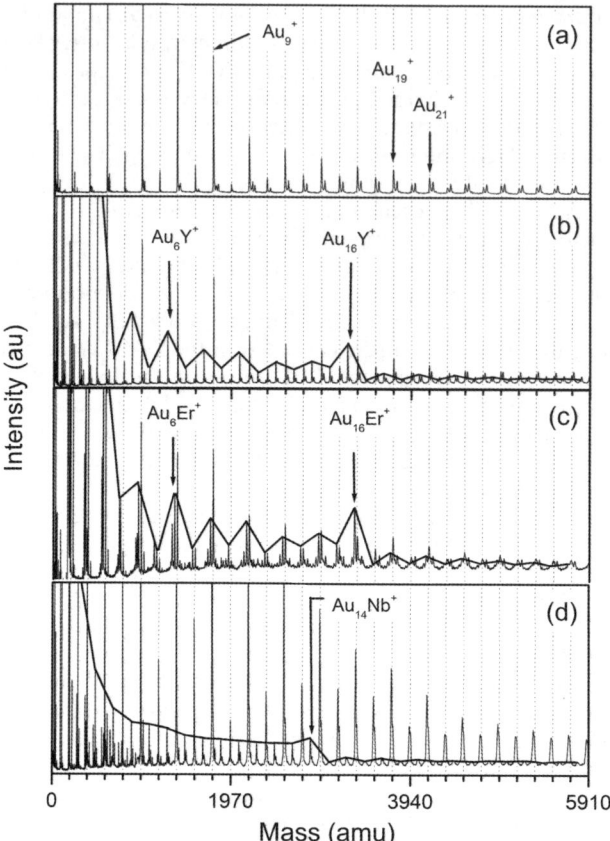

Fig. 1 Mass abundance spectra of photofragmented (a) Au_n^+, (b) Au_nY^+, (c) Au_nEr^+, and (d) Au_nNb^+ for $n = 1$–30. The grid lines mark the bare Au_n^+ clusters. Small peaks at the right side of the bare cluster peaks (mainly visible in panel (a)) correspond to metastable fragments stemming from one size up. Peaks corresponding to Au_nX^+ (X = Y, Er, and Nb) are connected by a solid line. Species after which there is a clear drop in abundance are labeled.

The delayed fragmentation pathways of mass selected Au_n^+ clusters (left column of Fig. 2) show a prominent peak, corresponding to the selected cluster (mother signal), and less intense peaks with a shorter flight time corresponding to metastable fragmentation (daughter peaks). The daughter clusters are created in the field free drift region and are separated in the reflectron when the electric field in the reflectron is detuned from the value which is optimal for mass spectrometric resolution. The fragmentation channels are indicated by labeled arrows. The preferred fragmentation channel is found to be size dependent.

The metastable fraction of monomer and dimer evaporation is shown as a function of the cluster size in Fig. 3a. For each mother cluster size n, the plotted values equal the integrated areas of respectively the monomer (■) and dimer (●) daughter peaks, divided by the total integrated area of the mother and daughter peaks.

For the smallest species ($n = 3$–5, 7) no delayed fragmentation is recorded. This is due to the combined results of three effects. First, the metastable fractions generally increase with cluster size. Secondly, the recoil energy upon dissociation is larger for small clusters, resulting in a poorer mass resolution, which makes the peaks more difficult to distinguish from the background. And thirdly, the beam path of daughter

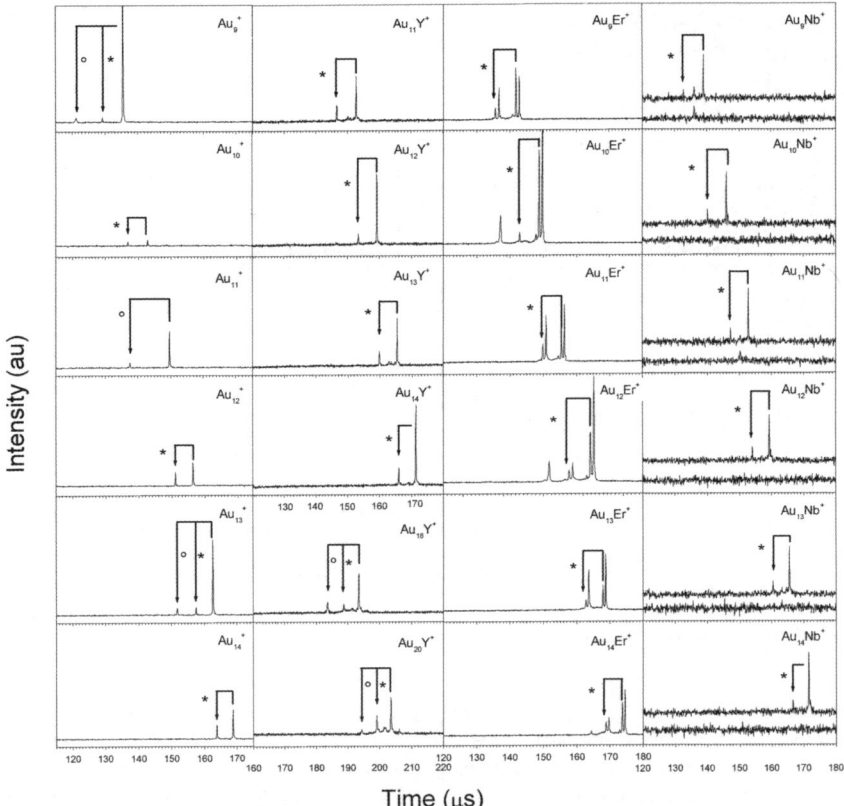

Fig. 2 Metastable fragmentation of Au_nX^+ for $n = 9$–14 (X = Au, Er, Nb) and $n = 11$–14, 18, 20 (X = Y). Because of the limited resolution of the mass selector for erbium doped gold clusters not only Au_nEr^+ was selected, but groups of Au_{n+1}^+, Au_nEr^+, $Au_{n-1}Er_2^+$ were transmitted. The dissociation channels (from parent to daughter) are indicated by arrows. Arrows labeled with * and ° correspond to neutral monomer and dimer evaporation, respectively. The bottom curves in the right column give the signal recorded as reference without any Nb in the clusters.

clusters deviates from the mother beam path after the reflectron. This deviation is larger for smaller clusters resulting in a loss of signal on the detector.

Au_n^+ ($n \geq 6$) clusters decay by neutral monomer dissociation with the exception of Au_9^+, Au_{11}^+, and Au_{13}^+, for which the main recorded dissociation channel is the evaporation of a gold dimer. For Au_{15}^+, dimer evaporation is observed but not as the main dissociation channel. Not only the preferred dissociation channel but also the dissociation probability strongly depends on the cluster size. The relative probability for monomer decay is higher for Au_n^+ with n even compared to their neighbors. Dimer evaporation however is more likely to occur for Au_n^+ with n odd. These results agree with photodissociation and collision induced dissociation experiments on size-selected Au_n^+ ($n = 2$–27) clusters in a Penning trap by Schweikhard et al.,[41–43] with the exception that we do not observe delayed fragmentation of the smallest clusters due to the above mentioned effects.

B. Au_nY^+

The photofragmentation spectrum for Au_nY^+ (Fig. 1b) exhibits strong finite size effects in agreement with an earlier study on yttrium doped gold clusters where the

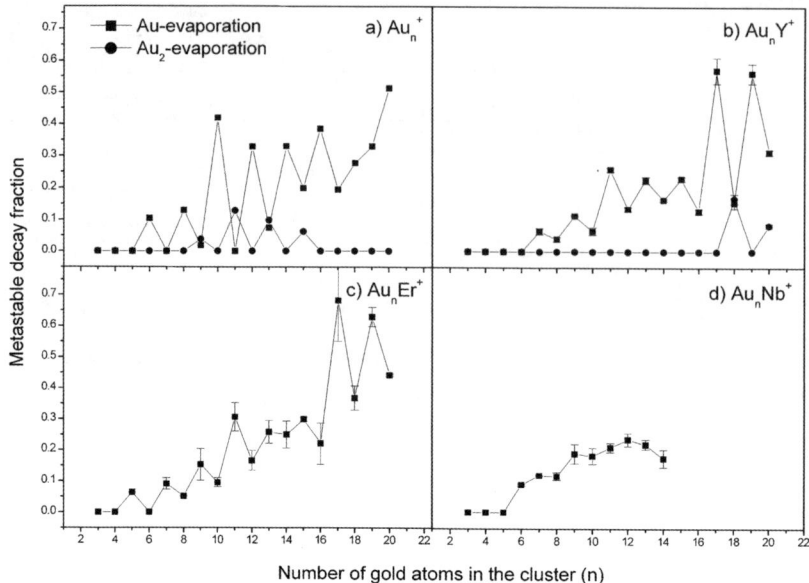

Fig. 3 Metastable decay fraction of neutral gold monomer and dimer evaporation pathways for (a) Au_n^+, (b) Au_nY^+, (c) Au_nEr^+, and (d) Au_nNb^+.

observed stability patterns, shell closures and odd–even effects were thoroughly discussed.[7] Prominent intensity steps can be observed at $n = 6$, 16 and 32 (32 not shown in the figure). Odd–even intensity alternations are present up to $n \approx 25$.

The evaporation channels recorded for metastable Au_nY^+ ($n = 11$–14, 18, 20) are shown in the second column of Fig. 2. From the labeled dissociation pathways it can be seen that the fragmentation channels, as for pure gold, are size dependent. The metastable fractions for monomer and dimer decay, deduced from the recorded delayed fragmentation spectra, are shown in Fig. 3b. The average is taken over three independent measurements, yielding reproducible ratios of the daughter-to-mother peak intensities. Error bars are statistically determined and correspond to the standard deviation of the average throughout. As for pure gold clusters, no delayed fragmentation is recorded for the smallest species ($n < 7$). Au_nY^+ ($n \geq 7$) clusters decay by monomer evaporation. For $Au_{18}Y^+$ and $Au_{20}Y^+$ also dimer loss is observed. Note also the odd–even alternations in relative decay probabilities: for odd n Au_nY^+ species a larger monomer metastable decay fraction is recorded than for their even n neighbours. Pronounced maxima in relative fragmentation probability appear at $Au_{17}Y^+$ and $Au_{19}Y^+$ for monomer loss and at $Au_{18}Y^+$ for dimer evaporation.

C. Au_nEr^+

The photofragmentation spectrum for Au_nEr^+ clusters (Fig. 1c) shows intensity steps at $n = 6$ and 16 in addition to an odd–even staggering. The measured dissociation pathways for Au_nEr^+ ($n = 9$–14) are shown in the third column of Fig. 2. The small mass difference between pure gold clusters and erbium doped species prohibited mass selection of Au_nEr^+ solely. Therefore, mass ranges corresponding to Au_{n+1}^+, Au_nEr^+, and $Au_{n-1}Er_2^+$ were selected for $n = 3$–20 (see ESI†), the bare gold being the heaviest. As a consequence, besides fragments stemming from Au_nEr^+, also delayed fragments of bare Au_{n+1}^+ are present. To guide the eye only the delayed fragments of Au_nEr^+ are indicated (labeled arrows). Dimer evaporation is not observed for any of the Au_nEr^+ clusters investigated. The intensities of $Au_{n-1}Er_2^+$ and their delayed fragments are too low to be able to analyze these species quantitatively. Therefore, these dissociation channels are not indicated and metastable fractions are only deduced for singly doped Au_nEr^+ (see Fig. 3c). Three independent measurements are averaged. Again odd–even alternations in the metastable fractions can be observed. As for Au_nY^+, neutral monomer loss is more likely for Au_nEr^+ containing an odd number of gold atoms and the metastable fraction increases with cluster size.

D. Au_nNb^+

The stability pattern of niobium doped gold clusters (Fig. 1d) shows a pronounced step in intensity at $Au_{14}Nb^+$. In contrast to the findings for Au_nY^+ and Au_nEr^+, no odd–even staggering in the intensity as a function of cluster size is seen for Au_nNb^+. The recorded dissociation pathways for Au_nNb^+ ($n = 9$–14) are shown in the right column of Fig. 2. The delayed fragmentation of Au_nNb^+ with $n > 14$ could not be monitored, because the cluster intensities were too small. In addition, a perfect mass selection was not always possible and a limited fraction of other species also passed the mass selector. Therefore, a reference signal of pure gold clusters (laser that vaporizes niobium switched off) in the cluster beam was recorded (bottom curves). For all investigated sizes, the only observed delayed fragmentation channel is the gold monomer evaporation, marked with arrows on Fig. 2. The metastable fractions plotted in Fig. 3d were deduced by averaging three independent measurements. In contrast to Au_n^+, Au_nY^+, and Au_nEr^+, no pronounced odd–even staggering or no clear maxima in the metastable fraction as a function of size could be identified.

4. Data analysis

A. Monomer and dimer evaporation rates

In order to establish a relation between mass spectra shaped by evaporation and dissociation energies, an expression for the evaporation rate is needed. Evaporative rate constants are functions primarily of the excitation energy, the evaporative activation energy, and a frequency factor. The main assumption used here is that the decays are activated processes on the experimental time scales of several tens of nanoseconds to hundreds of microseconds. Rate constants are generally given by:[49]

$$k_{n,\Delta n}(E) = \bar{\varpi}_{n,\Delta n} \frac{\rho_{n-\Delta n}(E - D_{n,\Delta n})}{\rho_n(E)} \qquad (1)$$

where ρ_n and $\rho_{n-\Delta n}$ are the mother and daughter level densities, respectively, $\Delta n = 1$, 2 for monomer and dimer decay, E is the excitation energy of the cluster, and $D_{n,\Delta n}$ the dissociation energy for the channel specified, i.e., the energy required to create the daughter cluster from the mother cluster, or the difference in ground state energies of the two:

$$D_{n,\Delta n} = E_{gs}(n - \Delta n) + E_{gs}(\Delta n) - E_{gs}(n) \qquad (2)$$

The subscript gs refers to ground state properties. The energy of the free monomer is used as the zero of energy; $E_{gs}(1) = 0$. It is assumed throughout that the dissociation energies equal the evaporative activation energies as is the case if there is no activation energy for the reverse reaction. This is expected to hold for most if not all metal clusters. The frequency factor, $\bar{\varpi}_{n,\Delta n}$, depends on the cluster size and the decay channel as indicated by the subscript. It varies with the surface area, i.e., with $n^{2/3}$, and slowly with energy (see ref. 49 for details). These variations are much slower than the variation of the ratio of the level densities, which, for large sizes, develops into a proper Boltzmann factor. The averaging of the frequency factor in eqn (1) refers to setting these slow variations to a constant.

Although the frequency factor is fairly insensitive to the cluster size, it does depend on whether the decay channel is monomer or dimer evaporation. Using eqn (2) a Born–Haber cycle relates two consecutive monomer evaporations to the dimer decay:

$$D_{n,2} - E_{gs}(2) = D_{n,1} + D_{n-1,1} \qquad (3)$$

with $-E_{gs}(2) = D_{2,1} = 2.306$ eV the binding energy of the gold dimer.[50]

We will use the level densities of the high energy limit of harmonic oscillators:

$$\rho_n(E) = \frac{(E + E_{0,n})^{3n-7}}{(3n - 7)!(\hbar\varpi_D)^{3n-6}} \qquad (4)$$

where the Debye frequency of gold, $\varpi_D = 165$ K $k_B/\hbar$, is used as a common vibrational frequency. $E_{0,n}$ is the sum of the zero point energies of the oscillators, i.e., with our use of the Debye frequency $E_{0,n} = 1/2\Sigma_i\hbar\varpi_i \approx 7.2 \times 10^{-3}(3n - 6)$ eV. Incorporation of (4) into (1) gives for the monomer dissociation rate:

$$k_{n,1}(E) \approx \bar{\varpi}_1 \left(\frac{E - D_{n,1} + E_{0,n-1}}{E + E_{0,n}}\right)^{3n-7} \qquad (5)$$

with

$$\bar{\varpi}_1' = \bar{\varpi}_1 \left[\frac{(3n - 8)\hbar\varpi_D}{E - D_{n,1} + E_{0,n-1}}\right]^3 \qquad (6)$$

In ref. 42 a numerical value of $\bar{\varpi}_1' = 3.6 \times 10^{16}$ s^{-1} was derived by setting the last factor in eqn (6) equal to the cube of $\hbar\varpi_D/0.1$ eV. The factor 0.1 eV is a typical daughter temperature. This was shown to be consistent with the experimental values.[42]

The main difference between the monomer and dimer evaporation frequency factors is related to the internal (particularly the rotational) degrees of freedom of the dimer. The rotational constant of Au$_2$ is $B = B_0 = 0.02636$ cm^{-1} = 3.3×10^{-6} eV.[50] The vibrational quantum energy is $\hbar\varpi_v = 190.9$ cm^{-1} = 0.02367 eV. In addition, one has a factor of 2 from the (reduced) mass of the channel, a factor of 1/2 from the symmetry number of the dimer, and a factor of 1/2 from the loss of the electronic degeneracy of the atom. Finally, applying a similar procedure as in eqn (6), one gets the dimer rate constant:

$$k_{n,2}(E) = \bar{\varpi}_2' \left(\frac{E - D_{n,2} + E_{0,n-2}}{E + E_{0,n}}\right)^{3n-7} \qquad (7)$$

with

$$\bar{\varpi}_2' = \bar{\varpi}_1' \frac{1}{2}\frac{0.1\,\text{eV}}{B}\frac{0.1\,\text{eV}}{\hbar\varpi_v}\left(\frac{\hbar\varpi_D}{0.1\,\text{eV}}\right)^3 = 186 \cdot \bar{\varpi}_1' \qquad (8)$$

This result is obtained by treating the vibrational and rotational degrees of freedom of the dimer as a perturbation, which permits integrating out the energy partitions and expressing the results in terms of canonical partition functions and the daughter microcanonical temperature.

B. Evaporative ensembles

The dependence of evaporative processes on binding energies endows ensembles with information on binding energies. When only one decay channel is present, the analysis is relatively simple but on the other hand only relative values can be found. The analysis can then be based either on the abundances, I_n, or on the amount of clusters of a certain size that decays in the first field free region of the RTOF mass spectrometer, *i.e.*, the metastable decay fraction, p_n. When both monomer and dimer evaporation occur, experimental branching ratios together with the energy constraint from the Born–Haber cycle allow one to extract absolute values.

To use the theoretical results of the evaporative ensemble each cluster must have undergone at least one evaporation. Furthermore it is required that the energy distribution right after the laser excitation is smooth on the scale set by the dissociation energies.[34] One contributing cause to a broad energy distribution is the Gaussian laser beam profile that gives rise to a broad range of the number of absorbed photons.

When just a single decay channel is active, for simplicity assumed here to be the monomer decay, we can find the highest energy present in the evaporating ensemble at any given time t as:

$$\frac{1}{t} = k_n(E_{\max,n}) \tag{9}$$

which is solved for the energy to give

$$E_{\max,n} = -E_{0,n} + \frac{D_n + E_{0,n} - E_{0,n-1}}{1 - (\bar{\varpi}'_1 t)^{-1/(3n-7)}} \tag{10}$$

If t is chosen as the time when the mass selection occurs during the initial acceleration in the time-of-flight, t_1, one can find the abundances to be proportional to the difference between the highest and lowest energy of size n:[51]

$$I_n \propto E_{\max,n} - E_{\min,n} = E_{\max,n} - (E_{\max,n+1} - D_{n+1})$$
$$= D'_{n+1} + D'_n \frac{1}{1 - (\bar{\varpi}'_1 t)^{-1/(3n-7)}} - D'_{n+1} \frac{1}{1 - (\bar{\varpi}'_1 t)^{-1/(3n-4)}} \tag{11}$$

where the definition $D'_n \equiv D_n + E_{0,n} - E_{0,n-1} \approx D_n + 0.02$ eV was used. The identification of the lowest energy of cluster size n with the highest energy of size $n + 1$ minus the evaporative activation energy only hinges on the assumption that only the dissociation energy is removed in the process. In other words, that the kinetic energy release is small and there is no reverse activation barrier for the process. The constant of proportionality in eqn (11) is a size dependent but smooth function of n, which contains instrumental parameters like detection efficiency and depends on the kinetics of the cluster formation, *etc.* It can be determined with a primitive application of the method known in nuclear physics as the Strutinsky procedure.[52] The smooth n dependence can be mocked up in a smoothed abundance function, which is then divided out. The procedure leaves an abundance distribution varying around unity, which is used to solve for the dissociation energies numerically. These values are defined only up to a scale factor with smooth size dependence.

It is also possible to calculate the amount of metastable fragmentation. It is given by the amount of decay occurring between the time of initial size selection relative to the time of the laser pulse, t_1, and the time of entry into the reflectron, t_2. With the same constant of proportionality as in eqn (11) it is:

$$p_n \propto E_{\max,n}(t_1) - E_{\max,n}(t_2) \tag{12}$$

The ratio of the metastable fraction to the cluster abundances is therefore:

$$P_n = \frac{p_n}{I_n} = \frac{E_{n,\max}(t_1) - E_{n,\max}(t_2)}{E_{n,\max} - (E_{n+1,\max} - D_{n+1})} \tag{13}$$

As in eqn (11) one may use the rate constants derived above directly in these formulae. It is more transparent to use the rate constant formulated in terms of an

Arrhenius expression. This implies an approximation of the rate constant as:

$$k_n(E) = \bar{\omega} \exp\left(-\frac{D'_n}{k_{\mathrm{B}}T - D'_n/2C_v}\right) \qquad (14)$$

where $k_{\mathrm{B}}T = (E + E_{0,n})/(3n - 7)$ and $C_v = 3n - 7$. The negative term in the denominator of the argument of the exponential is the so-called finite heat bath correction. Both expressions for the rate constants have been used in the analysis described below with practically identical results and we give equations for the Arrhenius expression only. With the above rate constant, the abundances become:

$$I_n \propto \frac{D'_n + D'_{n+1}}{2} - \frac{3D'_{n+1}}{G(t_1)} + \frac{3n - 7}{G(t_1)}(D'_n - D'_{n+1}) \qquad (15)$$

where G is the Gspann parameter defined as $G(t) = \ln(\bar{\omega}_1 t)$. The amount of metastable decay becomes:

$$p_n \propto D'_n(3n - 7)\left(\frac{1}{G(t_1)} - \frac{1}{G(t_2)}\right) = D'_n(3n - 7)\frac{\ln(t_2/t_1)}{(G(t_1)G(t_2))} \qquad (16)$$

Combining eqn (15) and (16) gives:

$$\frac{D'_{n+1}}{D'_n} = \left(\frac{(3n-7)\ln(t_2/t_1)}{G(t_1)G(t_2)} - P_n\left(\frac{1}{2} + \frac{3n-7}{G(t_1)}\right)\right) \times \left(P_n\left(\frac{1}{2} - \frac{3n-4}{G(t_1)}\right)\right)^{-1} \qquad (17)$$

In this analysis we have disregarded the possibility of radiative cooling. Radiative cooling would manifest itself as a reduced amount of metastable fragmentation, as observed, e.g., for fullerenes in similar experiments.[53] Because the amount of metastable fragmentation was found experimentally to be of the magnitude expected, we conclude that no appreciable radiative cooling was present on the experimental time scales.

C.   Dimer dissociation

The presence of competing decay channels complicates the determination of dissociation energies within the evaporative ensemble. The reason is that the branching between the two channels in general depends on the excitation energy. Therefore, the energy density is not constant across the distribution spanned by a single mass. This effect rules out a rigorous use of the concept of "highest and lowest possible excitation energy" used above for monomer decay.

On the other hand, there is an advantage with a competing dimer decay channel, viz. the possibility to introduce an absolute energy scale with the help of the Born–Haber cycle. The highest temperature is now determined as:

$$\frac{1}{t} = k_{n,1} + k_{n,2} \approx k_{n,1}(1 + B_n) \qquad (18)$$

where B_n is the experimentally determined branching ratio $p_{n,2}/p_{n,1}$. The approximate nature of the last identity is due to the fact that the branching ratio is energy dependent and the observed ratio is therefore a ratio of energy- (or time-) integrated values. However, the correction is small as is best seen if one evaluates the equation below at the two extreme times t_1 and t_2. We will therefore proceed with the identification of the highest temperature with the same relation as above but using a frequency factor modified by the factor $1 + B_n$. The associated Gspann parameter is denoted $G_{\mathrm{B}}(t) = \ln[\bar{\omega}_1 t(1 + B_n)]$.

Calculating the experimentally observed branching ratio as the ratio of the two rate constants and solving for the ratio of dissociation energies, one gets:

$$\frac{D'_{n,2}}{D'_{n,1}} = \frac{G_B(t_1)^{-1} + (2C_v)^{-1}}{(G_B(t_1) - \ln(B_n \bar{\varpi}_1 / \bar{\varpi}_2))^{-1} + (2C_v)^{-1}} \quad (19)$$

5. Dissociation energies

Eqn (11) and (15) are applied to extract dissociation energies from the experimental data, the result of which is shown in Fig. 4. Relative abundances $I'_n = I_n / \tilde{I}_n$ are used to correct for purely experimental intensity variations as described above. Hereto a simple fitting function was applied: $\tilde{I}_n = Ae^{-Bn}$. Estimated values of t_1 and t_2 were based on electrostatic simulations of the setup (using Simion 3D)[54] and the detection times of the clusters. The reference dissociation energy of the largest investigated size was set to $D_{21} = 1$. The dependence on this initial value is weak. The two expressions of the decay constant k_n, eqn (5) and (14), yield, via eqn (11) and (15), similar dissociation energies for all but the smallest sizes. As in the measured abundance spectra, clear steps in the extracted dissociation energies can be seen at specific sizes and an odd–even staggering is present for Au_nX^+ (X = Au, Y, Er).

The ratios of two consecutive dissociation energies, as given by eqn (17), are given in Fig. 5 for Au_n^+, Au_nY^+, Au_nEr^+, and Au_nNb^+ and are compared with the ratios of the dissociation energies obtained with eqn (11) and (15). For a range of Au_n^+ clusters and the high mass range of Au_nY^+ the admixture of dimer decay prevents an analysis in terms of one or both of these methods. The method based on the abundance spectra can be used for all measured Au_nEr^+ and Au_nNb^+ clusters, for Au_nY^+ (n = 3–15) and for Au_n^+ (n = 16–20). The metastable fraction can likewise be used for all measured Au_nEr^+ and Au_nNb^+ clusters, for Au_nY^+ (n = 3–15, 17), and for Au_n^+ (n = 6, 8, 10, 12, 14–19).

Generally, the values obtained with the two methods agree well. There are some features which should be commented on. We note that the uncertainties in the numbers extracted from the abundance spectra are smaller than those involving the

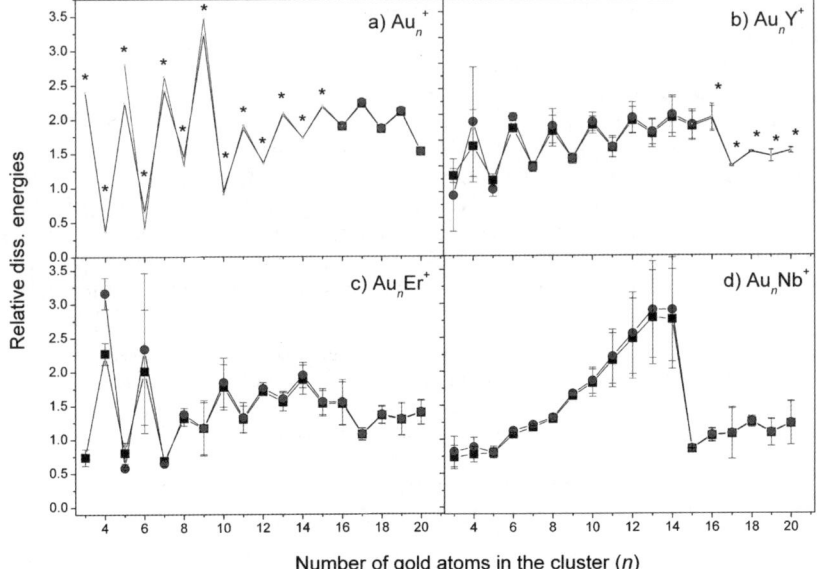

Fig. 4 Relative dissociation energies for (a) Au_n^+, (b) Au_nY^+, (c) Au_nEr^+, and (d) Au_nNb^+, obtained using eqn (11) (black squares) and eqn (15) (gray dots). Sizes marked with an asterisk are uncertain because of the competing dimer decay channel (see text for details).

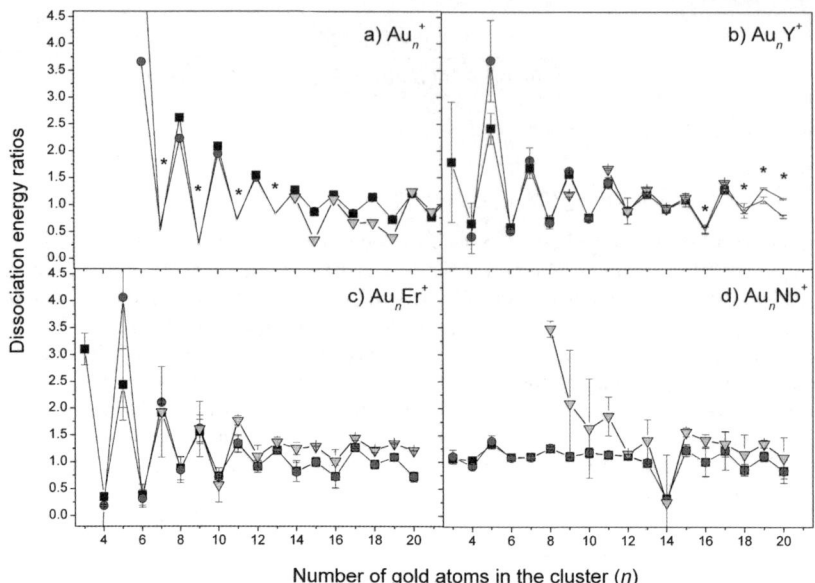

Fig. 5 Ratios of successive dissociation energies, D'_{n+1}/D'_n, for (a) Au_n^+, (b) Au_nY^+, (c) Au_nEr^+, and (d) Au_nNb^+, obtained using eqn (11) (black squares), (15) (gray dots), and (17) (light gray triangles). Sizes marked with an * are uncertain because of the competing dimer decay channel (see text for details).

metastable decay fraction. This is a purely statistical effect due to the small decay fractions. On the other hand there is some amount of arbitrariness involved in the use of the smoothing procedure for the abundances, which is expected to be small but which is difficult to quantify. The general agreement does indicate that the assumptions underlying the analysis are sound, *i.e.*, statistical decays, smooth energy distributions, and heat capacities are consistent with high temperature harmonic oscillator values.

Eqn (19) provides a straightforward way of calculating the fraction of dimer to monomer dissociation energies from the experimental branching ratios. Clear dimer peaks are recorded for Au_n^+ ($n = 9$, 11, 13, and 15) and Au_nY^+ ($n = 18$, 20) and the according results are presented in Table 1. The ratios for Au_n^+ are in good agreement with the model-free results presented in ref. 42, which for these cluster sizes are derived and have the relatively high uncertainties of about 10%. It should be noted that the agreement is obtained without any tuning of parameters from data of two distinctly different types of experiments. It should also be noted that a correct treatment of the dimer evaporation rate constant is essential for the agreement. If the

Table 1 Ratio of dimer to monomer dissociation energies, $D'_{n,2}/D'_{n,1}$, obtained with eqn (19) using the experimental branching ratios for Au_n^+ ($n = 9$, 11, 13, 15) and Au_nY^+ ($n = 18$, 20). For Au_n^+ also a comparison with model-free values obtained in ref. 42 is given. The * at $n = 11$ indicates an upper limit based on a monomer branching ratio of 5%. For a branching ratio of 1% the ratio is 1.01

N	$D_{n,2}/D_{n,1}$ (Au_n^+)	$D_{n,2}/D_{n,1}$ (Au_n^+) (ref. 42)	n	$D'_{n,2}/D'_{n,1}$ (Au_nY^+)
9	1.10	1.08	18	1.14
11	1.05*	1.14	20	1.19
13	1.13	1.22		
15	1.18	1.22		

monomer and dimer frequency factors were set equal, as in ref. 33, the ratio would be in error by almost 20%.

The determination of the dimer/monomer dissociation energy allows an absolute determination of the energy for a few cases. The agreement between the present results for pure gold clusters and those presented in ref. 42, shown in Table 1, does not require more comments. For $Au_{18}Y^+$ we use the relation:

$$D'_{n,1} = D'_{2,1} \left(1 + \frac{D'_{n-1,1}}{D'_{n,1}} - \frac{D'_{n,2}}{D'_{n,1}} \right)^{-1} \tag{20}$$

which is derived from the Born–Haber cycle, to find absolute dissociation energies. The values are $D'_{18,1} = 3.25$ eV, $D'_{18,2} = 3.73$ eV, and $D'_{17,1} = 2.78$ eV. As expected the monomer dissociation energy for $n = 18$ is higher than the one for $n = 17$.

6. Discussion

The mother abundances (Fig. 1), the preferred decay channel (Fig. 2), the metastable fractions (Fig. 3), and the calculated dissociation energies (Fig. 4) of the bare and doped gold clusters show a consistent picture that can be interpreted on the basis of an underlying electronic shell structure.

For bare gold clusters ($Au = [Xe] 4f^{14} 5d^{10} 6s^1$) pronounced steps in abundance and dissociation energies are found at $n = 9$ and to a smaller extent at $n = 19, 21$. These observations can be related to a shell structure and magic numbers 8, 18, and 20 that arise from the delocalization of the atomic valence 6s electrons. The electronic structure of the bare gold clusters is also reflected in the metastable fractions for monomer and dimer decay. The tendency of small Au_n^+ ($n < 16$) clusters to decay towards odd-sized clusters, Au_{n-1}^+, by monomer evaporation if n is even and to Au_{n-2}^+ by dimer evaporation if n is odd, can be related to the higher stability of odd-sized gold cluster ions due to their even number of delocalized valence s electrons. Moreover Au_{10}^+ and Au_{20}^+ give a local maximum in the monomer metastable fractions, since they decay to the closed shell Au_9^+ and Au_{19}^+ clusters, respectively.

For Au_nY^+ pronounced maxima in dissociation energies are obtained at Au_6Y^+ and especially $Au_{16}Y^+$ in addition to odd–even alternations. Assuming that each gold atom delocalizes its 6s valence electron, as for pure gold clusters, these cluster sizes correspond to 8 and 18 itinerant electrons, provided that the three yttrium ($Y = [Kr] 4d^1 5s^2$) valence electrons can be considered itinerant. Also the observation that odd n Au_nY^+ species give a larger monomer decay fraction can be explained as an electronic odd–even effect. The dimer decay that is only recorded for $Au_{18}Y^+$ and $Au_{20}Y^+$ leads to the formation of the closed shell $Au_{16}Y^+$ and $Au_{18}Y^+$ systems. The special stability of $Au_{16}Y^+$ might not only be due to the electronic closed shell (18 electrons) stabilization, but might also be related to a compact and symmetric geometry, similar to the doped golden fullerenes that have been recently predicted for $Au_{16}Cu^-$ and $Au_{17}Cu^-$.[28]

The intensity and dissociation energy (less pronounced) drops after Au_6Er^+ and $Au_{16}Er^+$ in the patterns of Au_nEr^+ clusters can be linked to the magic numbers 8 and 18 assuming that the erbium atom ($Er = [Xe] 4f^{12}5d^0 6s^2$) delocalizes three valence electrons. The trivalent character of erbium is not surprising, since erbium chemistry is dominated by the trivalent erbium ion Er^{3+} ($Er = [Xe] 4f^{11}5d^0 6s^0$). The observation of a larger monomer metastable fraction for Au_nEr^+ clusters having an odd number of gold atoms and a maximal metastable fraction for $Au_{17}Er^+$, can be related to the energetic preference for an even amount of itinerant electrons and the decay to the closed shell $Au_{16}Er^+$ system, respectively.

The abundance pattern of Au_nNb^+ is rather smooth and the dissociation energies provide less evidence for an electronic shell structure. No pronounced odd–even staggering is observed neither in the abundance spectra nor in the monomer metastable fractions. Nevertheless, the prominent peak at $Au_{14}Nb^+$ can be explained in terms of an electronic shell closing. In case the niobium atom (Nb = [Kr] $4d^4\,5s^1$) delocalizes both its 4d and its 5s electrons, $Au_{14}Nb^+$ has in total 18 delocalized electrons.

The appearance of the magic number 18 in the doped gold clusters, at the expense of the magic number 20, which is also present for bare gold, was noted earlier.[7,22] The electronic shell structure in a simple mean field potential with a flat bottom has 20 electrons as a prominent shell closing. However, if the mean field potential has a wine bottle shape, the 2s state is shifted up relative to the 1d level and the magic number 18 becomes more pronounced.[55] The formation of a wine bottle shaped potential can be induced by a centrally located electronegative dopant atom. However, the 18 *versus* 20 competition in gold and doped gold clusters might also have a geometrical origin. It was predicted that the neutral Au_{20} has a tetrahedral geometry and a large HOMO–LUMO gap, representing its electronically closed shell structure.[15,16] The isoelectronic Au_{21}^+ might be composed of tetrahedral neutral Au_{20} with an adatom on a facet. Since Au_{21}^+ is not more pronounced in the fragmentation spectra than Au_{19}^+, its stability is probably not as exceptional as the neutral Au_{20}, confirming that geometric structure does indeed play an important role in determining the cluster stability. Upon doping the clusters with a transition metal, any evidence for the magic number 20 disappears. This might indicate that a tetrahedral geometry (according to the magic number 20) is not stable for the doped clusters, whereas a more spherical caged structure with 18 itinerant electrons (as for $Au_{16}Cu^-$)[28] is.

The dependence of the odd–even effect on the dopant also deserves comment. The preference for dimer *versus* monomer decay in bare gold clusters is controlled by odd–even effects. Clusters with an even number of itinerant electrons are more stable than the ones with an odd number of delocalized electrons. If this stability variation is sufficiently strong, the preferred decay from an even electron numbered cluster will be the dimer decay. This is consistent with the observed correlation between the presence of dimer decay and the strength of the odd–even alternations in the abundances, which are strong for bare gold for which a number of dimer decay channels are observed, less strong for yttrium for which only two sizes evaporate dimers, again somewhat smaller for erbium doped clusters, and finally almost absent for the niobium doped species, the latter two having no observable dimer decay. In the simple mean field picture of shell structure, the odd–even effect is due to structural deviations from spherical symmetry: breaking the symmetry removes the angular momentum degeneracy of levels, leaving just the spin degeneracy of each level. *A priori*, there is no obvious reason why the amplitude of the odd–even effect should be reduced by introducing a dopant atom. Possibly the reason can be found in the specific character of the transition metal (or lanthanide) dopant atom. The valence d electrons are hybridized with the valence s electrons. Since the d electrons have a more localized character, itinerant electron counting is not straightforward. Moreover, the d electron behaviour can be size-dependent, this way reducing the odd or even electron character.[22]

The electronic behaviour of niobium and also of yttrium is very similar to that observed for light 3d transition metal dopant atoms (Sc, Ti, V) in gold clusters. A photofragmentation mass spectrometry study revealed that these dopant atoms contribute both their 4s and 3d electrons to the cloud of itinerant electrons because of the large spatial extent of the corresponding orbitals.[22,23] Since the 4d orbitals of yttrium and niobium are expected to be even larger than the scandium and titanium 3d orbitals, it is not surprising that they show up as delocalized electrons in the present experiments.

7. Conclusion

The dissociation pathways and size-dependent stability of cationic Au_nX^+ (X = Y, Er, and Nb, n = 3–20) clusters produced in a laser vaporization source have been investigated. Photofragmentation experiments were performed with a reflectron time-of-flight mass spectrometer incorporating a wire-type mass gate. Statistical models based on unimolecular decay rates are presented and applied to determine dissociation energies from both the recorded abundances and the fraction of mass selected clusters that undergo delayed fragmentation in the time window of the experiment. Gold monomer evaporation turns out to be the most likely fragmentation channel for doped Au_nX^+, with the exception of $Au_{18}Y^+$ and $Au_{20}Y^+$ for which gold dimer evaporation is a competing channel. The extracted dissociation energies show an odd–even staggering and enhanced stabilities for certain cluster sizes, in agreement with simple electronic shell model predictions.

Acknowledgements

The work in Leuven was supported by the Fund for Scientific Research-Flanders (FWO), the Flemish Concerted Action (GOA/2004/02), and the Belgian Interuniversity Poles of Attraction (IAP/P5/01) programs. E. J. is a postdoctoral researcher of the FWO. K. H. was supported by the Swedish National Research Council (VR).

References

1   M. Haruta, *Gold Bull.*, 2004, **37**, 27.
2   K. J. Taylor, C. L. Pettiette-Hall, O. Cheshnovsky and R. E. Smalley, *J. Chem. Phys.*, 1992, **96**, 3319.
3   C. Jackschath, I. Rabin and W. Schulze, *Ber. Bunsen-Ges. Phys. Chem.*, 1992, **96**, 1200.
4   I. Katakuse, T. Ichihara, Y. Fujita, T. Matsuo, T. Sakurai and H. Matsuda, *Int. J. Mass Spectrom. Ion Processes*, 1985, **67**, 229.
5   M. Lindinger, K. Dasgupta, G. Dietrich, S. Krückeberg, S. Kuznetsov, L. Lützenkirchen, L. Schweikhard, C. Walther and J. Ziegler, *Z. Phys. D*, 1997, **40**, 347.
6   S. Becker, G. Dietrich, H. U. Hasse, N. Klisch, H. J. Kluge, D. Kreisle, S. Krückeberg, M. Lindinger, K. Lützenkirchen, L. Schweikhard, H. Weidele and J. Ziegler, *Z. Phys. D*, 1994, **30**, 341.
7   W. Bouwen, F. Vanhoutte, F. Despa, S. Bouckaert, S. Neukermans, L. T. Kuhn, H. Weidele, P. Lievens and R. E. Silverans, *Chem. Phys. Lett.*, 1999, **314**, 227.
8   H. Häkkinen, M. Moseler, O. Kostko, N. Morgner, M. A. Hoffmann and B. von Issendorff, *Phys. Rev. Lett.*, 2004, **93**, 093401.
9   W. A. de Heer, *Rev. Mod. Phys.*, 1993, **65**, 611.
10 F. Furche, R. Ahlrichs, P. Weis, C. Jacob, S. Gilb, T. Bierweiler and M. M. Kappes, *J. Chem. Phys.*, 2002, **117**, 6982.
11 S. Gilb, P. Weis, F. Furche, R. Ahlrichs and M. M. Kappes, *J. Chem. Phys.*, 2002, **116**, 4094.
12 H. Häkkinen, M. Moseler and U. Landman, *Phys. Rev. Lett.*, 2002, **89**, 033401.
13 V. Bonačić-Koutecký, J. Burda, R. Mitrić, M. Ge, G. Zampella and P. Fantucci, *J. Chem. Phys.*, 2002, **117**, 3120.
14 J. Wang, G. H. Wang and J. Zhao, *Phys. Rev. B*, 2002, **66**, 035418.
15 J. Li, X. Li, H.-J. Zhai and L.-S. Wang, *Science*, 2003, **299**, 864.
16 E. Aprà, R. Ferrando and A. Fortunelli, *Phys. Rev. B*, 2006, **73**, 205414.
17 W. Fa, C. Luo and J. Dong, *Phys. Rev. B*, 2005, **72**, 205428.
18 M. P. Johansson, D. Sundholm and J. Vaara, *Angew. Chem., Int. Ed.*, 2004, **43**, 2678.
19 J. Wang, J. Jellinek, J. Zhao, Z. Chen, R. B. King and P. v. R. Schleyer, *J. Phys. Chem. A*, 2005, **109**, 9265.
20 S. Bulusu, X. Li, L.-S. Wang and X. C. Zeng, *Proc. Natl. Acad. Sci. USA*, 2006, **103**, 8326.
21 Y. Kondo and K. Takayanagi, *Science*, 2000, **289**, 606.
22 S. Neukermans, E. Janssens, H. Tanaka, R. E. Silverans and P. Lievens, *Phys. Rev. Lett.*, 2003, **90**, 33401.
23 E. Janssens, H. Tanaka, S. Neukermans, R. E. Silverans and P. Lievens, *Phys. Rev. B*, 2004, **69**, 085402.
24 E. Janssens, H. Tanaka, S. Neukermans, R. E. Silverans and P. Lievens, *New J. Phys.*, 2003, **5**, 46.

25 H. Tanaka, S. Neukermans, E. Janssens, R. E. Silverans and P. Lievens, *J. Am. Chem. Soc.*, 2003, **125**, 2862.
26 M. B. Torres, E. M. Fernandez and L. C. Balbas, *Phys. Rev. B*, 2005, **71**, 155412.
27 Y. Gao, S. Bulusu and X. C. Zeng, *J. Am. Chem. Soc.*, 2005, **127**, 156801.
28 L. M. Wang, S. Bulusu, H. J. Zhai, X. C. Zeng and L. S. Wang, *Angew. Chem., Int. Ed.*, 2007, **46**, 2915.
29 P. Pyykkö and N. Runeberg, *Angew. Chem., Int. Ed.*, 2002, **41**, 2174.
30 X. Li, B. Kiran, J. Li, H. J. Zhai and L. S. Wang, *Angew. Chem., Int. Ed.*, 2002, **41**, 4786.
31 K. Manninen, P. Pyykkö and H. Häkkinen, *Phys. Chem. Chem. Phys.*, 2005, **7**, 2208.
32 F. Naumkin, *Phys. Chem. Chem. Phys.*, 2006, **8**, 2539.
33 C. Bréchignac, Ph. Cahuzac, J. Leygnier and J. Weiner, *J. Chem. Phys.*, 1989, **90**, 1492.
34 C. E. Klots, *J. Chem. Phys.*, 1985, **83**, 5854.
35 C. Bréchignac, H. Busch, Ph. Cahuzac and J. Leygnier, *J. Chem. Phys.*, 1994, **101**, 6992.
36 C. Bréchignac, Ph. Cahuzac, F. Carlier, M. Defrutos and J. Leygnier, *J. Chem. Phys.*, 1990, **93**, 7449.
37 S. Krückeberg, L. Schweikhard, J. Ziegler, G. Dietrich, K. Lützenkirchen and C. Walther, *J. Chem. Phys.*, 2001, **114**, 2955.
38 V. A. Spasov, T.-H. Lee and K. M. Ervin, *J. Chem. Phys.*, 2000, **112**, 1713.
39 U. Hild, G. Dietrich, S. Krückeberg, M. Lindinger, K. Lützenkirchen, L. Schweikhard, C. Walther and J. Ziegler, *Phys. Rev. A*, 1998, **57**, 2786.
40 V. A. Spasov, T.-H. Lee, J. P. Maberry and K. M. Ervin, *J. Chem. Phys.*, 1999, **110**, 5208.
41 M. Vogel, K. Hansen, A. Herlert and L. Schweikhard, *Phys. Rev. Lett.*, 2001, **87**, 013401.
42 K. Hansen, A. Herlert, L. Schweikhard and M. Vogel, *Phys. Rev. A*, 2006, **73**, 063202.
43 M. Vogel, K. Hansen, A. Herlert and L. Schweikhard, *Eur. Phys. J. D*, 2001, **16**, 73.
44 V. A. Spasov, Y. Shi and K. M. Ervin, *Chem. Phys.*, 2000, **262**, 75.
45 E. Janssens, T. Van Hoof, N. Veldeman, S. Neukermans, M. Hou and P. Lievens, *Int. J. Mass Spectrom.*, 2006, **252**, 38.
46 J. B. Jaeger, T. D. Jaeger and M. A. Duncan, *J. Phys. Chem. A*, 2006, **110**, 9310.
47 W. Bouwen, P. Thoen, F. Vanhoutte, S. Bouckaert, F. Despa, H. Weidele, R. E. Silverans and P. Lievens, *Rev. Sci. Instrum.*, 2000, **71**, 54.
48 P. R. Vlasak, D. J. Beussman, M. R. Davenport and C. G. Enke, *Rev. Sci. Instrum.*, 1996, **67**, 68.
49 K. Hansen, *Philos. Mag. B*, 1999, **79**, 1413.
50 A. M. James, P. Kowalczyk, B. Simard, J. C. Pinegar and M. D. Morse, *J. Mol. Spectrosc.*, 1994, **168**, 248.
51 K. Hansen and U. Näher, *Phys. Rev. A*, 1999, **60**, 1240.
52 V. M. Strutinski, *Nucl. Phys. A*, 1968, **122**, 1.
53 K. Hansen and E. E. B. Campbell, *J. Chem. Phys.*, 1996, **104**, 5012.
54 D. A. Dahl, *Int. J. Mass Spectrom.*, 2000, **200**, 3 (Scientific Instrument Services, Inc., Ringoes, NJ, http://www.simion.com).
55 E. Janssens, S. Neukermans and P. Lievens, *Curr. Opin. Solid State Mater. Sci.*, 2004, **8**, 185.

Laser synthesis of bimetallic nanoalloys in the vapor and liquid phases and the magnetic properties of PdM and PtM nanoparticles (M = Fe, Co and Ni)

Victor Abdelsayed,[a] Garry Glaspell,[a] Minh Nguyen,[a] James M. Howe[b] and M. Samy El-Shall*[a]

Received 23rd April 2007, Accepted 11th May 2007
First published as an Advance Article on the web 8th November 2007
DOI: 10.1039/b706067j

In this work, we present several examples of the synthesis and characterization of bimetallic nanoparticle alloys using the Laser Vaporization Controlled Condensation (LVCC) method. In the first example, the vapor phase synthesis of Au–Ag, Au–Pd, and Au–Pt nanoparticle alloys are presented. The formation of nanoalloys is concluded from the observation of one plasmon absorption band at a wavelength that varies linearly with the gold mole fraction in the nanoalloy. Both XRD data and HRTEM-EDX data confirm the formation of nanoparticle alloys and not simply mixtures of the two metal nanoparticles. Irradiation of a mixture of Au/Ag nanoparticles dispersed in water with the 532 nm unfocused laser results in efficient alloying while the 1064 nm laser radiation results only in evaporation and size reduction of the unalloyed nanoparticles. Selective absorption of the femtosecond 780 nm radiation by large Au aggregates results in the formation of smaller aggregates with fractal structures, and no evidence for the Au–Ag alloy formation. The synthesis of palladium and platinum nanoparticles alloyed with transition metals such as iron and nickel using the LVCC method is also presented. The alloyed nanoparticles (FePd, FePt, NiPd, NiPt, and FeNi) are found to be superparamagnetic.

Introduction

The dependence of the properties of nanoscale materials on the size, shape and composition of the nanocrystal is a phenomenon of both fundamental scientific interest and many practical and technological applications.[1–5] These properties are often different, and sometimes superior, to those of the corresponding bulk materials. The origins of the differences can be explained by the high surface-to-volume ratio, dispersion factors and the quantum size effects unique to a specific length scale. Various effects that influence the new properties include the emergence of electronic and/or atom-packing shell structures, along with fundamentally altered interactions among the nanocrystals.[1–5] The characterization of the unique properties of nanocrystals can ultimately lead to identifying many potential uses and

[a] Department of Chemistry, Virginia Commonwealth University, Richmond, VA 23284-2006 USA. E-mail: mselshal@vcu.edu
[b] Department of Materials Science and Engineering, University of Virginia, Charlottesville, VA 22904 USA

applications, ranging from catalysis, ceramics, microelectronics, sensors, pigments, and magnetic storage to drug delivery and biomedical applications.[1–5] The applications of nanoparticles are thus expected to enhance many fields of advanced technology particularly in the areas of catalysis, chemical and biological sensors, optoelectronics, drug delivery, and media storage.

Recently, bimetallic alloy nanoparticles have gained much interest due to the additional new properties that arise from the combination of different compositions of metals on the nanoscale.[1–5] At a fundamental level, information on the evolution of electronic structures of bimetallic nanoparticles as a function of size and composition, and their effects on the optical absorption spectra continue to be a major goal of research in nanostructured materials. On a practical level, the unique optical properties of metallic and bimetallic nanoparticles are exploited for a variety of applications including optical markers for biomolecules, biological sensors, optical filters, surface enhancement in Raman spectroscopy and ultrafast nonlinear optical devices.

The presence of a surface plasmon resonance (SPR) band in the visible region of the absorption spectrum of noble metallic nanoparticles was recognized a long time ago and its origin is attributed to the collective oscillation of the free conduction electrons induced by an interacting electromagnetic field.[6–8] The oscillation frequency is determined by the metal electron density, the effective electron mass, and the shape and size of the charge distribution.[7] As the particle size becomes smaller than the mean free path of the free electrons, the plasmon band broadens until it disappears. For example, gold particles less than 1 nm in diameter had no plasmon absorption band. Nanocomposites, *i.e.* nanoalloys and core-shell nanoparticles are expected to exhibit different SPR characteristics. Indeed, core-shell nanoparticles of gold–silver exhibit two distinct SPR bands.[8] Nevertheless, gold–silver nanoalloys exhibit one single plasmon band and its absorption wavelength depends on the alloy composition.[9]

Gold and silver nanoparticles have been intensively studied because their SPR frequencies lie in the visible region of the electromagnetic radiation, usually centered around ∼520–540 nm and ∼400–420 nm, respectively.[6–17] Much of the interest in studying gold and silver nanoparticles as well as their alloys is due to a variety of biomedical applications.[10–16] Because of the efficient conversion of the optical excitation energies of the plasmon oscillation into heat, the excited gold and silver nanoparticles can generate a tremendous amount of heat following the plasmon absorption. These hot nanoparticles have been proposed for critical applications in medicine including, for example, the selective destruction of cancer cells following the irradiation of target cells. Silver nanoparticles have been used as antibacterial agents or bacteria sensors.[17] The alloying of gold and silver nanoparticles to form complexes with DNA ligands may promise new treatments for cell diseases and for cancer diagnosis and treatment.[10,13]

Several synthesis methods have been developed for the design of bimetallic nanocomposites (*i.e.*, alloy and/or core-shell structure) of gold–silver nanostructures.[6,8,18] Gold–silver alloy nanoparticles have been prepared chemically by co-reduction of silver and gold salts in a one-phase liquid system. For example, El-Sayed and co-workers have prepared Ag–Au nanoparticles in water by using sodium citrate as a reducing agent.[9] Lee and co-workers prepared the alloy nanoparticles in chloroform by using sodium borohydride ($NaBH_4$).[19] Monodispersed alloy nanoparticles have also been prepared in a two-phase liquid system. For example, He and co-workers prepared alloy nanoparticles passivated with alkyl thiol in a chloroform–water two liquid phase system.[20] Similar work by Kim and co-workers has been reported using dodecanethiol as a stabilizer.[21] A water-in-oil microemulsion method had been reported by Chen *et al.*, where hydrazine was used to co-reduce Au and Ag salts.[22] In addition to the chemical reduction methods, Papavassliou prepared colloidal Au–Ag alloy nanoparticles by applying an electric discharge between Ag–Au alloy and Pt electrodes immersed in 2-butanol.[23] More recently, laser ablation methods have been developed to prepare bimetallic and monometallic nanoparticles in solution. For example, Kimura and co-workers have prepared Au colloid in 2-propanol by the gas–flow solution–trap method.[24] Kondow and co-workers prepared Au and Ag nanoparticles by using pulsed laser ablation ($\lambda = 1064$ nm) of a gold (or a silver) plate in aqueous solution of the sodium dodecyl sulfate (SDS) surfactant.[25] Koda and co-workers, prepared Au nanoparticles by chemical reduction in water and then irradiating the solution by the second harmonic of a pulsed Nd:YAG laser (532 nm).[26] Hartland and co-workers studied the laser-induced interdiffusion of a core-shell structure to alloy nanoparticles in aqueous solution.[27] Zhang et al. prepared Ag–Au alloy nanoparticles by irradiating a metal powder mixture in an aqueous medium with the second harmonic of a pulsed Nd:YAG laser (532 nm).[28] Chen et al. have also prepared alloy nanoparticles by irradiating a colloidal mixture of pure Au and Ag by laser irradiation ($\lambda = 532$ nm).[29]

Most of the previous methods used to prepare bimetallic alloys between gold and silver were carried out in liquid media. In this work, we present the vapor phase synthesis of gold–silver alloy nanoparticles using the Laser–Vaporization Controlled–Condensation method (LVCC) method.[30–35] We also extend the application of this method to other bimetallic alloy nanoparticles such as gold–palladium and gold–platinum. The advantages of the vapor phase synthesis are the contamination-free products (as compared to chemical reductions in solutions), the elimination of the chemical precursors and solvents, and in most cases, the production of highly crystalline nanoparticles. Furthermore, by coupling the LVCC method with a differential mobility analyzer (DMA), size-selected nanoparticles can be prepared from the vapor phase.[35,36] The control of both particle size and composition are the major factors that determine the characteristic properties of alloy nanoparticles that are critical for optical devices and catalytic performance.

In this paper we also investigate the effects of laser irradiation in solution on the size, shape and composition of the alloyed nanoparticles. We also examine the extent of alloying of Au and Ag nanoparticles suspended in water by laser irradiation using different photon energies. We then introduce a fast and simple method of preparing monodispersed gold–silver and gold–palladium alloy nanoparticles stabilized by surface coatings with oleic acid–oleylamine mixtures. Finally, we present the vapor phase synthesis of the alloyed nanoparticles of PdM and PtM (where M = Co, Fe, and Ni) using the LVCC method.

Experimental

The nanoparticles were prepared using the LVCC method as described previously in several references.[30–36] Here, we only provide the necessary information relevant to the preparation of the alloy nanoparticles. A sketch of the chamber with the relevant components for the production of nanoparticles is shown in Fig. 1. The chamber consists of two horizontal, circular stainless steel plates, separated by a glass ring. A metal target of interest is set on the lower plate, and the chamber is filled with a pure carrier gas such as Ar (99.99% pure). The metal target and the lower plate are maintained at a temperature higher than that of the upper one. The top plate can be cooled to less than 150 K by circulating liquid nitrogen. The large temperature gradient between the bottom and top plates results in a steady convection current which can be enhanced by using a heavy carrier gas such as Ar under high pressure conditions (10^3 Torr). The metal vapor is generated by pulsed laser vaporization using the second harmonic (532 nm) of a Nd:YAG laser (15–30 mJ pulse^{-1}, 10^{-8} s pulse). The role of convection in the experiments is to remove the small particles away from the nucleation zone (once condensed out of the vapor phase) before they can grow into larger particles. The rate of convection increases with the temperature gradient in the chamber. Therefore, by controlling the temperature gradient, the total pressure and the laser power (which determines the number density of the metal

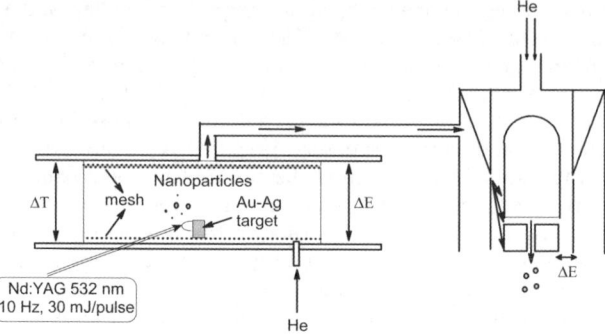

Fig. 1 Experimental set-up for the Laser–Vaporization Controlled–Condensation method (LVCC) coupled with a Differential Mobility Analyzer (DMA) for size selection.

atoms released in the vapor phase), it is possible to control the size of the condensing particles. No particles are found anywhere else in the chamber except on the top plate; this supports the assumption that nucleation takes place in the upper half of the chamber and that convection carries the particles to the top plate where deposition occurs. The charged nanoparticles produced through laser vaporization can be classified based on their electrical mobility in a dilute inert gas flow using a differential mobility analyzer (DMA).[36,37] To prepare size-selected nanoparticles and measure the size distribution produced under specific experimental conditions, the LVCC chamber is coupled to a DMA as shown in Fig. 1.

For the synthesis of bimetallic nanoparticles, a mixture of the appropriate elemental powder (micron size particles) was prepared in a specific molar ratio using a mortar and a pestle. The mixture was then pressed at 500 MPa, using a hydraulic press, in order to shape it into a cylindrical disk target. The pure metal nanoparticles were prepared as well, from bulk metals as reference materials, under the same experimental conditions.

Scanning electron microscopy (SEM) and energy dispersive X-ray spectroscopy (EDX) measurements were performed using a Quantum DS-130S Dual Stage Electron Microscope. A carbon substrate was placed inside the LVCC chamber (on the top plate) to observe the size and morphology of the as-deposited nano-particles under SEM. Additional transmission electron microscopy (TEM) analyses were performed using a JEOL 2010F field-emission gun operating at 200 kV and having an Oxford ultra-thin window EDX detector. The EDX spectra were analyzed using the NIST Desk-Top Spectrum Analyzer (DTSA) program to determine the counts in the Ag Kα, Ag Lα, Au Lα and Au Mα peaks, which were then input into the Thin Film Analysis (TFA) program to determine the compositions, using *k*-factors generated by the TFA software. During the acquisition of EDXS data averaged over many (order of thousands) particles, the beam was spread out at ∼ 50 k × magnification. The electron beam was focused to approximately 1 nm nominal diameter at high magnification (∼ 500 k ×) to obtain EDXS spectra from a single particle or a small region of the particle. The TEM specimens were prepared by placing a drop of aqueous dispersed nanoparticles on a thin carbon film supported on a copper grid, and then dried in a desiccator. Lattice parameter/composition measurements of the nanoparticles were performed using X-ray powder diffraction (XRD) on an X'Pert Philips Materials Research Diffractometer with Cu Kα radiation.

The elemental analyses were performed on an Inductively Coupled Plasma-Optical Emission Spectrometer (ICP-OES, Varian VISTA-MPX) to measure the amount of gold in both the starting targets and in the nanoparticles. For example, for the metal powder mixtures containing 51.8, 35.6 and 21.2 mol% Au, the nanoparticles were found to contain 47.3, 28.7 and 16.6 mol% Au, respectively.

This can be explained by the higher evaporation rate of Ag over Au. A similar effect was observed by Liu and co-workers in the synthesis of Fe–Al nanoparticles by a hydrogen plasma–metal reaction from a bulk alloy where the content of Al in nanoparticles was found to be higher than that in the bulk alloy.[38]

The UV-vis absorption spectra were obtained using a Hewlett-Packard HP 8453 diode array spectrometer. Colloidal solutions were prepared by dispersing the Ag–Au nanoparticles in triply deionized water (18 MΩ) using air ultra sound waves. In the alloying experiments in water, a mixture of nanoparticles with the desired composition was suspended in water and irradiated under agitation using a magnetic stirrer. The particles were irradiated using unfocused light either from the fundamental of a Nd:YAG pulsed laser (1064 nm, (6.0–7.6) $\times$ 10^7 W cm^{-2}, 5 ns, 10 Hz), the second harmonic of a Nd:YAG pulsed laser (532 nm, (4.3–5.4) $\times$ 10^7 W cm^{-2}, 5 ns, 10 Hz), or the femtosecond pulses from a titanium:sapphire laser (2 nJ pulse^{-1}, 150 fs, 100 MHz).

Temperature and magnetic field variations of the magnetization (M) of the PdM and PtM alloy nanoparticle samples were measured using a commercial superconducting quantum interference device magnetometer (SQUID, Quantum Design) with a remnant field of 7 Oe.

Results and discussion

I. Laser alloying of nanoparticles in the vapor phase

The DMA size distributions of the Au, Ag and Au–Ag alloy nanoparticles generated by the 532 nm laser vaporization are shown in Fig. 2. The average particle size for Ag (12 nm) appears to be significantly smaller than that of Au (20 nm) and of the Au–Ag alloy (22 nm). The stoichiometric coefficients given for the alloy nanoparticles represent the atomic ratio of the gold and silver in the nanoparticles as determined from the ICP analysis. The particles generated by 1064 nm have a smaller size compared to those generated by 532 nm. However, in all cases the size distributions exhibit significant broadening and tailing, characteristic of aggregated nanoparticles grown from the vapor phase.

Fig. 3 displays TEM images of the 25 nm selected Au–Ag and alloy nanoparticles with the Ag$_{0.75}$Au$_{0.25}$ and Ag$_{0.47}$Au$_{0.53}$ compositions prepared using the LVCC-DMA system. Although the nanoparticles are still aggregated, the primary particles appear to exhibit identical sizes. Individual monodisperse particles can be deposited only if the number density of the particles is kept very low to avoid the aggregation of the particles.

The UV-visible absorption spectra of dispersed Au, Ag and Au–Ag alloy nanoparticles in water are shown in Fig. 4. A single plasmon band at $\sim$533 and $\sim$420 nm was observed for the Au and Ag nanoparticles, respectively. The

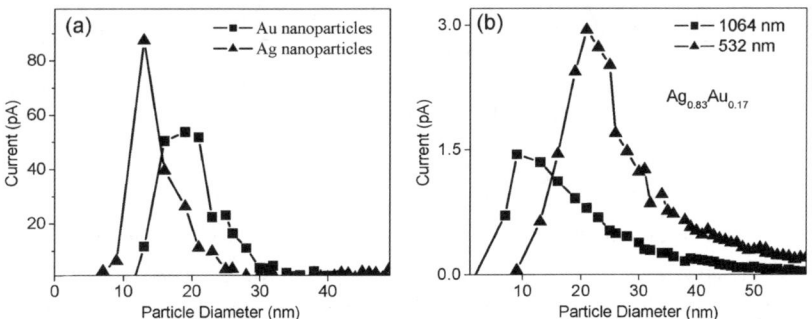

Fig. 2 (a) Size distributions of Au and Ag nanoparticles and (b) Ag$_{0.83}$Au$_{0.17}$ alloy nanoparticles prepared by laser vaporization of bulk targets using 532 and 1064 nm lasers.

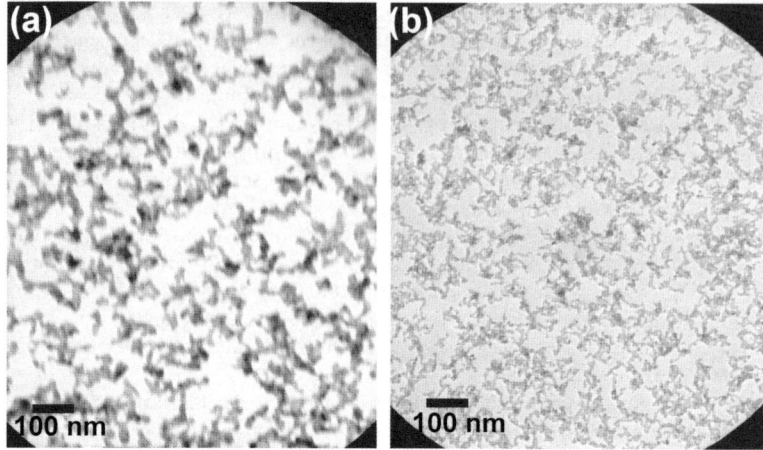

Fig. 3 TEM images of size-selected (a) $Ag_{0.75}Au_{0.25}$ and (b) $Ag_{0.47}Au_{0.53}$ alloy nanoparticles.

broadening in the absorption plasmon bands of Au and Ag observed in Fig. 4 is attributed to the broad particle size distributions shown in Fig. 2 and to the aggregation of the nanoparticles, where small particles have a high tendency to aggregate faster than big particles due to their high surface energy. Therefore, the SPR bands are inhomogeneously broadened by the different particle sizes and aggregates of different sizes and shapes.

In Fig. 4-a, the absorption spectrum of the alloy $Ag_{0.53}Au_{0.47}$ nanoparticles exhibits a distinct peak with a maximum observed at 487 nm. This plasmon peak is located at an intermediate position between the Au and the Ag surface plasmon bands. The single plasmon band implies that the particles are spherical rather than rods or triangles, which would have two or three plasmon peaks, respectively.[9,39,40] It also implies the formation of an alloy between Au and Ag nanoparticles rather than of a mixture, which would simply exhibit two distinct plasmon bands corresponding to Au and Ag. Fig. 4-c displays the UV-visible spectra of the alloy

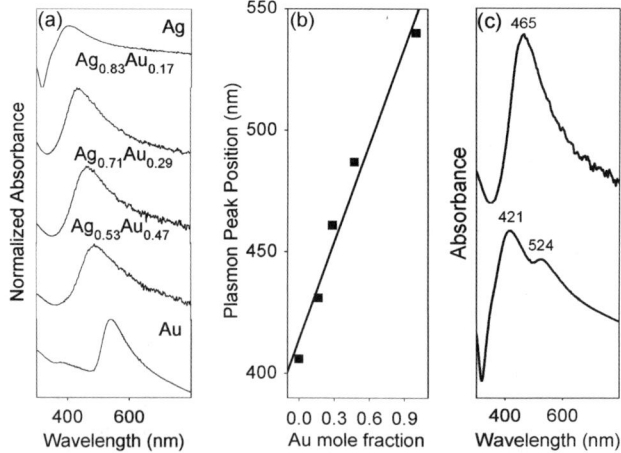

Fig. 4 (a) UV-visible absorption spectra of Ag–Au nanoparticles prepared by the LVCC method, (b) Plasmon peak position as a function of the Au content in the alloy nanoparticles and (c) UV-visible absorption spectra of $Ag_{0.71}Au_{0.29}$ alloy nanoparticles (upper graph) and a physical mixture of Au and Ag nanoparticles with the same molar ratio found in the alloy (lower graph).

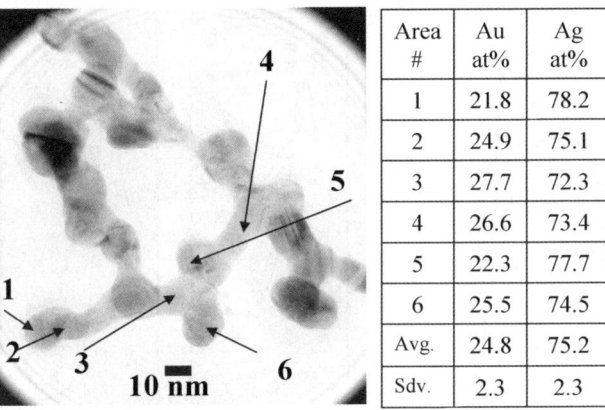

Area #	Au at%	Ag at%
1	21.8	78.2
2	24.9	75.1
3	27.7	72.3
4	26.6	73.4
5	22.3	77.7
6	25.5	74.5
Avg.	24.8	75.2
Sdv.	2.3	2.3

Fig. 5 High-resolution TEM image and EDX composition of the $Ag_{0.71}Au_{0.29}$ alloy nanoparticles measured in different regions of the aggregate.

and the mixture having the same molar ratio of Ag (71)/Au (29). The mixture had two distinct peaks at 421 and 524 nm corresponding to the absorption plasmon of the silver and the gold nanoparticles, respectively, while the alloy nanoparticles show only a single plasmon peak at 465 nm. The plasmon peak depends on the composition of the alloy prepared. It shifts linearly to higher energy with increased silver content in the nanocomposite alloy, as shown in Fig. 4-b. The linear relationship between silver content in gold–silver alloy and its surface plasmon energy has been extensively observed for many alloys prepared by different chemical reduction routes in liquids; for example, in single phase systems,[9,41] in binary phase systems,[20,21] in microemulsion systems,[22] and also by radiation chemistry[42] and laser ablation in solution.[28,43]

A high-resolution TEM micrograph for the $Ag_{0.71}Au_{0.29}$ alloy nanoparticles prepared by the LVCC method is shown in Fig. 5. Structural defects such as stacking faults and twins are visible within the particles. This can be explained by the different growth rates of silver on various planes of gold particles as well as by the anisotropy of the surface energy, which favors low-index {111} and {200} facets.[44] Stacking faults and twins are often found in nanoparticles of pure metals and metallic alloys, including Ag and Au.[45]

Further support for the formation of homogeneous Au–Ag alloy nanoparticles has been provided by the EDX spectroscopy of individual nanoparticles. The electron beam was focused on a number of particles in the strands, as well as in the necked regions between them, and it was found that all regions consisted of a Au–Ag alloy, without evidence of coring as evident from the EDX data shown in Fig. 5. The average composition $Ag_{0.75}Au_{0.25}$ determined from the EDX analysis is in excellent agreement with the $Ag_{0.71}Au_{0.29}$ composition determined from the ICP analysis, thus confirming that the alloyed nanoparticles made by the LVCC method have a uniform composition.

II. Laser irradiation of the alloy nanoparticles dispersed in water

The gold and silver nanoparticles dispersed in water were irradiated with the fundamental (1064 nm) as well as the second harmonic (532 nm) of a pulsed Nd:YAG laser (power density 4.3–7.6 × 10^7 W cm^{-2}) for 20 min. The UV-visible absorption spectra, as well as the TEM micrographs, showed clearly that after irradiation with the 532 nm light, the particle sizes have decreased and nearly monodispersed particles were formed, while the non-irradiated particles showed a broad size distribution with significant amounts of nonspherical particles. For the Au nanoparticles, the plasmon peak was shifted from ~533 to ~520 nm. Similar

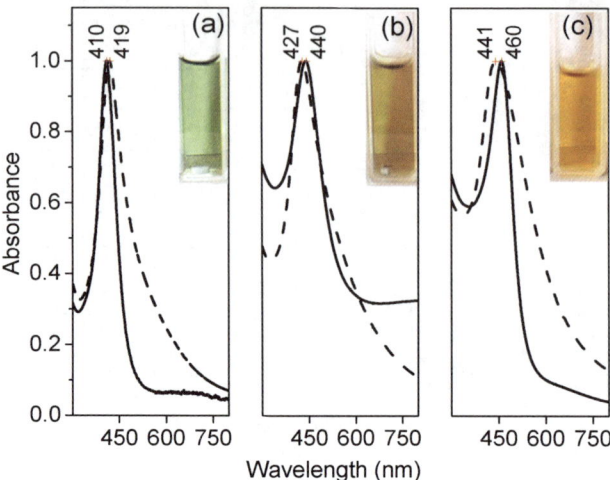

Fig. 6 UV-visible spectra of (a) $Ag_{0.83}Au_{0.17}$, (b) $Ag_{0.71}Au_{0.29}$ and (c) $Ag_{0.53}Au_{0.47}$ after irradiation with the 532 nm (solid line) and the 1064 nm (dotted line) lasers for 20 min.

results have been reported by Koda and co-workers where the 532 nm irradiation of a chemically-prepared gold colloid resulted in a significant reduction in the average particle size.[26] On the other hand, the decrease in the average particle size was less pronounced when the 1064 nm laser beam was used. Accordingly, the plasmon band was slightly blue shifted from $\sim$533 to 528 nm. The 532 nm radiation is more effective in reducing the particle size due to the strong plasmon absorption by gold which results in both melting and vaporization, whereas in the case of 1064 nm radiation only vaporization takes place. The melting of the nanoparticles leads to the shape change from nonspherical to spherical particles and the size reduction is due to the vaporization of the particles.

A comparison between the absorption spectra for the $Ag_{0.83}Au_{0.17}$, $Ag_{0.71}Au_{0.29}$ and $Ag_{0.53}Au_{0.47}$ nanoparticles irradiated with 532 nm and 1064 nm is shown in Fig. 6. The gradual change of color with changing Au–Ag compositions is clear in the photographs of the alloy solutions shown in Fig. 6. The surface plasmon bands for $Ag_{0.83}Au_{0.17}$, $Ag_{0.71}Au_{0.29}$ and $Ag_{0.53}Au_{0.47}$ nanoparticles were blue shifted from 431, 465 and 487 nm to 410, 440 and 460 nm, respectively after 20 min irradiation with the 532 nm laser. On the other hand, the plasmon bands were shifted to 419, 427 and 441 nm, respectively after 20 min irradiation with 1064 nm. The broadening in the UV-visible absorption spectra for Ag–Au particles decreased after irradiation, indicating a decrease in the size reduction of the nanoparticles which could lead to a narrow size distribution.

The effect of irradiation wavelength on the alloy nanoparticles dispersed in water was also examined using TEM. The TEM micrographs for $Ag_{0.71}Au_{0.29}$ nanoparticles are shown in Fig. 7 with and without irradiation in water medium. The as-prepared particles from the LVCC were nonspherical with facets and a broad size distribution. After being irradiated with 532 nm for 20 min (12 000 laser pulses) in water, size reduction was observed from the TEM results, where melting, reshaping, and vaporization took place to yield almost monodispersed spherical particles, as shown in Fig. 7-b. However, with the 1064 nm irradiation, only a small size reduction occurred in addition to reshaping to spherical particles as shown in Fig. 7-c. It was also observed that a broad size distribution still remained, even after the 1064 nm irradiation.

Several other bimetallic nanoparticles such as Au_xPd_{1-x}, Au_xPt_{1-x}, Pd_xCu_{1-x}, Au_xCu_{1-x} and Pt_xCu_{1-x}, with controlled compositions have been prepared by the

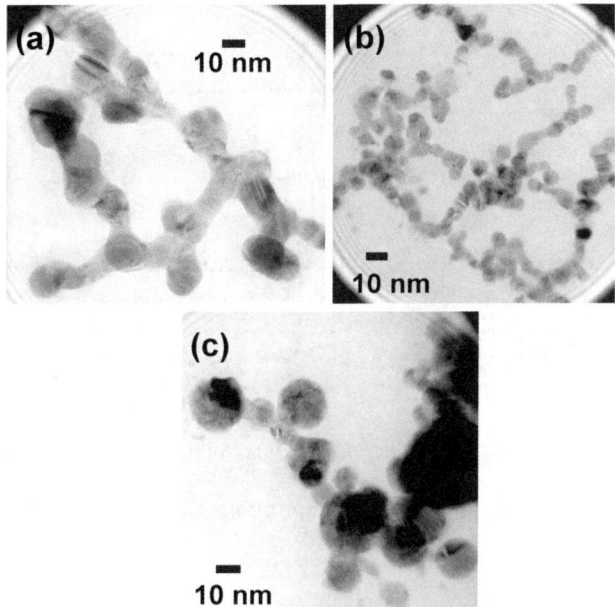

Fig. 7 TEM images of the $Ag_{0.71}Au_{0.29}$ nanoparticles (a) as-prepared, (b) after 20 min irradiation with 532 nm, and (c) after 20 min irradiation with 1064 nm. Scale bar = 10 nm.

LVCC method from mixed metal targets and the elemental compositions of the resulting alloy nanoparticles have been determined using the ICP technique. Fig. 8 and 9 display typical examples of the TEM images and the XRD patterns of the AuPd and AuPt alloy nanoparticles. The XRD of the alloy Au–Pd nanoparticles, shown in Fig. 8-b, matches well with the diffraction pattern for the AuPd alloy.[46] The observed peaks at 38.94, 45.33 and 66.19 (2θ), which correspond to reflections from the 111, 200, 220 planes, respectively, are significantly shifted from those of pure Au and Pd as shown in Fig. 8-b.

It is also significant to note that while the XRD of the LVCC target indicates a physical mixture of Au and Pd, the XRD diffraction pattern of the Au–Pd nanoparticles confirms the formation of the AuPd alloy in the vapor phase without any indication of the presence of its starting materials. Similar results were obtained for the AuPt alloy nanoparticles as shown in Fig. 9. These results demonstrate that

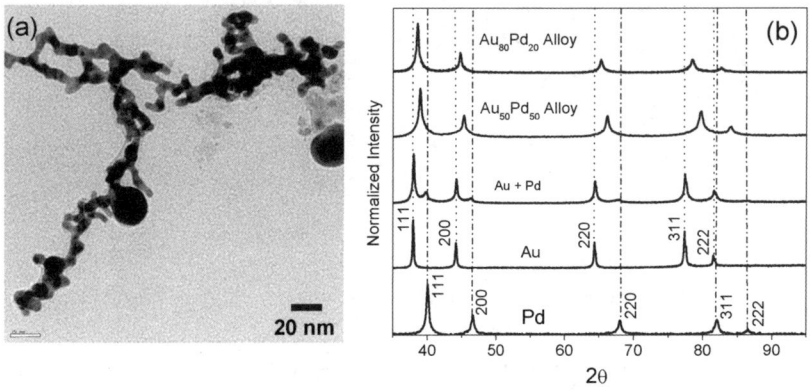

Fig. 8 (a) TEM image and (b) XRD the AuPd alloy nanoparticles.

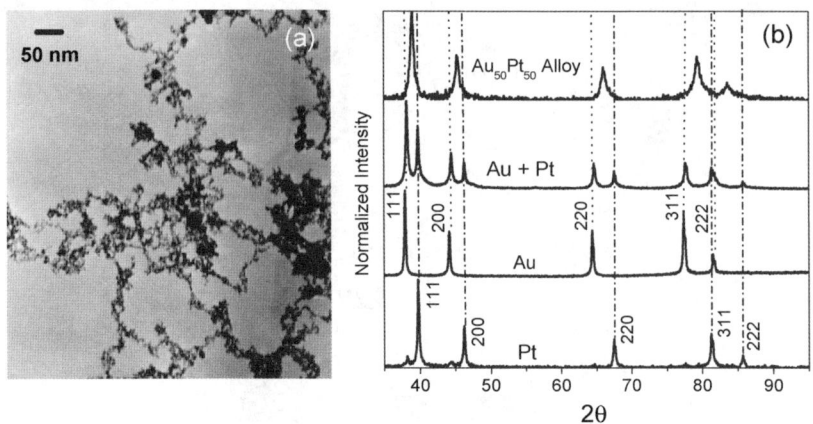

Fig. 9 (a) TEM image and (b) XRD the AuPt alloy nanoparticles.

alloy nanoparticles with well-defined compositions can be prepared by the LVCC method.

III. Laser alloying of nanoparticles in solution and hydrophobic passivation

In these experiments, Au and Ag nanoparticles were prepared separately under the same experimental conditions using the LVCC method. A mixture of Au and Ag nanoparticles corresponding to the molar composition of 52% Au–48% Ag was dispersed in water. Fig. 10-a shows the optical absorption spectra of the Au–Ag mixture before and after irradiation with the 532 nm laser. The absorption spectrum of the mixture before irradiation (Fig. 10-a (i)) is basically the sum of the absorption spectra of pure Au and pure Ag nanoparticles with the corresponding plasmon peaks at ~ 533 nm and ~ 429 nm, respectively. This confirms that both the Au and Ag nanoparticles are dispersed in water and, as expected no alloying takes place by simply mixing the Au and Ag nanoparticles in water. However, after 10 min irradiation with the 532 nm laser, the two plasmon peaks start to merge (Fig. 10-a

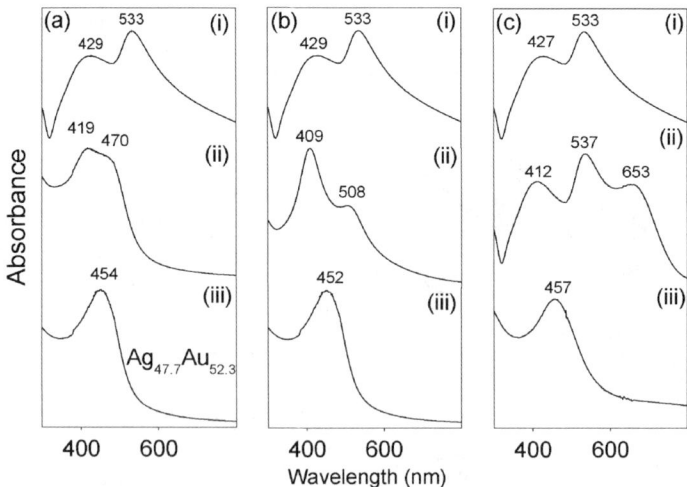

Fig. 10 UV-visible spectra of a mixture of Au (52%) and Ag (48%) nanoparticles dispersed in water obtained before and after irradiation with (a) 532 nm, (b) 1064 nm and (c) 780 nm lasers (see text for details).

(ii)) and with increasing the irradiation time to 40 min only one plasmon peak at ~454 nm is observed as shown in Fig. 10-c (iii). This indicates that the 532 nm irradiation results in alloying of the dispersed Au and Ag nanoparticles in water. This is consistent with the strong absorption of the 532 nm light with the Au nanoparticles which can lead to very rapid heating, melting and evaporation of the Au nanoparticles. The 1064 nm irradiation (20 min) of the mixture results in a significant decrease in the particle size as evident from the shift in the plasmon bands of the Au and Ag nanoparticles from ~533 nm and ~429 nm to ~508 nm and ~409 nm, respectively as shown in Fig. 10-b (ii). However, the irradiation of the resulting small Au and Ag nanoparticles with the 532 nm (20 min) clearly results in alloying the reduced size Au and Ag nanoparticles as evident from the appearance of a single plasmon band at ~452 nm as shown in Fig. 10-b (iii).

The results shown in Fig. 10 are consistent with the previous work on the laser synthesis of Au–Ag alloy nanoparticles using the 532 nm irradiation.[28,29] However, the irradiation of the Au–Ag nanoparticle mixture with the 780 nm femtosecond pulses appears to produce a new absorption band in the ~660 nm region. This band is attributed to the formation of small aggregates of Au nanoparticles with probably some elongated shapes. These small aggregates are formed as a result of the excitation of larger aggregates by the 780 nm femtosecond pulses which results in both size and shape changes in the large aggregates to produce the smaller aggregates characterized by the ~660 nm absorption. It should be pointed out that Au nanostructures that deviate from spherical shapes such as Au nanorods and Au nanoshells also exhibit absorption bands red shifted from the SPR of the spherical Au nanoparticles.

The absorption of strongly interacting Au nanoparticles within large aggregates in the 700–950 nm region has also been reported by Zhang and co-workers who used femtosecond spectroscopy to demonstrate the ultrafast electronic relaxation and coherent vibrational oscillation of these aggregates.[47] Using the hole burning technique, Zhang and co-workers showed that selective excitation of certain size and shape aggregates using 800 nm irradiation results in altering both the size and shape of these aggregates to produce smaller aggregates that absorb more in the blue region around 700 nm.[47] In our case, the smaller aggregates produced following the 780 nm irradiation exhibit absorption maximum near 653 nm as shown in Fig. 10-c (ii). These aggregates are clearly identified as shown in the TEM images of the Au–Ag nanoparticle mixture after laser irradiation with the 780 nm femtosecond pulses as shown in Fig. 11. Interestingly, irradiation of these small aggregates with the

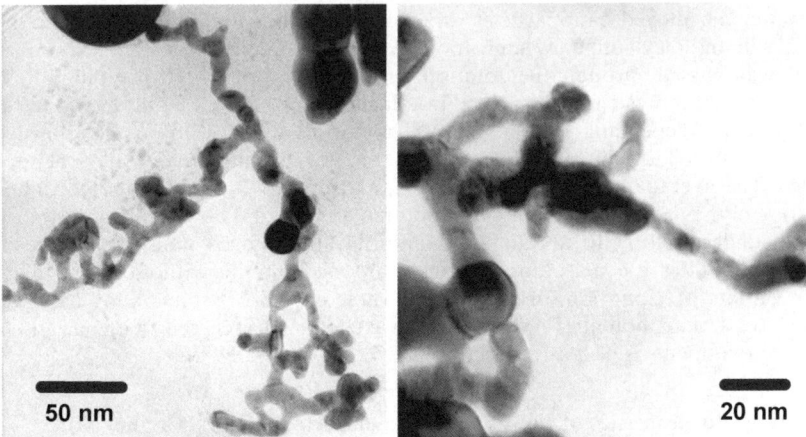

50 nm

20 nm

Fig. 11 TEM images of the Au–Ag aggregates obtained following the irradiation of the $Ag_{47.7}Au_{52.3}$ nanoparticle mixture with the 780 nm femtosecond pulses for 20 min.

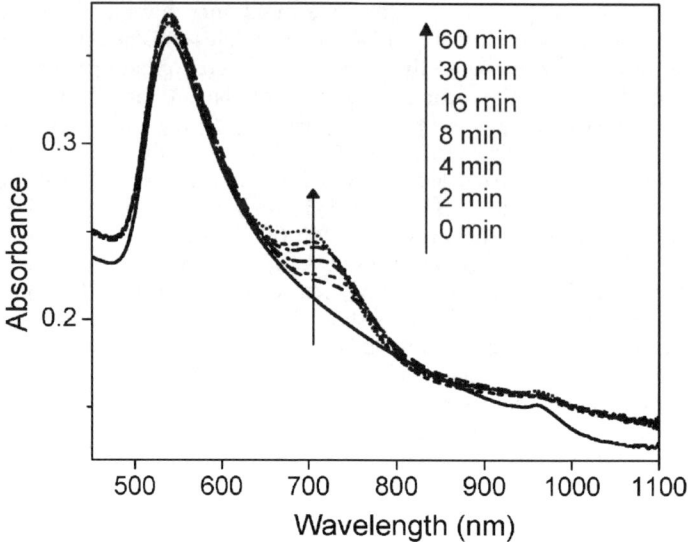

Fig. 12 UV-visible spectra of Au nanoparticles dispersed in water after irradiation with the 780 nm femtosecond pulses for different periods of time.

532 nm laser appears to destroy the aggregates and produce individual Au nanoparticles which are subsequently alloyed with the Ag nanoparticles within the nanosecond laser pulses. This result is clearly shown in Fig. 10-c (iii) which indicates the formation of a single plasmon band corresponding to the Au–Ag alloy following the 532 nm irradiation pulses.

In order to confirm the presence of the Au nanoparticle aggregates we measured the electronic absorption of the dispersed Au nanoparticles in water in the red region of the spectrum as shown in Fig. 12. In addition to the SPR band at ~530 nm, the spectrum shows a broad near-IR absorption band around 970 nm which is assigned to the large Au aggregates in agreement with the results of Zhang and co-workers.[47] The spectra recorded after irradiation of the solution containing the Au aggregates with the 780 nm femtosecond pulses for different periods of time clearly show the growth and blue shift of the maximum absorbance of the aggregates as shown in Fig. 12.

Following the alloying of the Au–Ag nanoparticles by the 532 nm laser irradiation in water, the alloyed nanoparticles were suspended in a 50 : 50 mixture containing oleic acid and oleylamine *via* sonication. While stirring, the unfocused 532 nm laser light was passed through the solution with a power of 30–40 mJ pulse^{-1}. The resulting solution was then dispersed in toluene which showed a purple–red to faint yellow color depending on the Au–Ag composition in the alloy nanoparticles. Fig. 13 shows TEM images of the coated Au, Au–Ag and Au–Pd alloy nanoparticles. The average size of the coated nanoparticles is 3–4 nm. The hydrophobic coating of the nanoparticles containing both oleic acid and oleylamine appears to be stable under heating to about 90 °C and also the alloyed nanoparticles remain suspended in the toluene solution for up to at least 3 months without any significant aggregation, precipitation, or separation. These coated bimetallic alloy nanoparticles have several potential applications in bacteria sensors, cell imaging, and cell disease treatment as well as in electronic devices.

IV. Magnetic properties of PdM and PtM nanoparticles (M = Co, Fe, Ni)

The PdM and PtM alloy nanoparticles represent rich systems with several important applications in catalysis and magnetic materials.[48–64] Both FePd and FePt are ideal

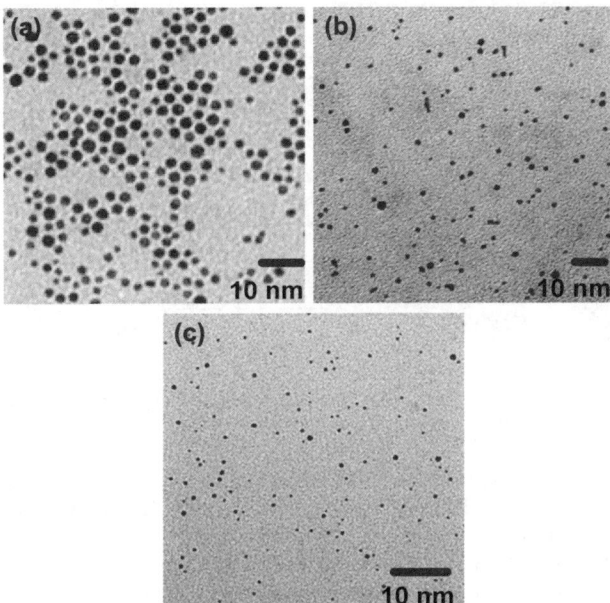

Fig. 13 TEM images of (a) Au, (b) Au–Ag and (c) Au–Pd alloy nanoparticles coated with a mixture of oleic acid–oleylamine.

materials for permanent magnetic applications because of their large uniaxial magnetocrystalline anisotropy and good chemical stability.[52,62–64] Magnetic enhancement has been expected in Pd nanoparticle alloys since Pd has a giant magnetic moment in the presence of ferromagnetic transition metals.[53] The giant moment was observed by Nunomura *et al.* in PdNi nanoparticles where the magnetization shows a significant increase when the Ni concentration exceeds 8%, where each Ni in the particle makes a ferromagnetic polarized region around itself.[54] Hori *et al.* have also observed the giant moment ferromagnetism in PdNi nanoparticles.[55] Choi *et al.* have shown that Pt has significant magnetization in various environments, including the PtNi system. Hou *et al.* have reported that FePd nanoparticles (16 nm) are also superparamagnetic, similar to PdNi and PtNi, and display a coercivity of 350 Oe at 2 K.[52] Chen *et al.* have also investigated the FePd system and have shown that the coercivity is affected by the annealing temperature with a maximum value observed at 550 °C.[57] This trend was also observed by Kang *et al.*[58] The FePt system also shows tunable control over the coercivity by controlling the annealing temperature, annealing time, and the ratio of Fe present.[59,62–64] Recently, the synthesis of FePt nanoparticles exhibiting high room temperature coercivity has been reported using a single-source molecular precursor.[64] In the present work, we demonstrate the synthesis of PtM and PdM alloy nanoparticles, where M = Co, Ni and Fe, from the corresponding elemental metallic powders using the LVCC method.

The optimum compositions of the vaporizing targets to produce the desired alloy nanoparticles were found to be 50 : 50 (PdCo), 50 : 50 (PtCo), 33 : 66 (PdNi), 33 : 66 (PtNi), 25 : 75 (PdFe) and 25 : 75 (PtFe), weight percent. Deviations from these values resulted in the presence of the pure components with the alloy nanoparticles as observed from the XRD patterns.

Fig. 14-a shows a TEM image of alloyed FePt nanoparticles which indicates that the average particle size is ∼8 nm. The XRD pattern for the nanocrystalline sample prepared from (75 : 25 wt%) of Fe and Pt metallic powder mixture is shown in Fig. 14-b, along with the diffraction patterns of pure Fe and Pt micron size particles and their mixture used to prepare the vaporizing target for the LVCC experiment. The

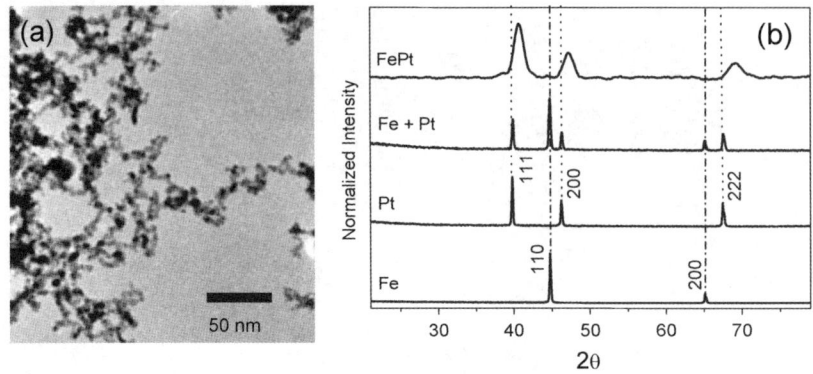

Fig. 14 (a) TEM image of FePt alloy nanoparticles and (b) XRD pattern of the FePt nanoparticle alloy (top) along with the patterns for Fe, Pt, and Fe + Pt bulk powders.

diffraction pattern of the prepared alloyed nanoparticles is in good agreement with the pattern of the bulk FePt alloy as obtained from the XRD data base (ICCD 00-029-0718). It is significant to note that the observed lines corresponding to reflections from the 111, 200, 220 planes from the FePt alloy nanoparticles are significantly shifted from those of pure Pt and Fe as shown in Fig. 14(b). Similarly, the XRD patterns of all the other PtM and PdM alloy nanoparticles are found to match the patterns of the corresponding bulk alloys. As examples, Fig. 15 displays the XRD patterns of the PdM alloyed nanoparticles prepared by the LVCC method.

It is significant to note that in all the XRD patterns of the six PtM and PdM alloyed nanoparticle samples prepared by the LVCC method, no evidence of the presence of the starting pure metals was found thus indicating that the target compositions used resulted in efficient alloying in the vapor phase.

The temperature variations of χ for both the field cooled (FC) and zero field cooled cases (ZFC) of the FePt nanoparticles are shown in Fig. 16-a. It is evident from the curves that the sample displays an overall superparamagnetic character. However, the width of the blocking temperature for the ZFC case is suggestive of a wide distribution of particle sizes. The convergence of the temperature variation of χ above the blocking temperature for the FC and ZFC data is characteristic of interacting particles. The temperature variation of χ for all the prepared samples also reveal a superparamagnetic character. The results of measuring the hysteresis loops for various temperatures are reported in Table 1, and an example of the

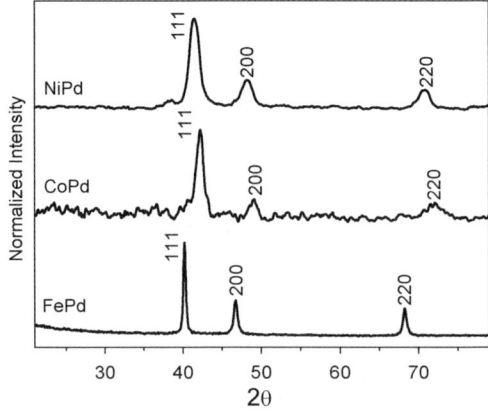

Fig. 15 XRD patterns of the FePd, CoPd and NiPd alloy nanoparticles prepared by the LVCC method.

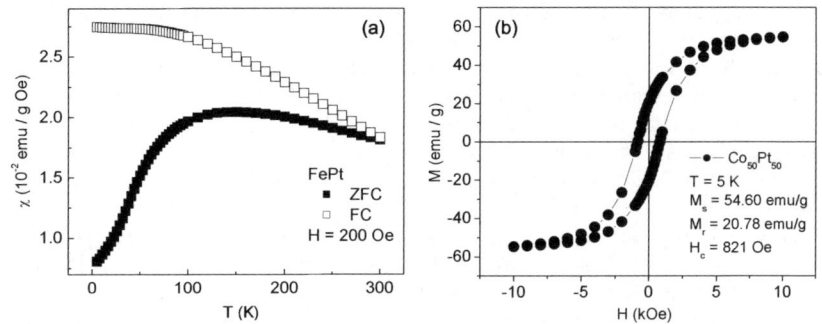

Fig. 16 (a) Temperature dependence of the magnetic susceptibility (χ) measured for the FePt alloy nanoparticles under field cooled (FC) and zero field cooled (ZFC) conditions, and (b) M vs. H variation measured at 5 K for the CoPt alloy nanoparticles.

hysteresis loops at 5 K and 300 K are reported in Table 1. The highly symmetrical shape of the loops indicates that other phases are not present in the sample.

The average magnetic size can be determined from the slope of the magnetization near zero with the major contributions arising from the largest particles. Using the equation:[35]

$$d_{\max} = \left[\frac{18 \, k \, T}{\pi} \frac{\frac{(dM)}{(dH)}}{\rho M_s^2} \right]^{\frac{1}{3}}$$

where k = Boltzamnn constant, T = temperature, dM/dH = the slope near zero field, M_s is the saturation magnetization and ρ = bulk density, an upper bound for the magnetic size can be estimated as shown in Table 1.

Table 1 Magnetic properties of the as-prepared (without thermal annealing) alloyed nano-particles

Sample	M_r/emu g^{-1}	H_c/Oe	M_s/emu g^{-1}	Magnetic size/nm	Critical temp./K
PdFe				1.8	403
T = 5 K	10.7	529	32.9		
T = 300 K	1.5	37	24.0		
PtFe				1.8	559
T = 5 K	8.5	661	27.2		
T = 300 K	1.4	55	20.9		
PdNi				1.5	806
T = 5 K	31.3	326	85.7		
T = 300 K	19.4	98	73.6		
PtNi				2.6	401
T = 5 K	5.1	614	11.8		
T = 300 K	1.6	74	7.8		
PdCo				1.4	480
T = 5 K	17.7	811	51.35		
T = 300 K	12.21	305	49.47		
PtCo				1.4	617
T = 5 K	20.8	821	54.6		
T = 300 K	14.3	384	51.9		

The small values obtained for the magnetic size in the prepared alloy nanoparticles indicate that the transition metals are well dispersed in the alloyed samples as a result of the uniform mixing in the vapor phase. The critical temperatures, saturation magnetization, remanence and coercivity data at 5 K and 300 K are reported in Table 1 for all the as-prepared alloyed samples without thermal annealing. Comparison of the remanence, coercivity at both 5 and 300 K and magnetic saturation reveals that the PdCo and PtCo alloy nanoparticles display the greatest ferromagnetic character of all the alloyed systems studied. Critical temperatures of 480 K and 617 K were found for the alloyed nanoparticles of PdCo and PtCo, respectively.

The results reported here clearly indicate that alloy nanoparticles of PdM and PtM (where M = Co, Fe, and Ni) can be synthesized in the vapor phase from mixtures of their bulk metal powders using the LVCC method. XRD analysis reveals that by carefully controlling the stoichiometry of the starting materials single phase alloys can be produced without any indication of the presence of the starting materials. Finally, SQUID measurements revealed that the nanoparticles are superparamagnetic.

Conclusions

Alloy nanoparticles of controlled size and composition can be prepared from the vapor phase using the LVCC method. The formation of Au–Ag alloy nanoparticles is concluded from the observation of only one plasmon band. The maximum of the plasmon absorption is found to vary linearly with the gold mole fraction. Irradiation of a mixture of Au/Ag nanoparticles dispersed in water with the 532 nm unfocused laser results in efficient alloying while the 1064 nm laser radiation results only in evaporation and size reduction of the nanoparticles. Selective absorption of the femtosecond 780 nm radiation by large Au aggregates results in the formation of smaller aggregates with fractal structures with no evidence of Au–Ag alloy formation. For the Au–Pd and Au–Pt systems, the XRD data confirm the formation of the alloy nanoparticles with no evidence of any of the pure components. Super paramagnetic alloy PdM and PtM nanoparticles (where M = Co, Fe, and Ni) nanoparticles are also synthesized *via* the LVCC method.

Acknowledgements

We thank the National Science Foundation (CHE-0414613) for support of this work. We thank Dr Darius Kuciaukas and Dr Chris Wohl for the use of the femtosecond titanium:sapphire laser. We also thank Dr Everett E. Carpenter for the SQUID measurements.

References

1   G. A. Ozin and A. C. Arsenault, *Nanochemistry: A Chemical Approach to Nanomaterials*, Royal Society of Chemistry, Cambridge, 2005.
2   L. M. Liz-Marzan, P. V. Kamat, and V. Prashant, *Nanoscale Materials*, Kluwer Academic Publishers Group, London, 2003.
3   A. S. Edelstein and R. C. Cammarata, *Nanomaterials: Synthesis, Properties and Applications*, Taylor & Francis Ltd., London, 1996.
4   G. C. Hadjipanayis and R. W. Siegel, *Nanophase Materials Synthesis – Properties – Applications*, Kluwer Academic Publications, London, 1994.
5   G. Schmid, *Nanoparticles: From Theory to Application*, Wiley-VCH, Weinheim, 2004.
6   S. Link and M. A. El-Sayed, *Int. Rev. Phys. Chem.*, 2000, **19**, 409.
7   K. L. Kelly, E. Coronado, L. L. Zhao and G. C. Schatz, *J. Phys. Chem. B*, 2003, **107**, 668.
8   M. Hu, J. Chen, Z.-Y. Li, L. Au, G. V. Hartland, X. Li, M. Marquez and Y. Xia, *Chem. Soc. Rev.*, 2006, **35**, 1084.
9   S. Link, Z. L. Wang and M. A. El-Sayed, *J. Phys. Chem. B*, 1999, **103**, 3529.
10 P. K. Jain, W. Qian and M. A. El-Sayed, *J. Am. Chem. Soc.*, 2006, **128**, 2426.
11 P. K. Jain, K. S. Lee, I. H. El-Sayed and M. A. El-Sayed, *J. Phys. Chem. B*, 2006, **110**, 7238.

12 X. Huang, I. H. El-Sayed, W. Qian and M. A. El-Sayed, *J. Am. Chem. Soc.*, 2006, **128**, 2115.
13 H. El-Sayed Ivan, X. Huang and A. El-Sayed Mostafa, *Cancer Lett.*, 2006, **239**, 129.
14 C. Loo, A. Lowery, N. Halas, J. West and R. Drezek, *Nano Lett.*, 2005, **5**, 709.
15 R. Hirsch Leon, M. Gobin Andre, R. Lowery Amanda, F. Tam, A. Drezek Rebekah, J. Halas Naomi and L. West Jennifer, *Ann. Biomed. Eng.*, 2006, **34**, 15.
16 X. Huang, K. Jain Prashant, H. El-Sayed Ivan and A. El-Sayed Mostafa, *Photochem. Photobiol.*, 2006, **82**, 412.
17 A. Panacek, L. Kvitek, R. Prucek, M. Kolar, R. Vecerova, N. Pizurova, V. K. Sharma, T. j. Nevecna and R. Zboril, *J. Phys. Chem. B*, 2006, **110**, 16248.
18 A. Henglein, *J. Phys. Chem. B*, 2000, **104**, 2201.
19 M.-J. Kim, H.-J. Na, K. C. Lee, E. A. Yoo and M. Lee, *J. Mater. Chem.*, 2003, **13**, 1789.
20 S. T. He, S. S. Xie, J. N. Yao, H. J. Gao and S. J. Pang, *Appl. Phys. Lett.*, 2002, **81**, 150.
21 S. W. Han, Y. Kim and K. Kim, *J. Colloid Interface Sci.*, 1998, **208**, 272.
22 D.-H. Chen and C.-J. Chen, *J. Mater. Chem.*, 2002, **12**, 1557.
23 G. C. Papavassiliou, *J. Phys. F*, 1976, **6**, L103.
24 Y. Takeuchi, T. Ida and K. Kimura, *J. Phys. Chem. B*, 1997, **101**, 1322.
25 F. Mafune, J. Kohno, Y. Takeda and T. Kondow, *J. Phys. Chem. B*, 2002, **106**, 7575.
26 A. Takami, H. Kurita and S. Koda, *J. Phys. Chem. B*, 1999, **103**, 1226.
27 J. H. Hodak, A. Henglein, M. Giersig and G. V. Hartland, *J. Phys. Chem. B*, 2000, **104**, 11708.
28 J. Zhang, J. Worley, S. Denommee, C. Kingston, Z. J. Jakubek, Y. Deslandes, M. Post, B. Simard, N. Braidy and G. A. Botton, *J. Phys. Chem. B*, 2003, **107**, 6920.
29 Y.-H. Chen and C.-S. Yeh, *Chem. Commun.*, 2001, 371.
30 M. S. El-Shall, W. Slack, W. Vann, D. Kane and D. Hanley, *J. Phys. Chem.*, 1994, **98**, 3067.
31 S. Li, S. J. Silvers and M. S. El-Shall, *J. Phys. Chem. B*, 1997, **101**, 1794.
32 M. A. Duncan, in *Advances in Metal and Semiconductor Clusters*, ed. M. A. Duncan, JAI Press Ltd., London, 1998, pp. 115–177.
33 Y. B. Pithawalla, M. S. El-Shall, S. C. Deevi, V. Stroem and K. V. Rao, *J. Phys. Chem. B*, 2001, **105**, 2085.
34 M. S. El-Shall, V. Abdelsayed, Y. B. Pithawalla, E. Alsharaeh and S. C. Deevi, *J. Phys. Chem. B*, 2003, **107**, 2882.
35 G. Glaspell, V. Abdelsayed, K. M. Saoud and M. S. El-Shall, *Pure Appl. Chem.*, 2006, **78**, 1667.
36 V. Abdelsayed, K. M. Saoud and M. S. El-Shall, *J. Nano Res.*, 2006, **8**, 519.
37 V. Abdelsayed, M. S. El-Shall and T. Seto, *J. Nano Res.*, 2006, **8**, 361.
38 T. Liu, Y. Leng and X. Li, *Solid State Commun.*, 2003, **125**, 391.
39 S. Link, C. Burda, B. Nikoobakht and M. A. El-Sayed, *Chem. Phys. Lett.*, 1999, **315**, 12.
40 Y. Sun and Y. Xia, *Science*, 2002, **298**, 2176.
41 M. P. Mallin and C. J. Murphy, *Nano Lett.*, 2002, **2**, 1235.
42 M. Treguer, C. de Cointet, H. Remita, J. Khatouri, M. Mostafavi, J. Amblard, J. Belloni and R. de Keyzer, *J. Phys. Chem. B*, 1998, **102**, 4310.
43 I. Lee, S. W. Han and K. Kim, *Chem. Commun.*, 2001, 1782.
44 J. C. Heyraud and J. J. Metois, *Acta Metall.*, 1980, **28**, 1789.
45 L. D. Marks, *Rep. Prog. Phys.*, 1994, **57**, 603.
46 C.-Y. Huang, H.-J. Chiang, J.-C. Huang and S.-R. Sheen, *Nanostruct. Mater.*, 1999, **10**, 1393.
47 C. D. Grant, A. M. Schwartzberg, T. J. Norman, Jr and J. Z. Zhang, *J. Am. Chem. Soc.*, 2003, **125**, 549.
48 P. Miegge, T. Rousset, B. Tardy, J. Massardier and J. Bertolini, *J. Catal.*, 1994, **149**, 404.
49 B. Khanra and M. Menon, *Chem. Phys. Lett.*, 1999, **305**, 89.
50 T. Deivaraj, W. Chen and J. Lee, *J. Mater. Chem.*, 2004, **13**, 2555.
51 K. Park, J. Choi, B. Kwon, S. Lee, Y. Sung, H. Ha, S. Hong, H. Kim and A. Weickowski, *J. Phys. Chem. B*, 2002, **106**, 1869.
52 Y. Hou, H. Kondoh, T. Kogure and T. Ohta, *Chem. Mater.*, 2004, **16**, 5149.
53 N. Nunomura, T. Teranishi, M. Miyake, A. Oki, S. Yamada, N. Toshima and H. Hori, *J. Magn. Magn. Mater.*, 1998, **177–181**, 947.
54 N. Nunomura, H. Hori, T. Teranishi, M. Miyake and S. Yamada, *Phys. Lett. A*, 1998, **249**, 524.
55 H. Hori, T. Teranishi, M. Taki, S. Yamada, M. Miyake and Y. Yamamoto, *J. Magn. Magn. Mater.*, 2001, **226–130**, 1910.
56 S. Choi, Y. Kwon, S. Hong, J. Lee and R. Wu, *J. Magn. Magn. Mater.*, 2001, **226–230**, 1662.
57 M. Chen and D. Nikles, *J. Appl. Phys.*, 2002, **91**(10), 8477.

58 S. Kang, Z. Jia, D. Nikles and J. Harrell, *J. Appl. Phys.*, 2004, **95**(11), 6744.
59 Q. Zeng, Y. Zhang, H. Wang, V. Papaefthymiou and G. Hadjipanayis, *J. Magn. Magn. Mater.*, 2004, **272–276**, e1223.
60 M. Chen and S. Sun, *J. Am. Chem. Soc.*, 2004, **126**, 8394.
61 Y. Ding, S. Majetich, J. Kim, K. Barmack, H. Rollins and P. Sides, *J. Magn. Magn. Mater.*, 2004, **284**, 336.
62 S. Sun, C. B. Murray, D. Weller, L. Folks and A. Moser, *Science*, 2000, **287**, 1989.
63 S. Sun, *Adv. Mater.*, 2006, **18**, 393.
64 R. D. Rutledge, W. H. Morris, M. S. Wellons, Z. Gai, J. Shen. J. Bentley, J. E. Wittig and C. M. Lukehart, *J. Am. Chem. Soc.*, 2006, **128**, 14210.

Magnetic properties of Co$_N$Rh$_M$ nanoparticles: experiment and theory†

M. Muñoz-Navia,‡[a] J. Dorantes-Dávila,[a] D. Zitoun,[b] C. Amiens,[b] B. Chaudret,[b] M.-J. Casanove,[c] P. Lecante,[c] N. Jaouen,[d] A. Rogalev,[d] M. Respaud[e] and G. M. Pastor*[f]

Received 4th April 2007, Accepted 9th May 2007
First published as an Advance Article on the web 28th September 2007
DOI: 10.1039/b705122k

The magnetism of Co–Rh nanoparticles is investigated experimentally and theoretically. The particles (≈ 2 nm) have been synthesized by decomposition of organometallic precursors in mild conditions of pressure and temperature, under hydrogen atmosphere and in the presence of a polymer matrix. The magnetic properties are determined by SQUID, Mössbauer spectroscopy, and X-ray magnetic circular dichroism (XMCD). The structural and chemical properties are characterized by wide angle X-ray scattering, transmission electronic microscopy and X-ray absorption near edge spectroscopy. All the studied Co–Rh clusters are magnetic with an average spin moment per atom μ that is larger than the one of macroscopic crystals or alloys with similar concentrations. The experimental results and comparison with theory suggest that the most likely chemical arrangement is a Rh core, with a Co-rich outer shell showing significant Co–Rh mixing at the interface. Measured and calculated magnetic anisotropy energies (MAEs) are found to be higher than in pure Co clusters. Moreover, one observes that the MAEs can be tuned to some extent by varying the Rh concentration. These trends are well accounted for by theory, which in addition reveals important spin and orbital moments induced at the Rh atoms as well as significant orbital moments at the Co atoms. These play a central role in the interpretation of experimental data as a function of Co–Rh content. A more detailed analysis from a local perspective shows that the orbital and spin moments at the Co–Rh interface are largely responsible for the enhancement of the magnetic moments and magnetic anisotropy.

[a] *Instituto de Física, Universidad Autónoma de San Luis Potosí, Alvaro Obregón 64, San Luis Potosí, Mexico. E-mail: jdd@ifisica.uaslp.mx*
[b] *Laboratoire de Chemie de Coordination, CNRS, 205 route de Narbonne, 31077 Toulouse, France. E-mail: chaudret@lcc-toulouse.fr*
[c] *Centre d'Elaboration de Matériaux et d'Etudes Structurales, CNRS, 29 rue Jeanne Marvig, 31077 Toulouse, France. E-mail: casanove@cemes.fr*
[d] *European Synchroton Radiation Facility, 6 rue Jules Horowitz, BP220, 38043 Grenoble, France. E-mail: rogalev@esrf.fr*
[e] *Laboratoire de Physique et Chimie des Nano-objets, INSA, 135 avenue de Rangueil, 31077 Toulouse, France. E-mail: respaud@insa-toulouse.fr*
[f] *Institut für Theoretische Physik der Universität Kassel, Heinrich Plett Str. 40, D-34132 Kassel, Germany. E-mail: pastor@uni-kassel.de*

† The HTML version of this article has been enhanced with colour images.
‡ Present address: Max-Planck-Institut für Mikrostrukturphysik, D-06120 Halle, Germany. E-mail: mnavia@mpi-halle.mpg.de.

1. Introduction

The magnetism in monometallic ferromagnetic 3d transition-metal (TM) nanoparticles containing less than 1000 atoms has been the subject of numerous experimental and theoretical studies. It is nowadays relatively well understood that the large surface-to-volume ratio induces an enhancement of the spin and orbital magnetic moments and magnetic anisotropy energy (MAE) as compared to the bulk materials.[1,2] In contrast very little is known about the behaviour of magnetic nanoalloys. This subject is currently attracting considerable attention both from fundamental and technological perspectives. For example, for material science applications one would like to be able to develop magnetic nanoparticles that combine both high saturation magnetization (M_S) and large MAE. This can indeed be achieved by starting from a ferromagnetic (FM) 3d metal and by associating it with a second heavier element that displays a stronger spin–obit coupling and a potentially significant contribution to the total magnetization. Quite generally, 4d and 5d metals appear as very good candidates for this purpose. Co–Rh clusters are particularly appealing since Rh shows non-saturated magnetism in small clusters despite being non-magnetic in bulk.[3–7] Alloying Co with Rh should be an effective way to combine large magnetic moments with large magnetic anisotropy energy. In addition, the diversity of local chemical environments present in these nanoalloys and the competition between Co–Co, Co–Rh and Rh–Rh effective exchange couplings lead us to expect very interesting size and structural dependence of the magnetic properties.

In the past years some experimental and theoretical studies on Co–Rh nanoparticles have been performed. For instance, CoRh particles have been synthesized by Zitoun *et al.* by decomposing organometallic precursors in the presence of a polymer.[6] Moreover, the measurements have shown that the average magnetic moment per Co atom for clusters of about 300–400 atoms is about 2.38 μ_B for a Co concentration $x_{Co} \simeq 0.5$. This value is much larger than the average magnetization found in bulk alloys of similar concentration. From the point of view of theory only a few studies have been concerned with Co–Rh clusters. The interplay between structural, chemical and magnetic properties of small free $Co_M Rh_N$ ($N + m \leq 13$) have been determined by Dennler *et al.* within the framework of spin-density-functional theory.[7] They found that all studied Co–Rh clusters are magnetic with average and local spin moments that are often a factor two larger than those of macroscopic crystals or alloys with similar concentrations. It is the purpose of this paper to report on recent experimental and theoretical progress on the study of the magneto-anisotropic properties of $Co_x Rh_{1-x}$ clusters as a function of size, composition and structure.

2. Experimental

In the following we briefly review the experimental method to study the magnetic properties of Co–Rh clusters. Details of both the synthesis and the structural studies have been published in ref. 6 and 8. The nanoparticle synthesis is performed in solution by decomposition of organometallic precursors in the presence of a stabilizing polymer. These precursors decompose under hydrogen pressure (3 bars) at room temperature to give zero-valency atoms. The nucleation–growth process leads to the formation of metallic clusters. The use of organometallics allows us to work under mild conditions of temperature compared to thermal decomposition of a metal carbonyl compound ($Fe(CO)_5$ or $Co_2(CO)_8$). These methods lead to nanoparticle assemblies of well-defined size and composition. However, a passivated surface by carbon monoxide and/or carbides is highly probable. The decomposition of both organometallic precursors (Co and Rh) does not release any contaminating by-products in contrast to the reduction of a metal salt by a borohydride. Only cyclooctane and pentane-2,4-diol could bind to the nanocrystals surface. Surface magnetism, which is crucial in small systems, is therefore not perturbed and the nanoparticle can be regarded as relatively close to the ideal free cluster from the magnetic point of view.

The organometallic approach combined with the use of polyvinylpyrrolidone K30 (PVP) at low metal concentration allows growth control of the nanoparticles. Three samples with Co final concentrations x_{Co} = 0.76, 0.49 and 0.25 were synthesized. The transmission electronic microscopy (TEM) studies evidence a regular dispersion of the clusters in the polymer matrix, with narrow log-normal size distributions, an average diameters around 2 nm, and a width below 15%. The analysis of the fine structure has been realized by using high resolution TEM and wide angle X-ray scattering (WAXS) techniques. The $Co_{0.25}Rh_{0.75}$ sample displays the bulk phase fcc structure with a first nearest neighbor (NN) distance of $d_{NN} \simeq$ 0.269 nm. $Co_{0.49}Rh_{0.51}$ and $Co_{0.76}Rh_{0.24}$ do not display any conventional crystalline phase. The WAXS pattern can be fitted with a polytetrahedral structure,[9] and the interatomic distance evolves from $d_{NN} \simeq$ 0.269 nm for $Co_{0.49}Rh_{0.51}$ to $d_{NN} \simeq$ 0.263 nm for $Co_{0.76}Rh_{0.24}$ which confirms the bimetallic character of the nanoparticles. We may conclude that the clusters probably present a Rh core and a Co rich shell since d_{NN} is very close to that of bulk Rh. In conclusion, we may assume that the nanoparticles adopt a close packed crystalline structure with $d_{NN} \simeq d_{NN}$ (Rh).

Fig. 1 shows the hysteresis loops measured at 2 K for the three samples. Data from mono-metallic Co particles of 1.5 nm diameter[10] are also included for comparison. All systems are found to be magnetic and display ferromagnetic behaviour with hysteresis. As a general tendency, on increasing the Rh concentration, the coercive field increases as well as the irreversible field up to and above 5 T. The differential high field susceptibility also increases, and none of the magnetization curves are saturated in this range of field. At 5 T, the magnetization per Co atom first increases and then decreases with increasing Rh concentration.

At higher temperature, the hysteresis disappears and the magnetization progressively decreases. This behaviour corresponds to the transition from a ferromagnetic behaviour to a super-paramagnetic one. This transition is confirmed by the behaviour of the susceptibility. The zero-field cooled and field-cooled susceptibility curves (measured at 1 mT) display the same typical shape (see Fig. 2 of ref. 6). The magnetization displays a narrow maximum which temperature corresponds to the blocking temperature, *i.e.* the transition from the superparamagnetic to the blocked ferromagnetic state. The same behaviour is observed for all the samples with

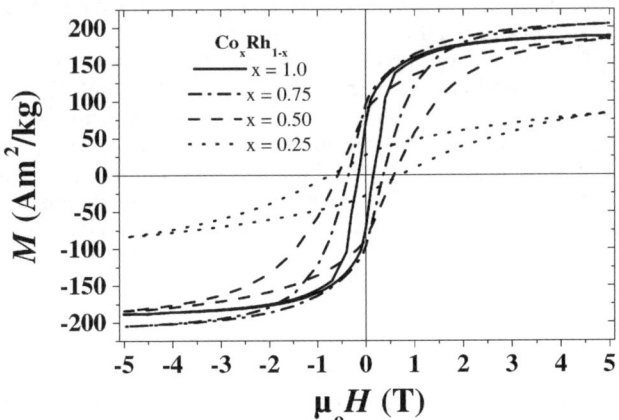

Fig. 1 Hysteresis loops measured at 2 K for bimetallic Co_xRh_{1-x} nanoparticles (diameter $\phi \simeq 2$ nm) for different Co concentrations x.

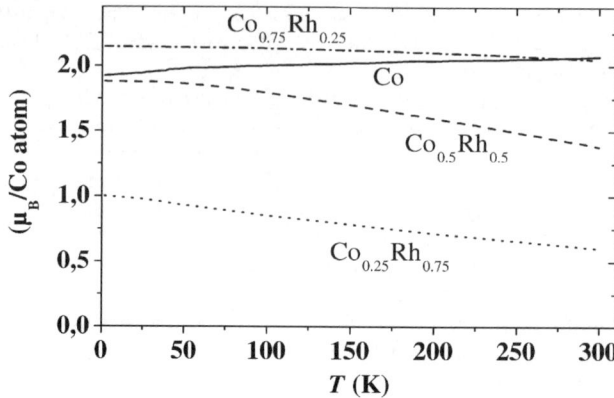

Fig. 2 Temperature dependence of the mean magnetic moment per Co atom in bimetallic Co_xRh_{1-x} nanoparticles ($\phi \simeq 2$ nm) for different Co concentrations x.

different blocking temperatures as a consequence of the different size distributions and effective anisotropy. We use the formalism described in ref. 6 and 10 to determine the size distributions, the variation of the spontaneous magnetization versus temperature $M_S(T)$ and the effective anisotropy K_{eff}.

The results plotted in Fig. 2 demonstrate that alloying with Rh induces a stronger dependence of M_S vs. temperature. The plot of the spontaneous magnetization with temperature can be fitted with a Bloch type law, the slope decreases while the Rh ratio increases. A surprisingly small increase of $M_S(T)$ is observed for pure Co as is also observed for free clusters by Billas et al.[11]

These data demonstrate that the Rh atoms in the $Co_{0.5}Rh_{0.5}$ clusters have a strong induced magnetic moment. The L_2 and L_3 XMCD signals are of the same order of magnitude, with opposite signs, indicating that the orbital contribution (L_z) is small compared to the spin one (S_z). Using Thole and Carra's sum rules,[12] we estimate the ratio $L_z/S_z = 0.066$.

The main experimental results can be summarized as follows: (i) For all the studied chemical compositions the clusters are magnetic. (ii) The cluster structure is closed packed with a tendency to fcc-like structure at least for low x_{Co} (iii) The mean NN distance in the cluster is very close to that of bulk Rh. (iv) The magnetic moment per Co atom is significantly enhanced with respect to the corresponding bulk alloy of similar concentration. (iv) The Rh atoms carry non vanishing spin and orbital magnetic moments. Despite all these interesting findings a number of questions still deserve to be clarified. For example, how does the cluster structure and distribution of Co and Rh within the cluster affect the magnetic behaviour? Can one infer any trends on the most likely chemical arrangement from the magnetic behaviour? What can be said about the role of segregation versus intermixing at the CoRh interface? How important are the orbital magnetic moments for the interpretation of experiment? And, is it possible to account for the measured magnetic moments quantitatively? In the following section we intend to address these kinds of questions from the point of view of theory by performing self-consistent calculations on Co_NRh_M as a function of size, structure and composition.

3. Theory

For the theoretical investigations we consider a d-band Hamiltonian given by ref. 13 and 14

$$H = H_0 + H_C + H_{SO}. \tag{1}$$

The first term

$$H_0 = \sum_{im,jm'} t_{ij}^{mm'} \, \hat{c}_{im\sigma}^{\dagger} \, \hat{c}_{jm'\sigma} \qquad (2)$$

takes into account inter-atomic hybridizations between the orbitals m and m' of atoms i and j. In the usual notation, $\hat{c}_{im\sigma}^{\dagger}$, $\hat{c}_{im\sigma}$, and $\hat{n}_{im\sigma} = \hat{c}_{im\sigma}^{\dagger}\hat{c}_{im\sigma}$ refers to the creation, annihilation, and number operator for a spin-σ electron at orbital im. The Coulomb interaction term H_C is treated in the unrestricted Hartree–Fock approximation and may be written as

$$H_C = \sum_{im\sigma} \Delta\varepsilon_{im\sigma} \, \hat{n}_{im\sigma}, \qquad (3)$$

where

$$\Delta\varepsilon_{im\sigma} = \sum_{m'} \left[\left(U_{mm'} - \frac{J_{mm'}}{2} \right) \nu_{im'} - \frac{\sigma}{2} J_{mm'} \mu_{im'} \right] \qquad (4)$$

are the orbital- and spin-dependent shifts of the d levels that depend on the local occupations $\nu_{im} = \langle\hat{n}_{im\uparrow}\rangle + \langle\hat{n}_{im\downarrow}\rangle$ and spin polarizations $\mu_{im} = \langle\hat{n}_{im\uparrow}\rangle - \langle\hat{n}_{im\downarrow}\rangle$. The direct and exchange Coulomb integrals $U_{mm'}$ and $J_{mm'}$ are expressed in terms of the three independent radial Coulomb integrals $F^{(0)}$, $F^{(2)}$, and $F^{(4)}$, whose values can be derived from the known atomic ratios.[15] In this way atomic symmetry is strictly respected and all Hund's rules are naturally fulfilled. The Coulomb interactions $U_{mm'}$ and $J_{mm'}$ that define the self-consistent equations [eqn (4)], may be treated in a simple way by the so-called orbital polarization (OP) approximation.[18,19,22] This approach has been proposed in the context of first principles studies of TMs in order to enhance phenomenologically the effects of Hund's rule orbital polarizations and to correct for the systematic underestimations of the orbital moments obtained with the usual exchange and correlation density functionals. In this way numerous important predictions of the orbital magnetic moments in bulk, surfaces and deposited atoms have been obtained.[19–21] Within the present model the OP term is given by $H_{OP} = -B\Sigma_i\hat{L}_{i\delta}^2$ where $B = (9F^{(0)} - 5F^{(4)})/441$ is the Racah coefficient. The corresponding mean-field energy levels are then written as[14]

$$\Delta\varepsilon_{im\sigma} = \left(U - \frac{J}{2} \right) \nu_i - \frac{\sigma}{2} J \mu_i - B\langle L_{i\delta}\rangle m. \qquad (5)$$

Here, $U = \overline{U_{mm'}} = F^{(0)}$ and $J = \overline{J_{mm'}} = (F^{(2)} + F^{(4)})/14$. In this context the OP approximation is equivalent to assuming that the orbital dependence of the Coulomb integrals has the form $U_{mm'} - U = J_{mm'} - J = -Bmm'$, which is the same for direct and exchange interactions. In practice, the OP calculations are much simpler than the rigorous orbital dependent treatment. In fact, they are not much more demanding than the simplest orbital independent approach,[24] since they require a self-consistent determination of only ν_i, μ_l and $\langle L_{i\delta}\rangle$ at all atoms. For the large clusters we use this approximation since it has been shown to yield similar results as the more demanding calculation in which the full orbital dependence of the Coulomb integrals are considered.[14] Notice that if $B = 0$ we obtain the simplest mean field approximation for eqn (5).[23–26]

Finally, the third term

$$H_{SO} = -\xi \sum_{i,m\sigma,m'\sigma'} (\vec{L}_i \cdot \vec{S}_i)_{m\sigma,m'\sigma'} \, \hat{c}_{im\sigma}^{+} \, \hat{c}_{im'\sigma'} \qquad (6)$$

takes into account spin–orbit interactions, where $(\vec{L}_i \cdot \vec{S}_i)_{m\sigma,m'\sigma'}$ refers to the intra-atomic matrix elements of $\vec{L} \cdot \vec{S}$. H_{SO} couples the up and down spin-manifolds and introduces the dependence of the magnetic properties on the relative orientation between the magnetization direction and the geometrical structure of the cluster.

The average local orbital moments $\langle L_{i\delta} \rangle$ at atom i are calculated from

$$\langle L_{i\delta} \rangle = \sum_{\sigma} \sum_{m=-2}^{2} \int_{-\infty}^{\varepsilon_F} m\rho_{im\sigma}^{\delta}(\varepsilon)\, \mathrm{d}\varepsilon, \tag{7}$$

where m indicates the magnetic quantum number. The quantization axis of the orbital momentum is thereby taken to be the same as the spin quantization axis.

The electronic energy per atom

$$E_{\delta} = \frac{1}{N} \sum_{i} E_{\delta}(i) \tag{8}$$

can be written as the sum of local contributions

$$E_{\delta}(i) = \sum_{m\sigma} \left[\int_{-\infty}^{\varepsilon_F} \varepsilon\rho_{im\sigma}^{\delta}(\varepsilon)\, \mathrm{d}\varepsilon - E_{im\sigma}^{dc} \right] \tag{9}$$

corresponding to each atom i of the cluster. Here $E_{im\sigma}^{dc} = (1/2)\Delta\varepsilon_{i\sigma} \langle \hat{n}_{im\sigma} \rangle$ stands for the double-counting correction. The MAE is defined as the change ΔE in the electronic energy E_{δ} associated to a change in the orientation of the magnetization. Thus, positive (negative) values of the anisotropy energy $\Delta E_{xz} = E_x - E_z$ indicate that the easy (hard) axis is along the z direction.

The parameters used in the calculations are then specified as follows. The two-center d-electron hopping integrals are given by the canonical expression in terms of the corresponding bulk d-band width. The intra-atomic Coulomb integrals $U_{mm'}$ and $J_{mm'}$ are expressed in terms of the three independent radial Coulomb integrals $F^{(0)}$, $F^{(2)}$, and $F^{(4)}$ allowed by atomic symmetry.[15] These are chosen by taking the ratios $F^{(0)}/F^{(2)}$ and $F^{(4)}/F^{(2)}$ from atomic calculations[27] and by fitting the value of $F^{(2)}$ to reproduce the bulk Co spin moment [$\overline{U}_{mm'} = F^{(0)} = 13.5$ eV and average exchange integral $\overline{J}_{mm'} = (F^{(2)} + F^{(4)})/14 = 0.74$ eV]. In the case of Rh we use $F^{(2)}$ such as $J_{av}^{Rh} = 0.48$ eV, which have been obtained from density functional calculations (Stoner theory) taking into account correlation effects beyond the local spin density approximation.[28] Notice that for properties like the spin and orbital magnetic moments, which derive directly from the spin-polarized density distribution, a size independent $U_{mm'}$ $(J_{mm'})$ has been proved to be a good approximation. Using these values for the radial integrals we estimate the Racah coefficient $B = (9F^{(2)} - 9F^{(4)})/441$ $(B_{Co} = 0.08$ eV and $B_{Rh} = 0.05$ eV). In this way, only one parameter $[F^{(2)}]$ is involved in the determination of the Coulomb integrals. The SO coupling constants $\xi(Co) = 0.088$ eV and $\xi(Rh) = 0.180$ eV are taken from ref. 16. As suggested by experiment,[6,9] we consider fcc-like clusters formed by a central atom and its successive shells of NN's. For Co–Rh this corresponds to a Co_N (Rh_M) core covered with a $Rh_M(Co_N)$ shell. The structural effects are discussed by considering additional geometries, namely octahedral-like fcc and polytetrahedral-like structures.

The local densities of electronic states (DOS) $\rho_{im\sigma}^{\delta}(\varepsilon)$ are determined self-consistently for each orientation δ of the spin magnetization $\vec{S}$. In this paper we consider $\delta = z$, along a principal C_n symmetry axis of the cluster, and $\delta = x$ along a nearest neighbor (NN) bond perpendicular to z. In the case of low-symmetry structures a full vectorial calculations as a function of both, the polar angle θ (between the magnetization $\vec{M}$ and the z axis) and the azimuthal angle ϕ were performed. The associated single-particle problem is solved by using Haydock–Heine–Kelly's recursion method.[17] The local orbital occupations ν_{im} and spin polarizations μ_{im} are determined with an accuracy $\varepsilon \leq 10^{-10}$ electrons per atom or better, which allows to derive the MAEs reliably.

4. Discussion

The calculated spin and orbital magnetic moments for Co–Rh clusters are shown in Table 1. The results for a 43-atom fcc-like spherical cluster formed by a central atom and its successive shells of NN's correspond to a Co_{19} (Rh_{19}) core covered with a Rh_{24}(Co_{24}) shell. The direction of magnetization $\delta = z$ is along a principal C_n symmetry axis of the cluster, and $\delta = x$ along a nearest neighbor (NN) bond perpendicular to z. In the case of Co_{19} the easy direction xy is along the central atom

Table 1 Theoretical results for the magnetic properties of $Co_N Rh_M$ clusters: total magnetic moment per cluster atom $\bar{\mu}_T = \langle L_T \rangle + 2 \langle S_T \rangle$, total magnetic moment per Co atom $\bar{\mu}_{N_{co}}$, orbital magnetic moment per atom $\langle L_T \rangle$ along the easy axis δ, and orbital-to-spin moment ratio $\frac{\langle L \rangle}{2 \langle S \rangle}$ (Rh) at the Rh atoms. The results are shown as a function of Co composition $x_{Co} = \frac{N}{(N+M)}$. The magnetic moments are in Bohr magnetons. The structure of the clusters is indicated: disordered alloys are denoted by (m); no indication refers to core-shell systems

Cluster	x_{Co}	$\bar{\mu}_T$	$\bar{\mu}_{N_{co}}$	$\langle L_T \rangle$	$\frac{\langle L \rangle}{2\langle S \rangle}$ (Rh)	δ
$N + M = 43$						
Co_{43}	1.00	1.98	1.98	0.29		x
$Co_{19}Rh_{24}$	0.44	0.80	1.44	0.06	0.07	z
$Rh_{19}Co_{24}$	0.56	1.40	2.51	0.30	0.14	x
$N + M = 79$						
Co_{79}	1.00	1.88	1.88	0.19		z
$Rh_{19}Co_{60}$	0.76	1.65	2.17	0.26	0.09	z
$Rh_{37}Co_{42}(m)$	0.53	1.40	2.62	0.22	0.10	z
$Rh_{55}Co_{24}$	0.30	0.67	2.22	0.11	0.08	z
$N + M = 87$						
			Spherical fcc			
Co_{87}	1.00	1.92	1.92	0.23		x
$Rh_{19}Co_{68}$	0.78	1.76	2.25	0.35	0.08	x
$Rh_{43}Co_{44}(m)$	0.51	1.29	2.56	0.26	0.08	z
$Rh_{67}Co_{20}(m)$	0.23	0.86	3.75	0.18	0.12	z
$N + M = 85$						
			Spherical fcc			
Co_{85}	1.00	1.89	1.89	0.20		xy
$Rh_{19}Co_{66}$	0.77	1.71	2.20	0.30	0.10	xy
$Rh_{43}Co_{42}(m)$	0.49	1.44	2.91	0.24	0.13	xy
$Rh_{61}Co_{24}$	0.28	0.62	2.20	0.11	0.17	z
$N + M = 70$						
			Polytetrahedral			
Co_{70}	1.00	2.00	2.00	0.31		x
$Rh_{17}Co_{53}$	0.76	1.84	2.43	0.47	0.03	x
$Rh_{38}Co_{32}$	0.46	1.28	2.80	0.35	0.20	z
$Rh_{52}Co_{18}$	0.26	0.85	3.30	0.25	0.14	z
$N + M = 72$						
			Polytetrahedral			
Co_{72}	1.00	2.07	2.07	0.41		x
$Rh_{22}Co_{50}$	0.69	1.91	2.76	0.60	0.11	x
$Rh_{40}Co_{32}$	0.44	1.64	3.70	0.52	0.21	x
$Rh_{52}Co_{20}$	0.28	1.31	4.73	0.47	0.23	x
$N + M = 405$						
			Spherical fcc			
Co_{405}	1.00	1.87	1.87	0.18		z
$Rh_{127}Co_{278}(m)$	0.68	1.36	1.98	0.19	0.02	z
$Rh_{199}Co_{206}(m)$	0.51	1.05	2.07	0.16	0.06	z
$Rh_{201}Co_{204}$	0.50	1.04	2.05	0.16	0.01	z

and one of the vertices in the plane x–y. The direction δ yielding the lowest-energy is indicated. For comparison, results for Co_{43} are also given. First one observes that the total magnetic moment per atom $\bar{\mu}_T$ decreases with Co concentration x_{Co} ($\bar{\mu}_T = 1.98\ \mu_B$ for $x_{Co} = 1.00$ and $\bar{\mu}_T = 0.80\ \mu_B$ for $x_{Co} = 0.44$ with a Co_{19} core and $\bar{\mu}_T = 1.40\ \mu_B$ for $x_{Co} = 0.56$ with a Rh_{19} core). Notice the large reduction of $\bar{\mu}_T$ if a Co_{19} core is assumed. In contrast, the cluster having a Rh_{19} core still has a significant $\bar{\mu}_T$. This is mainly due to two effects: the larger NN bond-length in the Rh_{19} core and to the orbital contribution ($\langle L_T \rangle = 0.30\ \mu_B$) of the Co surface atoms in $Rh_{19}Co_{24}$. Still, even in this case, $\bar{\mu}_T$ is smaller for $x_{Co} = 0.56$ than for $x_{Co} = 1.00$. This trend changes if the average magnetic moment per Co atom $\bar{\mu}_{N_{Co}}$ in Co–Rh clusters having a Rh core is considered instead of $\bar{\mu}_T$. In this case, $\bar{\mu}_{N_{Co}}$ increases with decreasing x_{Co} ($\bar{\mu}_{N_{Co}} = 1.98\ \mu_B$ for $x_{Co} = 1.00$ and $\bar{\mu}_{N_{Co}} = 2.51\ \mu_B$ for $x_{Co} = 0.56$), indicating that the Rh contribution to the total moment is significant [$\mu_{NCo}(Rh) \simeq 0.50\ \mu_B$]. The fact that the Co–Rh clusters are more likely to have a Rh core seems to be in agreement with the experimental findings of the NN's bond-length and magnetic moment per Co–Rh unit.[6] (In fact the experiments show that the NN's bond-length of the cluster is very similar to that of bulk Rh and that $\bar{\mu}_{N_{Co}} \simeq 2.38\ \mu_B$).[6] Furthermore, due to the induced magnetic moment in Rh atoms, the ratio $\frac{\langle \bar{L} \rangle}{2\langle S \rangle}(Rh) = 0.14$ is significant. As will be discussed later, the calculated $\frac{\langle \bar{L} \rangle}{2\langle S \rangle}(Rh)$ for clusters with size of about 400 atoms is smaller. At this stage, one concludes that Co–Rh clusters with a Rh core have a significant induced moment at Rh atoms.

Results for $N + M = 70, 72, 79, 85$ and 87 are shown in Table 1 in order to infer the effect of structure and Co composition on the magnetic properties. Three cases are analyzed: spherical fcc, octahedral fcc and polytetrahedral structures. An illustration of the considered structures is presented in Fig. 3. In the case of octahedral structures full vectorial calculations as a function of both, the polar angle θ between the magnetization $\vec{M}$ and the z axis and the azimuthal angle ϕ were performed. First, we compare the results of the spherical clusters with $N + M = 79$ and $N + M = 87$ [See Fig. 3(a) and (b)]. Notice that the 87-atom cluster is made up by taking the 79-atom cluster with an additional shell of 8 atoms. The results show that magnetic moments μ_T and $\mu_{N_{Co}}$ follow the same trends as those found in the smaller clusters, i.e., μ_T ($\mu_{N_{Co}}$) decreases (increases) with decreasing x_{Co} concentration and that the values of $\frac{\langle \bar{L} \rangle}{2\langle S \rangle}(Rh) = 0.09 - 0.12\ \mu_B$ are similar. One also observes that the average magnetic moments are larger in the 87-atom cluster than in the 79-atom cluster. This is mainly due to the larger orbital moment contribution of the low-coordinated Co surface atoms [the outermost 8 atoms in Fig. 3(b)]. Similar results are also found for the 85-atom cluster with octahedral structure. However, the results for $x_{Co} \simeq 0.5$ deserve special attention. Let us first notice that in this cluster there are 66 surface atoms. Thus, in order to have $x_{Co} \simeq 0.5$ and keeping a Rh core (with 19 atoms), 24 Rh atoms should be at the surface. The induced magnetic moments at the Rh atoms are larger since they have only low coordinated Co atoms as NNs. As a consequence, μ_T and $\mu_{N_{Co}}$ increase. Therefore, Co–Rh clusters having segregated Co atoms result in smaller average magnetic moments than in the case of mixed clusters.

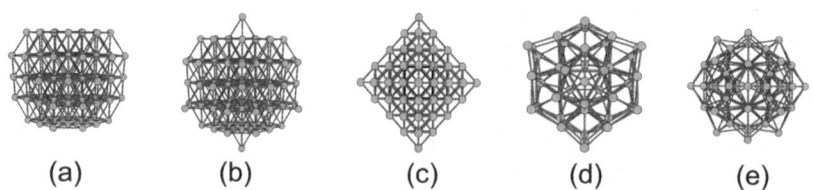

| (a) | (b) | (c) | (d) | (e) |

Fig. 3 Illustration of the cluster geometries considered in the calculations: (a) spherical fcc ($N + M = 79$), (b) spherical fcc ($N + M = 87$), (c) octahedral fcc ($N + M = 85$), (d) polytetrahedral ($N + M = 70$), and (e) polytetrahedral ($N + M = 72$).

Larger values of μ_T and $\mu_{N_{Co}}$ are expected if the symmetry is further reduced. In fact, it has been argued that the structure of Co–Rh clusters is polytetrahedral for large Co concentration.[9] Two types of polytetrahedral growth were considered in this case: the structure of the 70-atom (72-atom) cluster is made up by growing slightly strained face-sharing tetrahedra over a hexahedron (13-atom icosahedron) [see Fig. 3(d) and (e)]. Notice the larger values of μ_T and $\mu_{N_{Co}}$ as compared with those of clusters of similar sizes (see, Table 1 for $x_{Co} \simeq 0.75$). This is due to the large contribution of the Co orbital moments at the surface [i.e., for the 72-atom cluster, $\langle L \rangle (Co) \simeq 1.0 \ \mu_B$ at the surface].

The results for large fcc spherical clusters having 405 atoms are also shown in Table 1. Notice that μ_T and $\mu_{N_{Co}}$ follow the same trends as a function of x_{Co} as in the previous studied clusters. However, the value of $\frac{\langle L \rangle}{2 \langle S \rangle}(Rh) \simeq 0.06$ is smaller than the one obtained in smaller sizes.

The environment dependence of the local orbital moments $\langle L_\delta \rangle (j)$ provides further insight into the magnetic behaviour. Fig. 4 displays $\langle L_\delta \rangle (j)$ in the considered fcc-like clusters, where $j = 1$ refers to the central atom and $j > 1$ to the successive NN shells. The sites corresponding to the Co(Rh) atoms are indicated. One observes that $\langle L_\delta \rangle (j)$ generally increases with j, showing some oscillations as we move from the center to the surface of the cluster. Notice the particularly large values of $2 \langle S_\delta \rangle + \langle L_\delta \rangle$ which corresponds to the Rh atoms at $j = 12$ and $j = 14$. It is important to note that the Rh contributions to the total magnetic moment μ_T amount to about 20% [$\mu_{Rh} =$

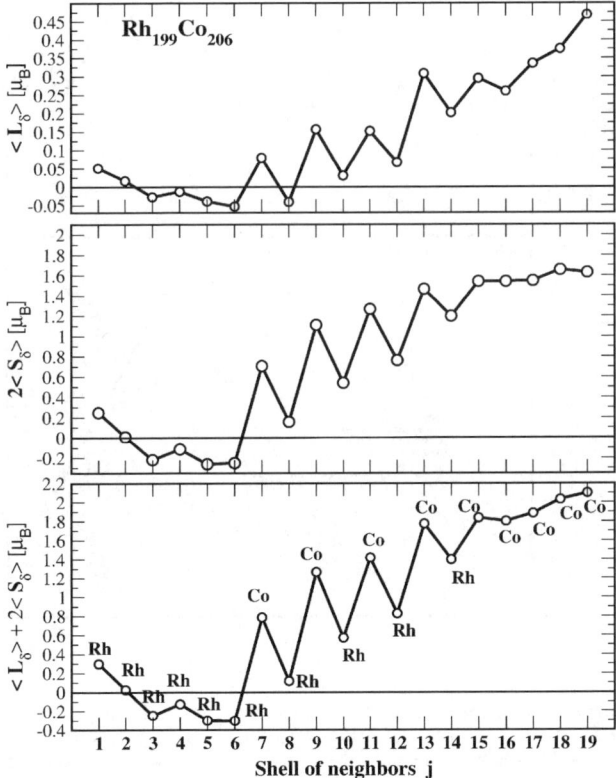

Fig. 4 Local orbital $\langle L(i) \rangle$, spin $2 \langle S(i) \rangle$ and total $2 \langle S(i) \rangle + \langle L(i) \rangle$ magnetic moment along the easy axis of a 405-atom fcc-like $Co_x Rh_{1-x}$ cluster with $x \simeq 0.5$. The results correspond to the average at each NN shell j surrounding the central atom $j = 1$. The shells correspond to a Rh core surrounded by successive Rh and Co shells (as indicated in the figure). The straight lines connecting the points are a guide to the eye.

Table 2 Shell average of the magnetic anisotropy energy $\Delta E_{zx}(j) = E_x(j) - E_z(j)$ of an fcc-like Co_NRh_M clusters having $N + M = 55$ atoms. The results refer to each NN shell j surrounding the central atom $j = 1$. $\Delta E_{zx} = E_x - E_z$ stands for the cluster average MAE. The direction z is taken along a C_n principal symmetry axis of the cluster and x along a NN bond perpendicular to z. The direction δ yielding the lowest-energy is indicated

Cluster	Shell j					ΔE_{xz}	δ
	1	2	3	4	5		
Co_{55}	1.00	−0.22	1.38	−1.33		−0.17	xy
$Co_{43}Rh_{12}$	1.00	0.02	0.35	0.35	−1.15	−0.43	x
$Rh_{13}Co_{42}$	0.44	−0.04	0.15	1.72	−1.70	0.13	z
$Rh_{43}Co_{12}$	0.56	−0.14	0.58	−0.12	−0.85	−0.53	x

$\langle L_\delta \rangle_{Rh} + 2\langle S_\delta \rangle_{Rh} = (0.2–0.5)\mu_B]$. The orbital and spin moments at the interfaces are mainly responsible for this increase of μ_{Rh}. These results demonstrate that it is the interface, rather than just the reduction of local coordination number, which is

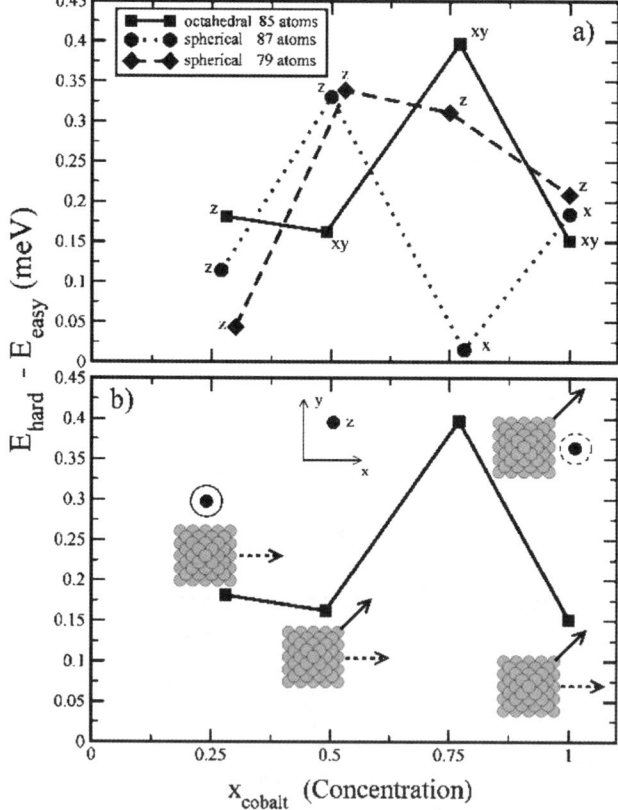

Fig. 5 (a) Magnetic anisotropy energy ΔE of Co_MRh_N clusters having $N + M = 79, 85$ and 87 atoms as a function of Co concentration. The corresponding structures are illustrated in Fig. 4. The labels x, xy, and z indicate the easy magnetization axis, where z refers to the direction along a principal C_n symmetry axis, x is along a NN bond perpendicular to z, and xy refers to the diagonal between the x and y axes [see the inset of figure (b)]. (b) Illustration of the easy axis (solid arrows for x and xy or solid circle for z) and hard axis (dashed arrows or dashed circle) of the 85-atom octahedral fcc cluster, as obtained from full vectorial calculations. A top view of the cluster structure is shown taken from the z principal C_4 symmetry axis. The directions of the magnetization x, y and z are illustrated in the inset.

responsible for the enhancement of magnetic moments. The calculated values well match the experimental ones, when considering both the spin and orbital contributions of the magnetic moments. It should be noticed that these calculations nicely reproduce the evolution of the magnetism with the Rh concentration and allow us to conclude that segregated nanoparticles containing a Rh core and a Co shell with some intermixing at the interface is the most likely scenario.

The interactions between magnetic ad-atoms and metallic substrates often lead to redistributions of the spin-polarized density and to changes in the electronic structure which affect sensitively the spin–orbit energies. The results shown in Table 2 indicate that the interface of Co–Rh plays the main role in determining the stable magnetization direction of the system. In fact, a rich and complex environment dependence of ΔE_{xz} is obtained by manipulating the interface. Replacing the Co core (Rh shell) by a Rh core (Co shell) changes the easy axis (see Table 2). Once again, the contribution of the local ΔE_{xz} to the MAE stabilizes the direction of magnetization. The environment dependence of ΔE_{xz} can be viewed as the result of two main contributions: the changes in the electronic structure of the Co cluster due to Co–Rh hybridizations and the local MAE of the interface Rh atoms which carry small magnetic moments.

Additional interesting magneto-anisotropic behaviour is expected if mixing of the 3d and 4d elements at the interface is allowed. Let us discuss in more detail the results for the MAE in larger clusters. A remarkable behaviour of the MAE as a function of x_{Co} is observed in Fig. 5. The results show that the MAE increases or decreases about 100% if the Co concentration changes from $x_{Co} = 1.00$ $x_{Co} = 0.75$ [see Fig. 5(a)]. Moreover, notice that the particular structure of the octahedral fcc cluster results in a large in-plane magnetic anisotropy energy (even larger than the usual off-plane MAE). In Fig. 5(b) results are shown for the MAE as a function of x_{Co}. Notice that only for $x_{Co} = 0.25$ the easy axis corresponds to the z direction. This shows the importance of performing full vectorial MAE calculations for structures having no spherical symmetry. A direct comparison of these results cannot be performed for several reasons, the main one being the strong dependence of the MAE with respect to the size, chemical order and structure. The effective anisotropy measured for the nanoparticles is in the range of 0.05–0.07 meV at^{-1}. It corresponds to the case of the less anisotropic clusters. Calculations for bigger nanoparticles are in progress.

5. Conclusion

In conclusion, it has been shown that alloying a 3d transition metal with a 4d element offers the possibility of tailoring new magnetic materials with optimized magneto-anisotropic properties for specific technological purposes. The magnetic moments of Co_xRh_{1-x} were determined using both experimental and theoretical approaches. The present discussion should encourage the development of new experimental work as well as further theoretical improvements. For example, it would be worthwhile to investigate more systematically the dependence of the magnetic properties on the geometry of the cluster and its immediate environment. This is relevant for the comparison between theory and experiment, since the morphology of the nanoparticles can be tuned, at least to some extent, by changing the growth and deposition conditions or by subsequent annealing. The well-known sensitivity of TM magnetism to the specific local atomic environments lets us expect a wide variety of interesting behaviours. From a theoretical perspective, it would be interesting to introduce a larger flexibility in the self-consistent calculations by allowing for non-collinear spin polarizations, since the SO interactions break the conservation of S_z, and since it has been shown that the magnetization direction giving the lowest local energy $E_\delta(i)$ is often different for different atoms i.[29]

References

1   See, for instance, *Nanoscale Materials*, ed. Prashant V. Kamat and Luis M. Liz Marzan, Kluwer Academic Press, Boston, USA, 2003.
2   J. Bansmann, S. H. Baker, C. Binns, J. A. Blackman, J.-P. Bucher, J. Dorantes-Dávila, V. Dupuis, L. Favre, D. Kechrakos, A. Kleibert, K.-H. Meiwes-Broer, G. M. Pastor, A. Perez, O. Toulemonde, K. N. Trohidou, J. Tuaillon and Y. Xie, *Surf. Sci. Rep.*, 2005, **56**, 189.
3   (*a*) A. J. Cox, J. G. Louderback and L. A. Bloomfield, *Phys. Rev. Lett.*, 1993, **71**, 923; (*b*) A. J. Cox, J. G. Louderback, S. E. Apsel and L. A. Bloomfield, *Phys. Rev. B*, 1994, **49**, 12295.
4   B. V. Reddy, S. N. Khanna and B. I. Dunlap, *Phys. Rev. Lett.*, 1993, **70**, 3323.
5   P. Villaseñor González, J. Dorantes-Dávila, G. M. Pastor and H. Dreyssé, *Phys. Rev. B*, 1997, **55**, 15084.
6   D. Zitoun, M. Respaud, M.-C. Fromen, M. J. Casanove, P. Lecante, C. Amiens and B. Chaudret, *Phys. Rev. Lett.*, 2002, **89**, 037203.
7   (*a*) S. Dennler, J. Morillo and G. M. Pastor, *J. Phys.: Condens. Matter*, 2004, **16**, S2263; (*b*) S. Dennler, J. Morillo and G. M. Pastor, *Surf. Sci.*, 2003, **532–535**, 334.
8   D. Zitoun, C. Amiens, B. Chaudret, M. Respaud, M.-C. Fromen, P. Lecante and M. J. Casanove, *New J. Phys.*, 2002, **4**, 77.1–77.11.
9   F. Dassenoy, M. J. Casanove, P. Lecante, M. Verelst, T. Ould Ely, C. Amiens and B. Chaudret, *J. Chem. Phys.*, 2000, **112**, 8137.
10   M. Respaud, J. M. Broto, H. Rakoto, A. R. Fert, L. Thomas, B. Barbara, M. Verelst, E. Snoeck, P. Lecante, A. Mosset, J. Osuna, T. Ould Ely, C. Amiens and B. Chaudret, *Phys. Rev. B*, 1998, **57**, 2925.
11   I. M. L. Billas, A. Châtelain and Walt A. de Heer, *Science*, 1994, **265**, 1682.
12   (*a*) B. T. Thole, Paolo Carra, F. Sette and G. van der Laan, *Phys. Rev. Lett.*, 1992, **68**, 1943; (*b*) Paolo Carra, B. T. Thole, Massimo Altarelli and Xindong Wang, *Phys. Rev. Lett.*, 1993, **70**, 694.
13   R. Guirado-López, J. Dorantes-Dávila and G. M. Pastor, *Phys. Rev. Lett.*, 2003, **90**, 226402.
14   G. Nicolas, J. Dorantes-Dávila and G. M. Pastor, *Phys. Rev. B*, 2006, **74**, 014415.
15   J. C. Slater, *Quantum Theory of Atomic Structure*, McGraw-Hill Book Co. Inc., New York, vol. I and II, 1960.
16   P. Bruno, *Magnetismus von Festkörpern und Grenzflächen*, 1993 Ferienkurse des Forschungszentrums Jülich (KFA Jülich), ch. 24 and references therein.
17   R. Haydock, in *Solid State Physics*, edited by H. Ehrenreich, F. Seitz and D. Turnbull, Academic, New York, 1980, vol. 35, p. 215.
18   M. Brooks, *Physica B*, 1985, **130**, 6.
19   O. Hjortstam, J. Trygg, J. M. Wills, B. Johansson and O. Eriksson, *Phys. Rev. Lett.*, 1996, **65**, 492.
20   (*a*) Olle Eriksson, Börje Johansson, R. C. Albers, A. M. Boring and M. S. S. Brooks, *Phys. Rev. B*, 1990, **42**, 2707; (*b*) Per Söderlind, Olle Eriksson, Börje Johansson, R. C. Albers and A. M. Boring, *Phys. Rev. B*, 1992, **45**, 12911.
21   P. Gambardella, S. Rusponi, M. Veronese, S. S. Dhesi, C. Grazioli, A. Dallmeyer, I. Cabria, R. Zeller, P. H. Dederichs, K. Kern, C. Carbone and H. Brune, *Science*, 2003, **300**, 1130.
22   I. V. Solovyev, A. I. Liechtenstein and K. Terakura, *Phys. Rev. Lett.*, 1998, **80**, 5758.
23   (*a*) G. M. Pastor, J. Dorantes-Dávila and K. H. Bennemann, *Physica B*, 1988, **149**, 22; (*b*) G. M. Pastor, J. Dorantes-Dávila and K. H. Bennemann, *Phys. Rev. B*, 1989, **40**, 7642; (*c*) J. Dorantes-Dávila, H. Dreyssé and G. M. Pastor, *Phys. Rev. B*, 1992, **46**, 10432.
24   G. M. Pastor, J. Dorantes-Dávila, S. Pick and H. Dreyssé, *Phys. Rev. Lett.*, 1995, **75**, 326.
25   J. Dorantes-Dávila and G. M. Pastor, *Phys. Rev. Lett.*, 1996, **77**, 4450.
26   J. Dorantes-Dávila and G. M. Pastor, *Phys. Rev. Lett.*, 1998, **81**, 208.
27   J. B. Mann, *Atomic Structure Calculations*, 1967, Los Alamos Sci. Lab. Rep. LA-3690.
28   (*a*) N. E. Christensen, O. Gunnarsson, O. Jepsen and O. K. Andersen, *J. Phys. (Paris)*, 1988, **49**, C8–17; (*b*) O. K. Andersen, O. Jepsen and D. Glötzel, *Highlights of Condensed Matter Theory*, ed. F. Bassani, F. Fumi and M. P. Tosi, North Holland, Amsterdam, 1985, p. 59.
29   (*a*) R. Félix-Medina, J. Dorantes-Dávila and G. M. Pastor, *Phys. Rev. B*, 2003, **67**, 094430; (*b*) R. Félix-Medina, J. Dorantes-Dávila and G. M. Pastor, *New J. Phys.*, 2002, **4**, 1.1–1.14.

Structure and chemical ordering in CoPt nanoalloys

G. Rossi,[a] R. Ferrando[a] and C. Mottet*[b]

Received 10th April 2007, Accepted 11th May 2007
First published as an Advance Article on the web 28th September 2007
DOI: 10.1039/b705415g

The structure and chemical ordering of CoPt nanoclusters in the size range of 1 to 3 nm are investigated by global optimization methods and Monte Carlo simulations using a many body potential derived from the tight binding model. For the smaller systems (number of atoms $N < 100$), the optimized clusters display a polyicosahedral-like atomic structure with a little core-shell chemical ordering characterized by a particular surface chemical configuration: some pentagonal or hexagonal Pt rings centered, respectively on a Co atom or a Co dimer. A transition to the decahedral symmetry occurs at about $N = 100$ atoms, with a pseudo $L1_0$ ordered phase in each tetrahedral unit. For larger cluster sizes, $201 < N < 1289$, the $L1_0$-ordered/disordered transition on the face centered cubic truncated octahedron is studied by canonical Monte Carlo simulations showing that the critical disordering temperature decreases with the cluster size. We also notice a Co surface segregation especially at edges and, possibly, (100) facets, depending on the cluster size, on both cubic and fivefold symmetry structures.

1. Introduction

Bimetallic clusters represent a large family of nanostructures where both the size and the chemical composition play a significant role in their structures and, as a consequence, in their chemical and physical properties. The order and phase stability in bulk alloys[1] together with the surface segregation in surface alloys[2] have been intensively studied over the last twenty years. The same issues are now investigated concerning finite-size systems.[3,4] The structure and chemical ordering of bimetallic clusters, now referred to as "nanoalloys", have given rise to many experimental[5–10] and theoretical[11–17] studies over the last ten years. Recently, global optimization approaches have been applied to resolve the structure of small Cu–Au nanoalloys,[18] and Ag–M, (M = Cu, Ni, Pd, Au) systems.[19] The core–shell polyicosahedral structures are particularly stable in systems with a large size-mismatch where the larger atom has the tendency to segregate at the surface.[20] The driving force for the surface segregation[3] being not only the size-mismatch but also the difference in the surface energy between the two elements and the tendency to phase separation, some systems such as Ag–Ni or Ag–Cu are particularly good candidates to form stable core-shell polyicosahedral structures.[19,20] Instead, systems like Cu–Au with large misfits but a much less surface segregation tendency (they have a chemical tendency

[a] INFM and IMEM/CNR, Dipartimento di Fisica dell'Università di Genova, via Dodecaneso 33, 16146 Genova, Italy. E-mail: ferrando@fisica.unige.it
[b] CRMCN-CNRS, Campus de Luminy, case 913, 13288 Marseille, France. E-mail: mottet@crmcn.univ-mrs.fr

to make ordered compounds instead of phase demixing), present less commonly the polyicosahedral structure.[19] It is worth noticing that such studies using semi-empirical potentials can be supplemented by *ab initio* density-functional theory calculations in order to check the stability of such structures (at least in what concerns the smallest sizes) and also characterize their electronic structure.[21,22]

In this study, we are interested in the Co–Pt alloy at equiconcentration, which can be classified in the second category of systems, *i.e.* in systems like Cu–Au, which form an $L1_0$-ordered phase at low temperature with an alternance of homogeneous atomic plans along a (100) axis. Among nanoalloys, Co–Pt nanoparticles are very promising systems in the domain of high density magnetic storage because of their magnetic properties[8,23] but they also present interesting catalytic properties as shown in an experimental study of CO adsorption on the PtCo(111) surface.[24] Density functional calculations have been performed on Pt-based bimetallic clusters concerning the adsorption of O, OH and H_2O[25] or the adsorption of specific intermediates of the oxygen reduction.[26] We propose to characterize the optimized mixed chemical and atomic structure and see how the $L1_0$ bulk ordering phase is accommodated in nanoclusters taking into account the symmetry change at small sizes. We show, using a semi-empirical potential, that the optimized structures, in the size range $N < 100$, have a weak core–shell chemical ordering as compared to the one obtained in the Ag–M, (M = Cu, Ni, Pd, Au) systems.[19] In particular, the surface is not homogeneous but displays a Co segregation on vertices, surrounded by Pt atoms forming pentagons or hexagons. For $N > 100$, the clusters adopt the decahedral structure accommodating a pseudo $L1_0$ ordering with an alternance of pure planes both in each (100) facet direction and along the fivefold symmetry axis. This chemical arrangement, which is not possible in the face centered cubic (fcc) symmetry (giving a well known competition between different variants) is allowed in the decahedral one, thanks to the particular stacking of the pentagonal layers along the fivefold symmetry axis. For larger sizes ($201 < N < 1289$), since the structure changes again to adopt the bulk-like fcc one, we have studied the $L1_0$-order/disorder transition by canonical Monte Carlo simulations. The article is organized as follows. In the next section we present the energetic and simulation models. Two different approaches are adopted: a global optimization method to resolve the ground state structure of the smaller clusters ($N < 150$), and a canonical Monte Carlo simulation to study the order/disorder transition and segregation effect on larger clusters ($201 < N < 1289$). The results are described in the third and forth sections and we give a brief conclusion in the last section.

2. The energetic and statistical models

2.1 Energetic model

We use the many-body tight binding potential based on the approximation of the density of states to its second moment.[27] In this framework, the energy at site i can be written as the sum of two terms:

—an attractive band term coming from the band structure which takes into account the many-body effect by the square root dependence of the neighboring atoms according to the following expression:

$$E_i^b = -\sqrt{\sum_{j,r_{ij} < r_{\alpha\beta}^{cut1}} \xi_{\alpha\beta}^2 e^{-2q_{\alpha\beta}\left(\frac{r_{ij}}{r_{\alpha\beta}^0}-1\right)}} \tag{1}$$

—a repulsive term of the Born–Mayer type which can be written as follows:

$$E_i^r = \sum_{j,r_{ij} < r_{\alpha\beta}^{cut1}} A_{\alpha\beta} e^{-p_{\alpha\beta}\left(\frac{r_{ij}}{r_{\alpha\beta}^0}-1\right)} \tag{2}$$

Table 1 Parameters for homo and hetero-atomic interactions and the solution energies fitted from the enthalpy formation of the alloys taken from ref. 30

α	β	$A_{\alpha\beta}/$ eV	$P_{\alpha\beta}$	$\xi_{\alpha\beta}/$ eV	$q_{\alpha\beta}$	$r_{\alpha\beta}^{\mathrm{cut1}}/$ nm	$r_{\alpha\beta}^{\mathrm{cut2}}/$ nm	E_{sol} (α in β)/ eV at.$^{-1}$	E_{sol} (β in α)/ eV at.$^{-1}$
Co	Co	0.1888	8.80	1.9066	2.96	0.354	0.433		
Pt	Pt	0.2424	11.14	2.5060	3.68	0.392	0.480		
Co	Pt	0.2447	9.97	2.3848	3.32	0.392	0.433	−0.472	−0.653

where r_{ij} is the distance between the atoms at sites i and j. Depending on the chemical nature of the atom at site i: α = A, B and at site j: β = A, B, the nearest-neighbors distance $r_{\alpha\beta}^{0}$ coincides with the homogeneous metal distance (if $\alpha = \beta$) or the average distance between the two homogeneous metal distances (if $\alpha \neq \beta$): $\frac{r_{AA}^{0}+r_{BB}^{0}}{2}$. The parameters ($A_{\alpha\beta}$, $P_{\alpha\beta}$, $q_{\alpha\beta}$, $\xi_{\alpha\beta}$) are fitted to the bulk experimental values: cohesive energy, lattice parameter[28] and elastic constants[29] for homoatomic interactions ($\alpha = \beta$); dissolution energy of one impurity of α in substitution in the bulk of β for heteroatomic interactions ($\alpha \neq \beta$) giving two equations to fit $A_{\alpha\beta}$ and $\xi_{\alpha\beta}$. $P_{\alpha\beta}$ and $q_{\alpha\beta}$ are taken as the average between the values of the homogeneous constituents. The misfit between the two constituents involves significant atomic relaxations around the impurity which are taken into account in the fitting procedure. The values of the parameters are reported in Table 1 together with the solution energies deduced from the formation enthalpies.[30] The cut-off distance $r_{\alpha\beta}^{\mathrm{cut1}}$ is the distance after which the potential is extended by a polynomial function that vanishes at $r_{\alpha\beta}^{\mathrm{cut2}}$. $r_{\alpha\beta}^{\mathrm{cut1}}$ and $r_{\alpha\beta}^{\mathrm{cut2}}$ are, respectively the second and third neighbor distance of the homogeneous metal for $\alpha = \beta$. When $\alpha \neq \beta$, $r_{\alpha\beta}^{\mathrm{cut1}}$ is the second neighbor distance of the metal with the larger lattice parameter and $r_{\alpha\beta}^{\mathrm{cut2}}$, the third neighbors distance of the metal with the smaller lattice parameter (see Table 1).

The total energy of a system of N atoms can then be written as follows:

$$E_N = \sum_{i=1}^{N}(E_i^b + E_i^r) \tag{3}$$

Within this model, the formation energies of the main known phases have been calculated (at 0 K) and listed in Table 2 together with the experimental values. We notice that our model reproduces well the main structures of the phase diagram of the Co–Pt system. The corresponding lattice parameters of the tetragonal $L1_0$ phase are also in good agreement with the experimental ones: $a = 0.39$ nm (compared to the experimental value of: 0.3793 nm[30]) and $c = 0.35$ nm (compared to the experimental value of: 0.3675 nm[30]). The tetragonalization factor of our model is then $c/a = 0.9$ compared to the experimental value of $c/a = 0.97$.

2.2 Simulation methods

Depending on the size of the system, we use different statistical methods to resolve the atomic structure and the associated chemical ordering. For small sizes, comprised between 20 and 150 atoms, the ground state structures (at 0 K) are determined

Table 2 Formation energies of the $A1$, $L1_0$ and $L1_2$ phases of the CoPt and CoPt$_3$ systems. The experimental values are taken from ref. 30

Phase	$\Delta E_{\mathrm{calc}}/\mathrm{eV}$ at.$^{-1}$	$\Delta E_{\mathrm{exp}}/\mathrm{eV}$ at.$^{-1}$
$A1$	−5.26	−5.16
$L1_0$	−5.35	−5.33
$L1_2$	−5.62	−5.63

by global optimization algorithms coupling standard Basin Hopping (BH) and Parallel Excitable Walkers (PEW).[31] For larger size systems, up to 1289 atoms and for bulk reference, we use Monte Carlo simulations[32] with the Metropolis algorithm[33] in the canonical ensemble where we keep constant the number of atoms N, the pressure ($P = 0$) and the temperature (T).

2.2.1 Global optimization.

BH and PEW have been used to globally optimize a variety of systems, like Lennard-Jones (LJ) clusters,[34–36] molecular clusters, peptides, polymers, and glass-forming solids.[37] Its effectiveness in dealing with multiple-funnel surfaces has also been proved, as in the case of LJ_{38} systems.

2.2.1 A BH algorithm.

BH method is described in detail elsewhere (as a general reference, see Wales[37]). Here we just remind that a BH search explores a transformed Potential Energy Surface (PES). Transformation consists in applying a local minimization procedure to all the points in the configuration space:

$$\tilde{E}(X) = \min\{E(X)\} \qquad (4)$$

During cluster optimization, the sampling of the configuration space is performed by means of a move upon atom coordinates. Moves from a starting local minimum configuration to a destination local minimum configuration can be accepted or refused according to a standard Metropolis algorithm (see below).

The moves implemented in the BH code are the following.

(i) **Shake move.** Every atom of the cluster is displaced within a spherical shell centered in its initial position. Minimum radius is usually fixed to 0, while the maximum one is related to the lattice parameter of the chemical species involved. It is usually chosen to be equal or slightly lower than half the first neighbor distance. Maximum radius has been fixed here to 1.3 Å.

(ii) **Shell move.** It is especially designed to get a better arrangement of the cluster surface. First, an atom i of the cluster is randomly chosen among the less coordinated, that is having less than 7 nearest neighbors. The maximum distance d_{max} between cluster atoms and the geometric center of the cluster is evaluated, and i is displaced to a randomly chosen position within a shell of fixed thickness (shell thickness being equal to 1.5 Å), centered in the geometric center and of minimum radius d_{max}. Moves that cause a negligible displacement of i are discarded.

(iii) **Exchange move.** The positions of two atoms of different species are swapped.

2.2.1 B PEW algorithm.

The BH search can be assisted by the definition and use of a geometrical order parameter, which helps in singling out all the competing structural families encountered during an optimization run. As an order parameter in PtCo clusters, we have used the signatures coming from Common Neighbor Analysis[38] (CNA). In particular, the percentage of (5,5,5) signatures, $p_{5,5,5}$, allows us to clearly distinguish fcc structures ($p_{5,5,5} = 0\%$), Dh structures ($0.5\% < p_{5,5,5} < 1.5\%$ in the size range considered), and Ih structures ($4\% < p_{5,5,5} < 10\%$ in the size range considered). In order to boost BH efficiency by the introduction of the order parameter, the large PtCo clusters ($N > 75$) were optimized by the PEW algorithm. PEW is particularly suitable for dealing with multiple funnel PES. In PEW, n_w walkers evolve in parallel on the BH-transformed PES. Two walkers, a and b, are neighbors in the order parameter space if the relation:

$$|p_{(5,5,5)}^a - p_{(5,5,5)}^b| < \delta \qquad (5)$$

is satisfied, δ being usually chosen in such a way that $2n_w\delta$ covers half the order parameter space. Moves are either accepted or refused again according to the Metropolis rule, but the energies attributed to each walker depend on the presence of neighbors in the order parameter space. Therefore, if a walker has no neighbors in the order parameter space, the choice is made just as in standard BH. On the other

Table 3 Input settings for BH and PEW global optimisations

	Temperature/K	n_w	δ (%)	E_{exc}/eV
BH	1500–2000	1	—	—
PEW	300–500	2–10	0.5–2	0.4–0.6

hand, if a walker lying in its starting position has at least one neighbor, its transformed potential energy $\tilde{E}$ is substituted by

$$E^* = \tilde{E} + E_{exc} \qquad (6)$$

with $E_{exc} > 0$. This means that the walker acts as being excited to an energy level placed higher than $\tilde{E}$ by the quantity E_{exc}. PEW reduces to BH for $n_w = 1$.

Table 3 reports the typical values of T, n_w, δ and E_{exc} used in BH and PEW global optimizations.

2.2.2 Monte Carlo simulations. Whereas the preceding method looks for the minimum energy structure, *i.e.* the ground state at 0 K, the Monte Carlo simulation describes the various chemical configurations as a function of the temperature on a given structure family with a local atomic relaxation of the interatomic distances. In one macrostep of the Monte Carlo simulation we propose N times (N being the atoms number) the successive trials:

(i) a random *displacement* of a randomly chosen atom; the maximum amplitude of the displacement is of the order of magnitude of the mean-square displacement at the corresponding temperature and must be fitted in order to keep an average acceptance percentage of about 50%;

(ii) an *exchange* between two randomly chosen atoms;

The acceptance of the trial is based on the standard Metropolis algorithm: the probability to keep the new configuration is equal to:

$$P_{new} = \text{Min}\left[1, \exp\left(-\frac{\Delta E}{kT}\right)\right] \qquad (7)$$

where k is the Boltzmann constant, and ΔE is the energy difference between the new and the old configurations. As a consequence, if $\Delta E \leq 0$, the trial is accepted because it lowers the energy of the system. If not, the trial is accepted according to the Boltzmann probability. In that case, we choose a random number η, with $0 < \eta \leq 1$, then if $\exp(-\frac{\Delta E}{kT}) \leq \eta$, the trial is accepted. We propose three displacements for one exchange in order to allow a better structure relaxation.

In the bulk case, where periodic conditions are applied to the simulation box, there is another trial which consists in the relaxation of the box size to ensure the pressure conservation (in our case, $P = 0$). Indeed, in the presence of a size mismatch between the two types of atoms, the alloy ordering leads to a deformation of the lattice parameter as compared to the disordered phase. This is for example a tetragonalization for the $L1_0$ structure. This must be introduced in order to modify the translation vector in the application of the periodic conditions. So each ten exchanges, we propose:

(iii) a *box expansion* in one direction; the probability of acceptance of this trial is equal to:

$$P_{new} = \text{Min}\left[1, \exp\left(-\frac{(\Delta E + P\Delta V - NkT\Delta \ln V)}{kT}\right)\right] \qquad (8)$$

where V is the volume of the box.

The number of macrosteps, which is independent of the size of the system, should be sufficiently high in order to reach the equilibrium at each temperature. Usually we take 2000 to 10 000 macrosteps to fulfil this condition. The total number of

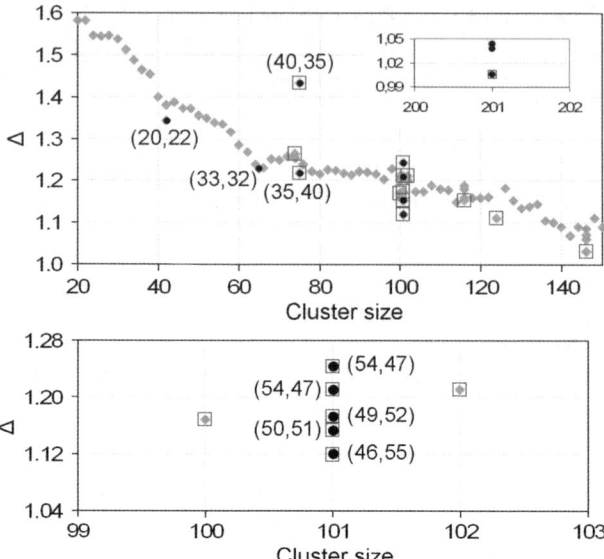

Fig. 1 Δ values for global minimum and local minimum clusters. Grey rhombi refer to clusters whose composition is $Pt_{0.5}Co_{0.5}$. Black dots refer to clusters with a different composition, always given in brackets. Open squares indicate decahedral clusters. In the low part of the sequence, the first minima of Δ corresponds to $Pt_{20}Co_{22}$, and the next minima corresponds to structures built as a packing of such pc6-like structures. From $N = 100$ on, decahedral structures get the lowest Δ both at magic and non-magic sizes.

elementary Monte Carlo steps N_{MC} then depends on the number of microsteps N_{micro} which depends on the system size according to the following expression: $N_{MC} = N_{macro}N_{micro} = N_{macro}(n_{disp} + n_{ex} + n_{box})N$ where N is the number of atoms, n_{disp}, n_{ex} and n_{box} are the number of different trials, (respectively the displacements, the exchanges and the box size).

3. Global optimization results

PtCo clusters were globally optimized by means of BH and PEW in the size range $10 < N < 150$. Within this size range, magic sizes for homogeneous clusters are known to adopt the icosahedral (Ih), decahedral (Dh), and fcc structures ($N = 38$ is magic for the fcc motif, 55 and 147 are magic for Mackay icosahedra, 75, 101 and 146 are magic sizes for the decahedral arrangement). In bimetallic equiconcentrated systems the global minimum clusters display a polyicosahedral-like symmetry for $N < 100$. The only exception to this trend is $Pt_{28}Co_{28}$, which adopts the icosahedral symmetry to recover the symmetry of the homogeneous magic cluster at 55 atoms. Decahedra are more and more common as the size increases beyond $N = 100$. On the whole, the sequence of global minimum structures can be split into two structural families.

As an index of the stability of a generic cluster A_nB_m, we refer to Δ, namely the difference between the energy of the cluster and the cohesive energy of n atoms of species A and m atoms of species B into the bulk metal. Δ for a Pt_nCo_m cluster is thus defined as:

$$\Delta = \frac{E(Pt_nCo_m) - nE_{coh}^{Pt} - mE_{coh}^{Co}}{N^{2/3}} \tag{9}$$

$N^{2/3}$ is introduced to take into account the increase of surface atoms with size. Fig. 1 shows the Δ value as a function of cluster size. Optimizations have first been applied to $Pt_{0.5}Co_{0.5}$ composition clusters. Then, in some cases, the composition has been slightly varied in order to achieve a better matching between the chemical and structural order.

3.1 Size range: $N < 100$

The first family characterizes sizes up to 100 atoms. Within this size range, most of the global minimum structures present common features:

(i) Very large (5,5,5) signature percentages.

(ii) A tendency of cobalt to segregate to the surface of the clusters, and an alternance between Co-rich and Pt-rich shells from the surface to the center.

(iii) The same surface pattern.

As regards to point (i), the order parameter $p_{5,5,5}$ is always larger than 35%. Concerning the (ii) statement, it derives both from the analysis of the distribution of homogeneous and heterogeneous bonds, and from the study of the atomic radial distributions. In fact, homogeneous Pt–Pt bonds cover from 20% to 30% of the total number of bonds in the clusters (which is comparable to the percentage of homogeneous bonds in the bulk $L1_0$ phase), Co–Co bonds are found to cover 10–20% of the total number of bonds, while heterogeneous Pt–Co bonds cover from 60% to 70% of the total number of bonds (which is also comparable to the percentage of heterogeneous bonds in the bulk $L1_0$ phase). This is consistent with a Co-rich cluster surface, the cobalt atoms being less coordinated than the platinum atoms, and with a possible ordering inside the cluster. As regards to the radial distributions, Fig. 2 shows the radial distribution of Pt and Co atoms in the global

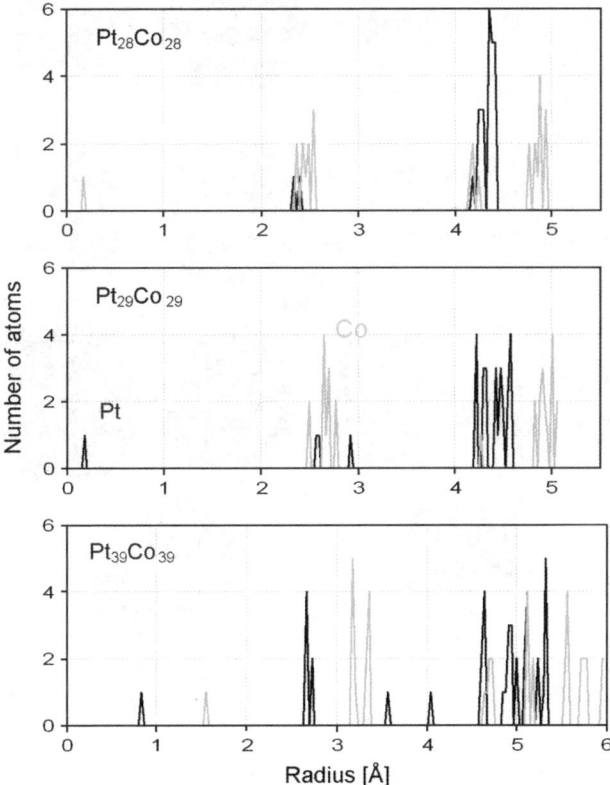

Fig. 2 Radial distribution of Co and Pt atoms in $Pt_{28}Co_{28}$ (Ih), $Pt_{29}Co_{29}$ and $Pt_{39}Co_{39}$ (pc6). Distances are calculated with respect to the geometrical center of the structure. Both clusters exhibit an alternance between Pt and Co shells, with the outer shell made up of Co atoms.

minimum structures found at compositions $Pt_{28}Co_{28}$, $Pt_{29}Co_{29}$ and $Pt_{39}Co_{39}$. All the clusters have a Pt atom close to the geometric center of the structure, and then exhibit an alternance of Co and Pt shells up to the Co-rich surface. The corresponding snapshots are shown in Fig. 3. Except for the $Pt_{28}Co_{28}$ configuration which adopts the icosahedral structure already predicted by Montejano-Carrizales[13] in the $Cu_{27}Pd_{28}$ system, all the clusters display a polyicosahedral-like structure which closely resembles the structure of the sixfold pancake (pc6), already found as the global minimum structure in a wide range of metal alloy nanoclusters, as AgCu, AgNi and AgPd.[19] We notice that the surface of the pc6 clusters is characterized by a combination of two chemically-ordered units. The first is made of a Pt pentagonal ring, having a Co atom at its center. The second is made of a Pt distorted hexagonal ring, surrounding a Co dimer. In both the surface structural units, the Co atoms do not lie in the same plane as the Pt ring, giving rise to the radial atomic distributions shown in Fig. 2. At variance with pc6, the $Pt_{20}Co_{22}$ global minimum has a cobalt

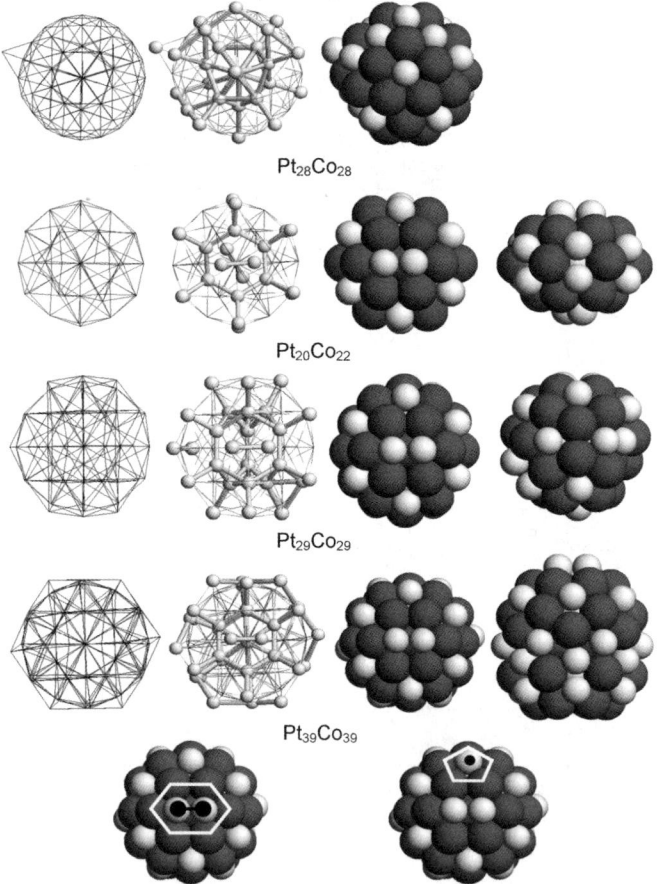

Fig. 3 Global minimum structures in the $N < 100$ size range. From the top: $Pt_{28}Co_{28}$ (Ih), $Pt_{20}Co_{22}$, $Pt_{29}Co_{29}$ and $Pt_{39}Co_{39}$ (pc6). $Pt_{20}Co_{22}$ (pc6) structure is quite similar to the sixfold pancake, found to be the global minimum structure in other metal systems. This structure can be considered the structural basic unit for many of the global minimum structures which one encounters in the sequence. Two examples are just $Pt_{29}Co_{29}$ and $Pt_{39}Co_{39}$. All these structures present some common surface patterns, as shown at the bottom. On the surface, in fact, both a pentagonal and an hexagonal Pt ring containing, respectively one and two Co atoms in the center are quite common. In all the snapshots, the Co atoms are represented as white spheres and the Pt atoms, as dark spheres.

dimer in place of a single cobalt atom on top of the two six-atom Pt rings. According to our model, the two dimers on the opposite sides of the clusters are not aligned, their directions forming a 60° angle. This structure can be considered the core unit for the global minimum clusters at the larger sizes. Both $Pt_{29}Co_{29}$ and $Pt_{39}Co_{39}$ replicate the $Pt_{20}Co_{22}$ structure along the sixfold symmetry axis, even if it is with a certain degree of chemical disorder.

3.2 Size range: $100 < N < 150$

The decahedral motif becomes more and more favorable as the size increases. Decahedral structures are highlighted by the open squares in Fig. 1.

At size $N = 74$ and 76, the global minimum structure at exact equiconcentration is not a decahedron, despite the fact that $N = 75$ is a magic size for such a motif. In principle, the decahedral arrangement allows accommodation of platinum and cobalt atoms by alternating cobalt and platinum planes along the (100)-like direction (see Fig. 4 and next section). This is just what happens in the ordered $L1_0$ bulk alloy. In the perfect Dh_{75}, moving from the central fivefold symmetry axis towards one of the (100) facets of the decahedron, one counts three such planes, and it is possible to artificially build the decahedron so as to alternate platinum and cobalt planes in the (100) direction. Two strategies are possible: Co (along the axis), Pt–Co–Pt (in the (100) facets), or Pt (along the axis), Co–Pt–Co (in the (100) facets). These two chemical orders correspond to the $Pt_{40}Co_{35}$ and $Pt_{35}Co_{40}$ compositions, respectively. The value of Δ for such clusters is reported in Fig. 1. The decahedron with Co on the (100) facets and around the central Pt axis is almost as stable as the global minimum, while the decahedron with Pt on the (100) facets and around the central Co axis is not stable at all. This is in agreement with the tendency of Co to segregate at the surface.

Around $N = 100$, the situation is different and the decahedral arrangement is the global minimum at size $N = 100$, 101, 102, both at the equiconcentration and in $Pt_{46}Co_{55}$, $Pt_{49}Co_{52}$ and $Pt_{54}Co_{47}$. At size $N = 101$, the global minimum cluster is $Pt_{46}Co_{55}$, which has the same alternations of Pt and Co planes as $Pt_{35}Co_{40}$. It is possible to state that starting from $N = 100$ global minimum structures enter a decahedral range, as this motif is also favorable at sizes which are not magic. As an example, one can consider the global minimum $Pt_{58}Co_{58}$ and $Pt_{62}Co_{62}$ clusters. Both present a decahedral arrangement (see again Fig. 4).

Increasing in size, another magic decahedral size can be found, namely $N = 146$. Only the $Pt_{73}Co_{73}$ cluster has been optimized, even if its composition does not allow the perfect arrangement of the cobalt and platinum planes in the same way as in $Pt_{35}Co_{40}$ or $Pt_{46}Co_{55}$. Cobalt atoms occupy all the sites on the fivefold symmetry axis, the intermediate plane along the (100) direction and (almost completely) the outer (100) facets.

According to semi-empirical calculations about homogeneous metal clusters,[39] as sizes increase the icosahedral range is dropped in favor of decahedral structures, and on approaching the bulk limit the fcc motifs become the most favorable. In the case of homogeneous Pt clusters, for example, stable decahedra cover a size range whose lower limit is placed at $N \sim 75$. In the case of homogeneous Co, the decahedral range starts at $N \sim 400$.

Dealing with heterogeneous particles, size is not the only element to affect the range of stability of the different structural motif. Composition and chemical order has to be taken into account, too. The global optimization results indicate that the decahedral range for the heterogeneous PtCo particles begins at $N \sim 100$. This could indicate that the possibility to accommodate the $L1_0$ chemical order in the Dh structure indeed favors these structures with respect to the Ih-like motifs. As regards to the transition from the Dh to fcc motif, calculations like those reported in ref. 39 suggest that decahedra and fcc motifs are competing already at $N \sim 600$, their Δ differing less than 0.001, so the transition can be thought to take place in the

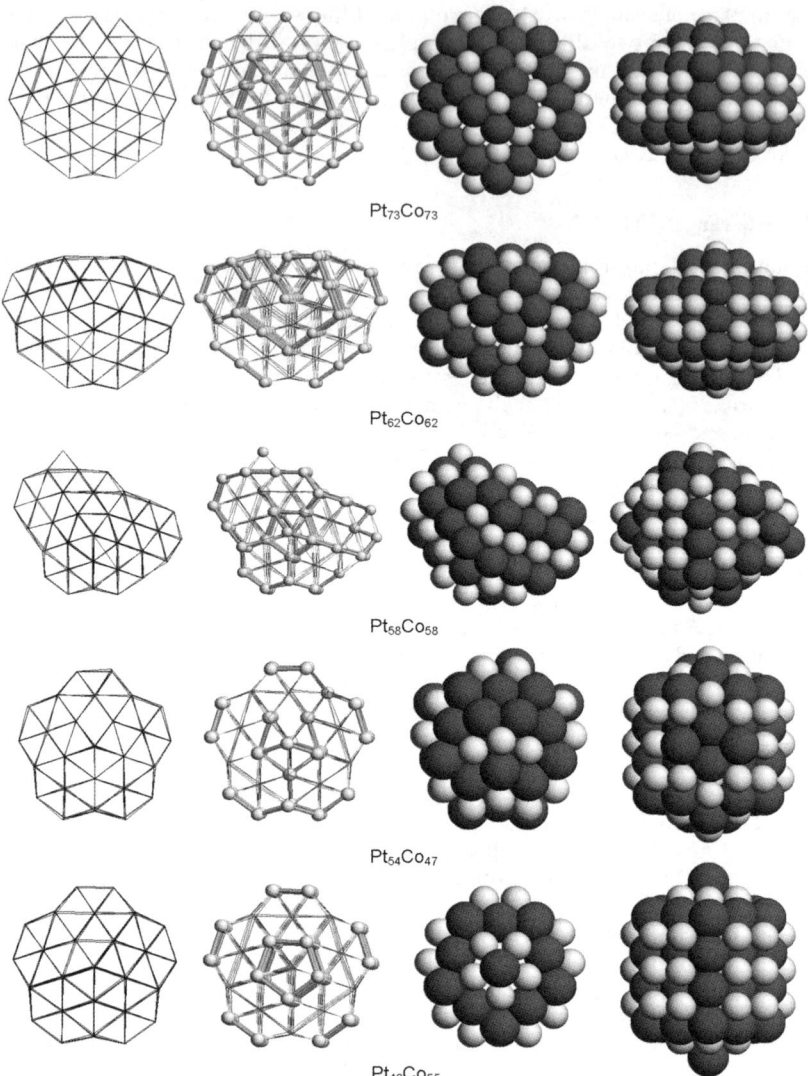

Fig. 4 Global minimum decahedral structures. Cobalt atoms occupy the sites of the outer (100) facets. Cobalt planes alternate with platinum planes along the (100) directions of the lateral facets, coherently with the chemical ordering of the $L1_0$ alloys. Moreover, there is another alternance of Co and Pt layers along the fivefold symmetry axis which consists in concentric pentagonal rings. As size increases beyond $N = 100$, global minima present a decahedral structure even at non-magic sizes (see for example $Pt_{62}Co_{62}$ and $Pt_{58}Co_{58}$ clusters). The color code is the same as in Fig. 3.

$600 < N < 1000$ size range. The fcc structures which will be analyzed later can thus be considered as low energy (and sometimes, lowest energy) cluster configurations.

4. Order/disorder phase transition

4.1 Bulk phase transition

The CoPt bulk alloy at equiconcentration is ordered at low temperature according to the tetragonal $L1_0$ structure which consists in an alternance of (100) planes which are pure in cobalt and platinum. The experimental critical temperature of order/disorder phase transition is equal to 1100 K. To determine such critical temperatures in our bulk theoretical system, we performed systematic canonical Monte Carlo simulations, increasing the temperature by 10 or 20 K. We start at a low temperature with the ordered $L1_0$ phase, using a sufficiently large box to prevent the effects of the periodic conditions. The convergence of the results is obtained for a box size of 512 atoms or more (the larger size we checked was a box of 768 atoms). In such cases, one point in the energy curve using 5000 Monte Carlo macrosteps takes 10 to 16 million elementary Monte Carlo steps. The energy curve, plotted in Fig. 5, displays a sharp break in the slope which is characteristic of the order/disorder transition. This transition is also well characterized by the variation of the order parameters (long and short range[40]) and by the variation of the lattice parameters due to the tetragonalization (ratio c/a) at low temperature, which disappears at high temperature. The long range order parameter (lro) is defined according to the definition of Cowley[40] by:

$$lro = \frac{n_A}{n}\left(\frac{p_A - c_A}{1 - c_A}\right) + \frac{n_B}{n}\left(\frac{p_B - c_B}{1 - c_B}\right) \tag{10}$$

where n_A, n_B and n are the numbers of sublattices which are occupied by A (resp. B) atoms and the total number of sublattices (equal to 4 in the fcc lattice), p_A and p_B are the occupation probabilities of each sublattices and c_A and c_B are their respective concentrations. In the $L1_0$ structure, this parameter is given by:

$$lro_{L10} = p_A + p_B - 1 \tag{11}$$

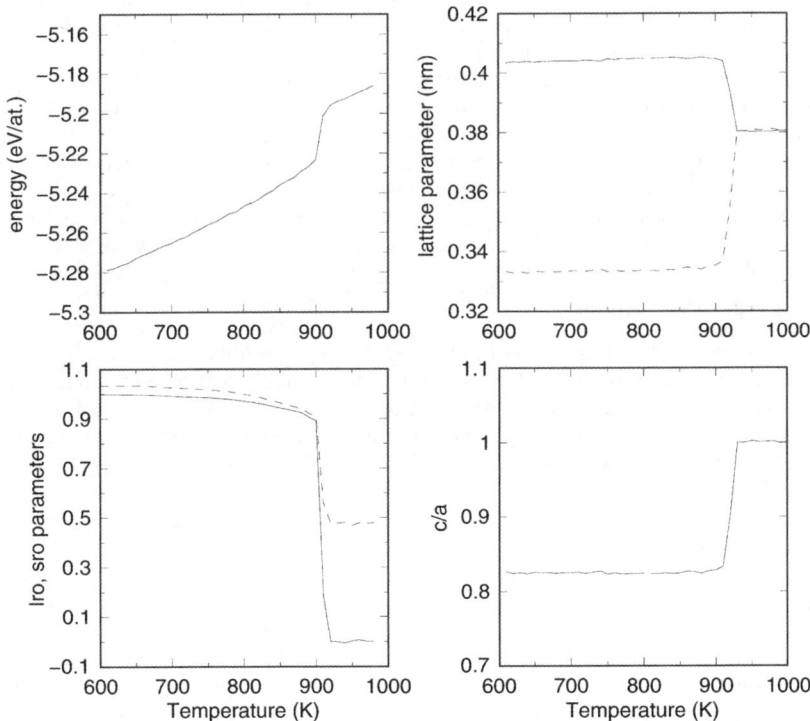

Fig. 5 Bulk order/disorder transition. In the left column, the calorimetric curve and the order parameters: lro with solid line and sro with dashed line. In the right column, the lattice parameters of the tetragonal lattice (a with solid line and c with dashed line) and the ratio (c/a).

The order/disorder transition in the bulk is entirely characterized by the variation of the long range order parameter. We have also determined a short range order (sro) parameter in the following way:

$$\text{sro} = 1 - \frac{m - m_0}{c_A - m_0} \tag{12}$$

where m is the number of mixed bonds (first neighbors bonds) in the actual system and m_0 is the number of mixed bonds in the ordered phase i.e. $m_0 = 8/12$ in the $L1_0$ structure. So we have:

$$\text{sro}_{L1_0} = 3(2m - 1) \tag{13}$$

By definition, these parameters are equal to one in the ordered phase and zero in the disordered phase.

We note in Fig. 5 that the lro parameter is equal to 1 at low temperature, then decreases abruptly to 0 at the temperature of the order/disorder transition. Besides, the sro parameter is also changing sharply but without reaching complete disorder because the structure keeps a local order in spite of the long range disordering. It is interesting to see however that the transition for the two parameters takes place at the same temperature, meaning that the sro parameter can be used instead of the lro one for the clusters where the lro parameter is basically less well defined. The lattice parameter of the disordered phase is equal to 0.38 nm which is in nice agreement with the experimental value of 0.375 nm. The ordered $L1_0$ phase, below 900 K displays a tetragonalization with $a = 0.4$ nm and $c = 0.33$ nm ($c/a = 0.83$). The deformation is larger at high temperature where we performed our calculations ($T > 600$ K) than at 0 K ($c/a = 0.9$ as mentioned in Section 2.1) in our model. In fact, the lattice parameters are sensitive to the temperature and the values of Fig. 5, which are taken at relatively high temperature close to the transition, include a thermal expansion of 2.5% laterally and nearly 6% vertically. Anyway, even at room temperature or near 0 K, our simulations have the tendency to overestimate the tetragonalization in the bulk phase (0.9 against 0.97 experimentally). The critical temperature of ordering is located around 900 K and presents a difference of about 20% as compared to the experimental one (1100 K). Finally, it is to be noticed that there is an important hysteresis (about 150 K) that we can evaluate when we cool the system down starting from the highest temperature configuration. So we should estimate the critical temperature of the order/disorder transition rather as being located between 750 K $< T_c^{\text{bulk}} < 900$ K.

4.2 Cluster phase transition

We apply the same kind of canonical Monte Carlo simulations to clusters of different sizes and structures. Let us begin with the fcc structure with sizes of 201 atoms (1.5 nm), 314 atoms (1.9 nm), 405 atoms (2 nm), 807 atoms (2.5 nm) and 1289 atoms (3 nm). The concentration is not exactly 0.5 in order to optimize the $L1_0$ type ordering. Indeed, as we have noticed in Section 3 concerning ordering on small clusters, the optimized stoichiometry with respect to the $L1_0$ bulk phase (as if we extract a fragment of the $L1_0$) deviates slightly from the equiconcentration in the clusters.[15] Such an exact equiconcentration, i.e. $c = N/2$ would induce anti-site defects as compared to the perfect $L1_0$ structure. So we have to take into account that the cluster stoichiometry depends on the cluster size. The systems of interest are the following: $(Co_{96}Pt_{105})_{201}$, $(Co_{176}Pt_{138})_{314}$, $(Co_{197}Pt_{208})_{405}$, $(Co_{399}Pt_{408})_{807}$ and $(Co_{632}Pt_{657})_{1289}$. In Fig. 6 the sro parameters of the different clusters are plotted. All the clusters display an order/disorder transition but with a variable degree of abruptness. As the size decreases, the transition is smoother and smoother until the smaller size of 201 atoms where the sro parameter varies quite continuously from 400 K to 800 K. As compared to the sharp transition in the bulk, the transitions in the clusters are continuous as has been predicted theoretically on the FePt system.[41]

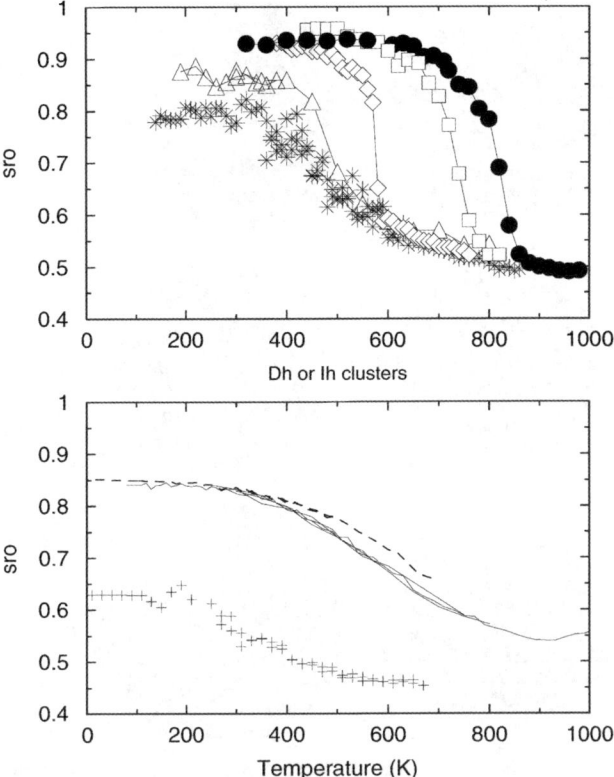

Fig. 6 sro parameter of clusters. Top graph: fcc clusters with 1289 at. (circles), 807 at. (squares), 405 at. (diamonds), 314 at. (triangles) and 201 at. (stars). Bottom graph: Dh and Ih clusters with Dh_{318} (solid line), Dh_{146} (dashed line) and Ih_{147} (plus).

We have checked that the hysteresis disappears for sizes smaller or equal to 405 atoms. The critical temperature is clearly decreasing as a function of cluster size as has been observed for Cu_3Au clusters and FePt nanoparticles.[42–45] In our case we observe that from 1289 atoms (3 nm) down to 314 atoms (1.9 nm), the disordering temperature goes from 820 K (1289 atoms) to 730 K (807 atoms), 550 K (405 atoms) and finally to 450 K (314 atoms) or less for the smaller sizes.

Some of the ordered and disordered fcc clusters are illustrated in Fig. 7. There are three main remarks relating to these structures:

(i) Even in the case of an optimized stoichiometry with respect to the $L1_0$ ordering, we notice that there are many anti-site defects, at the cluster surface, due to surface segregation. The Co surface concentration has been plotted in the left column of Fig. 8 together with the different site concentrations: edges, (100) and (111) facet sites present at the cluster surface. Co prefers the low coordinated sites, *i.e.* edges or (100) sites since their Co concentration is higher than 0.5 but Pt prefers the (111) sites since the Co concentration on the (111) sites is lower than 0.5. Co atoms have a tendency to segregate because the Co cohesive energy is lower than that for Pt. Moreover, the simple hierarchy based on broken-bonds arguments: preferential segregation on lower coordinated sites *i.e.* edges, then (100) and (111) facets is almost respected here, except for the (111) facets. On average, the surface is enriched with Co, but the segregation depends on the cluster size: it increases with cluster size from 201 at. to 1289 at. At larger sizes, we observe a slight transition in Co surface concentration corresponding to the order/disorder transition. Even if the edge segregation is still very strong at large sizes, its influence on the average surface segregation is weak

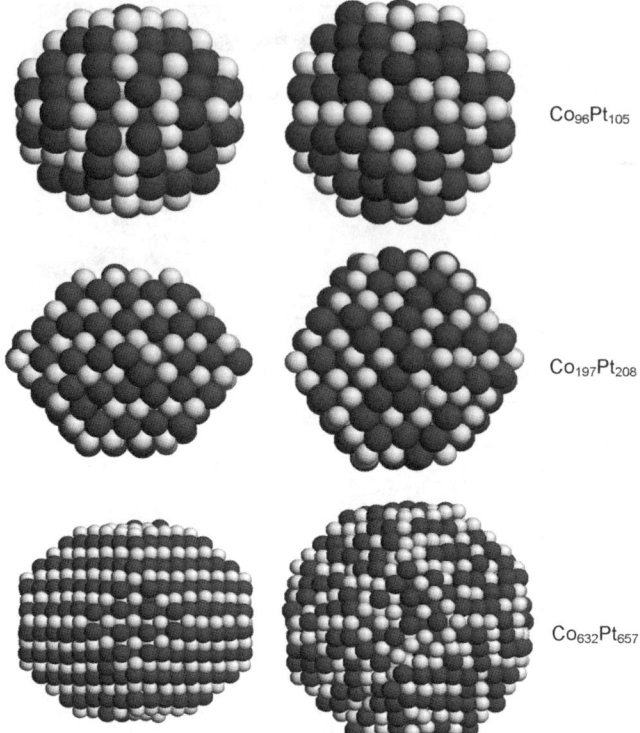

$Co_{96}Pt_{105}$

$Co_{197}Pt_{208}$

$Co_{632}Pt_{657}$

Fig. 7 Snapshots of the fcc clusters in the ordered low temperature (left column) and disordered high temperature (right column) phase: $(Co_{96}Pt_{105})_{201}$ at $T = 200$ K and $T = 500$ K, $(Co_{197}Pt_{208})_{405}$ at $T = 300$ K and $T = 650$ K and $(Co_{632}Pt_{657})_{1289}$ at $T = 500$ K and $T = 980$ K. The color code is the same as in Fig. 3.

because the proportion of edge sites is less and less important with increasing size. For the 1289 at. cluster, the average surface concentration of Co is similar to the (100) facet concentration, whereas at 201 and 405 at. sizes, the edge segregation is dominant and the (100) segregation is weak or even inversed in the 201 at. cluster (the (100) facets being reduced to one atom!)

(ii) From the ordering point of view, we notice that the 201 at. cluster is not well ordered with respect to the bulk $L1_0$ phase. It is geometrically frustrated because of a competition between two variants, *i.e.* the alternance of pure planes in two perpendicular (100) directions instead of only one as in the 405 at. and 1289 at. clusters (see Fig. 7).

(iii) The ordered clusters undergo a tetragonal deformation as observed in the bulk phase. We measured the aspect ratio of the cluster *i.e.* the height H along the axis of the pure planes alternance over the width W parallel to these planes. At low temperature, we obtain a ratio of H/W equal to 0.90 for the smaller clusters (<2 nm), that means no larger than in the bulk, and equal to 0.86 for the larger one (3 nm), that means a larger tetragonalization than in the simulated bulk. We checked that $H/W = 1$, *i.e.* the clusters are isotropic, at high temperature (right column of Fig. 7). There is another way to characterize the tetragonalization (c/a) by the ratio $d_{AB} \times 2/(d_{AA} + d_{BB})$ where d_{AB} are the average interatomic distances between atoms of type A = Co, Pt and B = Co, Pt. The different distances are plotted in the right column of Fig. 8. We see that the average interatomic distance is comparable to the one in the disordered bulk phase (0.38 nm) and slightly increases with increasing temperature. There is a clear modification of the homoatomic distances at least for cluster sizes larger or equal to 405 atoms: they decrease at the disordering transition

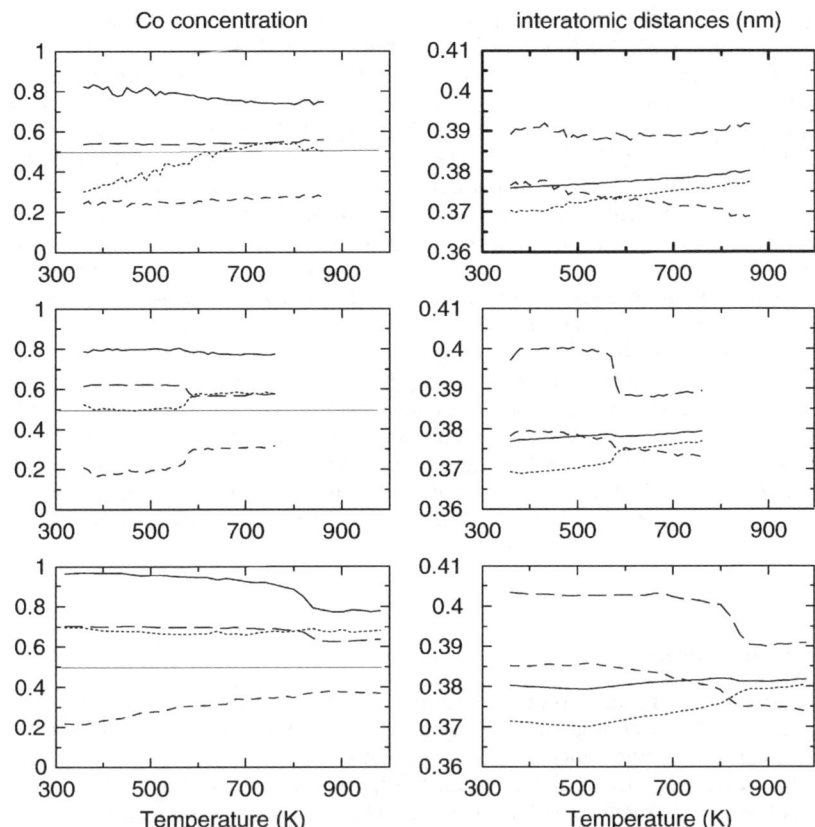

Co concentration | interatomic distances (nm)

Fig. 8 The different rows correspond to the clusters of Fig. 7. On the left column the surface and sites segregation curves of Co are plotted: average surface concentration (long-dashed line), edges concentration (solid line), (100) facet concentration (dotted line) and (111) facet concentration (dashed line). On the right column the different interatomic distances are plotted: the average distance (solid line), d_{CoPt} (dotted line), d_{CoCo} (dashed line), d_{PtPt} (long-dashed line).

whereas the heteroatomic distances increase slightly. The ratio c/a which would correspond to the ratio between the heteroatomic bonds and the homoatomic bonds is not exactly equal to 1 in the disordered clusters and is equal to about 0.95 (or 0.92 for the larger cluster) in their ordered state. Such results can be commented on in the context of a recent experiment[46] performed on CoPt clusters of 2 nm at 575 K which gave a measure of the interplanar distance $d_{111} = 0.221$ nm by high resolution transmission electron microscopy. It is larger than the bulk CoPt alloy where both the ordered $L1_0$ or disordered A_0 phase would give $d_{111} = 0.217$ nm. If we consider the a and c parameters as defined by $(d_{CoCo} + d_{PtPt})/2 = 0.39$ nm and $d_{CoPt} = 0.4$ nm in our ordered clusters, we get the same value of $d_{111} = 0.221$ nm (0.219 nm in the disordered state). The two different ways of characterizing the tetragonalization in the clusters are indicative of a certain profile of atomic relaxation in the clusters due to the surface relaxation. We note that the two methods give the same tendency as a function of the cluster size: the tetragonalization is amplified for sizes larger than 2 nm. This can be explained by the variation of the $L1_0$ ordering as a function of cluster size. At large size ($N > 400$ at.), the clusters select one variant whereas in the smaller size, the coexistence of two variants leads to a frustration of the tetragonalization.

Finally, we have considered the other preferential symmetries adopted by the clusters, *i.e.* the Dh and Ih symmetries. The sro parameters of the two quasi-isomers:

Dh_{146} and Ih_{146}, and a larger Dh_{318} are illustrated in Fig. 6. Whereas the Dh clusters display a quite well defined order/disorder transition with a pseudo-$L1_0$ ordering as described in Section 3, there is much less significant ordering transition on the Ih structure, although the sro changes from 0.6 at low temperature to 0.45 at high temperature. As mentioned in Section 3, there is probably a certain radial core-shell or multi-shell ordering in the Ih structure responsible for the slight variation in the sro parameter. However, it is worth noticing that on Pd–Pt clusters, Cheng et al.[47] found that the icosahedra of 147 and 309 atoms display a quasi perfect "onion-ring" structure. In our system, the bulk ordering tendency according to the $L1_0$ phase seems to play against a perfect radial shell ordering on the icosahedral structure. As a consequence, the perfect icosahedral structure is not stable, as seen in Section 3, except in the $N = 56$ atoms size, namely the $Pt_{28}Co_{28}$ structure illustrated in Fig. 3. On the other hand, the Dh with the pseudo $L1_0$ order is particularly stable, also against the fcc symmetry since it allows suppression of the geometrical frustration in the fcc structure arising from the competition between two possible variants. In the ordered Dh phase, these two variants are accommodated thanks to the pentagonal rings stacking along the fivefold symmetry axis which effectively alternates between pure Co and pure Pt layers while laterally the pseudo (100) planes alternate between pure Co and pure Pt planes, terminating with a Co plane to satisfy the Co surface segregation.

5. Conclusions

Using the global optimization method and Monte Carlo simulations, we have shown that the bulk $L1_0$ ordering tendency in the CoPt system induces particular structural and chemical ordering which are size-dependent:

—for $N < 100$ atoms, the optimized systems are polyicosahedral-like clusters (sixfold pancake structure) with a particular surface chemical ordering displaying the Pt pentagonal ring, having a Co atom in its center or a Pt hexagonal ring, surrounding a Co dimer. Such surface local ordering could be very promising for tuning catalytic properties. The Ih structure, with a weak radial shell chemical ordering, is energetically not stable except at the $N = 56$ size which corresponds to high stability of the magic Ih_{55} homogeneous cluster.

—for $N > 100$ atoms, the optimized systems are decahedral with a pseudo-$L1_0$ phase ordering in each tetrahedral unit together with an alternance of pure pentagonal rings layers along the fivefold symmetry axis. Such an ordered cluster is particularly stable since it removes the geometrical frustration existing in the fcc structure and we have observed this in small fcc clusters.

—for $N > N_c$ with $400 < N_c < 1000$ atoms, the fcc structures should be stabilized with the $L1_0$ ordering along one (100) axis, the tetragonalization of the lattice leading to a modification of the aspect ratio of the cluster.

The order/disorder transition in clusters has been investigated by canonical Monte Carlo simulations showing a clear tendency to lower the critical temperature of disordering as compared to the bulk disordering temperature. There is a clear tendency of the Co to segregate at edges and/or (100) facets, depending on cluster size and/or structure, having a close link with the cluster ordering.

Acknowledgments

We acknowledge financial support from the Italian CNR for the project SSA-TMN within the framework of the ESF EUROCORES SONS, and from European Community FP6 for the project GSOMEN (NMP4-CT-2004-001594). Giulia Rossi aknowledges financial support from L'Oréal Italia per le Donne e la Scienza.

References

1 F. Ducastelle, *Order and Phase Stability in Alloys, Cohesion and Structure*, vol. 3, North-Holland/Elsevier Science Publishers, Amsterdam, 1999.
2 G. Tréglia, B. Legrand, F. Ducastelle, A. Saùl, C. Gallis, I. Meunier, C. Mottet and A. Senhaji, *Comput. Mater. Sci.*, 1999, **15**, 196.
3 V. Moreno, J. Creuze, F. Berthier, C. Mottet, G. Tréglia and B. Legrand, *Surf. Sci.*, 2006, **600**, 5011.
4 F. Lequien, J. Creuze, F. Berthier and B. Legrand, *J. Chem. Phys.*, 2006, **125**, 094707.
5 S. Giorgio, C. Chapon and C. R. Henry, *Langmuir*, 1997, **13**, 2279.
6 J. L. Rousset, A. J. Renouprez and A. M. Cadrot, *Phys. Rev. B*, 1998, **58**, 2150.
7 J. L. Rousset, F. J. Cadete Santos Aires, B. R. Sehar, P. Mélinon, B. Prevel and M. Pellarin, *J. Phys. Chem. B*, 2000, **104**, 5430.
8 T. Ould Ely, C. Pan, C. Amiens, B. Chaudret, F. Dassenoy, P. Lecante, M.-J. Casanove, A. Mosset, M. Respaud and J.-M. Broto, *J. Phys. Chem. B*, 2000, **104**, 695.
9 B. Rellinghaus, O. Dmitrieva and S. Stappert, *J. Cryst. Growth*, 2004, **262**, 612.
10 B. Pauwels, G. Van Tendeloo, E. Zhurkin, M. Hou, G. Verschoren, L. Theil Kuhn, W. Bouwen and P. Lievens, *Phys. Rev. B*, 2001, **63**, 165406.
11 J. Jellinek and E. B. Krissinel, *Theory of Atomic and Molecular Clusters*, Springer, Berlin, 1999, 277.
12 L. Q. Yang and A. E. DePristo, *J. Catal.*, 1994, **148**, 575.
13 J. M. Montejano-Carrizales, M. P. Iiguez and J. A. Alonso, *Phys. Rev. B*, 1994, **49**, 16649.
14 E. Zhurkin and M. Hou, *J. Phys.: Condens. Matter*, 2000, **12**, 6735.
15 C. Mottet, G. Tréglia and B. Legrand, *Phys. Rev. B*, 2002, **66**, 045413.
16 F. Baletto, C. Mottet and R. Ferrando, *Phys. Rev. B*, 2002, **66**, 155420.
17 F. Baletto, C. Mottet and R. Ferrando, *Phys. Rev. Lett.*, 2003, **90**, 135504.
18 S. Darby, T. V. Mortimer-Jones, R. L. Johnston and C. Roberts, *J. Chem. Phys.*, 2002, **116**, 1536.
19 (a) A. Rapallo, G. Rossi, R. Ferrando, A. Fortunelli, B. C. Curley, L. D. Lloyd, G. M. Tarbuck and R. L. Johnston, *J. Chem. Phys.*, 2005, **122**, 194308; (b) G. Rossi, R. Ferrando, A. Rapallo, A. Fortunelli, B. C. Curley, L. D. Lloyd and R. L. Johnston, *J. Chem. Phys.*, 2005, **122**, 194309.
20 G. Rossi, A. Rapallo, C. Mottet, A. Fortunelli, F. Baletto and R. Ferrando, *Phys. Rev. Lett.*, 2004, **93**, 105503.
21 R. Ferrando, A. Fortunelli and G. Rossi, *Phys. Rev. B*, 2005, **72**, 085449.
22 G. Barcaro, A. Fortunelli, G. Rossi, F. Nita and R. Ferrando, *J. Phys. Chem. B*, 2006, **110**, 23197.
23 L. Favre, V. Dupuis, E. Bernstein, P. Melinon and A. Perez, *Phys. Rev. B*, 2006, **74**, 014439.
24 Y. Gauthier, M. Schmid, S. Padovani, E. Lundgren, V. Bus, G. Kresse, J. Redinger and P. Varga, *Phys. Rev. Lett.*, 2001, **87**, 036103.
25 P. B. Balbuena, D. Altomare, N. Vadlamani, S. Bingi, L. A. Agapito and J. M. Seminario, *J. Phys. Chem. A*, 2004, **108**, 6378.
26 S. R. Calvo and P. B. Balbuena, *Surf. Sci.*, 2007, **601**, 165.
27 V. Rosato, M. Guillopé and B. Legrand, *Philos. Mag. A*, 1989, **59**, 321.
28 C. Kittel, *Introduction to Solid State Physics*, 7th edn, Wiley, New York, 1996.
29 G. Simmons and H. Wang, *Single Crystal Elastic Constants and Calculated Aggregated Properties*, MIT, Cambridge, 1971.
30 R. Hultgren, P. D. Desai, D. T. Hawkins, M. Gleiser and K. K. Kelley, *Values of the thermodynamic properties of binary alloys*, American Society for Metals, Berkley, Jossey-Bass Publishers, 1981.
31 G. Rossi and R. Ferrando, *Chem. Phys. Lett.*, 2006, **423**, 17.
32 D. Frenkel and B. Smit, *Understanding Molecular Simulation*, Academic Press, 1996, p. 19.
33 N. Metropolis, A. W. Rosenbluth, M. N. Rosenbluth, A. N. Teller and E. Teller, *J. Chem. Phys.*, 1953, **21**, 1087.
34 D. J. Wales, J. P. K. Doye, A. Dullweber, M. P. Hodges, F. Y. Naumkin, F. Calvo, J. Hernandez-Rojas and T. F. Middleton, *The Cambridge Cluster Database*, http://www-wales.ch.cam.ac.uk/CCD.html.
35 D. J. Wales and J. P. K. Doye, *J. Phys. Chem. A*, 1997, **101**, 5111.
36 R. H. Leary and J. P. K. Doye, *Phys. Rev. E*, 1999, **60**, R6320.
37 D. Wales, *Energy Landscapes*, Cambridge University Press, Cambridge, 2003.
38 D. Faken and H. Jonsson, *Comput. Mater. Sci.*, 1994, **2**, 279.
39 F. Baletto, R. Ferrando, A. Fortunelli, F. Montalenti and C. Mottet, *J. Chem. Phys.*, 2002, **116**, 3856.
40 J. M. Cowley, *Phys. Rev.*, 1950, **77**, 669.

41 R. V. Chepulskii and W. H. Butler, *Phys. Rev. B*, 2005, **72**, 134205.
42 H. Yasuda and H. Mori, *Z. Phys. D*, 1996, **37**, 181.
43 T. Tadaki, T. Kinoshita, Y. Nakata, T. Ohkubo and Y. Hirotsu, *Z. Phys. D*, 1997, **40**, 493.
44 T. Schülli, J. Trenkler, I. Mönch, D. Le Bolloc'h and H. Dosch, *Europhys. Lett.*, 2002, **58**, 737.
45 T. Miyazaki, O. Kitakami, S. Okamoto, Y. Shimada, Z. Akase, Y. Murakami, D. Shindo, Y. K. Takahashi and K. Hono, *Phys. Rev. B*, 2005, **72**, 144419.
46 A. Hannour, L. Bardotti, B. Prével, E. Bernstein, P. Mélinon, A. Perez, J. Gierak, E. Bourhis and D. Mailly, *Surf. Sci.*, 2005, **594**, 1.
47 D. Cheng, W. Wang and S. Huang, *J. Phys. Chem. B*, 2006, **110**, 16193.

This journal is © The Royal Society of Chemistry 2008

General Discussion

Professor Johnson opened the discussion of Professor Broyer's paper: My question is concerned with the so-called 'Molecular Effect'. You argue that the 'anomalous' behaviour of NiAg alloy is due to the formation of Ag islands on the central Ni core, in other words there is no coherent film of Ag. This I accept but wonder why such a phenomenon also appears to happen for pure Ag particles and not for AuAg systems?

Professor Broyer answered: This effect is due to the fact that the number of silver atoms is not sufficient to form a complete shell. This effect does not exist for Au/Ag clusters because they form an alloy.

Dr Baletto remarked: Please comment on the optical spectra and on the signature of core–shell *versus* alloyed structures, in particular of AgNi.

Professor Broyer replied: Concerning the optical signature of core–shell *versus* alloy structure in mixed clusters, I think that this optical signature exists if the optical properties of alloys (namely the dielectric constants in the bulk) are known. This is clear for Au/Ag clusters for which the optical spectra of alloy clusters are different from those of the core–shell structure both experimentally and "theoretically". This means that we can compare experiments with model calculations for alloy and core–shell structure.

For Ni/Ag, we do not have dielectric constant of the Ni/Ag alloy which does not exist in the bulk. We only compare to a model cluster having a weighted averaged dielectric function of both pure materials. this "alloyed homogeneous model cluster" corresponds to a mixture of both materials like perhaps a solid solution. Clearly the spectrum of the core–shell cluster is not very different from the "model cluster". We use LEIS measurements to be sure of core–shell structure.

Dr Russell remarked: (1) Looking at Fig. 3 of your manuscript: If you were to have prepared a sample containing Ni clusters and Ag clusters (rather than the mixed or core/shell clusters), would you expect any difference in the optical spectra compared to what you have shown? (2) Follow up regarding LEIS: What size is the molecular beam morphology on the sample? Won't it be sampling the average surface composition of the matrix isolated sample rather than the surface of the clusters/particles?

Professor Broyer answered: We are sure from LEIS analysis that the clusters are core shell (see the text in the paper and ref. 7 of the paper). Moreover if we had independent Ni Clusters and independent Ag clusters, we would see the plasmon resonance of pure silver clusters which are slightly narrower and slightly red shifted. This is not possible.

Professor Henry asked: Does the variation of the electron–phonon relaxation time for the Pt–Ag cluster exhibit the same behaviour as a function of size as for NiAg clusters?

Professor Broyer answered: The behaviour of the electron–phonon relaxation time for PtAg clusters is the same as the behaviour of Ni/Ag clusters, both as a function of the size and as a function of the concentration. To my opinion, the interpretation is the same: Decrease of the e-ph time as a function of the size due to surface effects and "islands" of very small silver clusters (not complete shell of silver around the core of Pt/Ag alloy). These very small "islands" are responsible for the non-linear behavior of the e-ph relaxation time as a function of the Pt concentration for a given size.

Dr Schofield asked: What is the effect of the shape of the particles on the position and number of the plasmon resonances?

Professor Broyer responded: The spectrum corresponding to plasmon resonances is strongly dependent on the shape of the nanoparticle. We have recently developed a method for measuring the extinction spectrum of a single nanoparticle.[1] We have also studied the spectra as a function of the shape and we have shown that the aspect ratio of single ellipsoidal gold nanoparticles can be determined from spectra recorded in polarized light.[2]

1 A. Arbouet, D. Christofilos, N. Del Fatti, F. Vallée J. R. Huntzinger, L. Arnaud, P. Billaud, and M. Broyer, *Phys. Rev. Lett.* 2004, **93**, 127401.
2 O. Muskens, N. Del Fatti, F. Vallee, J. R. Huntzinger, P. Billaud, and M. Boyer, *Appl. Phys. Lett.*, 2006, **88**, 063109.

Professor Jellinek communicated: Could you say a few words about the effect(s) of the matrix environment on the positions of the measured absorption peaks?

Professor Broyer communicated in reply: The matrix environment has a clear influence on the measured peak positions through the matrix dielectric constant ε_m:

For pure noble metal clusters, the frequency of the surface plasmon resonance ω_s is given as a first approximation by:

$$\omega_s = \frac{\omega_p}{\sqrt{2\varepsilon_m + \varepsilon_d^{re}(\omega_s)}} \tag{1}$$

Where $\varepsilon_d(\omega) = 1 + \chi^d(\omega)$ is the interband part of the metal dielectric function, $\varepsilon_d^{re}(\omega)$ is the real part of $\varepsilon_d(\omega)$ and ω_p is the Drude bulk plasmon frequency.

It is clear from formula (1) that the absorption peaks (ω_s) shift to the red (ω_s decreases) when the dielectric constant ε_m of the matrix increases. The situation may be slightly more complicated for alloy or core–shell mixed clusters, but qualitatively the absorption peaks remain red-shifted when ε_m increases.

Professor Ferrando communicated: In Fig. 6 of your paper you show that the electron-lattice energy exchange time in 3.2 nm AgNi nanoparticles increases with Ni proportion. Essentially that time has a maximum when the proportion of Ni is 25% or 50%.

You tentatively attribute this maximum to an imperfect covering of Ni by Ag. However, when Ni is only 25%, it is unlikely that Ag will not cover it completely (unless separated Ag and Ni particles are formed). What about the possibility that the Ni nucleus is placed asymmetrically inside the Ag shell, so that the thickness of the Ag shell is not uniform?

Professor Broyer communicated in reply: I agree that my interpretation is only tentative and we cannot exclude that the Ni nucleus is placed asymmetrically inside the Ag shell.

Dr Mottet commented: My question is about the non-linear behaviour of the measured electron–phonon energy exchange time for Ni–Ag (Fig. 6) as compared to Au–Ag clusters (Fig. 5). I wonder if this non-linearity could not be due to a possible elastic strain as a function of Ni amount (and for a large amount, a relaxation of the strain by misfit interfacial dislocations) in the core–shell structure in the case of a strong size mismatch as it is in the Ni–Ag system, whereas the Au–Ag system, either in the solid solution or in the core–shell structure, would follow a kind of Vegard's law because of a weak size misfit?

Professor Broyer replied: I do not understand how the interfacial dislocations or defects can lead to an increase of the electron phonon relaxation time. I expect rather an acceleration of the electron–phonon relaxation by these phenomena. This effect could induce non-linearity, but not an increase of the electron–phonon relaxation time

Professor Fortunelli communicated: Do you think that PtAg nanoparticles are thermalised (in equilibrium shape)? I was surprised by the fact that you did not observe any plasmon.

Professor Broyer communicated in reply: The Pt/Ag clusters are thermalized at 300 K by the support (see E. Cottancin *et al.*[1]) I think that the fact that we do not observe any plasmon resonance means that we have an alloy structure at least in the center of the cluster and that the unknown optical properties of this alloy are such that no plasmon resonance arises. Clearly we consider that the absence of plasmon resonance in the optical spectra means that we do not have a core–shell structure. Indeed plasmon resonance must exist in PtAg core–shell clusters.

1   E. Cottancin, M. Gaudry, M. Pellarin, J. Lerme, L. Arnaud, J. R. Huntzinger, J. L. Vialle, M. Treilleux, P. Melinon, J. L. Rousset, and M. Broyer, *Eur. Phys. J. D*, 2003, **24**, 111–114.

Professor Johnston opened the discussion of Dr Veldeman's paper: As your dissociation energies (for monomer and dimer ions) are obtained from experiment, you have no value for *e.g.* $D_{17,2}$ as no dimer dissociation is observed. Are there any calculated dissociation energies to compare with experiment? Perhaps the calculated $D_{17,2}$ value would be very high.

Dr Veldeman responded: The dissociation energies were indeed extracted from the experiments, *i.e.* from metastable delayed fragmentation studies. When only one decay channel is present, only relative values can be obtained; when both monomer and dimer evaporation occur, experimental branching rations together with the energy constraint from the Born–Haber cycle allow extraction of absolute values. Dimer evaporation was however only observed for some pure gold clusters (Au_N; $N = 9, 11, 13, 15$) and for $Au_N Y$ ($N = 18, 20$). Therefore no absolute number for $D_{17,2}$ is deduced from the experiment. Moreover, to our knowledge dissociation energies are only computed for homogeneous gold clusters. Although neither an experimental nor a theoretical value of $D_{17,2}$ is available, the absence of dimer evaporation for $Au_{17}Y$ indeed indicates that a high value for $D_{17,2}$ might be expected.

Professor Johnson asked: You mention the stabilising effect of the interstitial atom within the Au cage. May I ask—and I am not familiar with the polyhedral forms of your gold polyhedra—would you not expect that given the asymmetry of the cage, major distortions might occur? I am inclined to ask, are there interstitial components of the boranes where no significant interaction is possible, in contrast to yours where interaction may or may not take place?

Dr Veldeman responded: A distinction should be made between "flat" and "hollow cages". Flat cages are cage-like structures with an internal diameter that is too small to accommodate a guest atom (< 5 Å). Hollow cages have a larger internal diameter (> 5 Å) and can accommodate a guest atom with little structural distortion. Of course, the bigger the cluster cage, the larger the interstitial atom may be. Structural distortion upon doping might influence the cluster stability compared to its hollow counterpart. Important is that by choosing the dopant (with the proper valence) so as to create a system with an electronic shell closure, the relative stability of the cluster is enhanced, irrespective of any eventual geometrical distortion, due to the more favorable electronic structure as is also commented on by Dr Fortunelli.

Professor Lievens responded: I don't know about boranes, but several other caged systems do exist, such as the well known endohedrally doped C_{60} systems. In that case interactions are limited. Other examples of centrally doped cage-like clusters are, *e.g.*, group 14 element clusters (Pb, Sn, Ge, Si) with a central (transition) metal atom (see for example ref. 1 and 2). In those cages, often a combination of geometrical fitting and optimal electron configuration (or chemical bonding) is of importance.

1 V. Kumar, *Comput. Mater. Sci.*, 2006, **36**, 1.
2 Z. Chen, *J. Am. Chem. Soc.*, 2006, **128**, 12829.

Professor Johnson said: For the record, I accept Peter Lievens' explanation.

Professor Hutchings asked: Based on the preceding discussion points, it is clear that your method of preparation produces clusters for which certain sizes, in terms of numbers of atoms, are preferred because they are more stable. If this is the case then it should be possible to make these structures using different methods *e.g.* vapour phase deposition. Is there any evidence in the literature to date that the clusters that you find are more stable, can be formed by alternative methods? If not, perhaps your finding could provide some impetus to investigate the possibility.

Professor Fortunelli responded: Another possibility would be to soft-land the stable clusters prepared in the gas phase onto a surface and see whether the cluster–surface interaction does not disrupt the peculiar stability of these structures. It would be interesting to soft-land both charged clusters and neutral clusters obtained *e.g.* by charge neutralisation of size-selected charged clusters. As far as I can understand, this has been attempted in the literature, but the experiments are complicated and the characterization step very difficult. It can however be expected that more of these kind of experiments will be performed in the future, and with more definite results.

Dr Veldeman answered: Indeed, it should be possible to make clusters with an enhanced stability for a particular size and composition with different methods. There are a few examples existing in the literature, for instance the well known fullerenes. Another recent example is an icosahedral Pb_{12} cluster that our group discovered as an Al-doped Pb_{12} cation with a 50 valence electron shell closure[1] and that was independently synthesized as a $PtPb_{12}$ dianion in a $[\{K(2,2,2\text{-crypt})\}_2]$-$[Pt@Pb_{12}]$ salt.[2]

1 S. Neukermans, E. Janssens, Z. F. Chen, R. E. Silverans, P. v. R. Schleyer, and P. Lievens, *Phys. Rev. Lett.*, 2004, **92**, 163401.
2 E. Esenturk, J. Fettinger, Y. Lam, and B. Eichhorn, *Angew. Chem., Int. Ed. Engl.*, 2004, **43**, 2132.

Professor El-Shall asked: Are the magic numbers observed in the photoionization of the neutral clusters similar to those found in the charged clusters directly generated by the laser vaporization of the metal target? Is there a dependence of the magic numbers on the fluence of the 193 nm photoionization laser?

Dr Veldeman answered: In our experimental setup clusters are produced with a laser vaporization source. Due to the low temperature of the inert gas (He), cluster production proceeds primarily by the successive addition of a single atom. Since the reverse process (*i.e.* evaporation) is negligible, the cluster abundance distribution does not reflect size-dependent thermodynamic stabilities. The mass spectra of charged clusters directly generated in the source are therefore, in general, smooth and structureless. Consequently, we do not obtain stability information on direct ions stemming from the source. Instead, we perform photofragmentation experiments in which a neutral cluster beam is irradiated with high fluence laser light, which results in stability information on cationic clusters. In a hot oven source, the equilibrium between growth and evaporation is however reached, and thus stability information on charged clusters directly generated in the source can be obtained. From experiments on clusters produced in a hot oven source, it is revealed that similar magic numbers are found for direct and photofragmented cluster ions (with one atom size difference for matching the total numbers of valence electrons).

Concerning the laser fluence, there is a minimal fluence needed to induce multiphoton absorption and to reach an excess in energy that is sufficiently large to locate enough internal energy in a single mode to overcome the binding energy of a fragment. Once this minimal laser fluence is exceeded, the same magic numbers are found for all laser fluences.

Professor Fortunelli commented: The extra stability of this $Au_{16}Y^+$ structure can be explained from a theoretical point of view in terms of a combination of good stability of Au cage structures with the insertion of Y^+ which gives two additional electrons and thus electronic shell closure. The rearrangement of the Au cage is compensated by this interaction.

Dr Veldeman answered: We fully agree with this comment.

Professor Jellinek commented: It would be of interest to assess the degree of stability of the gold cages without the dopants.

Professor Lievens commented: In the photofragmentation data, as discussed by Nele Veldeman *et al.*, there is strong evidence for enhanced stability for several doped gold clusters. Theory[1,2] predicts endohedral cage structures for several species consisting of 15 or 16 gold atoms and a central transition metal. The stability of species with such a compact geometry may be enhanced, for some dopant atoms, by the complete filling of electronic shells of delocalized valence electrons, both from gold and from the dopant atom. Examples are YAu_{16}^+, $ScAu_{16}^+$, $TiAu_{15}^+$,[3,4] that all have a closed 1s–1p–1d 18-electron closed shell structure.

Important for the occurrence of golden cage structures are the relativistic effects. Similar to the role these effects play for the planarity of gold anionic clusters up to large sizes (12, 13 atoms),[5] relativistic effects are responsible for the enhancement of the stability of cage structures. For the lighter coinage metals Cu and Ag, so far no empty cages have been reported, but we have reported evidence for endohedral systems, very similar to the doped gold cages[6,7]

1 Y. Gao, S. Bulusu, and X. C. Zeng, *J. Am. Chem. Soc.*, 2005, **127**, 156801.
2 K. Manninen, P. Pyykkö, H. Häkkinen, *Phys. Chem. Chem. Phys.*, 2005, **7**, 2208.
3 W. Bouwen, F. Vanhoutte, F. Despa, S. Bouckaert, S. Neukermans, L.T. Kuhn, H. Weidele, P. Lievens, and R.E. Silverans, *Chem. Phys. Lett.*, 1999, **314**, 227.
4 S. Neukermans, E. Janssens, H. Tanaka, R.E. Silverans, and P. Lievens, *Phys. Rev. Lett.*, 2003, **90**, 33401.
5 F. Furche, R. Ahlrichs, P. Weis, C. Jacob, S. Gilb, T. Bierweiler, and M.M. Kappes, *J. Chem. Phys.*, 2002, **117**, 6982.
6 E. Janssens, S. Neukermans, H. M. T. Nguyen, M. T. Nguyen, P. Lievens, *Phys. Rev. Lett.*, 2005, **94**, 113401.
7 N. Veldeman, T. Höltzl, S. Neukermans, T. Veszprémi, M. T. Nguyen, P. Lievens, *Phys. Rev. A*, 2007, **76**, 011201(R).

Professor Johnson asked: Julius (Professor Jellinek), did you say that the planar studies are governed (in part) by the formation of directed valence bonds? But why only planar, the polyhedral forms may also be rationalised in terms of directed valence bonds.

Professor Jellinek responded: Brian (Professor Johnson), your argument is well taken and the preferred structures of gold clusters are consistent with it. Whereas the smaller Au_n (a dozen of atoms or so) are planar, Au_{20} is a perfect tetrahedron, and

Au_{32} and Au_{50} form hollow cages. The cage form emerges as an energetically competitive structure for other medium size gold clusters as well, e.g., Au_{44} (cf., e.g., J. Wang et al.,[1] and references therein).

1   J. Wang, Jellinek, J. Zhao, Z. Chen, R. B. King, and P. von Rague Schleyer, *J. Phys. Chem. A*, 2005, **109**, 9265.

Dr Grönbeck remarked: Concerning the underlying reason for small gold clusters to adopt planar or tube-like structures, I do not think this can be understood from s–d hybridization. Although Cu clusters have the same degree of hybridization, such clusters are not planar for sizes $\geq Cu_6$. The reason for Au to adopt special structures is instead the d–d overlap and the electron delocalization that this overlap makes possible.

Professor Henry opened the discussion of Professor El-Shall's paper: Where does the aggregation of the naked particles occur: in the gas phase, on the cold plate or in solution?

Professor El-Shall replied: Some aggregation takes place in the gas phase, but probably significant aggregation takes place on the cold plate since under the conditions where a small number density of particles are deposited on the cold plate, much less aggregation is observed.

Dr Ellis asked: The TEM images of your AuAg nanoparticles show stacking faults. Were these seen in both monometallic and bimetallic samples?

Professor El-Shall responded: The stacking faults and twins were observed within the bimetallic nanoparticles which often appeared to be faceted rather than spherical. This can be explained by the different growth rates of silver on various planes of gold particles as well as by the anisotropy of the surface energy, which favors low-index {111} and {200} facets.

Professor Johnston asked: Have you carried out any ion-scattering studies to measure surface compositions of your nanoparticles? For example, in the Au–Pd case it is possible that Pd-enrichment of the surface may be driven by the stronger interaction of Pd with O in aqueous solution. However, the plasmon measurement alone may not be sensitive enough to see such surface enrichment.

Professor El-Shall answered: We have not carried out ion-scattering studies on the Au–Pd nanoalloys. However, we have observed surface enrichment in previous work on the formation of FeAl intermetallic nanoparticles. In this case, nano EDX analysis of single FeAl nanparticles and XPS results indicate that the surface composition is enriched in Al and this leads to the formation of an amorphous Al_2O_3 surface layer surrounding the FeAl nanocrystals.[1] The Au–Pd nanoalloys could show similar results and it would be interesting to carry out these studies on the Au–Pd nanoalloys.

1   Toru Egawa and Shigehiro Konaka, *J. Phys. Chem. B*. 2001, **105**, 2085–2090.

Professor Pastor asked: In your paper you report values for the critical temperature of mixed FePt, PdFe *etc.* nanoparticles. Could you explain how you measure this critical temperature and how the notion of critical temperature of a nanoparticle should be understood? In particular, I would like to know if you are referring to the temperature dependance of the saturation magnetization, in which case I wonder how strong your applied field is. Finally, I wonder if you see any signs of interaction between the nanoparticles in your measurement.

Professor El-Shall responded: The temperature variations of the remanance M_r (M at $H = 0$) and coercivity H_c (H at M $= 0$) were measured from 5 K to 350 K. Measurements for temperatures higher than 350 K could not be made due to experimental limitations. The critical temperatures were estimated from extrapolating the M_r and H_c data to zero. Although the extrapolations showed similar T_c obtained from the M_r and H_c data, the extrapolations are clearly very approximate since M_r and H_c are often non-linear approaching T_c. Our measurements do not show signs of interactions between the nanoparticles.

Professor Henry asked: What is the driving force for aggregated clusters to become more elongated after irradiation? Could they be changing during irradiation?

Professor El-Shall replied: For the Au nanoparticles the near IR absorption band around 900 nm is attributed to the presence of large aggregates. The excitation of these aggregates with the 780 nm fs pulses produces a new absorption band in the ~ 660 nm region which we attribute to smaller aggregates of probably elongated shapes based on the TEM results. It is possible that aggregates of different shapes and sizes are present and the 780 nm excitation selects certain aggregates and transforms them to those with absorption in the 660 nm region similar to elongated Au nanostructures such as nanorods. The elongated aggregates are produced from the break up of the larger multiply branched aggregates.

Professor Broyer commented: Have you measured the absorption spectra of the elongated aggregates using polarised light?

Professor El-Shall responded: No, but this is a very good suggestion since these experiments would provide information on the degree of elongation and the relative concentrations of the elongated aggregates.

Dr Cookson asked: Your nano-alloying works very well in solution when the laser frequency interacts with the gold surface plasmon resonance.
 (i) Can similar alloying be seen in the absence of gold; with other metals such as copper (where you will see resonances around 600 nm).
 (ii) Have you investigated the effect of irradiating different shaped particles? Rod-like particles would also provide higher wavelength longitudinal plasmon resonances that may allow the use of higher wavelength laser irradiation.

Professor El-Shall replied:
 (i) As long as the laser frequency matches the plasmon resonance of one of the components of the alloy it could result in the particle's melting through multiphoton absorptions followed by alloying with the other component in solution. This could also result in the core–shell structures.
 (ii) The melting of Au nanorods using femtosecond lasers has been studied by several groups and, depending on the plasmon resonance, near IR laser frequencies can be used for the melting followed by alloying.

Dr Baletto communicated:
 (1) How is the magnetic hysteresis affected by the size of the cluster?
 (2) What happens to Pt facets when some surface Pt atoms are substituted by Co atoms? What is the best composition? (compared to the composition of the whole cluster). Can we conclude that the finite size of clusters leads to new magnetic properties with respect to bulk materials and flat surfaces?

Professor El-Shall communicated in reply: We have not measured the magnetic hysteresis curves for different particle sizes of the PdM and PtM nanoalloys. However, from the measurements of the magnetic properties of ferrite nanoparticles, both the blocking temperatures and saturation magnetizations increase with particle size.

We have not measured the magnetic properties of PtCo nanoalloys of different compositions.

Professor Pastor communicated in reply: The coercive field, for example, is directly related to the magnetic anisotropy energy which in general increases with decreasing cluster size. This is a consequence of changes in the electronic structure and spin–orbit energies [see, for instance, G. M. Pastor et al.[1]]. Moreover, one often observes that small particles and ensembles of them have rather high saturation fields. This is shown, for example, by Fig. 1 of our paper. We have not done explicit calculations of Co atoms at Pt surfaces.

However, from a number of related studies, one can expect to find important induced magnetic moments at the Pt atoms which should be larger as the number of Co nearest neighbours increases.

We have seen such effects, for example, at the interface of Co–Pt and Co–Pd films as well as for Co clusters deposited on Pd [see, for instance, J. Dorantes-Davila et al.,[2] and R. Felix-Medina et al.[3]]. Moreover, there are very interesting recent results by Xie and Blackman[4] showing that the magnetic moments and anisotropy energy of the Co island on Pt(111) can be drastically enhanced by depositing Pt atoms on them.

1 G. M. Pastor, J. Dorantes-Dávila, S. Pick, and H. Dreyssé, *Phys. Rev. Lett.*, 1995, **75**, 326.
2 J. Dorantes-Davila, H. Dreysse, and G. M. Pastor, *Phys. Rev. Lett.*, 2003, **91**, 197206.
3 R. Félix-Medina, J. Dorantes-Dávila, and G. M. Pastor, *Phys. Rev. B*, 2003, **67**, 094430.
4 Y. Xie and J. A. Blackman, *Phys. Rev. B*, 2006, **74**, 054401.

Professor Johnson asked: Is dioxygen a magic number molecule? O_4 is, O_8 is but O_2! Let's not become too preoccupied with this term 'magic number'. Certainly many organo-metallic complexes possess an 18 electron configuration but many 'lightly stable' molecules do not.

Professor Sir Thomas remarked: We must not protest too much about the use of magic numbers. People began to use the word 'magic' when they saw (from mass spectrometry, computation and electron microscopy) that certain entities were more stable than other closely similar entities *e.g.* metal atoms (M) form a more stable structure when M is 40, 55 or 561 than when M is 41, 54 or 56. We are not reverting to a mystical, alchemical age when we use such terms as magic numbers. Clearly, certain significant discoveries have been made by noting the relative stability of certain significant numbers of atoms. This is how C_{60} became so prominent. It is not as thermodynamically stable as graphite, but it is readily preparable and chemically useful.

Professor Pastor stated: In relation to the point raised by Professor Johnson concerning the fact that O_2 is not 'magic' (*i.e.* $N = 2$ is not a magic number for O_N), I believe it is important to recall that in the case of finite clusters we are dealing with systems which are not thermodynamically stable. If one would consider thermodynamic stability as a criteria, then the only magic number for Na_N clusters would be $N = \infty$ and not $N = 20$, 40 *etc.* as beam mass spectra and electronic shell closings have shown. In the case of clusters we are often interested in the relative stability for a given size N, as compared to $N - 1$ and $N + 1$, rather than in the stability of cluster size N in equilibrium.

Professor Lievens remarked: The term "magic numbers", as used by several of the speakers and delegates at this meeting, has a well defined meaning in the cluster science community. While the term historically has been used to indicate certain sizes with an enhanced abundance in mass spectra, its meaning usually is restricted for indicating clusters with a size that corresponds to either a closed "shell of electrons" or a closed "shell of atoms". Experimental evidence for shells of electrons in clusters was first reported by Knight and coworkers[1] in 1984 when they published a mass spectrum of Na clusters with remarkable peaks and steps for certain cluster sizes. This finding was explained in terms of a phenomenological shell model for delocalized elcctrons—one valence electron for each atom—very analogous to the description of neutron and proton shell closures in nuclear physics where the use of "magic number" to indicate closed shells is widely accepted. Ample evidence has been reported for the existence of electronic shell structure in simple metal clusters, for instance also in other physical properties such as ionization energies and electron affinities.[2] Another reason for peaks and steps in mass abundance spectra is related to enhanced stabilities for clusters with a closed outer atom shell. A historically important example of experimental evidence for the existence of shells of atoms was the observation of peculiar size dependence in the mass spectra of van der Waals bonded Xe clusters,[3] which could be explained in terms of geometrical packing in icosahedral symmetry and enhanced stability for complete atom shells. Several other examples, also for metal clusters, were reviewed in ref. 4.

Another well established magic number, also in pseudo-sciences, is the so-called "golden ratio" or "*sectio divina*". This number phi, with a value of 1.618..., was already identified by the ancient Egyptians and Greeks as a mathematical proportion generally recognized to be aesthetically pleasing: two quantities are in "golden mean" if the whole is to the larger as the larger is to the smaller. The same number also relates to the Fibonacci series: the ratio of consecutive Fibonacci numbers converges to phi. The golden ratio is found in nature in for example nautilus shells, (flower and tree) leaves, bird feather patterns, or the human body and face, and is implied in man-made "constructions" like paintings, monuments, buildings, cars or simply the rectangular shape of a credit card. Interestingly, symmetry and perfect proportions are joined in the highest possible symmetry occurring in nature: icosahedral symmetry; the smallest of which is the icosahedron composed out of 20 identical equilateral triangles meeting in 12 equivalent vertices. Such an icosahedron indeed is constructed from three identical mutually orthogonal rectangles obeying the golden ratio.[5]

Both shells of electrons and shells of atoms are particularly relevant for binary or bimetallic clusters. Indeed, by changing the cluster composition, the number of valence electrons and the number of constituent atoms can be tuned independently, which allows constructing clusters that combine an electronic shell closure with a closed shell of atoms. A number of examples of such species are reviewed in ref. 6.

1 W. D. Knight, K. Clemenger, W. A. de Heer, W. A. Saunders, M. Y. Chou, M. L. Cohen, *Phys. Rev. Lett.*, 1984, **52**, 2141.
2 W. A. de Heer, *Rev. Mod. Phys.*, 1993, **65**, 611.
3 O. Echt, K. Sattler, E. Recknagel, *Phys. Rev. Lett.*, 1981, **47**, 1121.
4 T. P. Martin, *Phys. Rep.*, 1996, **273**, 199.
5 S. Neukermans, E. Janssens, X. Wang, N. Veldeman, Z. Chen, P. v. R. Schleyer, R. E. Silverans, P. Lievens, *Phys. Mag.*, 2004, **26**, 319.
6 E. Janssens, S. Neukermans, P. Lievens, *Curr. Opin. Solid State Mater. Sci.*, 2004, **8**, 185.

Professor Henry stated: I am not sure that the word 'magic' (numbers) is well chosen. Taking the case of atomic shell closure: the clusters correspond to a lower reactivity. However, this observation applies simply to the smoothness of the cluster surface. Take the case of a crystal growth: if you have a crystal with a smooth facet the growth rate is very low except if you have defects like dislocation but this is completely independent of the size of the crystal.

Professor Fortunelli answered: To be more precise, the smaller interaction energies with incoming species that 'magic' clusters often exhibit are due not only to their smoother surface, but also to a smaller reactivity of their closed electronic shells or to charge distribution effects as can be found *e.g.* the paper by Oliver Hampe.[1] Anyway, I agree that the effect of their smoother surface is often more important, as I mentioned in my response to Calvo: we found[2] that $Ag_8/DV/MgO(100)$ has a much lower interaction energy with the oxygen molecule than the neighboring sizes, basically due to the compact character of its 'magic' structure and also to the effects that electronic shell closure has on its chemical properties.

1 M. Neumaier, F. Weigend, O. Hampe and M. M. Kappes, *Faraday Discuss.*, 2008, **138**, DOI: 10.1039/b705043g
2 G. Barcaro and A. Fortunelli, *Phys. Rev. B*, 2007, **76**, 165412.

Professor Lievens answered: In terms of reactivity this viewpoint is correct. However, there is more to it than smoothness of the surface. For instance the binding energy of one atom to a cluster with a closed shell (of electrons or of atoms) will be significantly lower. In any case, the use of the work 'magic' is historical, and follows the terminology used for fully occupied quantum levels in other fields, such as nuclear physics.

Professor Ferrando remarked: There has been recently interest in the so-called "magic melters", for example in relation to free sodium clusters. Magic melters correspond to peaks in size dependence of the cluster melting temperature.

As is well known, gas-phase sodium clusters of a few ten or hundred atoms present abundances related to electronic magic numbers (see Knight *et al.*[1]). However, the sizes of sodium magic melters correspond to geometric shell closures, so that magic melters have a geometric origin. (see H. Haberland *et al.*[2] and K. Joshi *et al.*[3]). This is an example in which the concept of magic sizes is a fruitful guideline in relation to cluster equilibrium properties.

1 W. D. Knight, Keith Clemenger, Walt A. de Heer, Winston A. Saunders, M. Y. Chou and Marvin L. Cohen, *Phys. Rev. Lett.*, 1984, **52**, 2141.
2 Hellmut Haberland, Thomas Hippler, Jörn Donges, Oleg Kostko, Martin Schmidt and Bernd von Issendorff, *Phys. Rev. Lett.*, 2005, **94**, 035701.
3 K. Joshi, S. Krishnamurty, and D. G. Kanhere, *Phys. Rev. Lett.*, 2006, **96**, 135703.

Professor Bond addressed Professor Lievens: Arising from Professor Lievens' explanation of the use of the term 'magic number', I would draw attention to the fact that the occurrence of mass numbers less than 40 (*i.e.* in the 30–39 range) is not so very different from that of mass 40 itself. Perhaps therefore we should speak of a 'magic range' rather than a single magic number.

Professor Lievens replied: Indeed the measured intensities of several clusters are similar in certain size ranges, such as for the specified range with numbers of atoms lower than 40. However, the differential abundances are not, *e.g.* following M = 40, the abundance of Na_{41} is very small. The differential abundance can be related to the cluster binding energy, as is explained, for instance, by W. A. de Heer in his review article.[1]

1 W. A. de Heer, *Rev. Mod. Phys.*, 1993, **65**, 611–76

Professor Broyer said: The discussion on magic numbers is interesting but a little bit semantic and puzzling—the point is the specific properties of clusters. To my opinion, the fascinating subject of this conference is to find new clusters with specific new and interesting properties; optical properties, magnetic properties, catalytic reactivities and so on. Working with alloys, mixed clusters or doped clusters open new opportunities as compared to homogeneous clusters. This is a fascinating program for the future, to be discussed here.

Professor Fortunelli communicated: In response to the comment by Professor Broyer: it is true that this conference is about new properties of alloys and not about magic clusters, but the hope is to find new properties by producing magic nanoalloyed clusters.

Professor Broyer communicated in reply: You are right but the important features are new properties, not magic numbers. The interpretation of high intensities on mass spectra is not easy.

Professor Broyer opened the discussion of Professor Pastor's paper: What is the best system relatively to the blocking temperature; Co–Sm or others?

Professor Pastor responded: Your question addresses one of the main motivations of current research on magnetic nanoalloys. I'm afraid we don't have the answer to it yet. Co–Sm clusters are certainly good candidates for high anisotropy and blocking temperature. In fact, the Lyon group (V. Dupuis and coworkers) has shown that Co–Sm clusters produced with a laser vaporization source and deposited in a matrix show very large blocking temperatures. As far as I know a post-deposition thermal treatment was necessary to obtain the best results. From the point of view of theory very little is known on Co–Sm clusters. They are particularly challenging due to the strong correlation effects at the 4f orbitals.

Professor Jellinek commented: Very nice results. It would be of interest to explore in a more systematic way the size-evolution of the magnetic properties of the Co/Rh bimetallic particles.

Professor Ferrando asked: In your paper, you show the nanocluster structures that you use in the calculations of magnetic properties. These structures are either pieces of fcc crystals or polytetrahedra. The cuts of fcc crystals seem indeed quite far from what I would expect as the best fcc CoRh clusters. Moreover, a specific chemical ordering is assumed.

Can you comment on how the magnetic properties that you calculate depend on the specific cluster structure and chemical ordering?

Professor Pastor answered: The point you raise is one of the central issues in magnetic nanoalloys.

We have investigated the problem by considering different representative structures and chemical arrangements like Co-core and Rh-shell or *vice versa*, with various degrees of intermixing at the interface. The experiments actually indicate that these particles have a Rh-rich core and a Co-rich surrounding shell. Concerning the theoretical results they are very sensitive to the chemical order and in particular on whether Rh is at the core or at the surface. In the first case, large average magnetics are obtained with an important Rh contribution to the magnetic behaviour. In contrast, in the second case the Co moments and the induced Rh moments are both much smaller.

This also results in important changes in the magnetic anisotropy.

In the case of CoRh, these general trends are not significantly affected by the details of the most stable cluster structures.

Professor Sir Thomas asked: Professor Pastor's fascinating results concerning the influence of Rh on Co remind us that we have still a great deal to learn about magnetism, especially ferromagnetism. There have been recent reports showing that minute globules of potassium inside a zeolite exhibits ferromagnetism. The same is true about nanoparticles of silver wrapped in layers of graphene. Does he have any general guidance to give us to enable us to make sense of these somewhat surprising facts?

Secondly, if one wants to evaluate the influence of particle size and shape on magnetic properties, a good way of doing so is to use electron-wave-holography.

Professor Pastor replied: I agree completely with your remark that a lot has still to be learned on magnetism in general, and on ferromagnetism in particular. There are in fact different sources of magnetism and, of course, a too large variety of magnetic behaviors. General simple rules that would cover most of this large and subtle field are not available to my knowledge. However, there are a number of more or less intuitive rules for specific physical situations.

For instance, in the context of clusters and low-dimensional systems it is well established that a reduction of cluster size or dimensionality generally leads to an enhancement of both spin and orbital magnetic moments. This reflects the increasing importance of exchange and correlations, with respect to electron delocalization and bonding, as the coordination number decreases. Moreover, for 3d–4d nanoalloys we have found a general trend to an enhancement of the 4d induced magnetic moments when the species intermix. This is a consequence of the increasing number of magnetic 3d neighbours at the 4d atoms. In this case we also observe an enhancement of the magnetic anisotropy energy that can be explained by the increasing importance of spin–orbit interactions at the now magnetic 4d sites.

Professor Hou opened the discussion of Dr Mottet's paper: We performed similar Metropolis Monte Carlo simulations of CoPt clusters and compared them with the experimental results of the Lyon group.[1,2] The simulations were repeated with two slightly different parameterizations of an EAM potential and two different parameterizations of a Modified EAM accounting for directional dependencies. We found that phase stability, the bulk order–disorder transition temperature and the segregation state of clusters are much dependent on the potential. Moreover, we did not predict the $L1_0$ ordered phase in clusters, which you find using a second moment tight binding potential. Experimentally, the Lyon group found a fcc solid solution in their clusters deposited on α-C while Langlois *et al.* do find the $L1_0$ ordered phase in CoPt clusters that they synthesized on α-C.[3]

Regarding this situation, I think it is necessary to identify the experimental factors that govern the phase formation in clusters and then to reconsider the potential models and parameterizations that we use accordingly, in order to make predictions beyond experimental possibilities. My question to the discussion paper by Langlois *et al.* is directly related to this.

1 P. Moskovkin and M. Hou, *JALCOMP*, 2007, **434–435**, 550.
2 P. Moskovkin, S. Pisov, M. Hou, C. Raufast, F. Tournus, L. Favre, V. Dupuis, *Eur. J. Phys. D*, 2007, **43**, 27.
3 C. Langlois, D. Alloyeau, Y. Le Bouar, A. Loiseau, T. Oikawa, C. Mottet, C. Ricolleau, *Faraday Discuss.*, 2008, **138**, DOI: 10.1039/b705912b

Professor Ferrando commented: In the ordered $L1_0$ phase, "segregation" is intended simply to mean that the outmost facets (and edges) are cobalt rich.

Dr Mottet responded: The $L1_0$ ordered phase on the FCC symmetry leads to a frustration in surface segregation because clusters expose only 2 pure Co facets over the 6 (100) facets, the other being mixed. In the decahedral structure, the pseudo $L1_0$ ordering allows exposing only pure Co facets, which removes a certain frustration between ordering and surface segregation. However, even in the FCC clusters, we show that there is globally a Co surface segregation (see Fig. 8 of our paper). However the surface segregation is anisotropic: it concerns preferentially the edges and the (100) facets whereas we observe an inversion of the segregation (enrichment in Pt) on the (111) facets.

Professor Evans asked: Unlike the preference shown by calculations of the structures of clusters with fewer than 100 atoms for polytetrahedral and icosahedral growth patterns, the structures adopted by isolable transition metal cluster complexes are mostly fragments of closed packed lattices. Is this discrepancy the fault of the ligands?

Dr Mottet answered: Probably yes but I didn't do the calculations which would imply the use of quantum chemistry methods (density functional theory). The effect of the ligands, by modifying the surface energy of the clusters, can probably influence the internal structure of the clusters. Indeed, the icosahedral structure is stabilized because of its surface energy gain which overtakes the core strain. If the ligands can reduce the surface energy without core distortion, the close packed bulk structure can be stabilized even at small sizes.

Professor Johnston commented: Just a short follow up to Professor Evans' comment. A few years ago we performed some simple calculations on thiol-passivated gold clusters.[1] We found they reduce the size at which non-crystalline structures (e.g. icosahedra) convert to fcc-like (cuboctahedra or truncated octahedra). This is in part driven by the ligands stabilizing the more-open (100)-type facets. I am sure that CO ligands will cause surface restructuring and even (for small clusters) global cluster rearrangement. So I agree with his comment.

1   N. T. Wilson and R. L. Johnston, *Phys. Chem. Chem. Phys.*, 2002, **4**, 4168.

Dr Schofield remarked: The whole discussion of computational modelling of clusters would move into a new dimension of usefulness, particularly for catalysis, if it were possible to look at segregation processes and particle structure in an external environment—oxidising or reducing. While this has been done for bulk alloys, the point of this discussion is that nanoalloys would behave differently. How far are we away from being able to do calculations based on a nanoalloy cluster in an oxidising atmosphere?

Dr Mottet replied: I understand that the interest concerning surface segregation in clusters could be extended to more realistic systems than bimetallic clusters in ultra-vacuum, particularly in the domain of the catalysis where the metal oxide tends to play a major role as recent studies have shown, but the introduction of an atmosphere (oxygen or hydrogen) in our model would increase dramatically the computational task. Indeed, as the oxide formation is necessarily accompanied by a charge transfer, the simple semi-empirical potentials used in the bimetallic system are no more relevant in the presence of oxygen (or hydrogen) thus it is required to use quantum chemistry methods (the more appropriate would be the density functional theory) which is so time consuming that instead of doing relevant statistics, we could only try some of the possible configurations and compare their energy at 0 K.

The more successful attempts I know are related to semi-infinite pure systems (pure metallic surfaces) in the group of M. Scheffler.[1] The complexity of a bimetallic system (the preference of one of the components to make the oxide at the surface) and of the finite size with a multiplicity of possible structures and morphologies make this aim quite difficult to access at this time. But the studies of model nanoalloys even in vacuum but supported on oxide surfaces could give some appreciable insights in the first steps of the comprehension of the catalytic process.

1   J. Rogal, K. Reuter, and M. Scheffler, *Phys. Rev. Lett.*, 2007, **98**, 046101.

Direct synthesis of hydrogen peroxide from H_2 and O_2 using supported Au–Pd catalysts

Jennifer K. Edwards,[a] Albert F. Carley,[a] Andrew A. Herzing,[b] Christopher J. Kiely[b] and Graham J. Hutchings*[a]

Received 18th April 2007, Accepted 4th May 2007
First published as an Advance Article on the web 20th September 2007
DOI: 10.1039/b705915a

The direct synthesis of H_2O_2 at low temperature (2 °C) from H_2 and O_2 using carbon-supported Au, Pd and Au–Pd catalysts is described and contrasted with data for TiO_2, Al_2O_3 and Fe_2O_3 as supports. The Au–Pd catalysts all perform significantly better than the pure Pd/TiO_2 and Au/TiO_2 materials. The Au–Pd/carbon catalysts gave the highest rate of H_2O_2 production, and the order of reactivity observed is: carbon > TiO_2 > Al_2O_3. Catalysts were prepared by co-impregnation of the supports using incipient wetness with aqueous solutions of $PdCl_2$ and $HAuCl_4$, and following calcination at 400 °C the catalysts were stable and could be re-used several time without loss of metal. The method of preparation is critical, however, to achieve stable catalysts. No promoters are required (e.g. halides) to achieve the high rates of hydrogen peroxide synthesis. The surface and bulk composition of the gold palladium nanoparticles was investigated by STEM-XEDS spectrum imaging. For TiO_2 and Al_2O_3 as supports the Au–Pd particles were found to exhibit a core-shell structure, Pd being concentrated on the surface. In contrast, the Au–Pd/carbon catalyst exhibited Au–Pd nanoparticles which were homogeneous alloys and X-ray photoelectron studies were consistent with these observations. The origin of the enhanced activity for the carbon supported catalysts is a result of higher H_2 selectivity for the formation of hydrogen peroxide which is due to the surface composition and size distribution of the nanoparticles. The key problem remaining is the sequential hydrogenation of hydrogen peroxide which limits the utilisation of the direct synthesis methodology and this is discussed in detail.

Introduction

The activation of hydrocarbons by oxidation or hydrogenation are well established procedures in the commercial exploitation of fossil fuels. Of the two procedures, hydrogenation is by far both better established and understood. Indeed, hydrogenation is practised on a very small scale in laboratories and in the production of fine chemicals using a range of well-established catalysts. Using hydrogen in this way seems to present no problems for undergraduates or commercial companies. The

[a] School of Chemistry, Cardiff University, Main Building, Park Place, Cardiff, UK CF10 3AT. E-mail hutch@cardiff.ac.uk, edwardsjk@cf.ac.uk, carley@cf.ac.uk; Fax: +44 29 2087 4030; Tel: +44 29 2087 4805
[b] Center for Advanced Materials and Nanotechnology, Lehigh University, 5 East Packer Avenue, Bethlehem PA 18015-3195, USA. E-mail: aah5@lehigh.edu, chk5@ehigh.edu

situation is not the same with oxidation using dioxygen as the oxidant and small scale oxidations in the laboratory are not generally carried out, and at the commercial scale, often, stoichiometric oxygen donors are used which entail a considerable environmental burden due to their low atom efficiency. The reason for this difference in the use of hydrogen and dioxygen as reagents lies in their reactivity. Hydrogen, to be effective in hydrogenation requires activation on the surface of a heterogeneous catalyst. Consequently, non-catalysed hydrogenations do not compete with the catalysed process and this affords great control over the reaction selectivity. This is not the case for oxidations, since dioxygen in its ground state is a di-radical and so homogeneous non-catalysed oxidations can compete readily, especially at elevated temperatures, with the catalysed process. This is a major reason why small scale oxidations using dioxygen or air have not been exploited, as control of selectivity presents a major experimental problem.[1] Hence, there remain many grand challenges in the field of oxidation chemistry which tantalise scientists, yet their solution always seems just out of reach. Two in particular are the selective oxidation of methane to methanol[2] and the direct reaction of hydrogen and oxygen to form hydrogen peroxide, rather than water. These two reactions have dogged scientists for almost a century. Every generation rises to the challenge but in the case of methane oxidation no one has yet succeeded. Nature has shown how these reactions can be induced, since although natural enzymes do not manufacture hydrogen peroxide (rather specific systems are designed to deal with peroxides and destroy them should they be formed outside of a specified environment), natural systems do utilise hydroperoxy and peroxy species as oxidising agents. The second of these demanding reactions, *i.e.* the direct synthesis of hydrogen peroxide is the subject of this paper.

The recent discovery of the catalytic efficacy of gold for oxidation and reduction reactions has given a new impetus in the search for selective redox catalysts,[3-9] and new discoveries are being made on a regular basis. Supported gold catalysts have been shown to be effective for low temperature oxidation of CO,[10] especially in the presence of H_2, H_2O and CO_2 and this reaction[11,12] is finding some potential application in fuel cells. Gold has also been found effective for the selective oxidation of alkenes[13,14] and alcohols,[15] and for the selective hydrogenation of unsaturated carbonyl compounds and nitroso groups.[16] In all these applications it is the selectivity that is coupled with high activity that is viewed as important. Recently, we have shown that supported gold palladium alloys are very active and selective for the oxidation of alcohols,[17] being 25 times more active than the corresponding gold or palladium monometallic catalysts. In addition, we have shown that these catalysts are effective for the direct formation of hydrogen peroxide but as yet selectivity at high conversion remains a problem.[18-22]

Hydrogen peroxide is produced currently by the sequential hydrogenation and oxidation of an alkyl anthraquinone, which has been designed so that explosive mixtures of hydrogen and oxygen can be avoided.[23] However, there are problems associated with the anthraquinone route and these include the cost of the quinone solvent system and the requirement for periodic replacement of anthraquinone due to hydrogenation. In addition, the process is only economically viable on a relatively large scale and this necessitates the transportation and storage of concentrated solutions of hydrogen peroxide. In view of this, most hydrogen peroxide is used for cleaning or in the paper and textile industries in large scale applications.

The direct reaction between hydrogen and oxygen has been a research target for many years with the first reported study in 1914,[24] using a Pd catalyst. Since then, there have been a number of investigations and virtually all of these have been pioneered in industrial laboratories.[24-37] Early studies used H_2/O_2 mixtures in the explosive region but, more recently, studies have concentrated on carrying out the reaction with dilute H_2/O_2 mixtures well away from the explosive region.[33,36] It is reported that the hydrogen peroxide yield is improved by the addition of acid and bromide,[29,30] and solutions of over 35 wt% hydrogen peroxide have been made by reacting H_2/O_2 over Pd catalysts at elevated pressures.[29] Against this background, we have shown that supported gold palladium alloys can be very effective for this reaction. In our earlier studies[20-22] we have found that the gold–palladium alloys spontaneously form core-shell structures with palladium-rich shells. This raises the question as to whether these structures are essential for the establishment of the active catalytic sites. An additional question concerns the factors controlling hydrogen selectivity and whether high hydrogen selectivity can be achieved at hydrogen conversions of commercial interest. The purpose of this discussion paper is to present new data concerning the use of carbon as a catalyst support and to address these two key questions.

Experimental

Catalyst preparation

5 wt% Pd, 5 wt% Au and a range of Au–Pd catalysts were prepared by impregnation of three supports (carbon (Aldrich G60)), TiO_2 (Degussa P25, mainly anatase), Al_2O_3 (Aldrich) *via* an incipient wetness method using aqueous solutions of $PdCl_2$ (Johnson Matthey) and/or $HAuCl_4 \cdot 3H_2O$ (Johnson Matthey). For the 2.5%Au–2.5%Pd/TiO_2 catalyst the detailed procedure used was as described below. An aqueous solution of $HAuCl_4 \cdot 3H_2O$ (10 ml, 5 g dissolved in water (250 ml)) and an aqueous solution of $PdCl_2$ (4.15 ml, 1 g in water (25 ml)) were simultaneously added to TiO_2 (3.8 g). The paste formed was ground and dried at 80 °C for 16 h and calcined in static air, typically at 400 °C for 3 h, although other heat treatment conditions have been investigated. Other Au–Pd ratios were prepared by varying the amounts of starting reagents accordingly.

An additional series of catalysts containing 5 wt% Au on TiO_2 were prepared by deposition precipitation. For 5 wt% Au/TiO_2 an aqueous solution of $HAuCl_4 \cdot 3H_2O$ (20 ml, 5 g dissolved in water (250 ml) was added to a solution of TiO_2 in water (3.9 g, 150 ml). This solution was stirred at room temperature for 10 min, and then a solution of NaOH in water (1 g, 25 ml) was added drop-wise until the pH of the solution reached 9. The solution was stirred and maintained at pH 9 for 60 min. The catalyst was then washed, filtered and dried (80 °C, 16 h).

The catalysts prepared by impregnation or deposition-precipitation were pre-treated using a range of conditions: drying at 120 °C in air, calcination in static air at 200 °C for 3 h, calcination in static air at 400 °C for 3 h and reduction in flowing H_2 (5 wt% H_2 in Ar) at 500 °C.

Hydrogen peroxide synthesis and hydrogenation

Catalyst testing was performed using a stainless steel autoclave (Parr Instruments) with a nominal volume of 50 ml and a maximum working pressure of 14 MPa. The autoclave was equipped with an overhead stirrer (0–2000 rpm) and provision for measurement of temperature and pressure. Typically, the autoclave was charged with the catalyst (0.01 g unless otherwise stated), solvent (5.6 g MeOH and 2.9 g H_2O), purged three times with 5%H_2/CO_2 (3 MPa) and then filled with 5% H_2/CO_2 and 25% O_2/CO_2 to give a hydrogen to oxygen ratio of 1:2 at a total pressure of 3.7 MPa. Stirring (1200 rpm unless otherwise stated) was commenced on reaching the desired temperature (2 °C), and experiments were carried out for 30 min unless otherwise stated. Gas analysis for H_2 and O_2 was performed by gas chromatography using a thermal conductivity detector and a CP—Carboplot P7 column (25 m, 0.53 mm id). Conversion of H_2 was calculated by gas analysis before and after reaction. H_2O_2 yield was determined by titration of aliquots of the final filtered solution with acidified $Ce(SO_4)_2$ (7×10^{-3} mol l^{-1}). $Ce(SO_4)_2$ solutions were standardised against $(NH_4)_2Fe(SO_4)_2 \cdot 6H_2O$ using ferroin as indicator. Hydrogen peroxide hydrogenation was carried out in an identical manner but in this case the

25% O_2/CO_2 was not added so that hydrogen peroxide could not be synthesised, but part of the water was replaced by hydrogen peroxide (50 vol%).

Catalyst characterization

Atomic Absorption Spectroscopy (AAS) was performed with a Perkin-Elmer 2100 Atomic Absorption spectrometer using an air–acetylene flame. Gold/palladium samples were run at wavelengths 242.8 nm (Au) and 247.6 nm (Pd). Samples for analysis were prepared by dissolving 0.1 g of the dried catalyst in an aqua regia solution followed by the addition of 250 ml deionised water to dilute the sample. AAS was used to determine the wt% of the metal incorporated into the support after impregnation and also the concentration (ppm) of Au or Pd that had leached out into solution during reaction, by determining the Au and Pd content of the used catalyst and comparing it to the fresh catalyst.

Samples for examination by scanning transmission electron microscopy (STEM) were prepared by dispersing the catalyst powder in high purity ethanol, then allowing a drop of the suspension to evaporate on to a holey carbon film supported by a 300 mesh copper TEM grid. Samples were then subjected to chemical microanalysis and annular dark-field imaging in a VG Systems HB603 STEM operating at 300 kV equipped with a Nion C_s corrector. The instrument was also fitted with an Oxford Instruments INCA TEM 300 system for energy dispersive X-ray (XEDS) analysis.

X-ray photoelectron spectra were recorded on one of two different instruments: a VG EscaLab 220i spectrometer, using a standard Al-Kα X-ray source (300 W) and an analyser pass energy of 100 eV (survey scans) or 20 eV (detailed scans), and a Kratos Axis Ultra DLD spectrometer employing a monochromatic Al-Kα X-ray source (75–150 W) and analyser pass energies of 160 eV (survey scans) or 40 eV (detailed scans). Samples were mounted using double-sided adhesive tape and binding energies referenced to the C(1s) binding energy of adventitious carbon contamination which was taken to be 284.7 eV.

Results and discussion

Evaluation of calcined Au, Au–Pd and Pd supported catalysts

In our previous studies[18–22] we have examined three supports in detail for Au–Pd nanoparticles, namely TiO_2 and Al_2O_3 and Fe_2O_3. These materials were selected as they have been found to be effective as supports for gold nanoparticles for the low temperature oxidation of CO. However, all the Au–Pd catalysts were found to be inactive for CO oxidation whilst being highly effective for the direct synthesis of hydrogen peroxide from the hydrogenation of oxygen. Recently, we have found that Au nanocrystals supported on activated carbon are very active for selective oxidation of glycerol[38] and alkenes;[14] these Au/carbon catalysts are also inactive for CO oxidation at ambient conditions. This observation has prompted us to study carbon as a support for Au–Pd nanoparticles for the direct hydrogen peroxide synthesis reaction. A series of catalysts, prepared by either impregnation or co-impregnation using incipient wetness with carbon, TiO_2 and Al_2O_3 as supports, were investigated for the direct synthesis of hydrogen peroxide and the results for reaction at 2 °C are shown in Table 1. Prior to testing, all the catalysts were calcined at 400 °C. In all cases, the pure Au catalysts generate H_2O_2 but at low rates, and at an insufficient level to enable the hydrogen selectivity to be determined with any accuracy. The addition of Pd to Au, to give catalysts comprising 2.5 wt% Au–2.5 wt% Pd, significantly enhances the catalytic performance for the synthesis of H_2O_2, and the rate of H_2O_2 production is much higher than for the pure Pd catalyst, which in itself is significantly more active than pure gold. The highest rates of hydrogen peroxide formation and hydrogen selectivities are observed for the 2.5 wt% Au/2.5 wt%Pd/ carbon catalyst, and these are significantly higher than those observed with the other

Table 1 Formation of hydrogen peroxide using Au, Pd and Au–Pd supported catalysts[a]

Catalyst	Hydrogen peroxide formation/mol H_2O_2 h^{-1} kg_{cat}^{-1}	Hydrogen selectivity (%)
5%Au/carbon	1	nd
2.5%Au–2.5%Pd/carbon	110	80
5%Pd/carbon	55	34
5%Au/Al$_2$O$_3$	2.6	nd
2.5%Au–2.5%Pd/ Al$_2$O$_3$	15	14
5%Pd/ Al$_2$O$_3$	9	nd
5%Au/TiO$_2$	7	nd
2.5%Au–2.5%Pd/TiO$_2$	64	70
5%Pd/TiO$_2$	30	21

[a] Standard reaction conditions, nd = not determined as the yield is too low for reliable measurement.

supports. Furthermore, the rates of hydrogen peroxide formation with the carbon-supported catalysts are almost a factor of two higher than the corresponding TiO$_2$-supported catalysts, and almost an order of magnitude greater than the Al$_2$O$_3$-supported catalysts, demonstrating that the nature of the support plays an important role in the direct oxidation reaction. However, the synergistic effect observed for the rate of hydrogen peroxide formation on addition of Au to Pd is similar for the carbon- and titania-supported catalysts, being roughly a factor of 2, and the synergistic effect is less marked in the alumina-supported catalysts (Table 1).

Catalyst stability

For any catalyst that is used in a batch process it is important to determine if the catalyst can be successfully re-used: with precious metal catalysts it is essential that the metals are not lost from the catalyst during use. Indeed, one of the key factors that must be considered for heterogeneous catalysts operating in three phase systems, as is the case in the direct synthesis of hydrogen peroxide, is the possibility that active components can leach into the reaction mixture, thereby leading to catalyst deactivation or, in the worst case, leading to the formation of an active homogeneous catalyst.[39] We have investigated the effect of calcination on the stability of catalysts used in the direct synthesis reaction since, in general, catalysts that are not calcined prior to use are highly active *e.g.* 2.5% Au/2.5% Pd/TiO$_2$ dried at 25 °C, gives a rate of H$_2$O$_2$ synthesis of 202 mol H$_2$O$_2$ h^{-1} kg$_{cat}^{-1}$, H$_2$ conversion of 46% and H$_2$O$_2$ selectivity of 89%. However, these materials are highly unstable due to loss of metals during use and representative data for Al$_2$O$_3$- and TiO$_2$-supported catalysts are shown in Table 2. In contrast, catalysts that are pre-calcined at 400 °C prior to use are very stable and do not leach any Au or Pd into solution. Data for consecutive reuse of the carbon and TiO$_2$-supported catalysts are shown in Fig. 1 and it is apparent that these catalysts can be re-used many times. It is important to note that the way in which the calcination is carried out is critical. Both the temperature and gas environment need to be carefully controlled to achieve the stable catalysts that we report in this study. Use of lower temperatures than 400 °C, even for a longer time, leads to catalysts that leach Au and Pd on use. Furthermore, we have found that samples calcined in a tube furnace produce superior catalysts when compared with those prepared using a muffle furnace. Hence we wish to emphasise that the precise way in which these catalysts are prepared is critical to achieving stable catalysts and great care has to be taken with the calcination step of the preparation.

Table 2 Reuse of catalysts and catalyst stability

Catalyst	Pre-treatment	Run	Au remaining[a] (%)	Pd remaining[a] (%)
2.5wt%Au–2.5wt%Pd/TiO$_2$	Air 25 °C	1	20	10
	Air 25 °C	2	8	5
2.5wt%Au–2.5wt%Pd/TiO$_2$	Air 400 °C	1	100	100
	Air 400 °C	2	100	100
2.5wt%Au–2.5wt%Pd/Al$_2$O$_3$	Air 25 °C	1	25	21
	Air 25 °C	2	20	15
2.5wt%Au–2.5wt%Pd/Al$_2$O$_3$	Air 400 °C	1	100	100
	Air 400 °C	2	100	100

[a] (1-(Au or Pd in fresh catalyst) – (Au or Pd after run 1 or 2))/(Au or Pd in fresh catalyst) × 100.

In many studies in which monometallic supported palladium catalysts are evaluated for the direct synthesis of hydrogen peroxide, promoters (typically halides) are added to ensure enhanced selectivity for hydrogen.[40–42] In view of this we have carried out some initial studies of the addition of bromide as a potential promoter with our Au–Pd catalysts. These initial studies have shown that the addition of these additives is deleterious for our catalysts and this emphasises a key difference between supported Pd-only and supported Au–Pd catalysts for this reaction.

Catalyst characterisation

To determine the nature of the supported Au–Pd catalysts a detailed structural and chemical characterization was carried out using scanning transmission electron microscopy (STEM) and X-ray photoelectron spectroscopy.

XPS characterisation. Fig. 2 shows the combined Au(4d) and Pd(3d) spectra for a 2.5 wt% Au–2.5 wt%Pd/carbon, 2.5 wt% Au–2.5 wt%Pd/TiO$_2$ and 4.2 wt% Au–0.8 wt%Pd/Al$_2$O$_3$ catalysts following drying (Fig. 2a) and calcination at 400 °C (Fig. 2b). For the dried samples, which all exhibit higher rates of H$_2$O$_2$ production

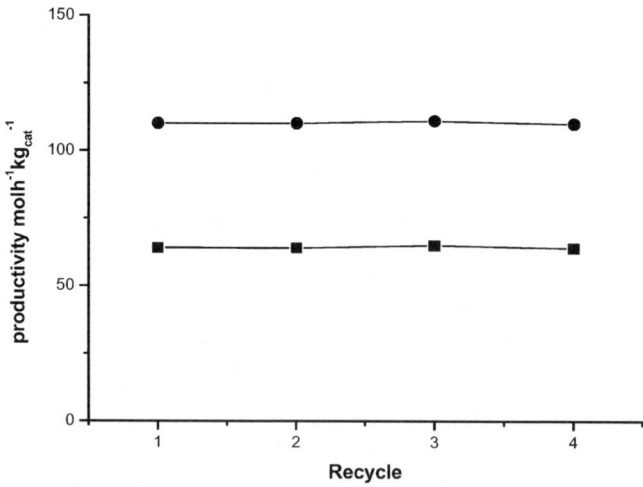

Fig. 1 The effect of the number of uses on the productivity for H$_2$O$_2$ formation: (●) 2.5 wt% Au–2.5 wt%Pd/carbon catalyst calcined in air at 400 °C; (■) 2.5 wt% Au–2.5 wt%Pd/TiO$_2$ catalyst calcined in air at 400 °C.

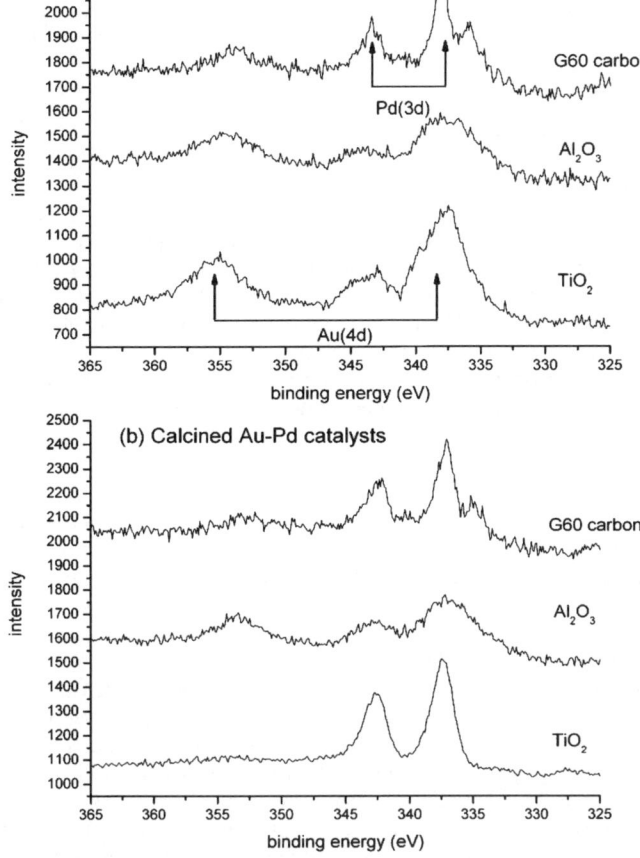

Fig. 2 Au(4d) and Pd(3d) spectra for 2.5 wt% Au–2.5 wt% Pd catalysts after different heat treatments: (a) uncalcined, (b) calcined at 400 °C in air.

when compared with the calcined catalysts, there are clear spectral contributions from both Au and Pd leading to severe overlap of peaks. For the dried TiO_2 supported catalyst the Pd : Au surface ratio (by weight) is significantly different from the bulk value of 1 : 1 (Table 3), whereas the Pd : Au surface ratio for the Al_2O_3- and carbon supported catalysts is similar to that of the corresponding bulk composition. After calcination at 400 °C, for the TiO_2- and Al_2O_3-supported catalysts there is a dramatic decrease in the intensity of the Au(4d) peaks, and for the TiO_2-supported catalyst the $Au(4d_{3/2})$ feature is not readily apparent. It is clear that for these materials the surface has become enhanced in Pd on calcination at 400 °C (Table 3), which is consistent with the formation of core-shell structures for the Au–Pd nanoparticles. The formation of Au–Pd core–shell structures has been reported previously for ligand-stabilized bimetallic Au–Pd colloids,[43] and in the preparation of bimetallic nanoparticles by the sonochemical reduction of solutions containing gold and palladium ions.[44] Such core-shell particles were found to exhibit superior catalytic activity compared with Au–Pd alloy particles exhibiting the same overall Au : Pd ratio.[44] Detailed [197]Au Mössbauer measurements have confirmed the presence of a pure Au core, and also identified a thin alloy region at the interface between the Au-core and Pd-shell.[44,45] Inverted Pd core/Au shell particles may be prepared, but with difficulty;[46] even if Au is deposited on already-formed Pd particles, Au-core/Pd-shell structures are formed.[47] Interestingly, the core-shell particles reported in our study are stable at temperatures up to at least 500 °C, in

Table 3 Pd : Au ratios measured by XPS for titania, alumina and carbon supported Au–Pd catalysts

Catalyst	Measured Pd : Au by weight	
	Dried	Calcined
2.5 wt% Pd–2.5 wt% Au/TiO$_2$	0.3	5.1
0.8 wt% Pd–4.2 wt% Au/Al$_2$O$_3$	0.2	0.5
2.5 wt% Pd–2.5 wt% Au/G60 carbon	1.0	1.1

contrast to core-shell nanoparticles prepared ultrasonically in a porous silica support, where transformation to a random alloy was observed at 300 °C.[48]

In contrast, the carbon catalyst shows almost no differences in the surface Pd : Au ratio which remains close to the bulk value of 1 : 1 by weight (Table 3). This shows that the carbon-supported material, which is by far the most active catalyst for the direct synthesis reaction, is significantly different from the oxide supported catalysts. All these materials have been examined in detail using STEM characterisation.

STEM characterisation. High angle annular dark field (HAADF) images of the calcined Au–Pd/TiO$_2$ and the Au–Pd/carbon (2.5% Au–2.5% Pd) catalysts have been obtained and the measured particle size distributions are shown in Fig. 3 and 4 (Fig. 4 permits more detail to be observed for the particles in the 1–10 nm size range). It is apparent that both catalysts comprise mainly small particles (*ca.* 2–10 nm), but both also contain larger particles and it is clear that the carbon-supported material, which is the more effective catalyst, has a higher number density of the larger particles (Fig. 3 and 5). This finding may be significant with respect to the origin of enhanced catalytic activity, *i.e.* that higher catalytic activity in calcined catalysts is associated with the presence of larger particles. This is consistent with our earlier observations concerning the relationship between the particle size distribution and activity of Au–Pd/Al$_2$O$_3$ catalysts.[20] There it was observed that fresh catalysts comprised Au–Pd nanoparticles with a 3–10 nm size distribution, whereas, the same catalysts that had aged on storage at ambient temperature exhibited a bimodal size distribution with some larger particles (> 10 nm) being observed. Most significantly, the aged Au–Pd/Al$_2$O$_3$ catalysts that contained the larger particles showed an enhancement in activity by a factor of three.

X-Ray energy dispersive spectroscopy (STEM-XEDS) of the larger particles has been carried out for the Au–Pd (2.5 wt% Au–2.5 wt% Pd) catalysts supported on carbon, TiO$_2$ and Al$_2$O$_3$ calcined at 400 °C and the results are shown in Fig. 6. XEDS maps of the larger particles show that the Au–M$_2$ (9.712 keV) and the Pd Lα (2.838 keV) signals are spatially coincident, indicating that the metal nanoparticles in the field of view are in fact Au–Pd alloys in all these samples. Multivariate statistical analysis[49] (MSA) was performed on the XEDS spectrum image data sets obtained from the metal particles. MSA is a group of processing techniques that can be used to identify specific features and to reduce random noise components in the large STEM-XEDS datasets in a statistical manner. The MSA algorithm performs a data smoothing calculation by portioning the XEDS data using a probability density function and has recently been shown to be a particularly useful analysis tool for identifying core–shell morphologies in Au–Pd nanoparticles. This suggests that there is a tendency for Pd surface segregation to occur in the alloy particles supported on TiO$_2$ and Al$_2$O$_3$, as indicated by the XPS analysis (*qv*). The strong tendency for palladium surface segregation, observed in this study and for bulk alloys,[48] is not expected. We consider that it is presumably brought about by the preferential formation of Pd–O bonds at the alloy surface since in this temperature range palladium oxidizes more readily than gold.

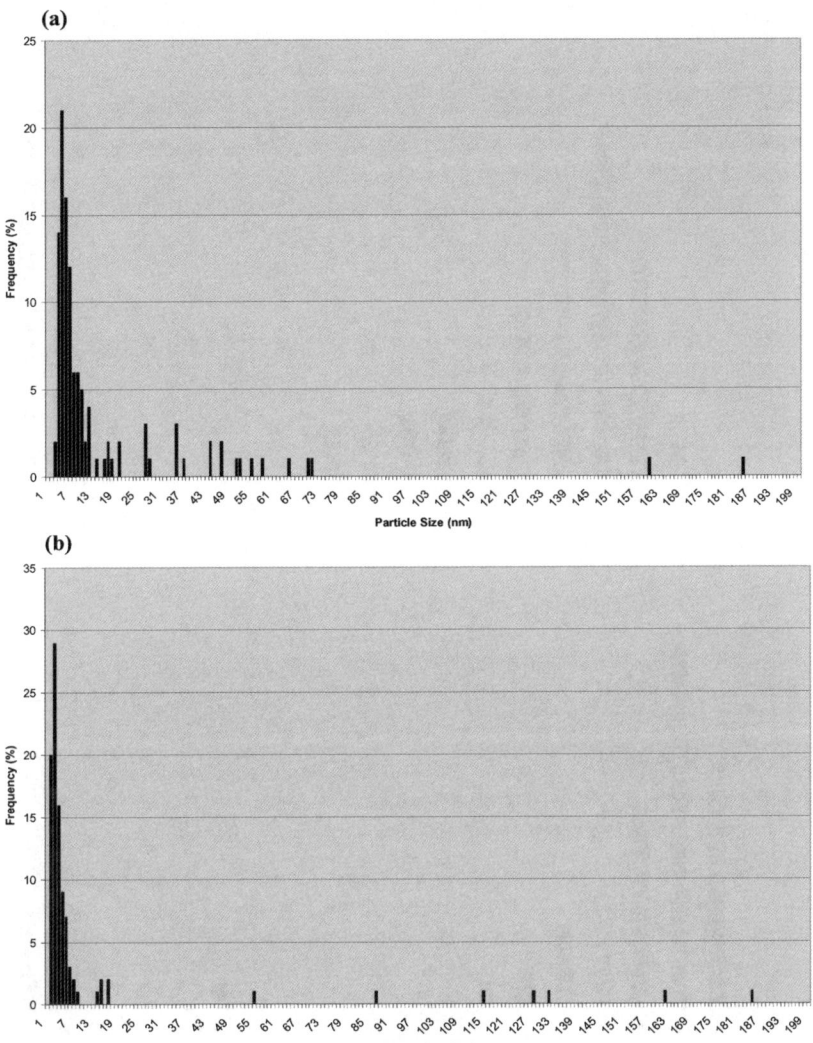

Fig. 3 Particle size distributions for Au–Pd catalysts (a) Au–Pd/carbon, (b) Au–Pd/TiO$_2$.

However, the structure of the Au–Pd nanoparticles supported on carbon is significantly different, since the core–shell morphology is not observed and, rather, homogeneous Au–Pd alloys are now observed (Fig. 6). Again, this is wholly consistent with the XPS evidence (Fig. 2b). Furthermore, the composition of the Au–Pd alloy nanocrystals changes markedly with the particle size (these data are presented in a related discussion paper[50]) and the amount of gold present in the nanocrystals increases with the particle diameter. These are significant observations, since it demonstrates that the core–shell structures, which spontaneously form on TiO$_2$ and Al$_2$O$_3$ supports (Fig. 6) (and have also been observed with Fe$_2$O$_3$-supported catalysts[21]), are not essential for the observation of high activity for the direct synthesis of hydrogen peroxide. Rather, we suggest that the particle size of the Au–Pd alloy is the important factor, with larger particles appearing to be associated with high activity and in this respect the variable composition of the Au–Pd particles may be an important factor. Given these factors, the active sites for the direct formation of hydrogen peroxide may well be very different from the isolated sites

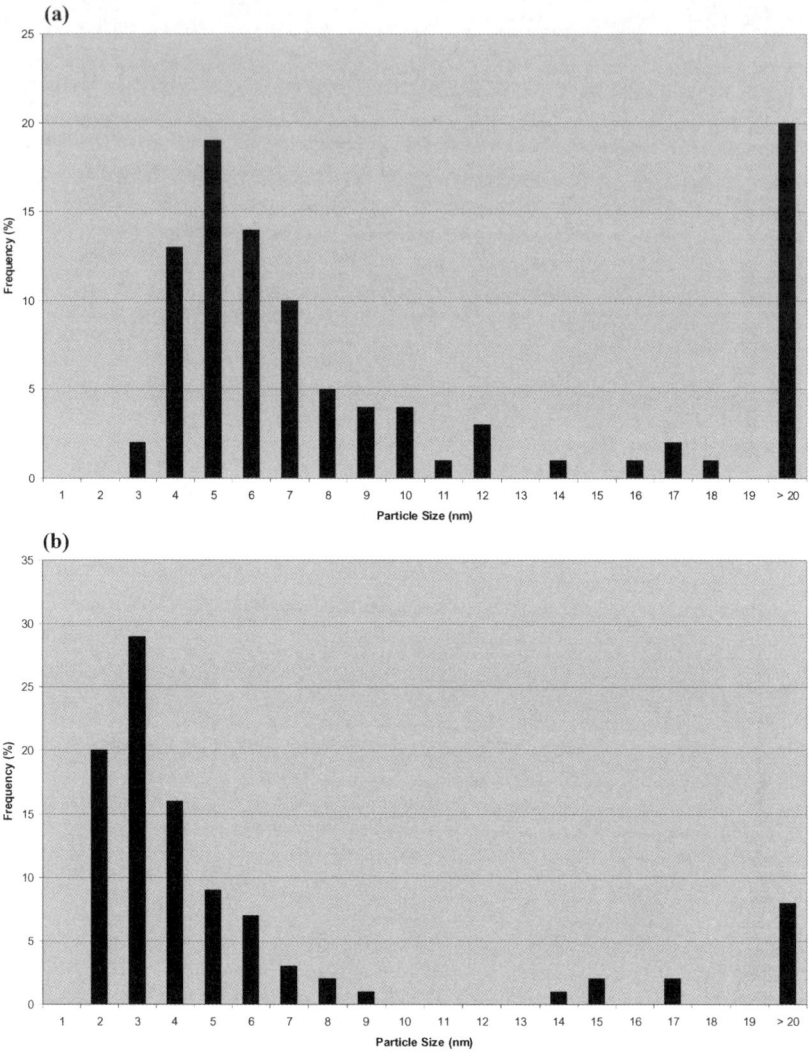

Fig. 4 Particle size distributions for Au–Pd catalysts showing more detail of small particle sizes (a) Au–Pd/carbon, (b) Au–Pd/TiO$_2$.

described by Chen *et al.*[51] for Au–Pd alloy catalysts that are effective for the acetoxylation of ethene in the production of vinyl acetate. The question arises as to why core–shell structures form spontaneously on the oxide supports but not on carbon. This may be due to the oxidation efficacy of the support surface, since carbon is a reducing support. As noted earlier, the formation of palladium shell structures may be due to the surface of the nanocrystals being oxidised and PdO, therefore, phase separates to the surface. This process is hindered by the reducing nature of the carbon support. Indeed, carbon is used in the extraction of Au^{3+} salts during the commercial production of gold and during this process the Au^{3+} is reduced to metallic gold. This process may also occur during the preparation of the Au–Pd alloys used in this study and this may stabilise the random alloys rather than core–shell structures.

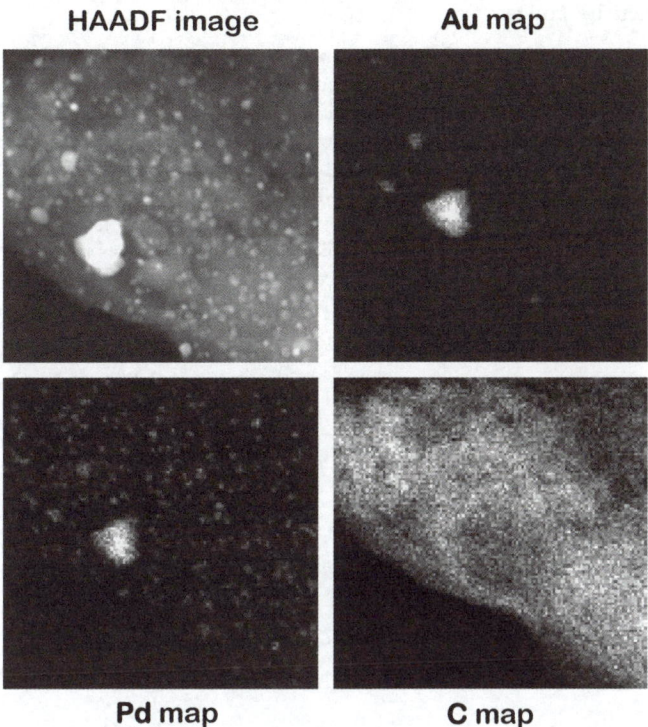

Fig. 5 HAADF STEM image and Au, Pd and C STEM XEDS maps showing the spatial and chemical distribution of alloy particles in the uncalcined 2.5wt% Au–2.5wt% Pd/carbon sample.

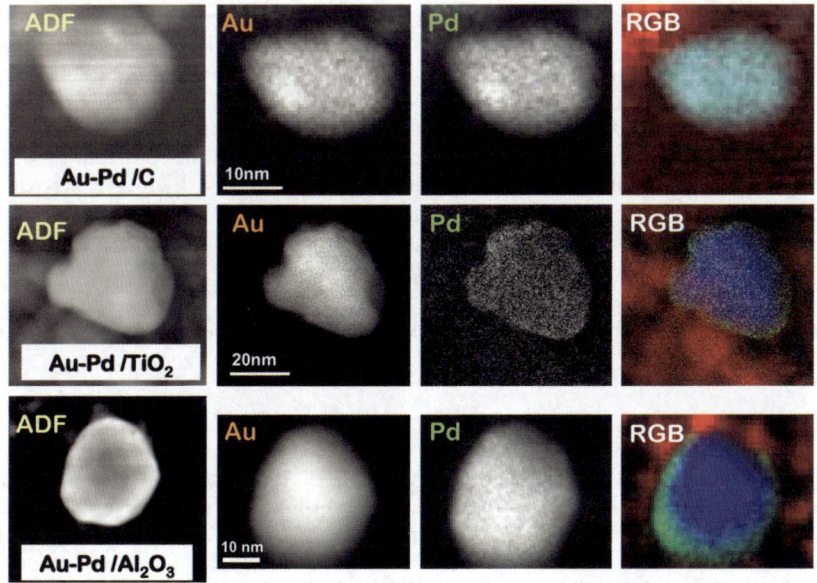

Fig. 6 Montage of HAADF image (*column* 1), Au map (*column* 2), Pd map (*column* 3) and RGB reconstructed overlay map (*column* 4) [(Au–blue: Pd–green)] for calcined AuPd/C (*row* 1), calcined AuPd/TiO₂ (*row* 2) and calcined AuPd/Al₂O₃ (*row* 3). Note that the calcined AuPd particles on TiO₂ and Al₂O₃ supports show a Au rich-core/Pd-rich shell morphology, whereas calcined AuPd particles on activated C are homogeneous alloys.

The H_2 selectivity problem

Whilst many may view the direct reaction between H_2 and O_2 to form hydrogen peroxide as an oxidation reaction, this is not really the case, since it is the H–H bond that is broken and the O–O bond is retained. The competing reaction in which water is directly formed is an oxidation reaction. In this sense the direct formation of hydrogen peroxide is better viewed as the hydrogenation of oxygen. For this reason, catalysts that are effective for the reaction, *e.g.* supported Pd catalysts, are also effective hydrogenation catalysts, as their primary role is the activation of H_2 at low temperatures. This then leads to a major problem, since catalysts that are effective in the initial hydrogenation of oxygen to give hydrogen peroxide selectively are also equally effective at the hydrogenation of hydrogen peroxide to water in a subsequent hydrogenation reaction:

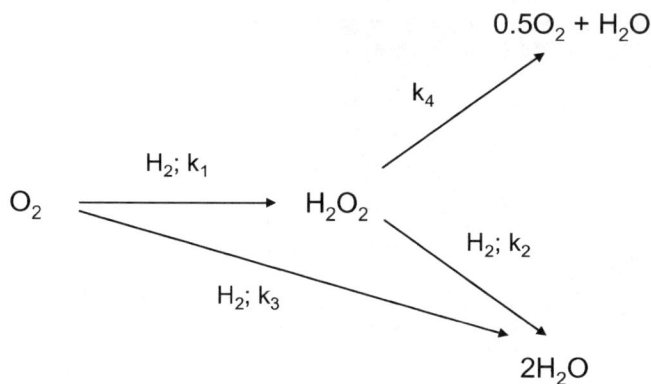

The decomposition of hydrogen peroxide to give oxygen and water is not observed during the direct reaction when hydrogen is present, and hence we shall ignore the contribution of this reaction when analysing the selectivity of hydrogen for the formation of hydrogen peroxide. In catalytic reactions it is important to consider the primary selectivity for the desired product, denoted S_o. S_o is the selectivity obtained by extrapolating selectivity to zero conversion and is therefore the limiting selectivity of the catalytic system. It is therefore essential that S_o is high, since if it is not then very little of the desired product can be expected at higher conversions. At very low conversions of H_2, *i.e.* as the conversion approaches zero, then the contribution of k_2 to the observed selectivity (*i.e.* the sequential hydrogenation of hydrogen peroxide) becomes negligible, and hence we can write:

$$S_o = \frac{k_1}{k_3}$$

and at higher conversions, since the contribution from k_4 is negligible, the selectivity, S, can be written as:

$$S = \frac{k_1}{k_2 + k_3}$$

The data in Table 1 gives the selectivities at 30 min reaction time when an appreciable conversion of H_2 is observed. To determine the primary selectivity we have carried out reactions at very short reaction times (2 min) when the H_2 conversion was $<1\%$. At these short reaction times very high rates of reaction are observed (Au–Pd/carbon = 391 mol H_2O_2 h^{-1} $kg(cat)^{-1}$, compared with 110 mol H_2O_2 h^{-1} $kg(cat)^{-1}$ observed at 30 min reaction time; Au–Pd/TiO_2 = 205 mol H_2O_2 h^{-1} $kg(cat)^{-1}$, compared with 64 mol H_2O_2 h^{-1} $kg(cat)^{-1}$ observed at 30 min reaction time) and these could be beneficial if the hydrogen peroxide could be captured *in situ* under these reaction conditions (*i.e.* at low temperature in aqueous solution and in the presence of high pressures of H_2 and O_2). The primary selectivity for hydrogen utilisation determined at 2 min reaction time was $>99\%$ for both the TiO_2- and the carbon-supported catalysts. This demonstrates that the direct oxidation of hydrogen to form water is negligible for the Au–Pd catalysts presented in this study. This means that the active sites for the total oxidation reaction, which involves the scission of the dioxygen bond, must be different from the active sites for the selective hydrogenation reaction of dioxygen. Hence the reaction scheme can be simplified as follows to consider just two reactions:

$$O_2 \xrightarrow{\ H_2;\ k_1\ } H_2O_2 \xrightarrow{\ H_2;\ k_2\ } 2H_2O$$

The loss in selectivity observed during the reaction for 30 min (Table 1) is therefore due to the influence of k_2 as the sequential hydrogenation becomes important since the concentration of hydrogen peroxide increases as the reaction proceeds. It is apparent that the carbon-supported catalyst exhibits a higher selectivity than the other catalysts, in particular TiO_2, and this may be due to both the support and the nature of the Au–Pd nanoparticles present on the surface. The effect of the over-hydrogenation of hydrogen peroxide has been examined for the carbon-supported catalysts by stirring a solution of hydrogen peroxide with the catalyst in the absence of oxygen but otherwise using identical experimental conditions to those used in the synthesis reaction and carrying out the reaction for 30 min. In these experiments (Fig. 7) we investigated the hydrogenation of hydrogen peroxide over the carbon support without the addition of any metal, as well as the calcined 5 wt%Pd/carbon and 2.5wt% Au–2.5 wt%Pd/carbon catalysts. These data show that all three materials hydrogenate hydrogen peroxide to an equal extent. This is an unexpected result, since one could expect that the supported metal catalysts would be more effective for the sequential hydrogenation of hydrogen peroxide. Hence, we consider that the surface of the support is a critical factor that determines the selectivity for the direct hydrogenation reaction. The support, therefore, has two important parameters that control reactivity; namely (i) the dispersion of the metal and whether core–shell or homogeneous alloys are obtained, and (ii) minimising the sequential hydrogenation of hydrogen peroxide, which is the main route by which hydrogen peroxide is lost. These data also demonstrate that the sites for the sequential hydrogenation of hydrogen peroxide are different from the active

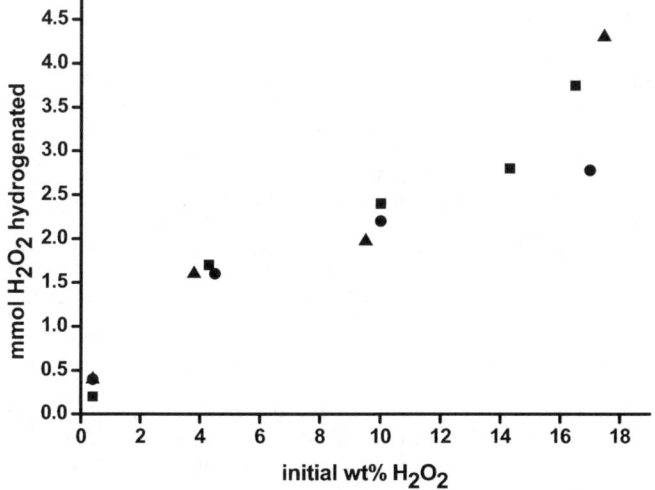

Fig. 7 Hydrogen peroxide hydrogenation under standard reaction conditions at 2 °C in the absence of O_2. Key: ▲ carbon support; ● 5 wt%Pd/carbon; ■ 2.5wt%Au–2.5wt%Pd/carbon.

sites for the direct synthesis reaction, since although the carbon support is as effective for the hydrogenation reaction as the Pd and Au–Pd-supported catalysts (Fig. 7), the carbon support is not active for the direct reaction. However, taken overall, these data clearly show that the hydrogen selectivity and the sequential hydrogenation of hydrogen peroxide remain the key problems that require a solution if the Au–Pd-supported catalysts are to find commercial exploitation.

Conclusions

In the direct synthesis of hydrogen peroxide from the reaction of H_2 and O_2 the addition of Au (which is relatively inert for this reaction) to Pd to produce Au–Pd nanoalloy particles enhances the activity of the Pd markedly. On TiO_2 and Al_2O_3 the nanoalloy exhibit core–shell structures, with Pd-rich surfaces, which form spontaneously in the calcined Au–Pd nanoparticles. However, these core–shell morphologies clearly are not essential for the observation of high reaction rates and selectivities because the calcined carbon supported Au–Pd nanoparticles exhibit homogeneous alloys of Au and Pd and these give the highest activities. The calcined catalysts are stable and re-usable, and no loss of Au or Pd is observed as long as care is taken in the preparation of the catalysts, and in particular with the calcination step. Furthermore, no additives are required to observe high reaction rates and this may have significant advantages with respect to the synthesis of hydrogen peroxide for medical applications. However, the key problem that still remains is the sequential hydrogenation of hydrogen peroxide and this appears to be dependent mainly on the support. This is a crucial problem that urgently requires solution for these catalysts to be commercially exploited.

Acknowledgements

This work formed part of the EU AURICAT project (Contract HPRN-CT-2002-00174) and the EPSRC/Johnson Matthey funded ATHENA project and we thank them for funding this research. We also thank the World Gold Council (through the GROW scheme), and Cardiff University (AA Reed studentship) for providing support for J. K. E. C. J. K. and A. H. would also like to acknowledge the generous support of NSF Materials Research Science and Engineering Center (NSF DMR-0079996).

References

1   M. S. Scurrell and G. J. Hutchings, *Cattech*, 2003, **7**, 90.
2   G. J. Hutchings, M. S. Scurrell and J. R. Woodhouse, *Chem. Soc. Rev.*, 1989, **18**, 251.
3   G. C. Bond and D. T. Thompson, *Catal. Rev. Sci. Eng.*, 1999, **41**, 319.
4   G. C. Bond and D. T. Thompson, *Gold Bull.*, 2000, **33**, 41.
5   M. Haruta, *Gold Bull.*, 2004, **37**, 27.
6   A. S. K. Hashmi, *Gold Bull.*, 2004, **37**, 51.
7   R. Meyer, C. Lemaire, Sh. K. Shaikutdinov and H.-J. Freund, *Gold Bull.*, 2004, **37**, 72.
8   G. J. Hutchings, *Gold Bull.*, 2004, **37**, 37.
9   A. S. K. Hashmi and G. J. Hutchings, *Angew. Chem., Int. Ed.*, 2006, **45**, 7896.
10 M. Haruta, T. Kobayashi, H. Sano and N. Yamada, *Chem. Lett.*, 1987, **16**, 405.
11 P. Landon, J. Ferguson, B. E. Solsona, T. Garcia, A. F. Carley, A. A. Herzing, C. J. Kiely, S. E. Golunski and G. J. Hutchings, *Chem. Commun.*, 2005, 3385.
12 P. Landon, J. Ferguson, B. E. Solsona, T. Garcia, S. Al-Sayari, A. F. Carley, A. Herzing, C. J. Kiely, M. Makkee, J. A. Moulijn, A. Overweg, S. E. Golunski and G. J. Hutchings, *J. Mater. Chem.*, 2006, **16**, 199.
13 A. K. Sinha, S. Seelan, S. Tsubota and M. Haruta, *Angew. Chem., Int. Ed.*, 2004, **43**, 1546.
14 M. D. Hughes, Y.-J. Xu, P. Jenkins, P. McMorn, P. Landon, D. I. Enache, A. F. Carley, G. A. Attard, G. J. Hutchings, F. King, E. H. Stitt, P. Johnston, K. Griffin and C. J. Kiely, *Nature*, 2005, **437**, 1132.
15 A. Abad, P. Conception, A. Corma and H. Garcia, *Angew. Chem., Int. Ed.*, 2005, **44**, 4066.
16 A. Corma and P. Serna, *Science*, 2006, **313**, 332.

17 D. I. Enache, J. K. Edwards, P. Landon, B. Solsona-Espriu, A. F. Carley, A. A. Herzing, M. Watanabe, C. J. Kiely, D. W. Knight and G. J. Hutchings, *Science*, 2006, **311**, 362.

18 P. Landon, P. J. Collier, A. J. Papworth, C. J. Kiely and G. J. Hutchings, *Chem. Commun.*, 2002, 2058.

19 P. Landon, P. J. Collier, A. F. Carley, D. Chadwick, A. J. Papworth, A. Burrows, C. J. Kiely and G. J. Hutchings, *Phys. Chem. Chem. Phys.*, 2003, **5**, 1917.

20 B. E. Solsona, J. K. Edwards, P. Landon, A. F. Carley, A. Herzing, C. J. Kiely and G. J. Hutchings, *Chem. Mater.*, 2006, **18**, 2689.

21 J. K. Edwards, B. Solsona, P. Landon, A. F. Carley, A. Herzing, M. Watanabe, C. J. Kiely and G. J. Hutchings, *J. Mater. Chem.*, 2005, **15**, 4595.

22 J. K. Edwards, B. Solsona, P. Landon, A. F. Carley, A. Herzing, C. J. Kiely and G. J. Hutchings, *J. Catal.*, 2005, **236**, 69.

23 H. T. Hess, in *Kirk-Othmer Encyclopaedia of Chemical Engineering*, ed. I. Kroschwitz and M. Howe-Grant, Wiley, New York, 1995, vol. 13, p. 961.

24 H. Henkel and W. Weber, *US Pat., 1108752*, 1914.

25 G. A. Cook, *US Pat., 2368640*, 1945.

26 Y. Izumi, H. Miyazaki and S. Kawahara, *US Pat., 4009252*, 1977.

27 Y. Izumi, H. Miyazaki and S. Kawahara, *US Pat., 4279883*, 1981.

28 H. Sun, J. J. Leonard and H. Shalit, *US Pat., 4393038*, 1981.

29 L. W. Gosser and J.-A. T. Schwartz, *US Pat., 4772458*, 1988.

30 L. W. Gosser, *US Pat., 4889705*, 1989.

31 C. Pralins and J.-P. Schirmann, *US Pat., 4996039*, 1991.

32 T. Kanada, K. Nagai and T. Nawata, *US Pat., 5104635*, 1992.

33 J. van Weynbergh, J.-P. Schoebrechts and J.-C. Colery, *US Pat., 5447706*, 1995.

34 S.-E. Park, J. W. Yoo, W. J. Lee, J.-S. Chang, U. K. Park and C. W. Lee, *US Pat., 5972305*, 1999.

35 G. Paparatto, R. d'Aloisio, G. De Alberti, P. Furlan, V. Arca, R. Buzzoni and L. Meda, *EP Pat., 0978316A1*, 1999.

36 B. Zhou and L.-K. Lee, *US Pat., 6168775*, 2001.

37 M. Nystrom, J. Wangard and W. Herrmann, *US Pat., 6210651*, 2001.

38 S. Carrettin, P. McMorn, P. Johnston, K. Griffin and G. J. Hutchings, *Chem. Commun.*, 2002, 696.

39 R. A. Sheldon, I. Arends, G.-J. ten Brink and A. Dijksman, *Acc. Chem. Res.*, 2002, **35**, 774.

40 V. R. Choudhary and C. Samanta, *J. Catal.*, 2006, **238**, 28.

41 C. Samanta and V. R. Choudhary, *Catal. Commun.*, 2006, **8**, 73.

42 V. R. Choudhary, C. Samanta and P. Jana, *Appl. Catal., A*, 2007, **317**, 234.

43 A. F. Lee, C. J. Baddeley, C. Hardacre, M. R. Ormerod, R. M. Lambert, G. Schmid and H. J. West, *Phys. Chem.*, 1995, **99**, 6096.

44 H. Takatani, H. Kago, Y. Kobayashi, F. Hori and R. Oshima, *Trans. Mater. Res. Soc. Jpn.*, 2003, **28**, 871.

45 Y. Kobayashi, S. Kiao, M. Seto, H. Takatani, M. Nakanishi and T. Oshima, *Hyperfine Interact.*, 2004, **156–157**, 75.

46 H. Takatani, F. Hori, M. Nakaishi and R. Oshima, *Aust. J. Chem.*, 2003, **53**, 1025.

47 C. Kan, W. Cai, C. Li, L. Zhang and H. J. Hofmeister, *J. Phys. D*, 2003, **36**, 1609.

48 T. Nakagawa, H. Nitani, S. Tanabe, K. Okitsu, S. Seino, Y. Mizukoshi and T. A. Yamamoto, *Ultrason. Sonochem.*, 2004, **12**, 249.

49 N. Bonnet, *J. Microsc.*, 1998, **190**, 2.

50 A. A. Herzing, M. Watanabe, J. K. Edwards, M. Conte, G. J. Hutchings and C. J. Kiely, *Faraday Discuss.*, 2008, **138**, DOI: 10.1039/b706293c.

51 M. Chen, D. Kumar, C.-W. Yi and D. W. Goodman, *Science*, 2005, **310**, 291.

Structures and associated catalytic properties of well-defined nanoparticles produced by laser vaporisation of alloy rods†

Valérie Caps, Sandrine Arrii, Franck Morfin, Gérard Bergeret and Jean-Luc Rousset*

Received 23rd April 2007, Accepted 15th May 2007
First published as an Advance Article on the web 19th September 2007
DOI: 10.1039/b706131e

Bimetallic clusters, all containing gold, have been produced by laser vaporisation of bulk alloys followed by deposition of the formed clusters onto Al_2O_3 and TiO_2 powders or flat silica supports. This technique allows a narrow size distribution of highly dispersed gold-based nanoparticles on powders and nanocrystalline structured thin films on 2D supports to be obtained. The catalytic performances of the as-obtained AuFe, AuNi, AuTi powdery catalysts have been studied in the PROX reaction and compared with those obtained in the oxidation of CO in the temperature range 25–300 °C. By comparing the activities of the different catalysts, it is concluded that the nature of the gold partner directly affects the activity of gold. The following tendency is observed: AuFe and AuNi have rather similar activities, significantly lower than that of AuTi. In this paper, we also present a first attempt to study reactivity of original self-supported systems. We show that significant CO oxidation reactivity can be obtained over unsupported nanoporous AuTi and PdAu thin films. By completely excluding the support effect, unsupported catalysts could provide a way of understanding the relevant catalytic mechanisms more easily.

1. Introduction

Metallic clusters constitute an exciting field of research, and over at least the past 30 years a great deal of effort has been devoted to them. A very attractive issue in this field is the transition in electronic and related properties on going from the bulk to small clusters. These differences should be reflected in numerous fields such as optics, magnetism or catalysis. More particularly, these particle size effects are of interest for those investigating chemisorptive or catalytic properties since catalysts in practice always consist of small supported particles. Moreover, in the field of heterogeneous catalysis, bimetallic particles constitute a promising class of catalysts that are found to exhibit superior properties, compared to single metals, in terms of activity, selectivity,[1] stability, and resistance to poisoning.[2]

It is generally recognised that the development of a new catalyst requires the control of its properties, and, in particular, of its surface composition and local order

Institut de Recherches sur la Catalyse et l'Environnement de Lyon (IRCELYON, CNRS/ Université de Lyon), 2 Avenue Albert Einstein, F-69626, Villeurbanne Cedex, France. E-mail: jean-luc.rousset@ircelyon.univ-lyon1.fr

† The HTML version of this article has been enhanced with colour images.

at the atomic scale that directs its electronic and hence chemical properties. Generally, most bimetallic clusters are prepared in a chemical way, *i.e.*, by co-impregnation or co-exchange techniques but large distributions of composition are generally observed and it is thus difficult to study the relation between surface composition and reactivity. In the present study, alloyed particles and related systems, are synthesised by laser vaporisation followed by low energy cluster beam deposition. The laser vaporisation technique allows the generation of ligand-free metal clusters and it can be used to evaporate even the most refractory metals.

Since clusters are pre-formed before deposition and do not fragment upon impact on the substrate one has already been able to deposit the same active phase on different supports to prove the direct contribution of the supports.[3] Conversely, in the case of bimetallic systems, the advantage of this technique becomes essential since the composition of each produced particle is identical to that of the vaporised rod. This allows us to deposit clusters with similar size but different composition onto the same substrate and to study precisely the alloying effect on reactivity.[4,5] In this paper, we also present a first attempt to study the reactivity of the original self-supported system or, in other words, nanocrystalline structured thin films. The growth of these thin films are obtained by random stacking of the incident free clusters. These cluster assembled films are highly porous with densities as low as about one half of the corresponding bulk material densities and are characterised by numerous cavities of the nanometre size that may be interconnected and form nanopores.[6]

2. Experimental

2.1 Preparation

All materials were prepared using low-energy cluster beam deposition.[7] The laser vaporisation source available in Lyon (Fig. 1) uses a Nd:YAG pulsed laser operating at 532 nm to vaporise the metal(s) from a bimetallic rod in order to create a plasma within a high vacuum chamber. Cluster nucleation and growth occur when an inert gas is introduced in synchronisation with the laser. Differential pumping extraction of the clusters through a skimmer yields a cluster beam that carries neutral and ionised species, that are ejected at a rate of about 1000 m s^{-1}. After removal of the ionised clusters by electrostatic deflection, the neutral ones are deposited on the desired substrates with no fragmentation, since the kinetic energy of the clusters (~ 0.1 eV atom^{-1}) is lower than the cohesive energy in face centered cubic bulk metals/alloys (~ 4–5 eV atom^{-1}). Deposition rates of typically 1 Å (equivalent thickness) s^{-1} are achieved, as measured with a quartz microbalance. For these experiments, a monometallic rod of gold (99.99% purity), as well as bimetallic rods

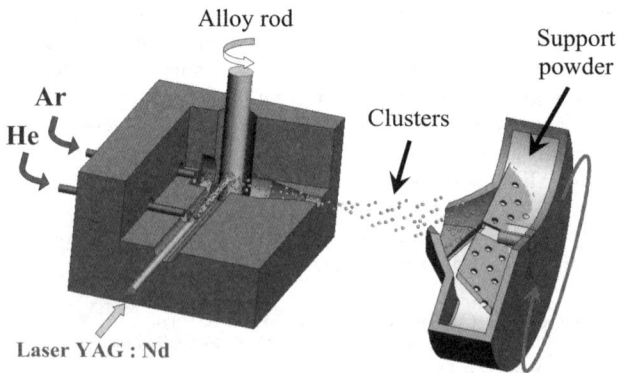

Fig. 1 Schematic description of the laser vaporisation source.

of $Au_{20}Pd_{80}$, $Au_{25}Ti_{75}$, $Au_{25}Fe_{75}$ and $Au_{50}Ni_{50}$ (obtained by melting of Au 99.99% and the other metal 99.95%) from Goodfellow are used. The substrates used can be copper grids for transmission electron microscopy analysis or oxide powders (γ-alumina Condea Puralox SCFa-215, 221 m^2 g^{-1} and titania Hombikat T100 20-S20 from Sachtleben Chemie, 60 m^2 g^{-1} anatase) for catalytic evaluation in a fixed-bed flow reactor. Substrates can also be Suprasil™ silica disks, sintered glass disks or disks obtained by pressing the oxide powder supports for studies with X-ray diffraction and X-ray photoelectron spectroscopy.

These systems are exposed to air before characterisation and reactivity studies.

2.2   Chemical analysis

The metal content of the powder materials are determined in-house using inductively-coupled plasma mass spectroscopy.

2.3   Transmission electron microscopy

A JEOL JEM 2010 Transmission Electron Microscope (TEM) operating at 200 kV is used to characterise nanoalloy deposits on copper grids as well as nanoalloys deposited on powders. In the latter case, the powders are dispersed in ethanol by ultrasonication. A drop of the solution is deposited onto a thin holey-carbon film supported on a copper microscopy grid (200 mesh, 3.05 mm) and left to dry in order to get TEM imaging of the shape and distribution of the clusters over the support. Size distributions are determined from TEM imaging of extractive replicas.

EDX (Energy Dispersive X-ray) analysis is carried out in a JEOL 2010 F microscope equipped with a field emission gun for high-performance measurements. The Pentafet LinK Isis (Oxford Instruments) EDX spectrometer allows detection of elements with atomic numbers $Z > 4$ amu. The field emission gun allows one to obtain small electron probes that are sufficiently bright to perform EDX analysis on individual particles and achieve spatial resolution better than 1 nm. In this case, analysis is performed on very thin deposits (~ 0.4 nm equivalent thickness) of the bimetallic clusters on copper grids in order to probe the chemical composition of single particles.

2.4   X-Ray diffraction

X-Ray diffraction powder patterns are recorded using a Panalytical X'Pert Pro MPD diffractometer (Bragg-Brentano parafocussing geometry, Cu-Kα radiation, X'Celerator detector). The samples (cluster deposits on Suprasil™ silica disks with ~ 300 nm equivalent thickness) are put on a X–Y–Z stage. Data are collected in continuous scanning mode at $0.002068°$ s^{-1}, *i.e.* 16 s per step of $0.033°$ (2θ). The determination of the particle size cannot be obtained from the classical Scherrer formula because of the overlapping of the broad reflection lines. The average crystallite size is thus determined by the Rietveld method (Fullprof code)[8] assuming various possible particle compositions and fitting the whole experimental diffraction pattern.

2.5   X-Ray photoelectron spectroscopy

XPS experiments are carried out in an ESCALAB 200R spectrometer from Fisons Instruments equipped with a hemispherical analyzer, using the Al-Kα line (1486.6 eV) of a dual anode and a pass energy of 50 eV in ultra-high vacuum ($P < 3 \times 10^{-10}$ mbar). The peaks are referenced to the C–(C,H) components of the C 1s band at 284.6 eV. The data are processed using Shirley background subtraction and peak decomposition into Gaussian–Lorentzian products. A degree of asymmetry is added to fit the Au 4f peaks. For these experiments, the bimetallic clusters are deposited on Suprasil™ silica disks ($\sim$400 nm equivalent thickness).

2.6 Catalytic evaluation

Bimetallic clusters deposited on oxide powders and on Suprasil™ silica disks (or sintered glass disks) are tested for CO oxidation, both in the absence and in the presence of H_2, and H_2 oxidation in a typical continuous flow fixed-bed reactor at atmospheric pressure. For alumina and titania powders, 800 mg and 1200 mg of the as-synthesised material are used, respectively, except for $Au_{25}Fe_{75}/Al_2O_3$ and $Au_{50}Ni_{50}/Al_2O_3$ where lower amounts are used (320 and 267 mg, respectively). The latter are diluted in γ-alumina (inactive under all reaction conditions used) in order to keep the bed lengths constant. The reactant flows consist of 2% CO + 2% O_2 + 96% He for the H_2-free oxidation of CO, 2% CO + 2% O_2 + 48% H_2 + 48% He for the preferential oxidation of CO (PROX) and 48% H_2 + 2% O_2 + 50% He for the oxidation of H_2 (all percentages in vol%), using high purity (>99.95%) gases from Air Liquide. A total flow rate of 50 mL min^{-1} (STP) is generated. The reactor is heated to 300 °C then cooled down to 20 °C with a heating (and cooling) rate of 1 °C min^{-1} several times in order to reach steady-state conditions. Product analysis is carried out on-line with a Varian Micro GC (CP2003) equipped with a TCD detector. Two columns, operated at 50 °C, are used in parallel: the Molsieve 5A column (Ar as carrier gas) is used to quantify O_2 and CO, and the poraPLOT Q column (He as carrier gas) to quantify CO_2.

3. Results and discussion

3.1 AuTi, AuFe and AuNi

The characteristics of the powder-supported clusters are shown in Table 1. Micrographs of alumina-supported $Au_{25}Ti_{75}$ and $Au_{50}Ni_{50}$ clusters are shown in Fig. 2; micrographs of $Au_{25}Ti_{75}$ and $Au_{25}Fe_{75}$ clusters (40 nm and 0.4 nm-thick films, respectively on copper grids) are shown in Fig. 3.

   The mean sizes of gold particles (as determined on imaging of replicas) were found to be similar in the fresh powder samples, between 2 and 3 nm. Gold particles produced from vaporisation of alloy rods have a slightly lower diameter than those produced from vaporisation of a monometallic gold rod. This is attributed to the lower gold contents in the bimetallic rods. Gold nanoparticles arising from the same $Au_{25}Ti_{75}$ rod exhibit identical mean diameters whether supported over alumina (2.1 nm) or titania (2.2 nm). This was previously observed with gold nanoparticles

Table 1 Gold loadings of the supported materials and mean diameters of supported gold particles as deduced from chemical analysis and TEM experiments, respectively

| Rod | Powder support | Metal content (wt%) | | Mean diameter/nm[b] | Catalytic CO oxidation | |
		Au	M[a]		$T_{1/2}$/°C	Rate at 80 °C[c]
Au	TiO_2	0.023	—	2.9	118	0.54
$Au_{25}Ti_{75}$	TiO_2	0.017	N.D.	2.2	116	0.73
$Au_{25}Ti_{75}$	γ-Al_2O_3	0.06	0.07	2.1	84	0.71
$Au_{25}Fe_{75}$	γ-Al_2O_3	0.05	0.08	N.D.	>300	<0.1
$Au_{50}Ni_{50}$	γ-Al_2O_3	0.06	0.02	N.D.	>300	<0.1

[a] M stands for the metal associated with gold in the vaporised rod. [b] As determined by imaging of replicas of the powder-supported clusters in fresh samples (i.e. before catalytic reaction). [c] Reaction rates (mmol CO g_{Au}^{-1} s^{-1}) calculated at 80 °C from Fig. 5. $T_{1/2}$ is the temperature at which the conversion of CO attains 50%.

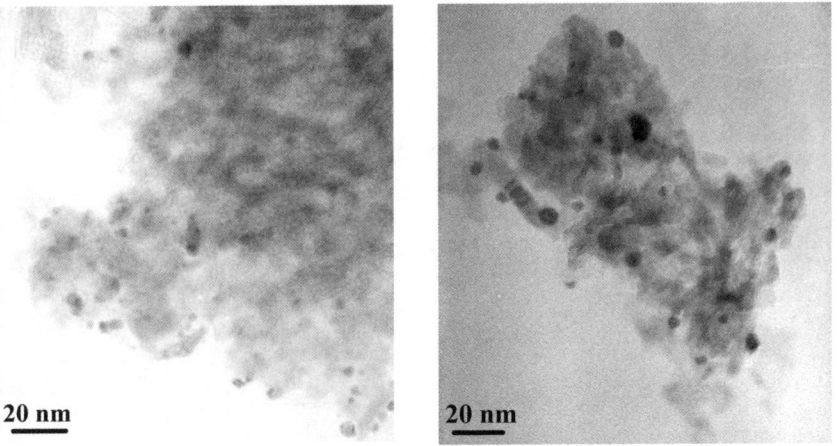

Fig. 2 Transmission electron micrographs of fresh $Au_{25}Ti_{75}/Al_2O_3$ (left) and $Au_{50}Ni_{50}/Al_2O_3$ (right).

vaporised from a monometallic gold rod.[3] It illustrates once more the ability of the laser vaporisation technique to generate similar metal particle sizes over various supports and substrates. A size histogram of gold particles in $Au_{25}Ti_{75}/Al_2O_3$ is given in Fig. 4 to show that, like in monometallic systems,[9] the particles produced from bimetallic rods have a narrow size distribution centred at about 2 nm.

The gold contents of the powder samples are low, typically below 0.1 wt%, but sufficient to allow testing of these materials in a typical fixed-bed flow reactor. Their catalytic behaviours in the oxidation of CO as a function of temperature are shown in Fig. 5.

3.1.1 Influence of the titania morphology on gold/titania-catalysed CO oxidation.
The CO oxidation rate observed over alumina-supported $Au_{25}Ti_{75}$ clusters (Table 1) is similar to that observed over Au/TiO_2 (note: the apparent superior catalytic activity of $Au_{25}Ti_{75}/Al_2O_3$ in Fig. 5 is due to the higher metallic loading of the catalytic bed, see Experimental section and Table 1). This can be explained by the

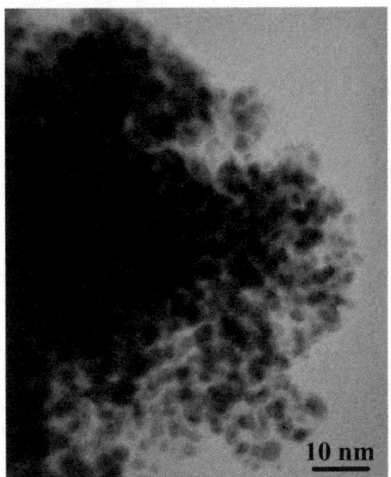
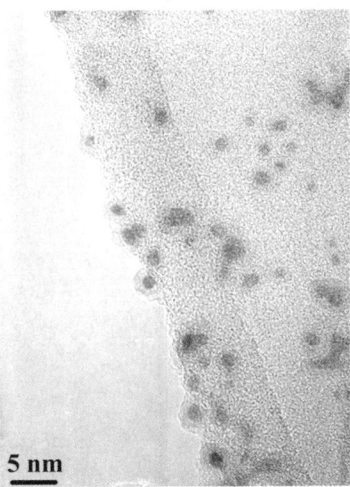

Fig. 3 Transmission electron micrographs of $Au_{25}Ti_{75}$ (40 nm-thick film, left) and $Au_{25}Fe_{75}$ (4 nm-thick film, right).

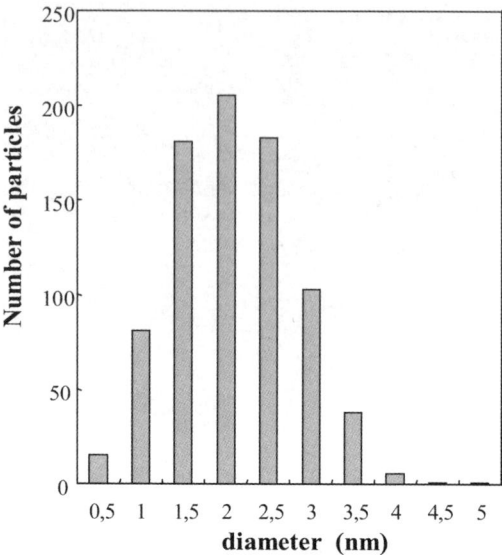

Fig. 4 Size histogram of gold particles in $Au_{25}Ti_{75}/Al_2O_3$ as determined from micrographs of an extractive replica of the sample.

fact that the $Au_{25}Ti_{75}$ clusters actually consist of metallic gold nanoparticles and titania nanostructures, as shown by XRD (Fig. 6 and Table 2) and XPS (Fig. 7 and Table 2) experiments.

The diffraction pattern of the self-supported $Au_{25}Ti_{75}$ system (after catalytic reaction) indeed exhibits very broad peaks (Fig. 6) corresponding to Au(111) and Au(200) reflections. The large widths of the peaks are attributed to the small size of the particles and to a possible distribution in compositions with the potential alloying of gold and titanium. The determination of the particle size is obtained by fitting the whole pattern using the Rietveld method, assuming that all the particles have the same pure gold (100%) composition. A fairly good agreement with the experimental data is obtained for a particle size of 1.8 ± 0.2 nm (Table 2). No other peak, that could be attributed to Ti-containing crystallites, is detected.

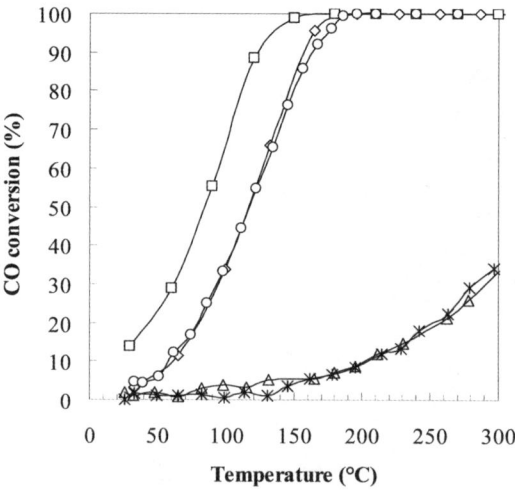

Fig. 5 CO conversion as a function of temperature over $Au_{25}Ti_{75}/Al_2O_3$ ($\square$), Au/TiO_2 ($\diamond$), $Au_{25}Ti_{75}/TiO_2$ ($\bigcirc$), $Au_{25}Fe_{75}/Al_2O_3$ (*) and $Au_{50}Ni_{50}/Al_2O_3$ ($\triangle$).

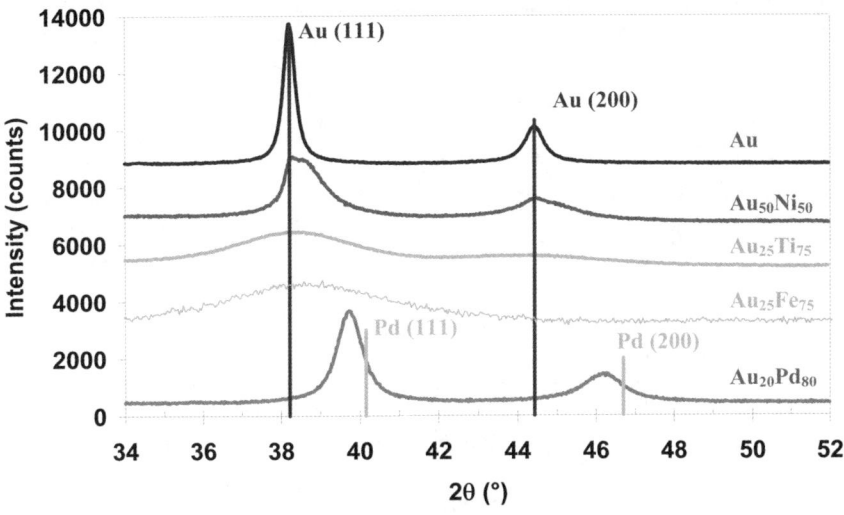

Fig. 6 X-Ray diffraction patterns of thick films of Au, $Au_{50}Ni_{50}$, $Au_{25}Ti_{75}$, $Au_{25}Fe_{75}$, $Au_{20}Pd_{80}$ clusters, after catalytic reaction (except for $Au_{25}Fe_{75}$).

The XPS spectra of the Au 4f and Ti 2p regions (Fig. 7) exhibit doublet peaks. The binding energies of 83.7 eV and 458.2 eV for Au $4f_{7/2}$ and Ti $2p_{3/2}$, respectively correspond to those of metallic gold (84.0 eV)[10] and TiO_2 (458.8 eV).[11] The slight shift towards lower binding energies for the Au 4f peaks could be attributed to surface-core level shift due to small particle size[12] or electron transfer from the TiO_2 to the gold particles.[3] The fact that these TiO_2 phases are invisible in XRD patterns indicates that they are very small and not well-crystallised.

This shows that the presence of nano-oxides in the direct proximity of the gold particles is enough to warrant oxygen activation and subsequent CO oxidation on gold (or at the gold/titania interface).[13] It also demonstrates that the synergy between Au and Ti is only slightly dependant on the intimate structure between the gold atoms and the TiO_2 phase. The influence of long-distance mechanisms, such as oxygen reverse spill-over sometimes used to explain CO oxidation over supported gold catalysts,[14] seems to be minimal in this reaction.

Table 2 Average gold crystallite size of the self-supported clusters before and after reaction and binding energies of the two metals in the clusters as deduced from XRD analysis and XPS experiments, respectively

	Average Au crystallite size/nm[a]		Binding energies/eV[b]		
Rod	Fresh sample	After reaction	Au $4f_{7/2}$	M $2p_{3/2}$	Pd $3d_{5/2}$
Au	3	16	N.D.	—	—
$Au_{25}Ti_{75}$	N.D.	1.8	83.6	458.2	—
$Au_{25}Fe_{75}$	~1.5	N.D.	83.5	710.8	—
$Au_{50}Ni_{50}$	~1.5	3.7	83.6	855.4	—
$Au_{20}Pd_{80}$	N.D.	6.2 ($Au_{20}Pd_{80}$)	83.9	—	335.5

[a] Determined on 300 nm—thick films of clusters directly deposited on Suprasil™ silica disks (sintered glass disks for $Au_{50}Ni_{50}$ and $Au_{20}Pd_{80}$) by laser vaporisation. [b] Determined on fresh 300 nm—thick films of clusters directly deposited on sintered glass disk by laser vaporisation (0.4 nm C/Cu microscopy grid). It is noted that the binding energies measured on cluster deposits (~0.4 nm equivalent thickness) over C/Cu microscopy grid are similar (*e.g.* values of 83.8 eV and 458.2 eV are found for Au $4f_{7/2}$ and Ti $2p_{3/2}$, respectively in $Au_{25}Ti_{75}$).

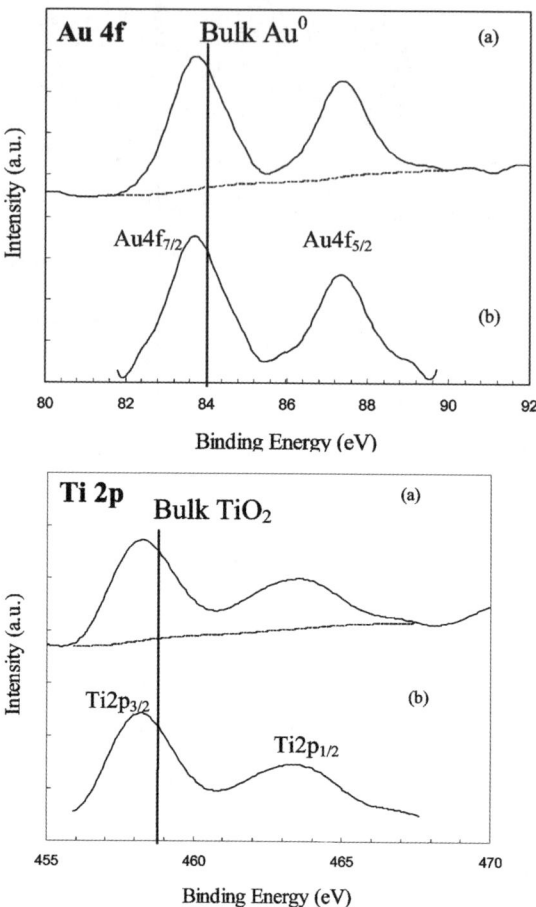

Fig. 7 Au 4f and Ti 2p photoemission spectra obtained from a fresh 0.4 nm-thick $Au_{25}Ti_{75}$ deposit on a C/Cu microscopy grid. (a) Raw spectra and Shirley background, (b) processed spectra.

3.1.2 Support effect on $Au_{25}Ti_{75}$ clusters. Nevertheless, we found it interesting to prepare and test titania-supported $Au_{25}Ti_{75}$ clusters in order to probe the influence of possible long-range oxygen reverse spill-over on the $Au_{25}Ti_{75}$ system. The reaction rate found for $Au_{25}Ti_{75}/TiO_2$ is however similar to that obtained over $Au_{25}Ti_{75}/Al_2O_3$ (Table 1), showing either that the extent of reverse oxygen spill-over over the large titania crystallites is very small or that its contribution to the catalysed reaction is very small (*i.e.* not rate determining).

3.1.3 Alloy effect over alumina. The alumina-supported clusters produced from $Au_{50}Ni_{50}$ and $Au_{25}Fe_{75}$ rods display similar catalytic activities in the oxidation of CO (Fig. 5) and only slightly better than that already observed for alumina supported pure gold catalyst.[3] The reaction rates observed over these materials are also much lower than that obtained over $Au_{25}Ti_{75}/Al_2O_3$ (Table 1). In order to understand these catalytic behaviours, characterization of thick films of the clusters has been performed.

The diffraction pattern of a fresh $Au_{50}Ni_{50}$ thick film (Fig. 8a) displays only one very small and wide peak that could correspond to the Au(111) reflection. The Au(200) reflection is however apparently absent from the diffractogram, which makes it impossible to find a suitable fit for the whole pattern, considering furthermore that background subtraction is complicated by the presence of a large hump probably due to amorphous (nickel) oxide phases. A size of 1.5 nm for gold-rich particles can however be estimated (Table 2). On the other hand, the XRD pattern of the self-supported $Au_{50}Ni_{50}$ system after catalytic reaction exhibits several broad peaks (Fig. 8b). The main peaks correspond to Au(111), Au(200) (with possible contribution from Ni(111)), Au(220) and Au(311) reflections. The peak with a very low intensity near 52° corresponds to the Ni(200) reflection. No crystallised NiO_x phase is detected. Again, the large widths of the peaks are attributed to the small size of the particles and also to a possible distribution in compositions with the apparent alloying of a fraction of gold and nickel. The determination of the particle size is obtained by fitting the whole pattern using the Rietveld method, assuming that most particles have the same gold-rich (90–95%) composition, with a fraction of pure gold and pure nickel particles. Agreement with the experimental data is obtained for a particle size of 3.7 ± 0.3 nm (Table 2), showing that catalytic reaction (*i.e.* heating to 300 °C under reaction conditions) induces a small extent of sintering of the gold particles. It is interesting that the gold crystallites are larger than those resulting from the vaporisation of an $Au_{25}Ti_{75}$ rod and heated to 300 °C. It is likely that the higher titanium content (as compared with nickel content) in the vaporised rod, which provides the oxide phase embedding the gold particles (Fig. 3), prevents their agglomeration. A similar nickel oxide phase must however also be present in the $Au_{50}Ni_{50}$ system and somehow limit the sintering, as the gold particle size in the monometallic self-supported system increases from 3 to 16 nm under identical reaction conditions (Table 2), while it reaches only 3.7 nm in the self-supported $Au_{50}Ni_{50}$ system.

The Au 4f and Ni 2p regions of the XPS spectra (not shown) give binding energies of 83.6 eV and 855.4 eV for Au $4f_{7/2}$ and Ni $2p_{3/2}$, respectively (Table 2). These can be attributed to metallic gold (84.0 eV) and oxidised states of nickel lying between NiO (854.4 eV)[15] and Ni_2O_3 (855.5 eV).[16] The XRD-silent NiO_x phases are probably very small and not well-crystallised.

The diffraction pattern of a fresh $Au_{25}Fe_{75}$ film exhibits only one very broad peak (Fig. 6) corresponding to the Au(111) reflection. Like in the diffractogram of the self-supported $Au_{50}Ni_{50}$ system (Fig. 8a), the Au(200) reflection is absent and it is still impossible to fit the diffraction pattern and get an accurate average crystallite size for

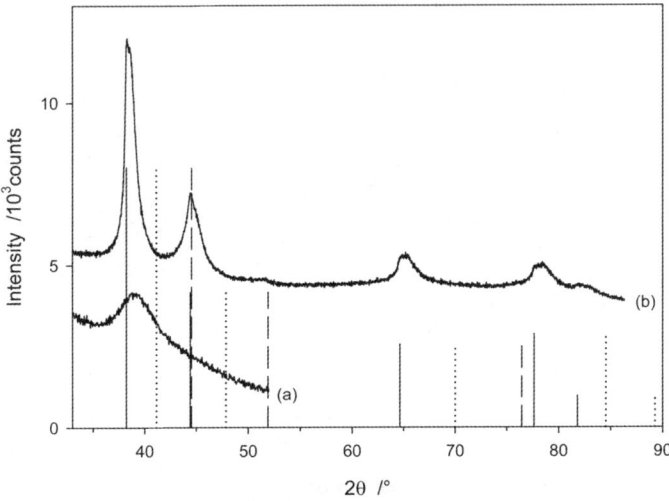

Fig. 8 X-Ray diffraction patterns of a 300 nm-thick film of $Au_{50}Ni_{50}$ clusters before (a) and after (b) catalytic reaction. Vertical lines indicate Bragg lines of Au (solid lines), Ni (dashed lines), and $Au_{50}Ni_{50}$ (dotted lines).

gold. Several attempts with the Rietveld method, assuming various compositions of the particles, including partial alloying of gold and iron, failed. A particle size of 1–1.5 nm can however be estimated (Table 2). No crystallised phase of Fe or FeO_x is detected.

The Au 4f and Fe 2p regions of the XPS spectra (Fig. 9) give binding energies of 83.5 eV and 710.8 eV for Au $4f_{7/2}$ and Fe $2p_{3/2}$, respectively, which are assigned to metallic gold (84.0 eV) and Fe_2O_3 (710.6–711.2 eV).[17] The shift towards lower binding energies for the Au 4f peaks is attributed to a surface–core level shift due to the small particle size but it could also include a contribution from charge transfer from the iron oxide to the gold particle, as discussed in ref. 3. The XRD-silent Fe_2O_3 phases are likely to be very small and not well-crystallised.

Both $Au_{50}Ni_{50}$ and $Au_{25}Fe_{75}$ systems thus consist of small gold particles (of 1–1.5 nm before and up to 3.7 nm after catalytic reaction) intimately mixed with nano-oxides of nickel and iron, respectively. EDX analyses of single particles (see Experimental section) of these bimetallic systems indeed show the presence of both elements within the 10 nm area probed, in relative concentrations close to that of the vaporised rod.

The catalytic results (Fig. 5, Table 2) show that the synergy between gold and these oxides is less efficient than the gold–titanium oxides synergy for the oxidation of CO. It seems that oxygen activation is performed better over TiO_2 than over NiO_x

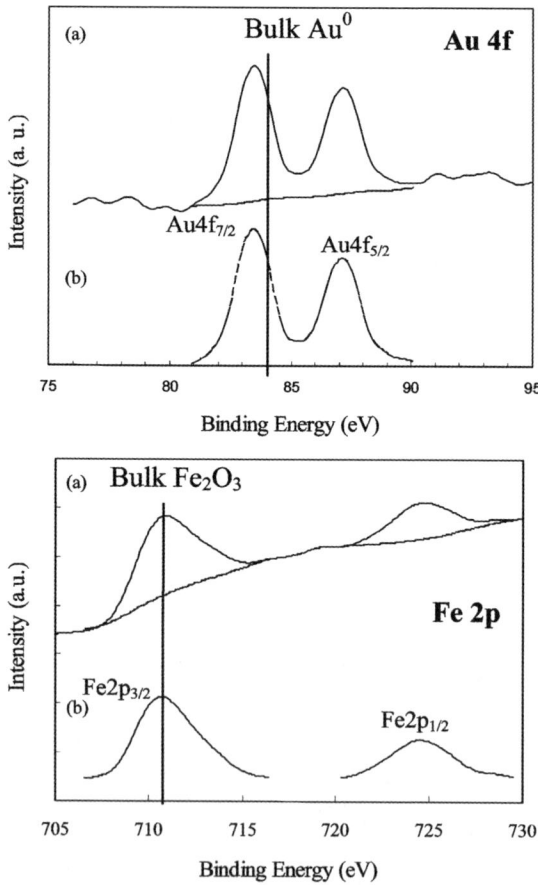

Fig. 9 Au 4f and Fe 2p photoemission spectra obtained from a fresh 0.4 nm-thick $Au_{25}Fe_{75}$ deposit on a C/Cu microscopy grid. (a) Raw spectra and Shirley background, (b) processed spectra.

and Fe_2O_3. However, one cannot exclude the possibility that during reaction over $Au_{50}Ni_{50}/Al_2O_3$ and $Au_{25}Fe_{75}/Al_2O_3$ gold nanoparticles diffuse on the alumina support and are no longer in the direct proximity of the Ni or Fe oxide phases. Further EDX experiments on these catalysts after reaction should be useful to answer this question.

3.2 The preferential oxidation of CO over alumina-supported $Au_{50}Ni_{50}$ and $Au_{25}Fe_{75}$

When hydrogen is present in the reaction feed, the conversion of CO is however markedly enhanced over both systems (Fig. 10). This phenomenon has already been observed over Au/Al_2O_3[18] (and, to a lesser extent, Au/ZrO_2) while Au/TiO_2 exhibited similar activities for CO oxidation in the absence and in the presence of hydrogen.[18] The CO oxidation rates at 80 °C in PROX reach 0.7 and 0.6 mmol CO $g_{Au}^{-1} s^{-1}$ for $Au_{50}Ni_{50}/Al_2O_3$ and $Au_{25}Fe_{75}/Al_2O_3$, respectively. These values are now close to those obtained over Au/TiO_2 (0.54 mmol CO $g_{Au}^{-1} s^{-1}$) and $Au_{25}Ti_{75}/Al_2O_3$ (0.7 mmol CO $g_{Au}^{-1} s^{-1}$) in CO oxidation (Table 1), confirming that, when H_2 is present, O_2 activation does not depend on the type of oxide any more. It seems that hydrogen can activate oxygen, just as efficiently as titanium oxides.

The promotion of the oxidation of CO has already been achieved using water in the reactant feed over Au/Al_2O_3[19] and Au/Fe_2O_3.[20] In the first case, where the active sites for oxygen activation are proposed to be hydroxyl groups, the beneficial effect

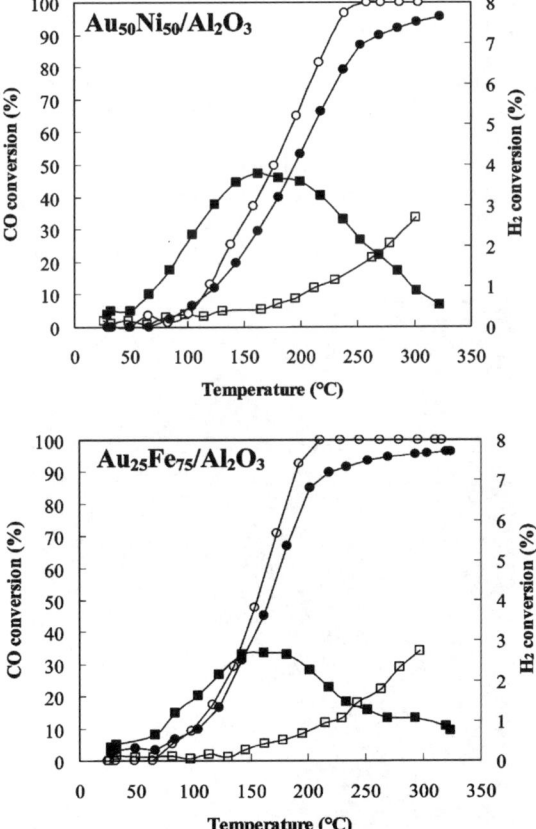

Fig. 10 CO and H_2 conversion as a function of temperature over $Au_{50}Ni_{50}/Al_2O_3$ and $Au_{25}Fe_{75}/Al_2O_3$: CO conversion in H_2 free mixture (□); H_2 conversion in CO free mixture (○); CO conversion in PROX (■); H_2 conversion in PROX (●).

of moisture was related to hydroxylation of the gold/alumina catalyst and the decomposition of hydroxycarbonyls (formed by reaction of CO with OH), which block and consume the active sites.[19] In the second case, the positive influence of water was attributed essentially to the decomposition of carbonates (formed by reaction of CO with lattice oxygen), which regenerates the active sites for oxygen activation.[20]

Over our alumina-supported gold/nano-oxides materials, the promotion of CO oxidation by hydrogen could proceed *via* similar pathways. It is however interesting that promotion is readily achieved at low temperature where no hydrogen is converted and thus no water is present in the reaction mixture. This would indicate that hydrogen alone (*i.e.* in the presence of oxygen but without the formation of H_2O) is capable of promoting the oxidation of CO. It is likely that, instead of participating in the hydroxylation of the gold/oxide catalyst, hydrogen reacts directly with oxygen to generate oxidising species that are more reactive than O_2 towards CO, as we have suggested recently.[18]

3.3 Self-supported $Au_{25}Ti_{75}$ and $Au_{20}Pd_{80}$ in PROX

Fig. 11 shows a TEM picture of a 40 nm-thick film of $Au_{25}Ti_{75}$, where gold particles are clearly separated by an inorganic/amorphous phase of TiO_2 (as discussed in Section 3.1.1). Catalytic activity (Fig. 12) is thus unsurprisingly consistent with that of Au/TiO_2 powder,[3] with CO conversion reaching a maximum at low temperature, then decreasing due to competition with H_2 oxidation in an O_2-limited gas mixture (see Experimental section).

On the other hand, the catalytic behaviour of $Au_{20}Pd_{80}$ clusters (Fig. 13) resembles more that of palladium in Pd/Al_2O_3,[21] with a CO conversion maximum at higher temperature and a poor selectivity to CO_2 formation due to competition with highly efficient H_2 oxidation. This is consistent with TEM imaging of the fresh sample (Fig. 14) which shows a very dense metallic phase. This contrasts with the mixed metal/oxide phases of the $Au_{25}Ti_{75}$ system (Fig. 11); in the $Au_{20}Pd_{80}$ system, the absence of amorphous/inorganic phases from TEM pictures is striking.

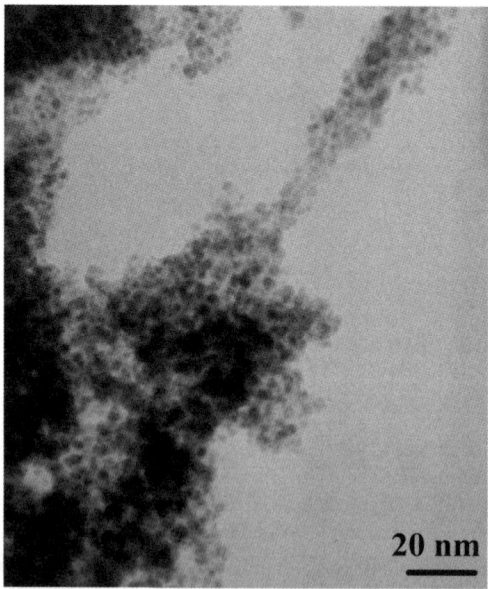

20 nm

Fig. 11   Transmission electron micrograph of a fresh 40 nm-thick film of $Au_{25}Ti_{75}$.

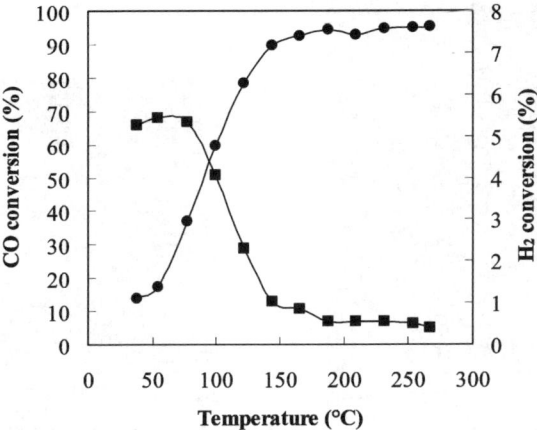

Fig. 12 CO (■) and H_2 (●) conversions in PROX as functions of temperature over a 300 nm-thick film of $Au_{25}Ti_{75}$ clusters on a sintered glass disk.

It is supported by XPS studies which do not detect any oxide phase. The Au 4f and Pd 3d regions of the XPS spectra give binding energies of 83.9 eV and 335.5 eV for $Au4f_{7/2}$ and Pd $3d_{5/2}$, respectively (Table 2). These are consistent with Au^0 and Pd^0 (335.0 eV).[22] The shift towards higher binding energies for the Pd 3d peaks is too small to justify the presence of palladium oxides (PdO $3d_{5/2}$ at 336.3 eV).[22] It is more likely to be due to alloying with gold atoms, as shown by XRD studies.

The diffraction pattern of an $Au_{20}Pd_{80}$ film (after catalytic evaluation) indeed exhibits essentially one peak between the Au(111) and the Pd (111) reflections and one peak between the Au(200) and Pd(200) reflections (Fig. 6). Simulation of the whole pattern (Fig. 15a) with the Rietveld method is achieved by assuming that all the particles have the same composition (Fig. 15b). The absence of amorphous phases perturbating the background allows good agreement with the experimental data (Fig. 15d) using a $Au_{21.2}Pd_{78.8}$ alloy composition and anisotropic crystallite sizes of 8.4, 4.9, 5.2 and 5.3 nm for (111), (200), (220) and (311) directions, respectively. An average crystallite size of 6.2 nm (Table 2) is found for these AuPd nanoalloys. It is interesting that this alloy particle is larger than the gold particle (1.8 nm) produced from an alloy rod of close stoichiometry, such as $Au_{25}Ti_{75}$.

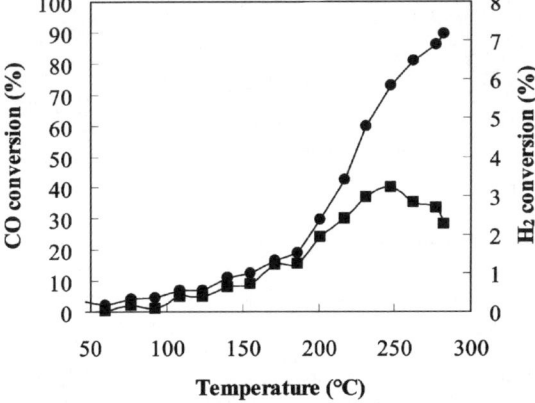

Fig. 13 CO (■) and H_2 (●) conversions in PROX as functions of temperature over a 300 nm-thick film of $Au_{20}Pd_{80}$ clusters on Suprasil™.

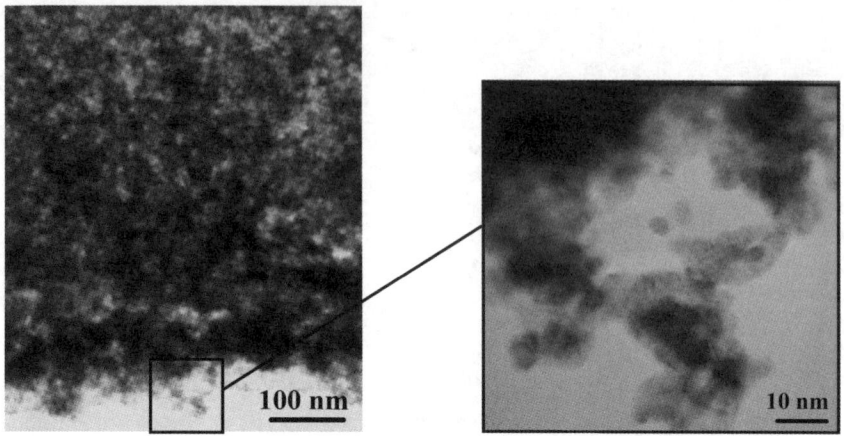

Fig. 14 Transmission electron micrographs of a fresh 40 nm-thick film of $Au_{20}Pd_{80}$.

This supports the fact that amorphous oxide phases act as a barrier for sintering (Section 3.1.3). It indicates that, in the absence of such a phase around the metal/ alloy particles, sintering is determined by the melting point of the metals and the resulting Tammann temperature of the nanoalloys, as the gold particle size in the monometallic self-supported system reaches 16 nm under the reaction conditions (Table 2).

Although a meaningful and representative signature of the catalytic properties of the clusters is obtained over Suprasil™ silica disks, the use of sintered glass disks, which allows a better accessibility of the gas phase to the clusters, is probably more appropriate for the systematic study of bimetallic nanoparticles and will thus be used for future studies.

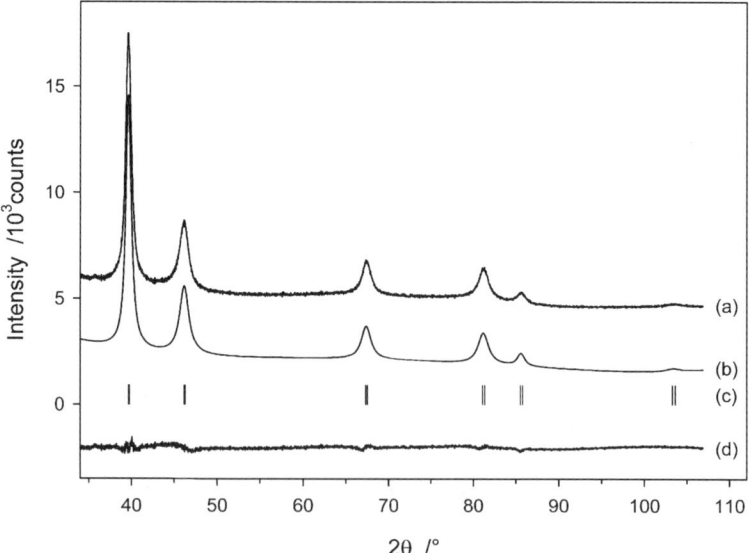

Fig. 15 X-Ray diffraction patterns of a 300 nm-thick film of $Au_{20}Pd_{80}$ clusters (after catalytic reaction): (a) $Au_{20}Pd_{80}$ alloy (observed); (b) calculated using the Rietveld method with anisotropic crystallite sizes; (d) difference between calculated and observed. Vertical lines (c) indicate Bragg lines of a $Au_{21.2}Pd_{78.8}$ alloy.

4. Summary and conclusions

We have shown that the laser vaporisation source can produce bimetallic nanosystems that can be applied to catalysis and thoroughly characterized with the usual methods used for characterising solid catalysts, such as X-ray diffraction and X-ray photoelectron spectroscopy. This allowed us particularly to look into some challenges of gold catalysis. Depending on the oxidisability of the metal associated with gold in the vaporised alloy rod, gold nanoparticles intimately mixed with nano-oxides or true nanoalloys (solid solutions) are obtained after exposure to air. In the first case, both the size of the gold particle and its thermal stability (for example under reaction conditions) are determined by the stoichiometry of the alloy rod. It is possible that thermal stability is also further influenced by the nature/stability of the oxide phase.

We have also shown that a self-supported system can be synthesised using laser vaporisation and low energy cluster beam deposition. A first attempt to study the reactivity of these self-supported systems has been presented. We have shown that significant CO oxidation reactivity can be obtained over unsupported nanoporous AuTi and PdAu thin films. These systems constitute a precious tool to investigate the extent of the support effects on a catalytic reaction.

Acknowledgements

The authors wish to thank Dr M. Aouine for TEM and EDX analysis, Dr P. Delichère and L. Massin for XPS experiment, N. Cristin and P. Mascunan for chemical analyses and the LPMCN team (O. Boisron and G. Guiraud) for its technical help with laser experiments.

References

1   J. H. Sinfelt, *Bimetallic Catalysts*, Wiley, New York, 1983.
2   H. Yasuda, T. Kameoka, T. Sato, N. Kijima and Y. Yoshimura, *Appl. Catal., A*, 1999, **185**, L199–L201.
3   S. Arrii, F. Morfin, A. J. Renouprez and J. L. Rousset, *J. Am. Chem. Soc.*, 2004, **126**, 1199–1205.
4   J. L. Rousset, L. Stievano, F. J. Cadete Santos Aires, C. Geantet, A. J. Renouprez and M. Pellarin, *J. Catal.*, 2001, **197**, 335–343.
5   J. L. Rousset, L. Stievano, F. J. Cadete Santos Aires, C. Geantet, A. J. Renouprez and M. Pellarin, *J. Catal.*, 2001, **202**, 163–168.
6   L. Bardotti, B. Prevel, P. Melinon, A. Perez, Q. Hou and M. Hou, *Phys. Rev. B*, 2000, **62**, 2835–2842.
7   J. L. Rousset, A. J. Renouprez and A. M. Cadrot, *Phys. Rev. B*, 1998, **58**, 2150–2156.
8   J. Rodríguez-Carvajal, *Physica B*, 1993, **192**, 55–69, The manual of FullProf can be obtained from a Web browser at http://www-llb.cea.fr/fullweb/.
9   J. L. Rousset, F. J. Cadete Santos Aires, B. R. Sekhar, P. Mélinon, B. Prevel and M. Pellarin, *J. Phys. Chem. B*, 2000, **104**, 5430–5435.
10 Y. F. Han, Z. Zhong, K. Ramesh, F. Chen and L. Chen, *J. Phys. Chem. C*, 2007, **111**, 3163–3170.
11 L. Óvári and J. Kiss, *Appl. Surf. Sci.*, 2006, **252**, 8624–8629.
12 P. Heimann, J. F Van der Veen and D. E. Eastman, *Solid State Commun.*, 1981, **38**, 595–598.
13 K. Tanaka, Y. Moro-oka, K. Ishigure, T. Yajima, Y. Okabe, Y. Kato, H. Hamanoa, S. Sekiyaa, H. Tanakaa, Y. Matsumoto, H. Koinuma, H. Hee, C. Zhang and Q. Feng, *Catal. Lett.*, 2004, **92**, 115–123.
14 M. M. Schubert, S. Hackenberg, A. C. Van Veen, M. Muhler, V. Plzak and R. J. Behm, *J. Catal.*, 2001, **197**, 113–122.
15 M. Walker, C. R. Parkinson, M. Draxler, M. G. Brown and C. F. McConville, *Surf. Sci.*, 2006, **600**, 3327–3336.
16 W. Zhao, W. Ma, C. Chen, J. Zhao and Z. Shuai, *J. Am. Chem. Soc.*, 2004, **126**, 4782–4783.
17 T. Yamashita and P. Hayes, *J. Electron Spectrosc. Relat. Phenom.*, 2006, **152**, 6–11.

18   C. Rossignol, S. Arrii, F. Morfin, L. Piccolo, V. Caps and J.-L. Rousset, *J. Catal.*, 2005, **230**, 476–483.
19   C. K. Costello, J. H Yang, H. Y. Law, Y. Wang, J.-N. Lin, L. D. Marks, M. C. Kung and H. H. Kung, *Appl. Catal., A*, 2003, **243**, 15–24.
20   M. M. Schubert, A. Venugopal, M. J. Kahlich, V. Plzak and R. J. Behm, *J. Catal.*, 2004, **222**, 32–40.
21   F. Marino, C. Descorme and D. Duprez, *Appl. Catal., B*, 2004, **54**, 59–66.
22   A. I Titkov, A. N. Salanov, S. V. Koscheev and A. I. Boronin, *Surf. Sci.*, 2006, **600**, 4119–4125.

In situ investigation of $Pt_{100-x}Au_x$ and $Pt_{100-y}Sn_y$ nanoalloys

Ken A. Grant, Kelei M. Keryou and Paul A. Sermon*

Received 11th June 2007, Accepted 28th June 2007
First published as an Advance Article on the web 15th October 2007
DOI: 10.1039/b708810h

Dispersed sols of 1–10 nm sized $Pt_{100-x}Au_x$ and $Pt_{100-y}Sn_y$ nanoalloys have been prepared separately at various x and y above and below the miscibility limit in the bulk metals. $Pt_{100-x}Au_x$ was derived from trisodium citrate reduction of aqueous solutions of H_2PtCl_6 and $HAuCl_4$. $Pt_{100-y}–Sn_y$ was produced by (i) complexing Sn^{2+} with glucose at 323 K at pH > 7, (ii) neutralising this with H_2PtCl_6 addition and (iii) reducing the bimetallic precursor with glucose on raising the temperature to 373 K. For $Pt_{100-x}Au_x$ (where both metals were zero-valent) as x increased the average size of nanoalloy particles increased. These particles adsorbed onto graphite, where the extent of hydrogen chemisorption at 298 K decreased by 67% at 9 at% Au. Pt/SnO_2 nanoparticles (<3 nm in size) were adsorbed onto alumina. The Pt interacted with and catalysed the reduction of SnO_2, with some $Pt_{100-y}Sn_y$ nanoalloy formation at about 673 K which even in the bulk occurs over a wider range of compositions than Pt–Au) and enhanced H_2 chemisorption at 17–33 at% Sn. Nevertheless some Sn must remain in a positive oxidation state on the alumina surface. The ratio of rates of 2MP/3MP formation from MCP and *n*-hexane may be informative in chemically fingerprinting (and revealing fundamental differences in) these nanoalloy surfaces. The reasons for this are seen in terms of the surface structures on these two types of nanoalloy particles (*i.e.* the availability of contiguous asymmetric pairs of active surface atoms *, which, as expected, is found to pass through a maximum or decrease beyond specific values of x and y).

1. Introduction

Colloidal solutions of catalytic metals are readily prepared by reduction in solution of appropriate salts and as such are part of bottom-up nanotechnology. Some have seen similarities between their high catalytic activity and that of enzymes. Despite their instability,[1] metal nanoparticles have many biological, chemical and physical applications.[2] Therefore attention has now turned to the optimisation of the surface composition and properties of binary metallic nanoparticles; some can be modelled[3–9] (*e.g.* segregation,[10–14] melting point,[15,16] ordering[17] and stability[18]), but others are best followed experimentally *via* high-resolution transmission electron microscopy,[19] X-ray diffraction[20] or catalytic/adsorptive probes.[21–23] This is the subject of this Discussion.

Au catalyses CO oxidation at low temperature[24] and, intriguingly, Pd–Au nanoparticles are 70 times more active[23] than Pd alone and may be relevant to bimetallic nanoparticle catalysts for aqueous-phase trichloroethene hydro-

Chemistry, FHMS, University of Surrey, Guildford, Surrey, UK GU2 7XH

dechlorination (*i.e.* groundwater remediation). Here we have chosen Pt–Au[21] to test because of their potential immiscibility and previous work.[25]

Monodisperse PtRu nanoalloy can be prepared from acac precursors using 1,2-hexadecanediol reducing agent deposited on carbon.[26] Here PtSn nanoalloys have also been chosen because they have *greater* electro-oxidation activity towards ethanol than Pt–Ru/C[27] which have a resistance to CO poisoning that is important in a fuel cell context and is not possessed by Pt alone. A Pt–Sn alloy may be formed on alumina above 0.6 wt%, but Pt and Sn[28] species will need to be related to the surface chemistry of Pt and Sn in reforming catalysts.[29] Using ethylene glycol reductant $Pt_{50}Sn_{50}$ ($d = 2.7$ nm) has been produced and supported on carbon;[30] this nanoalloy appears to have a bright future in that it is a *better* electrocatalyst than PtRu/C or PtSnRu/C in CH_3OH oxidation. Specifically, addition of Sn allows oxidation of ethanol at lower potentials than Pt alone.[31] For 0.4–0.9 mg PtSn nanoparticles per cm^2 of graphite, anodic stripping was used to determine the surface Pt : Sn ratio (3.6); this was higher than the interior ratio (2.5)[32] possibly as a result of a surface stabilisation by tetraoctylammonium triethylhydroborate.

Here then we have selected $Pt_{100-x}Au_x$ and $Pt_{100-y}Sn_y$ (the former less mutually miscible than the other, but the latter with greater ease of oxidation of one component) as examples of nanoalloys to explore.

It seemed appropriate to try to ascertain the mode of sol-formation.[33] In addition it seemed that *in situ*[34] characterisation methods would also be needed under conditions of use. Active sites and surface properties of bimetallic solids have often been investigated using catalysis of carbon–hydrogen bonds[35] (*e.g.* in *n*-hexane (*n*Hex) and 2-methylpentane (2MP)) as a function of temperature at low (<0.133 mPa) and high (≥ 133 Pa) pressures in the absence and presence of hydrogen. Gaseous *n*-hexane can be in equilibrium with surface-held hexylidene ($\equiv$C–CH_2–CH_2–CH_2–CH_2–CH_3) and π-bound methylcyclopentane (MCP) over Pt–Pd. Turnover frequencies (TOFs) and product selectivities for *n*-hexane (*n*Hex)[36] and methylcyclopentane (MCP) conversion at 623 K–773 K at a H_2/MCP = 20 after pre-reduction in H_2 to 773 K have been reported,[37] although it was recognised that carbon accumulation and deactivation/reactivation occur.[38] Ring opening of methylcyclopropane (an intermediate in *n*-hexane conversion) appeared to be highly sensitive to changes in surface structure of catalysts (as is hydrogenolysis[39]). Non-selective MCP ring opening will give 2MP/3MP = 2. *n*-Hexane (one of the simplest alkanes to be able to undergo skeletal rearrangement (hydrogenolysis, isomerisation, dehydrocyclisation and aromatisation) over Pt catalysts[40]) and methylcyclopentane (MCP, that converts to *n*-hexane, 3-methylpentane (3MP) or 2-methylpentane (2MP)[41–43]) were selected as the probe reactants here. In the past[44] ethane hydrogenolysis was studied and found to decrease in rate on alloying of a low activity metal with a high activity one,[45] while facile dehydrogenation[46] changed less dramatically.

2. Experimental

Preparation

One can extrapolate directly from the work on $AuCl_4^-$ reduction to give Au nanoparticles undertaken by Faraday[47] to Frens.[48] Tri-sodium citrate is effective at the solution boiling point developing a blue and then a red colour.[49] Now more intricate core–shell Au–Cu nanocomposites[50,51] and Au–Cu nanoparticles embedded in SiO_2[52] have been prepared. Here $Pt_{100-x}Au_x$ (0 < x < 100, where x was 0 and 9 and also 25, 50 and 100 at% Au) was prepared in pre-cleaned glassware by reducing a mixed aqueous (HPLC grade water) solutions (30 mg dm^{-3}) of H_2PtCl_6 (Johnson Matthey) and $HAuCl_4$ (Johnson Matthey) with a 1% trisodium citrate solution and then stirring for 2–4 h at 373 K.[53] The product sol was purified by addition of Amberlite MBI ion-exchange resin with stirring until constant conductivity and pH were obtained. Such sols were adsorbed on graphite (Fluka; 99.9% purity; 11 m^2 g^{-1}), alumina (Degussa Al$_2$O$_3$–C; 100 m^2 g^{-1}), silica (Degussa Aerosil 200; 200 m^2 g^{-1}), titania (Degussa P25, 50 m^2 g^{-1}) and MgO (BDH AnalaR, 16 m^2 g^{-1}). Concentrations were deduced by atomic absorption after digestion.

Pt$_{100-y}$Sn$_y$ ($0 < y < 100$, where y was 0, 17, 33, 43, 52 and 55 at%) was prepared by dissolving SnCl$_2$ (Aldrich; 99%) and NaOH in HPLC-grade water and then complexing this with glucose (BDH, AnalaR). H$_2$PtCl$_6$ was then added and reduced while refluxing at 373 K. The product darkened from amber as the glucose polymerised, the SnCl$_2$ hydrolysed and H$_2$PtCl$_6$ reduced. After cooling the sols were purified by addition of Amberlite MB1 resin, that was then removed by filtering. These sols were also adsorbed to a level of 1 wt% Pt on alumina (Degussa Al$_2$O$_3$–C; 100 m^2 g^{-1}). The Sn^{2+}–glucose interaction in solution (that occurred during sol synthesis) was followed at 373 K by polarographic (Metrohm Ploracord 626 with dropping Hg and Ag/AgCl electrodes) measurements of limiting currents in solutions under N$_2$ (i_d in nA) and half-wave potentials ($E_{1/2}$ in V that were -0.63 to -0.65 V (anodic wave) and -1.09 to -1.15 V (cathodic wave) for Sn^{2+}). Clearly Fig. 1 and Table 1 show that Sn^{2+} is consumed within 5 min at 373 K (and longer at lower temperatures) and a new electroactive species, presumably produced from the Sn^{2+}–glucose complex, is seen (at $-E_{1/2} = 1.64$–1.68 V) to form to an increasing

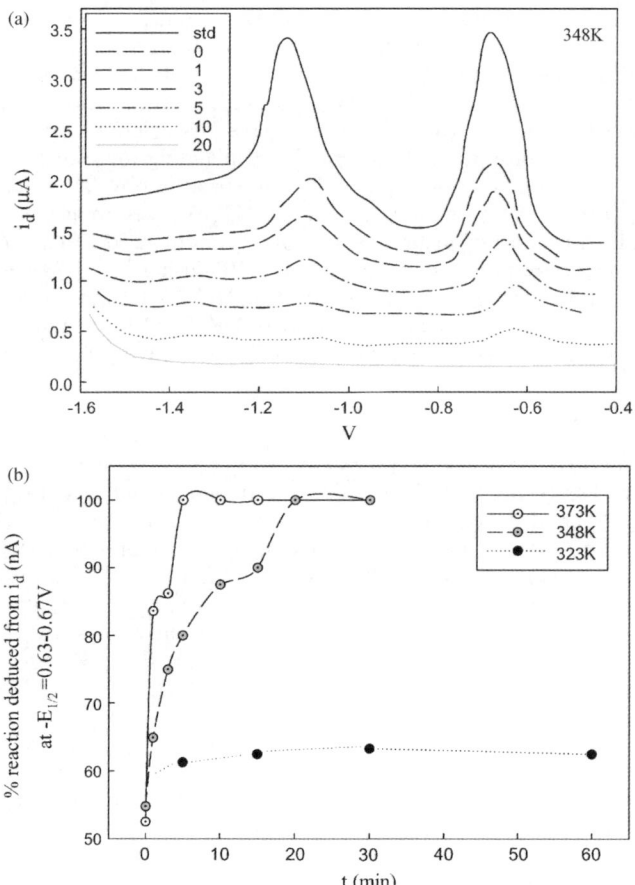

Fig. 1 (a) Polarographs of Sn^{2+}–glucose reaction at 348 K after 0, 1, 3, 5, 10 and 20 min (and in the absence of glucose (standard: std)). (b) % decrease in i_d peak at $-E_{1/2} = 0.63$–0.67 V (relative to the standard value in the absence of glucose) at 323 K, 348 K and 373 K measured as a function of reaction time.

Table 1 Polarographic analysis of the Sn^{2+}–glucose reaction at 373 K

t/min	i_d/nA		
	Sn^{2+} anodic wave[a]	Sn^{2+} cathodic wave[b]	Unknown[c]
0	3040	2500	0
0.2	1440	1200	660
1.0	500	280	1140
3.0	420	160	1840
5.0	0	0	2010
20.0	0	0	2540

[a] For Sn^{2+} $E_{1/2}$ = −0.63 top −0.65 V. [b] For Sn^{2+} $E_{1/2}$ = −1.09 to −1.15 V. [c] $E_{1/2}$ = −1.64 to −1.68 V.

extent as the Sn^{2+} is lost. Hence, sol formation at 373 K was selected, rather than a slower process at a lower temperature (*i.e.* at 348 K a reaction time increased by a factor of 2 would have been required (see Fig. 1)).

Characterisation methods

Extents of hydrogen chemisorption (n) at 298 K were measured volumetrically for $Pt_{100-x}Au_x/C$ and $Pt_{100-y}Sn_y/Al_2O_3$ (pre-calcined in air at 673 K for 2 h, reduced *in situ* at 673 K in H_2 for 2 h and finally outgassed to 0.1 mPa for 1 h) by extrapolation of n–p isothermal data to p = 0 Pa. High-resolution transmission electron microscopy (TEM)[54–56] and electron diffraction[57] have characterised the structure, size and morphology of unsupported nanoparticles, including those of Pt–Sn;[32] here a Jeol 200CX[25] was used.[32] X-Ray photoelectron spectroscopy (XPS; Kratos E3500 using AlK$_\alpha$ (1486.6 eV) normalised to Al_{2p} at 74.7 eV) was applied to $Pt_{100-y}Sn_y/Al_2O_3$ samples (pre-calcined in air at 673 K for 2 h and reduced *in situ* at 673 K in H_2 for 2 h). Temperature-programmed reduction (TPR) using 5% H_2/Ar flowing at 40 cm^3 min^{-1} during programming from 103–900 K at 5 K min^{-1} was applied to $Pt_{100-y}Sn_y/Al_2O_3$ (also pre-calcined in air at 673 K for 2 h but not pre-reduced). AAS data for digested samples were used to calculate x and y on an atomic basis.

Catalytic methods

Surface atomic arrangements and ratios are always critical to nanoalloys; these have been assessed here using a catalytic probe.[58] Catalytic activities and selectivities of $Pt_{100-x}Au_x/C$ were deduced using H_2 : MCP = 19 : 1 or H_2 : hexane = 18 : 1 with H_2 flow rates of 27 cm^3 min^{-1} with catalysts that had been pre-reduced at 573 K for 1 h. Catalytic activities and selectivities were measured for $Pt_{100-x}Sn_x/Al_2O_3$ using MCP (Fluka AG Chemicals) and n-hexane (Alfa Products) with catalyst samples (0.05 g for MCP conversion and 0.25 g for n-hexane) precalcined (673 K; 2 h) and then pre-reduced (673 K; 2 h). H_2 (30 cm^3 min^{-1} for MCP and 90 cm^3 min^{-1} for n-hexane) flowed through a saturator giving 5 kPa MCP (H_2/MCP = 17.1) or 6.1 kPa n-hexane (H_2/n-hexane = 15.6). This reactant stream flowed over the catalyst samples as the catalyst temperature was controlled at 576–621 K (MCP) or 623–663 K (n-hexane). In such tests all supports were entirely unreactive in both MCP and n-hexane conversions below 723 K. All products were analysed using gas chromatography with a flame ionisation detector.

3. Characterisation results

The micrograph in Fig. 2a shows the TEM-defined 1.9 nm Pt particles produced here by reaction with glucose were monodispersed, unlike the 2.3 nm particles produced by glucose–SnCl$_2$ reaction that were aggregated (see Fig. 2b), in the same manner as

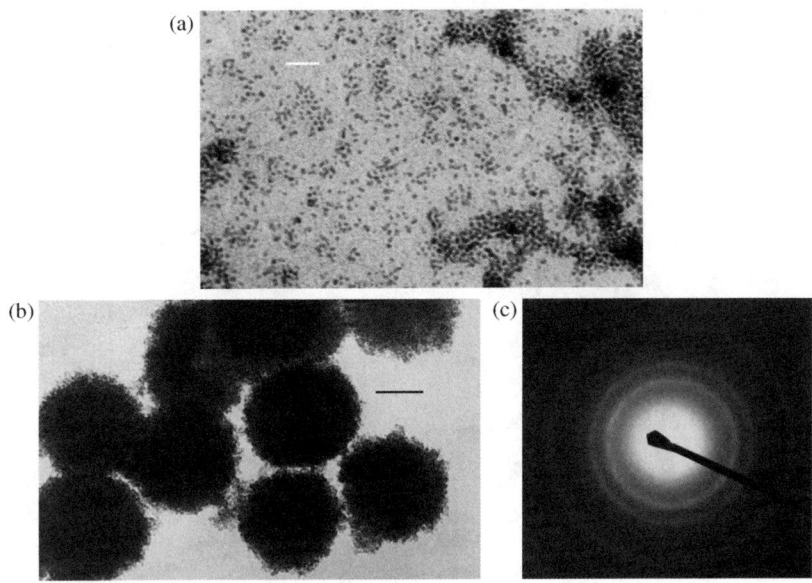

Fig. 2 Analysis of glucose-produced nanoparticles. (a) Pt sol after 40 min (scale bar = 50 nm); (b) SnO_2 sol after 45 min (scale bar = 50 nm); (c) electron diffraction of SnO_2 sol (d = 0.335 nm (100%), 0.264 nm (80%), 0.236 nm (25%) and 0.176 nm (65%)) that compares well with cassiterite (0.340 nm, 0.266 nm, 0.230 nm and 0.178 nm).

were those produced previously[59] on $SnCl_4$ hydrolysis in the presence of HCl. The product in Fig. 2b gave electron diffraction patterns (see Fig. 2c) consistent with cassiterite SnO_2. When Pt and SnO_2 were produced they were combined in Pt/SnO_2 aggregates that were at the end of the synthesis adsorbed onto Al_2O_3. In TPR Pt species reduced at 73–273 K and Sn species at 440–773 K. For TPR of SnO_2 alone the temperature for the maximum rate of reduction (T_{max}) was seen at 1051 K with 13.6 μmol H_2 g^{-1} consumed overall, which corresponds to 103% reduction. Fig. 3 and Table 2 show TPR data for $Pt_{100-y}Sn_yO_{2y}/Al_2O_3$-$Pt_{100-y}Sn_y/Al_2O_3$. As y increases the contribution of reduction peaks above 273 K and associated with Sn^{y+} reduction increased and corresponding T_{max} values increased from 595 K to 664 K. Further evidence of a Pt–Sn interaction is seen in the maximum H_2 adsorption capacity (see Table 2) that was noted at intermediate y (17–33 at% Sn). For the moment it appears that Pt is reducing at much lower temperatures (<273 K) and then catalyses reduction of the Sn^{y+} species to an extent that increases with lower Sn concentrations relative to the Pt (despite suggestions previously[60] that Pt does not catalyse Sn^{4+} reduction). It may be that some Sn^{4+} reduces to Sn^{2+} (that is then stabilised by the alumina surface). Although, there is no clear evidence of the precise Sn oxidation state from XPS (see Fig. 3 and Table 2) or TPR (see Fig. 3 and Table 2 since $Sn_{3d\ 5/2}$ was always at 486.2 ± 0.2 eV and less than for SnO_2), it is clear that the Pt is in close contact with the Sn and that, after 673 K reduction in H_2, catalyst reduction is complete. Here the Pt-enhancement of Sn^{4+} reduction is affected by the [Sn]. Indeed it is likely that the low temperature TPR shoulder ≈480 K in Fig. 3 is associated with Pt-catalysed Sn^0 formation which is especially evident at low [Sn]. Certainly XRD has found[61] PtSn alloy formation.

Previously[25] nanoalloy particle sizes were shown to increase as x increased. Table 2 confirms that as x increased the average size of particles seen in transmission electron microscopy (d_{TEM}) increased,[25] but in addition as x increased the extent of hydrogen chemisorption decreased by 67% at 9 at% Au (and d_{chem} increased). Although both the Pt and the Au were zero-valent at all compositions, the Au may reduce faster than Pt and lead to some surface enrichment by Pt.

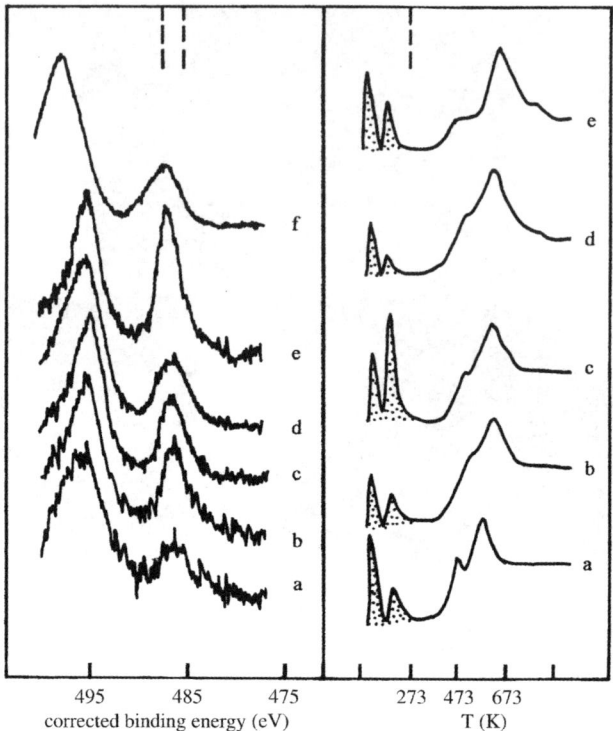

Fig. 3 X-Ray photoelectron spectra (XPS; left) and temperature programmed reduction (TPR; right) of alumina-supported $Pt_{100-y}Sn_yO_{2y}/Pt_{100-y}Sn_y$ containing 1 wt% Pt and varying at% Sn relative to Pt: a: 17% Sn; b: 33% Sn; c: 43% Sn; d: 52% Sn and e: 55% Sn compared with 1% Sn/Al_2O_3 (f).

Thus zero-valent $Pt_{100-x}Au_x$ nanoalloy particles (whose size increased with x) adsorbed onto graphite surfaces, and showed a level of hydrogen chemisorption at 298 K depressed by more than half when $x = 9$ at% Au. Pt/SnO_2 nanoparticles (<3 nm in size) adsorbed onto alumina where after reduction at 673 K for 2 h was best represented as $Pt_{100-y}Sn_y/Al_2O_3$ (despite some Sn^{y+} remaining on the support surface) that showed enhanced levels of H_2 chemisorption at $y = 17$–33 at% Sn. The surfaces of such nanoalloy particles were now probed *in situ* in terms of their reactivity.

4.   Characterisation through catalytic activity and selectivity

In terms of thermodynamic favourability of MCP conversion Fig. 4 shows that the reaction products are *n*-hexane (*n*Hex) > 2-methylpentane (2MP) > 3-methylpentane (3MP).

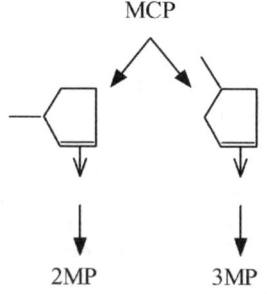

Table 2 Average particle size determined by transmission electron microscopy (d_{TEM}), extent of monolayer H_2 adsorption (n_{H2}), extent of H_2 reduction and temperature of maximum rates of reduction in temperature-programmed reduction (TPR) and X-ray photoelectron spectroscopic (XPS) binding energies in $Pt_{100-y}Sn_yO_{2y}/Pt_{100-y}Sn_y$

			TPR^c		
	$d_{TEM}/$ nm^a	$n_{H2}{}^b$ μmol H_2/g cat	% of H_2 consumed for Sn^{y+} above 273 K	T_{max}/K for most Sn redn	$Sn_{3d\ 5/2}/$ eV^d
1.00% Pt	1.9	12.8	0.00	—	—
1 wt% Pt with					
17 at% Sn		14.9	10.3	595 K	486.1
33 at%Sn		14.6	49.6	614 K	486.1
43 at% Sn		9.1	43.8	621 K	486.4
52 at% Sn			67.1	634 K	486.2
55 at% Sn	2.3		67.8	664 K	486.4

^{a} TEM after calcination at 673 K and reduction in H_2 at 673 K. ^{b} Amount of H_2 adsorption at 298 K, $p = 0$ kPa after pre-calcination and reduction *in situ* at 673 K and outgassing to 0.1 mPa. ^{c} The extent of H_2 reduction of Pt and Sn species in samples pre-calcined at 673 K in air was at $T < 273$ K and $T > 273$ K. The percentage of H_2 consumed in Sn reduction above 273 K and the temperature (T_{max}) of the dominant TPR peak for Sn reduction are shown. For SnO_2 13.6 μmol H_2 g^{-1} was consumed (corresponding to 103% reduction) with T_{max} at 1051 K. ^{d} $Sn_{3d\ 5/2}$ binding energy in samples pre-calcined in air at 673 K and then reduced *in situ* in H_2 at 673 K for 2 h. These should be compared with the values for Sn^0 (484.9 eV), SnO_2 (487.0 eV) and 1% Sn/Al_2O_3 (486.7 eV) and so it is difficult to differentiate $0/2 + /4 +$ oxidation states of Sn.

Since below 723 K under present conditions C, Al_2O_3 and Sn/Al_2O_3 showed no activity in MCP conversion, reactions are assumed to be primarily initiated on the surfaces of the supported nanoalloy particles. Again Fig. 4 suggests that *n*-hexane conversion is more favourable to 2-methylpentane (2MP) than 3-methylpentane (3MP), but for the same reason the reaction is also thought to be associated with (and therefore to probe) the $Pt_{100-x}Au_x$ and $Pt_{100-y}Sn_y$ nanoparticle surfaces rather than the supports.

Fig. 5 shows that the conversion of MCP and *n*-hexane *decreased* with reaction time (when tested isothermally), presumably with the build-up of a carbonaceous deposit on these surfaces.

However, as a function of isothermal reaction time (see Fig. 6) the 2MP/3MP (R) product ratio
(i) increased over $Pt_{100-x}Au_x/C$

Table 3 Average particle size of the sol $Pt_{100-x}Au_x$ seen in TEM (d_{TEM}), extent of monolayer adsorption of H_2 at 298 K (n^*_{H2}) and the average particle size estimated from chemisorption (d_{chem})

	n^*_{H2}/μmol H_2 g^{-1} cat	$d_{chem}{}^a$/nm	$d_{TEM}{}^b$/nm
$Pt_{100}Au_0/C$	2.55	8.0	2.3
$Pt_{91}Au_9/C$	0.66	31.1	4.8

^{a} Measured after adsorption on Fluka graphite by hydrogen chemisorption at 298 K and extrapolation to $p = 0$ Pa. ^{b} Previously[25] d_{TEM} was measured for samples at different x (2.3 nm at $x = 0$; 4.8 nm at $x = 10$, 12.3 nm at $x = 50$; 14.3 nm at $x = 90$; 24.7 nm at $x = 100$).

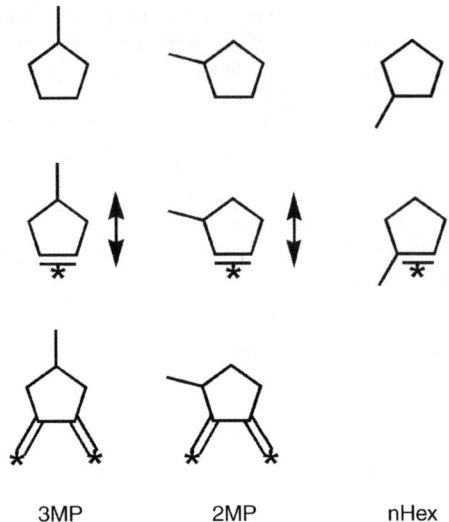

Fig. 4 Inter-relationship of 3-methylpentane (3MP; $\Delta_f H° = -171.9$ kJ mol^{-1} at 298 K), 2-methylpentane (2MP; $\Delta_f H° = -174.6$ kJ mol^{-1} at 298 K), π- and multiply-bound species on surface atoms (*), n-hexane (nHex; $\Delta_f H° = -166.9$ kJ mol^{-1} at 298 K) and methylcyclopentane (MCP; $\Delta_f H° = -106$ kJ mol^{-1} at 298 K).

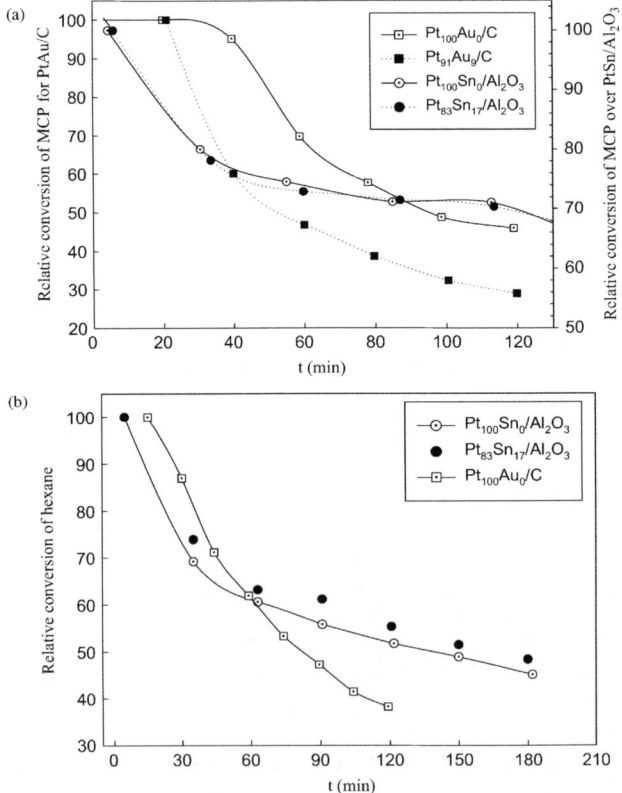

Fig. 5 Decrease in the relative conversion of (a) MCP over PtAu/C at 623 K and PtSn/Al$_2$O$_3$ at 603 K and (b) n-hexane over PtAu/C at 623 K and PtSn/Al$_2$O$_3$ at 643 K with increasing reaction time.

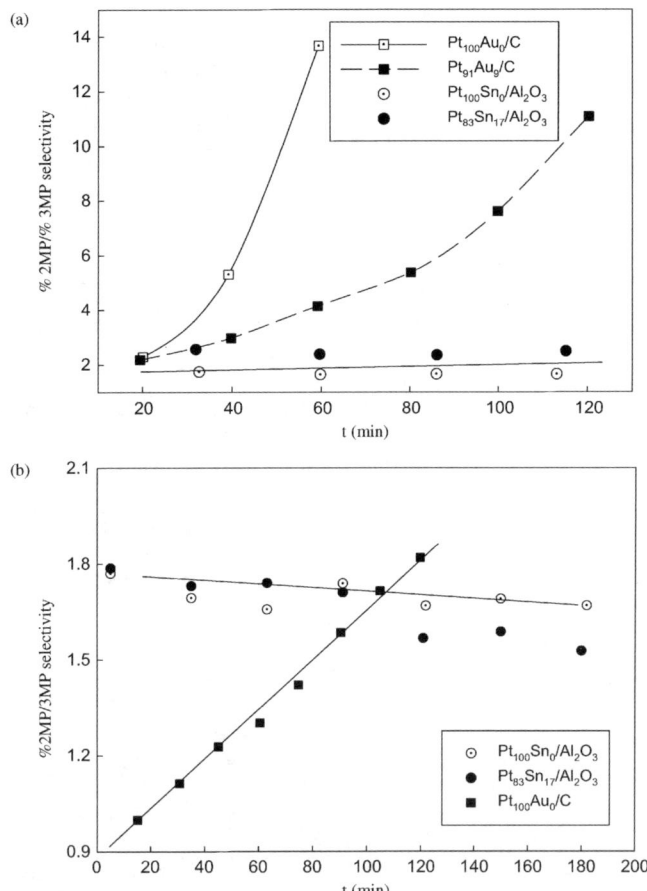

Fig. 6 2MP/3MP product ratio seen in (a) MCP conversion over $Pt_{100-x}Au_x/C$ at 623 K and $Pt_{100-y}Sn_y/Al_2O_3$ at 603 K and (b) n-hexane conversion over $Pt_{100-x}Au_x/C$ and $Pt_{100-y}Sn_y/Al_2O_3$ at 643 K. Using n-hexane reactant MCP was also a significant product over $Pt_{100-y}Sn_y/Al_2O_3$.

(ii) was close to constant on $Pt_{100-y}Sn_y/Al_2O_3$ from *both* MCP *and* n-hexane conversion. Of course MCP was another significant product from n-hexane, as were traces of C_{1-5} hydrogenolysis products (predominantly C_3).

Fig. 7 suggests that the 2MP/3MP (*R*) product ratio decreased with increasing reaction temperature with both MCP and n-hexane reactants over either $Pt_{100-x}Au_x/$ C or $Pt_{100-y}Sn_y/Al_2O_3$, but most especially with n-hexane over $Pt_{100-x}Au_x/C$.

Consider then (in Fig. 8 and 9) what happens as y increases in $Pt_{100-y}Sn_y/Al_2O_3$. With MCP, the initial activity decreases at 603 K as y increases. In parallel the product 2MP/3PM ratio rises and then above $y = 40$ at% Sn drops. With n-hexane, first, the % conversion of n-hexane also drops as y increases at 643 K. Second, selectivity towards isomerisation to 2MP, 3MP and MCP is dominant until over 50 at% Sn is present, at which point dehydrogenation ($-H_2$) becomes dominant. Third, the 2MP/3MP product ratio is constant at just below 2 until it drops at and above 50 at% added Sn. Therefore, beyond specific y some reactive surface atoms pairs are not available. Further, HRTEM is required to investigate this further.

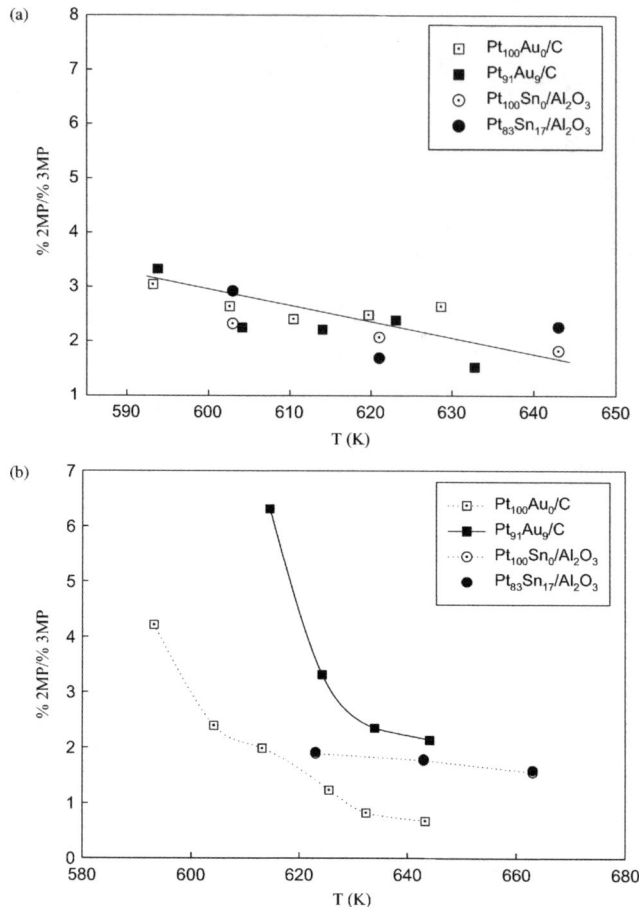

Fig. 7 Effect of temperature (T) on the 2MP/3MP product ratio seen in (a) MCP and (b) n-hexane conversion over $Pt_{100-x}Au_x/C$ and $Pt_{100-y}Sn_y/Al_2O_3$.

5. Discussion and conclusions

Here trisodium citrate produces zero-valent $Pt_{100-x}Au_x$ particles; as x increased the average size increased and the extent of hydrogen chemisorption *decreased*. These adsorbed onto graphite. Glucose produces composite aggregates of monodispersed Pt particles (1.9 nm) and 2.3 nm SnO_2 particles (similar to those seen previously[59]) that adsorbed onto alumina surfaces. Pt species reduced below 273 K and Sn species at 440–773 K (*i.e.* temperatures much lower than for SnO_2 alone (1051 K) and so there was a good interaction between Pt and Sn components). Further evidence of this interaction was seen in the effect of y on T_{max} values and H_2 adsorption capacities (*i.e.* a maximum at y = 17–33 at% Sn), possibly with Pt^0 formed at <273 K then catalysing reduction of the Sn^{y+} species especially at lower [Sn]/[Pt] (despite previous suggestions[60] to the contrary), possibly with a low temperature TPR shoulder suggesting some PtSn nanoalloy formation. However, some Sn^{y+} must remain on the alumina surface.

Although MCP conversion to n-hexane, 2-methylpentane and 3-methylpentane is thermodynamically favourable, below 723 K no supports used were active. Gault envisaged that *if* there was equal probability of attack on all MCP bond positions in

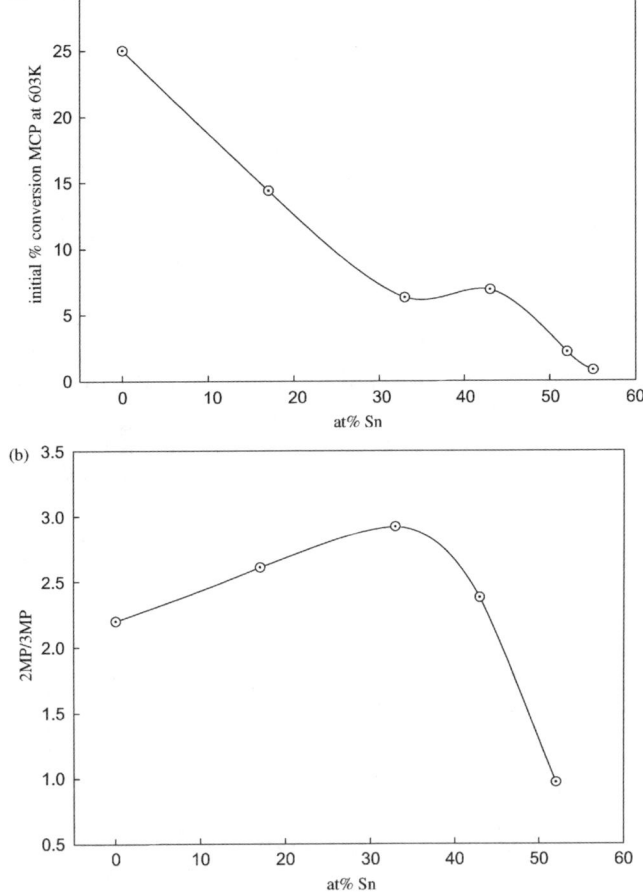

Fig. 8 (a) Decrease in initial activity in MCP conversion at 603 K as the at% Sn added to 1 wt% Pt on Al_2O_3 increases. (b) 2MP/3MP product ratio in MCP ring opening at 603 K over catalysts containing 1 wt% Pt but varying at% Sn as $Pt_{100-y}Sn_y/Al_2O_3$.

the ring on the surface of nanoalloy particles then the product 2MP/3MP ratio (R) would be 2.

Over $Pt_{100-x}Au_x$ and $Pt_{100-y}Sn_y$ nanoalloy particles the product R ratio for MCP conversion dropped as

(i) the reaction temperature rose,

(ii) as x or y increased (suggesting a similarity in these two nanoalloys) and was stable with time for PtSn, but rose for PtAu with catalyst restructuring (or deactivation).

2MP and 3MP are also favourable products of n-hexane conversion over metals. With n-hexane over $Pt_{100-y}Sn_y/Al_2O_3$ when held isothermally:

(i) initial conversion of n-hexane decreased as y increased

(ii) conversion decreased with increasing reaction time due to build-up of carbonaceous deposits

(iii) isomerisation was the dominant reaction over hydrogenolysis,dehydrogenation or isomerisation except at the highest levels of Sn addition (when it became replaced by dehydrogenation).

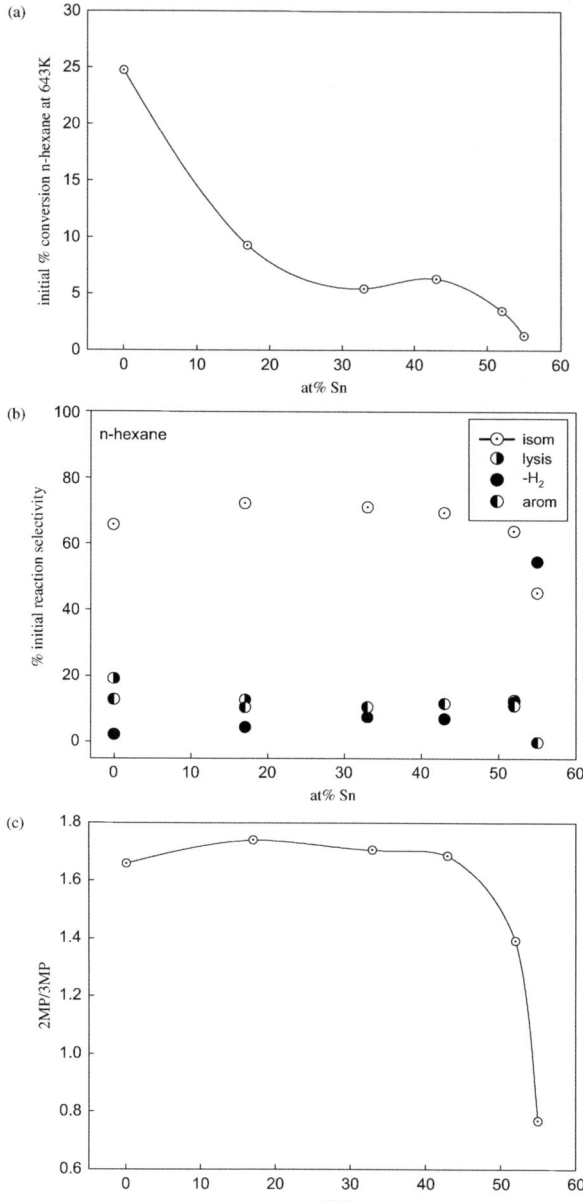

Fig. 9 (a) Decrease in initial activity in hexane conversion at 643 K as the at% Sn added to 1 wt% Pt on Al_2O_3 increases. (b) Variation in initial selectivity in hexane conversion over catalysts containing 1 wt% Pt but varying at% Sn as $Pt_{100-y}Sn_y$ at 643 K to ring-opening isomerisation in preference to hydrogenolysis to CH_4–C_5H_{12}, dehydrogenation to methylcyclopentenes or aromatisation to benzene. (c) 2MP/3MP product ratio in n-hexane isomerisation at 643 K over catalysts containing 1 wt% Pt but varying at% Sn as $Pt_{100-y}Sn_y/Al_2O_3$ at 643 K.

(iv) 2MP/3MP selectivity within isomerisation decreased with reaction time with either $y = 0$ or 17 at% Sn (although the initial 2MP/3MP product ratio at 643 K was independent of $y < 50$ at% Sn and then decreased)

(v) the minor hydrogenolysis reactions produced predominantly C_3 products irrespective of y ($y < 40$–50 at% Sn).

For n-hexane reaction over $Pt_{100-x}Au_x/C$ at 590–645 K

(i) the rate of conversion was higher for Pt/C than for $Pt_{91}Au_9$/C, although the activation energies (98 ± 1 kJ mol^{-1}) were similar, but

(ii) the 2MP/3MP product ratio is higher after gold addition to 9 at%, despite falling with increasing temperature.

Many years ago catalytic probes were used to elucidate alloy structures (*e.g.* the rate of ethane hydrogenolysis was noted to drop faster on Cu addition to Ni than did the rate of dehydrogenation[45]). Here we have prepared two different nanoalloys by low energy bottom-up nanotechnology and then used adsorption and catalysis to probe their surface states. The reactive probes selected here (*i.e.* methylpentanes formation on MCP ring opening and *n*-hexane isomerisation) seem appropriate and useful.

It is known that ring opening of MCP to methylpentanes occurs over EuroPt-1 silica-supported Pt,[62] where at 473 K and $8 < p_{H2} < 101$ kPa the steady-state product ratio 2MP/3MP (*R*) ≈ 3.0. It seems that the products adsorb more weakly on a vacant active site * than reactant MCP (*i.e.* $C_6H_{12} + * = C_6H^*_{12-2a} + aH_2$, where $2a$ is the average number of H atoms lost per reactant MCP molecule). In addition it has been suggested[63] that a partially selective ring-opening might prevail at higher temperatures. Over ZSM-5 it converts predominantly to cyclohexane with hydride transfer[64] while hexane isomers are formed more on metals,[65] where larger Pt particles (*i.e.* $d_{Pt} = 17$ nm) can lead to more selective MCP hydrogenolysis (*i.e.* 85% selective MCP hydrogenolysis giving higher 2MP/3MP product ratios than 2) than smaller particles (*i.e.* when at $d_{Pt} = 8.5$ nm there was 0% selective MCP hydrogenolysis leading to 2MP/3MP = 2 (*i.e.* the statistical and thermodynamic equilibrium value[66])).[67] Not surprisingly then sulfation of Pt/ZrO$_2$ does not raise the 2MP/3MP (*R*) product ratio in MCP conversion, even though this is 3.3–4.2 at 313–380 K over sulfated Pt/ZrO$_2$; rather the ratio is thought to be defined by the nature of surface sites.[68] Thus pairs of contiguous surface atoms (where one is electron deficient and an asymmetric site pair exists) are said[68] to enhance rates of formation of 2MP. The drop in product *R* ratio at higher *y* must indicate a change in the availability of contiguous Pt atom * pairs at this surface composition; HRTEM is being used to investigate this further.

As with alumina-supported Pd–Re surfaces, it seems that the 2MP/3MP product ratio arising from *n*-hexane and MCP conversion is a useful 'chemical fingerprint' for such bimetallic catalysts.[69] Certainly Pd–Pt surfaces have been studied using the 2MP/3MP ratio arising from conversion of *cis* or *trans* methyl-ethyl-cyclopropane (MECP) and MCP.[70,71] Of course one could also consider 2MP/hexane product ratios for MCP conversion, but to enable *both* reactions to be followed in the same way only 2MP/3MP ratios in the products of MCP and *n*-hexane conversion were considered here.

The present results suggest that catalytic probes *do* provide another insight into the nature of nanoalloy surfaces.

1–5 nm bimetallic nanoparticles[72] are of increasing importance. They have useful optical,[73,74] catalytic,[75] superparamagnetic,[76] fuel cell[77] and bio-[78] properties and applications. It may be that catalytic probes will enhance this potential and allow their potential to be realised that much faster.

Acknowledgements

The authors thank EPSRC for support of K. A. G. through the provision of a studentship.

References

1   E. K. Rideal, *Concepts in Catalysis*, Academic Press, New York, 1968, p. 8.
2   D. I. Gittins and F. Caruso, *ChemPhysChem*, 2002, **3**, 111.
3   S. H. Overbury, P. A. Bertrand and G. A. Somorjai, *Chem. Rev.*, 1975, **75**, 547.
4   F. G. Meng, H. S. Liu, L. B. Liu and Z. P. Jin, *J. Alloys Compd.*, 2007, **431**, 292.

5 L. O. Paz-Borbon, R. L. Johnston, G. Barcaro and A. Fortunelli, *J. Phys. Chem. C*, 2007, **111**, 2936.
6 D. Cheng, S. Huang and W. Wang, *Eur. Phys. J. D*, 2006, **39**, 41.
7 A. Rapallo, G. Rossi, R. Ferrando, A. Fortunelli, B. C. Curley, L. D. Lloyd, G. M. Tarbuck and R. L. J. Ohnston, *J. Chem. Phys.*, 2005, **122**, 194308.
8 S. Darby, T. V. Mortimer-Jones, R. L. Johnston and C. J. Roberts, *J. Chem. Phys.*, 2002, **116**, 1536.
9 X. D. Dai, Y. Kong and J. H. Li, *Phys. Rev. B*, 2007, **75**, 104101.
10 S. Sahoo, G. Rollmann and P. Entel, *Phase Transitions*, 2006, **79**, 693.
11 D. J. Cheng, W. C. Wang and S. P. Huang, *J. Phys. Chem. B*, 2006, **110**, 16193.
12 (*a*) N. A. Zarkevich, T. L. Tan and D. D. Johnson, *Phys. Rev. B*, 2007, **75**, 104203; (*b*) V. S. K. Balagurusamy, R. Streitel, O. G. Shpyrko, P. S. Pershan, M. Meron and B. H. Lin, *Phys. Rev. B: Condens. Matter*, 2007, **75**, 104209.
13 C. Massen, T. V. Mortimer-Jones and R. L. Johnston, *J. Chem. Soc., Dalton Trans.*, 2002, 4375.
14 A. S. Shirinyan and M. Wautelet, *Nanotechnology*, 2004, **15**, 1720.
15 D. J. Cheng, S. P. Huang and W. C. Wang, *Phys. Rev. B: Condens. Matter*, 2006, **74**, 064117.
16 A. Aguado and J. M. Lopez, *J. Chem. Theor. Comput.*, 2005, **1**, 299.
17 N. T. Wilson and R. L. Johnston, *J. Mater. Chem.*, 2002, **12**, 2913.
18 L. D. Lloyd, R. L. Johnston, S. Salhi and N. T. Wilson, *J. Mater. Chem.*, 2004 **14**, 1691.
19 L. D. Menard, H. P. Xu, S. P. Gao, R. D. Twesten, A. S. Harper, Y. Song, G. L. Wang, A. D. Douglas, J. C. Yang, A. I. Frenkel, R. W. Murray and R. G. Nuzzo, *J. Phys. Chem. B*, 2006, **110**, 14564.
20 S. Pal and G. De, *J. Mater. Chem.*, 2007, **17**, 493.
21 Y. B. Lou, M. M. Maye, L. Han, J. Luo and C. J. Zhong, *Chem. Commun.*, 2001, 473.
22 T. Montanari, O. Marie, M. Daturi and G. Busca, *Appl. Catal., B*, 2007, **71**, 216.
23 M. O. Nutt, K. N. Heck, P. Alvarez and M. S. Wong, *Appl. Catal., B*, 2006, **69**, 115.
24 E. G. Szabo, A. Tompos, M. Hegedus, A. Szegedi and J. L. Margitfalvi, *Appl. Catal., A*, 2007, **320**, 114.
25 P. A. Sermon, J. M. Thomas, K. Keryou and G. R. Millward, *Angew. Chem., Int. Ed. Engl.*, 1987, **26**, 918.
26 Y. H. Lee, G. Lee, J. H. Shim, S. Hwang, J. Kwak, K. Lee, H. Song and J. T. Park, *Chem. Mater.*, 2006, **18**, 4209.
27 E. Antolini, F. Colmati and E. R. Gonzalez, *Electrochem. Commun.*, 2007, **9**, 398.
28 R. Srinivasan, R. J. de Angelis and B. H. Davis, *J. Catal.*, 1987, **106**, 449.
29 H. Lieske and J. Volter, *J. Catal.*, 1984, **90**, 96.
30 A. O. Neto, R. R. Dias, M. M. Tusi, M. Linardi and E. V. Spinace, *J. Power Sources*, 2007, **166**, 87.
31 C. Lamy, E. M. Belgsir and J. M. Leger, *J. Appl. Electrochem.*, 2001, **31**, 799.
32 D. R. Lycke and E. L. Gyenge, *Electrochim. Acta*, 2007, **52**, 4287.
33 O. Trapp, *Electrophoresis*, 2007, **28**, 691.
34 M. Bruncko, I. Anzel and A. Kneissl, *Corros. Sci.*, 2007, **49**, 1228.
35 G. A. Somorjai and A. L. Marsh, *Philos. Trans. R. Soc. London, Ser. A*, 2005, **363**, 879.
36 Z. Paal, A. Wootsch, I. Bakos, S. Szabo, H. Sauer, U. Wild and R. Schlogl, *Appl. Catal., A*, 2006, **309**, 1.
37 C. Dossi, A. Pozzi, S. Recchia, A. Fusi, R. Psaro and V. Dal Santo, *J. Mol. Catal. A: Chem.*, 2003, **204**, 465.
38 Z. Paal, A. Wootsch, R. Schlogl and U. Wild, *Appl. Catal., A*, 2005, **282**, 135.
39 (*a*) S. M. Davis, F. Zaera and G. A. Somorjai, *J. Am. Chem. Soc.*, 1982, **104**, 7453; (*b*) J. A. Dalmon and G. A. Martin, *J. Catal.*, 1999, **66**, 214.
40 A. Wootsch and Z. Paal, *J. Catal.*, 1999, **185**, 192.
41 H. Matsumoto, Y. Saito and Y. Yoneda, *J. Catal.*, 1970, **19**, 101.
42 J. R. Anderson and Y. Shimoyam, *Proc. 5th Int. Cong. Catal.*, 1972, 48.
43 (*a*) G. Maire, G. Plouidy, J. C. Prudhom and F. G. Gault, *J. Catal.*, 1965, **4**, 556; (*b*) J. R. Anderson, *Adv. Catal.*, 1974, **23**, 1.
44 J. H. Sinfelt, *Catal. Rev. Sci. Eng.*, 1969, **3**, 175.
45 J. H. Sinfelt, D. J. C. Yates and J. L. Carter, *J. Catal.*, 1972, **24**, 283.
46 M. Boudart, A. W. Aldag, L. D. Ptak and J. E. Benson, *J. Catal.*, 1968, **11**, 35.
47 M. Faraday, *Philos. Trans. R. Soc. London*, 1857, **147**, 145.
48 G. Frens, *Nat. Phys. Sci.*, 1973, **241**, 20.
49 S. Pande, S. K. Ghosh, S. Praharaj, S. Panigrahi, S. Basu, S. Jana, A. Pal, T. Tsukuda and T. Pal, *J. Phys. Chem. C*, 2007, **111**, 4596.

50 G. Barcaro, A. Fortunelli, G. Rossi, F. Nita and R. Ferrando, *J. Phys. Chem. B*, 2006, **110**, 23197.
51 P. W. Zheng, X. W. Jiang, X. Zhang, W. Q. Zhang and L. Q. Shi, *Langmuir*, 2006, **22**, 9393.
52 Y. Suchorski, J. Beben, A. Frac, V. K. Medvedev and H. Weiss, *Surf. Interface Anal.*, 2007, **39**, 161.
53 (*a*) J. Turkevich, P. C. Stevenson and J. Hillier, *Discuss. Faraday Soc.*, 1951, **11**, 55; (*b*) L. G. Tejuca, K. Aika, S. Namba and J. Turkevich, *J. Phys. Chem.*, 1977, **81**, 1399; (*c*) D. N. Furlong, A. Launikonis, W. H. F. Sasse and J. V. Sanders, *J. Chem. Soc., Faraday Trans. 1*, 1984, **80**, 571.
54 (*a*) M. J. Yacaman and J. M. Domingueze, *J. Catal.*, 1980, **64**, 213; (*b*) D. A. Jefferson, J. M. Thomas, G. R. Millward, K. Tsuno, A. Harriman and R. D. Brydson, *Nature*, 1986, **323**, 428.
55 R. T. K. Baker, *Catal. Rev. Sci. Eng.*, 1979, **19**, 161.
56 J. M. Thomas, *Faraday Discuss.*, 1996, **105**, 1.
57 J. M. Cowley and R. J. Plano, *J. Catal.*, 1987, **108**, 199.
58 S. M. Augustine and W. M. H. Sachtler, *J. Catal.*, 1987, **106**, 417.
59 M. Ocana, C. J. Serna and E. Matijevic, *Colloid Polym. Sci.*, 1995, **273**, 681–686.
60 R. Burch, *J. Catal.*, 1981, **71**, 348.
61 S. R. Adkins and B. H. Davis, *J. Catal.*, 1984, **89**, 371.
62 Y. P. Zhuang and A. Frennet, *Appl. Catal., A*, 1999, **177**, 205.
63 F. G. Gault, *Adv. Catal.*, 1981, **30**, 1.
64 K. H. Lee and D. Farcasiu, *J. Chem. Eng. Jpn.*, 2001, **34**, 1557.
65 T. J. McCarthy, G. D. Lei and W. M. H. Sachtler, *J. Catal.*, 1996, **159**, 90.
66 B. A. Lerner, B. T. Carvill and W. M. H. Sachtler, *J. Mol. Catal.*, 1992, **77**, 99.
67 J. M. Dartigues, A. Chambellan and F. G. Gault, *J. Am. Chem. Soc.*, 1976, **98**, 856.
68 M. R. Smith, J. K. A. Clarke, G. Fitzsimons and J. J. Rooney, *Appl. Catal., A*, 1997, **165**, 357.
69 W. Juszczyk and Z. Karpinski, *Appl. Catal., A*, 2001, **206**, 67.
70 N. Gyorrfy, L. Toth, M. Bartok, J. Ocsko, U. Wild, R. Schlogl, D. Teschner and Z. Paal, *J. Mol. Catal. A: Chem.*, 2005, **238**, 102.
71 A. Barrera, J. A. Montoya, M. Viniegra, J. Navarrete, G. Espinosa, A. Vargas, P. del Angel and G. Perez, *Appl. Catal., A*, 2005, **290**, 97.
72 *Metal Nanoparticles Synthesis Characterisation and Applications*, ed. D. L. Feldheim and C. A. Foss, Marcel Dekker, New York, 2002.
73 G. L. Hornyak, C. J. Patrissi, E. B. Oberhauser, C. R. Martin, J. C. Valmalette, L. Lemaire, J. Dutta and H. Hofmann, *Nanostruct. Mater.*, 1997, **9**, 571.
74 G. Mattei, G. Battaglin, E. Cattaruzza, C. Maurizio, P. Mazzoldi, C. Sada and B. F. Scremin, *J. Non-Cryst. Solids*, 2007, **353**, 697.
75 I. Pastoriza-Santos, J. Perez-Juste, S. Carregal-Romero, P. Herves and L. M. Liz-Marzan, *Chem.–Asian J.*, 2006, **1**, 730.
76 M. Mandal, S. Kundu, T. K. Sau, S. M. Yusuf and T. Pal, *Chem. Mater.*, 2003, **15**, 3710.
77 N. M. Galea, D. Knapp and T. Ziegler, *J. Catal.*, 2007, **247**, 20.
78 M. Noyong, K. Gloddek, J. Mayer, T. Weirich and U. Simon, *J. Cluster Sci.*, 2007, **18**, 193.

To alloy or not to alloy?
Cr modified Pt/C cathode catalysts for PEM fuel cells†

Peter P. Wells,[a] Yangdong Qian,[b] Colin R. King,[a]
Richard J. K. Wiltshire,[a] Eleanor M. Crabb,[b] Lesley E. Smart,[b]
David Thompsett[c] and Andrea E. Russell*[a]

Received 15th May 2007, Accepted 29th June 2007
First published as an Advance Article on the web 13th September 2007
DOI: 10.1039/b707353b

The cathode electrocatalysts for proton exchange membrane (PEM) fuel cells are commonly platinum and platinum based alloy nanoparticles dispersed on a carbon support. Control over the particle size and composition has, historically, been attained empirically, making systematic studies of the effects of various structural parameters difficult. The controlled surface modification methodology used in this work has enabled the controlled modification of carbon supported Pt nanoparticles by Cr so as to yield nanoalloy particles with defined compositions. Subsequent heat treatment in 5% H_2 in N_2 resulted in the formation of a distinct Pt_3Cr alloy phase which was either restricted to the surface of the particles or present throughout the bulk of the particle structure. Measurement of the oxygen reduction activity of the catalysts was accomplished using the rotating thin film electrode method and the activities obtained were related to the structure of the nanoalloy catalyst particles, largely determined using Cr K edge and Pt L_3 edge XAS.

1. Introduction

Commercialisation of proton exchange membrane fuel cells (PEM FCs) for demanding applications, such as transport, requires a reduction of the platinum content from the currently attainable 0.85–1.1 g_{Pt} kW^{-1} to less than 0.2 g_{Pt} kW^{-1}.[1] The Pt loading at the cathode represents nearly 90% of the total of today's catalysts, reduction of which is limited by the activity of Pt for the oxygen reduction reaction (ORR).[2] The Pt content of the cathode can be reduced by either improving the utilisation of the catalyst or by the use of catalysts that are more active than Pt with respect to the mass of Pt in the electrode (enhanced mass activity). The former is realised by the use of highly dispersed nanoparticle catalysts supported on a conductive support, whilst the latter may be accomplished via alloying Pt with a second component to yield nanoalloy particles with enhanced activity.

[a] School of Chemistry, University of Southampton, Highfield, Southampton, UK SO17 1BJ.
E-mail: A.E.Russell@soton.ac.uk
[b] Department of Chemistry, The Open University, Walton Hall, Milton Keynes,
UK MK7 6AA
[c] Johnson Matthey Technology Centre, Blounts Court, Sonning Common, Reading,
UK RG4 9NH

† The HTML version of this article has been enhanced with colour images.

PtCr catalysts of various Pt : Cr ratios have consistently been shown to be amongst the most active for the ORR and, as such, they have received considerable attention in the literature as cathode catalysts for both PEM and phosphoric acid fuel cells (PAFC), with an aim to understanding the origins of the enhanced catalytic activity over that of Pt and other bimetallic Pt based catalysts. For example, in a study of catalysts for the PAFC in 1983 Jalan and Taylor[3] suggested that the enhanced activity of Pt bimetallic catalysts was due to geometrical considerations. They proposed that by introducing a second metal into the Pt lattice, the Pt–Pt interatomic distance on the surface of the catalyst was shortened, enhancing the rate of cleavage of the O–O bond. A range of Pt bimetallic alloys was investigated and a linear correlation between interatomic separation and specific activity towards the ORR was found, with PtCr catalysts exhibiting the highest specific activity towards the ORR compared to the other Pt alloys studied. However, other research groups later contested this simple explanation of the enhancement towards the ORR offered by Pt based bimetallic catalysts.

Glass et al.[4] investigated a range of PtCr alloys of different compositions, suggesting that increased Cr content should decrease the Pt–Pt spacing. The performance of these catalysts towards the ORR did not correlate with the Pt–Pt interatomic spacing and furthermore showed no improved performance over pure Pt. The work did show some differences in electrochemical behaviour between the Pt and PtCr systems, most notably differences in OH coverage and open circuit potential, suggesting an electronic effect.

Paffet et al.[5] attributed the enhanced activity of PtCr catalysts to a roughening of the surface. They proposed that during the experiments, Cr(III) at the surface, which is present as an oxide or hydroxide, is dissolved as a Cr(VI) solution species at operating potentials, causing a roughening of the surface, and a corresponding increase in the number of active sites.

Mukerjee and Srinivasan[6] studied carbon supported Pt and Cr, Co, or Ni bimetallic catalysts and found that all of the bimetallic catalysts gave a two to three fold enhancement towards the ORR under PEMFC operating conditions over Pt alone. The work showed that the specific activity (activity per unit surface area of the catalyst particles) of the bimetallic catalysts was greater, even though they had an increased particle size, and thus a smaller available active area than carbon supported Pt for similar metal loadings expressed as wt%. This improvement was attributed to the formation of superlattices of the type Pt_3Cr, Pt_3Co, and Pt_3Ni. These superlattices exhibited the same cubic fcc structure of Pt but with lattice contractions. The work also showed that the secondary component was not evenly distributed, with some of the secondary component (in excess to the Pt_3X phase) remaining on the surface of the support or the bimetallic particles in an oxide form.

Mukerjee et al.[7] subsequently used in situ XAS measurements to relate structural and electronic properties of the bimetallic/alloy catalysts to their activity towards the ORR. Structural information in the form of the Pt–Pt bond distance was obtained by analysis of the EXAFS data collected at the Pt L_3 edge. XANES data collected at both the Pt L_3 and L_2 edges was used to calculate the Pt d-orbital vacancy per atom, using the method of Mansour et al.[8] The work showed that both the Pt d-orbital vacancy per atom and Pt–Pt distance exhibit a volcano type behaviour with respect to electrocatalytic activity towards the ORR, suggesting that there is a symbiotic relationship between Pt–Pt distance and d-orbital vacancy. The PtCr/C catalyst was shown to be the most active and at the top of both volcano plots. In addition, comparison of the potential dependence of the d-band vacancies of Pt and the bimetallic catalysts provided evidence that OH adsorption is weaker on the bimetallics, thereby suppressing oxide formation.

Most recently, Pt bimetallic catalysts that have a structure described as a Pt-skin,[9,10] i.e. having a pure-Pt surface atomic layer, have been shown to be the most active ORR catalysts.[11,12] In particular, in a study of single crystal surfaces, Stamenkovic et al.[12] have shown that the (111) face of Pt_3Ni with a Pt-skin layer is 10 times more active than the corresponding Pt(111) surface and 90 times more than the current state-of-the-art carbon supported catalysts. The effect of the underlying alloy on the Pt-skin layer was shown to be predominantly electronic in nature, altering Pt–OH chemical bonding. In an earlier study of Pt and Ni, Co, or Fe bimetallic catalysts prepared by sputtering, Toda et al.[13,14] had also found that catalysts with this Pt-skin structure exhibited an increased d-band vacancy without a contraction in the Pt–Pt distance. They suggested that this change in electronic property, assuming a lateral interaction between O_2 and Pt, promotes the 2π donation from O_2 to surface Pt, resulting in stronger O_2 adsorption and a weakening of the O–O bond.

As indicated in the brief review presented above, the roles of alloy phase formation and surface composition in determining the ORR activities of Pt bimetallic catalysts are not fully understood. Whilst investigations of single crystal surfaces can provide model systems in which individual structural parameters may be isolated, exploitation of such results requires translation of the findings to practical supported catalysts. However, traditional methods for producing bimetallic catalysts such as impregnation, electrochemical deposition and precipitation lack the selectivity to control the surface composition of the resulting particles. In the study presented here, a controlled surface modification procedure developed by Crabb et al.[15–19] was used to prepare Pt/C catalysts modified by a controlled amount of Cr (denoted Cr/Pt/C). The modified catalysts were characterised using XANES, EXAFS, cyclic voltammetry, XRD and TEM and their ORR was compared to conventionally prepared Pt_3Cr/C catalysts.

2. Experimental

Catalyst preparation

Cr/Pt/C catalysts were prepared by modification of a 20 wt% Pt/C (XC-72R) catalyst supplied by Johnson Matthey using the CSR method described previously.[15–19] Briefly, 3 g of the Pt/C substrate catalyst was first reduced in the reactor under flowing $H_2(g)$ (flow 60–100 cm^3 min^{-1}) at 200 °C for 3 h and then cooled to room temperature under flowing $N_2(g)$. The required amount of either chromocene ($Cr(Cp)_2$) or bisbenzene chromium ($Cr(Bz)_2$), corresponding to 0.167 or 0.66 monolayers of Cr atoms based on the dispersion of the substrate catalyst (0.48), was dissolved in n-heptane, purged with N_2 and added to the reactor. $H_2(g)$ was then passed through the reactor whilst stirring and heating at 90 °C for 8 h. The reactor was then allowed to cool and flushed with $N_2(g)$ for 30 min. The contents of the reactor were discharged, filtered, and washed with n-heptane. The filtered catalyst was allowed to dry in air and then returned to the cleaned reactor and the initial reduction step at 200 °C was repeated. Final catalyst compositions were confirmed by ICP-AES and found to be within 10% of the predicted values. Portions of the catalyst were subjected to additional annealing treatments in 5% H_2 in N_2 for 1 h at 750 °C and 900 °C.

Conventionally prepared 20 wt% and 40 wt% Pt_3Cr/C catalysts were supplied by Johnson Matthey.

Physical characterisation

The catalysts were characterised by XRD using a Bruker AXS D-500 diffractometer with Cu K_α X-ray radiation and TEM using a Tecnai F20 Transmission Electron Microscope. Both bright field and high-resolution electron microscopy modes were used. Powder samples for TEM EDX were crushed between two glass slides and samples positioned onto a lacey carbon coated copper 'finder' grid with the aid of a micromanipulator.

All catalysts were prepared as BN pellets for XAS measurements and examined without further treatment. Cr K edge XAS spectra were acquired on station 7.1 at the SRS, Daresbury Laboratory, using a double crystal Si(111) monochromator.

Data were acquired in fluorescence mode using a liquid nitrogen cooled 9-element Ge solid-state detector. Reference EXAFS of metal foils and oxide standards were acquired in transmission mode. Pt L_2 and L_3 edge XAS spectra were acquired on the 6 T wiggler station 16.5 at the SRS with a double crystal Si(220) monochromator tuned to 50% intensity to reject the higher harmonics. Spectra of the catalyst materials were acquired in fluorescence mode using a liquid nitrogen cooled 30-element Ge solid-state detector. Reference EXAFS of metal foils and oxide standards were acquired in transmission mode.

The EXAFS data were analysed using the Daresbury suite of analysis programs including, EXCURV98, a least squares fitting program based on curved-wave theory, using a $Z + 1$ core hole approximation.[20–23]

Electrochemical characterization

The ORR activity of each catalyst was assessed using the thin film rotating disc electrode method described by Schmidt *et al.*[24] Briefly a glassy carbon rotating disc electrode (Pine) was polished to a 0.05 μm mirror finish and sonicated in triply distilled water to remove any particulates affixed to the disk. 10 mg of the catalyst was dispersed in 10 mL chloroform and mixed using an ultrasonic bath for 30 min to form a well-dispersed suspension. A 5 μL aliquot of this solution was deposited onto the surface of a glassy carbon rotating disk electrode (area = 0.196 cm^2) and allowed to dry. The dried catalyst layer was then covered by 5 μL of a 5 wt% solution of Nafion® in low weight alcohol and allowed to dry in air at room temperature for 30 min. A Pine AFMSRX modulated speed rotator was used with an Autolab PGSTAT30 potentiostat. A standard 3 electrode RDE cell (Pine) was used with a Pt gauze counter electrode. Potentials were measured with respect to a Hg/Hg$_2$SO4 (MMS) reference electrode and corrected to RHE by calibration of the MMS electrode against a dynamic hydrogen electrode in the same solution; 1 mol dm^{-3} H$_2$SO$_4$ prepared using 18 MΩ cm water from a Barnstead Nanopure system.

Voltammograms were acquired between 1.0 V and 0.05 V *vs.* RHE with a scan rate of 2 mV s^{-1} whilst flowing O$_2$(g) over the surface of the previously O$_2$ saturated electrolyte for rotation rates of 900, 1600, 2500, and 3600 rpm, as shown in Fig. 1. In between measurements, the electrolyte was purged with O$_2$(g) whilst rotating the electrode at 1000 rpm. Diffusion controlled kinetics were confirmed by the Levich plot as shown in Fig. 1.

The forward scan of the voltammograms acquired at 2500 rpm was used in each case to assess the performance of the catalysts. The kinetically limited current as a function of the potential, I_k, was extracted from the data and the results presented as Tafel plots, potential *vs.* Log $> I_k$]. The currents were normalised to the available Pt surface area, calculated using the area determined from stripping of a monolayer of

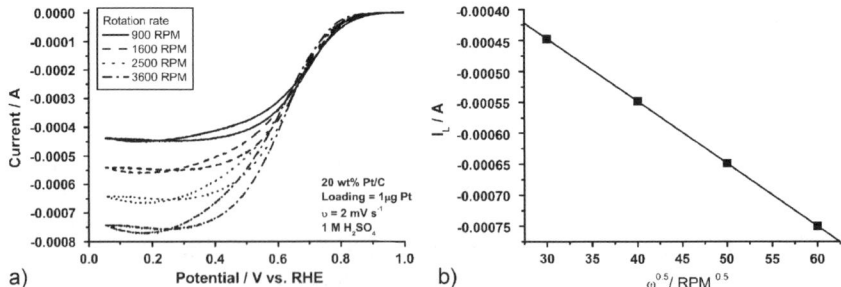

Fig. 1 (left) Voltammograms of 20 wt% Pt/C from 1.0 to 0.05 V in 1 M H$_2$SO$_4$ with a scan rate of 2 mV s^{-1} at rotation rates as indicated. (right) The limiting currents at each rotation rate plotted against $\omega^{0.5}$. The line of best fit is also included, with the straight line indicating good mass transport control.

adsorbed CO, to provide specific activities for the ORR, reported as $\mu A \text{ cm}^{-2}$ Pt. The CO adlayer was prepared by holding the potential at 0.15 V vs. RHE for 60 min, during which CO was purged through the solution for 30 min, followed by N_2 for 30 min. The area under the CO oxidation peak was integrated, using the Autolab software, to give the charge passed and this converted to the active electrochemical surface area of the electrode using the value $420 \times 10^{-6} \text{ C cm}^{-2}$.

3. Results and discussion

Fig. 2 shows the normalised XANES spectra obtained at the Cr K edge for the Cr/Pt/C catalysts, Cr foil, and Cr_2O_3. The data were acquired under atmospheres of H_2 and air, although no discernable differences were apparent.

Features in the XANES region arise due to electronic transitions and this is exemplified by the data obtained at the Cr K edge. A pre-edge feature can be observed at approximately 5980 eV which corresponds to the dipole forbidden 1s → 3d transition. This pre-edge feature is highly sensitive to the coordination geometry of the Cr as the t_2 orbitals have some p-orbital character. It has been found that tetrahedrally coordinated Cr species (e.g. CrO_3) exhibit a sharp white line in this pre-edge region and that there is, on average, a shift in edge position of 1.6 eV per unit change in valency.[25]

As can be seen in the figure, the two preparatory methods give rise to quite different XANES spectra. The Cr/Pt/C catalyst prepared by the $Cr(Bz)_2$ route all exhibit an edge position around 5990 eV, similar to that of a Cr foil. This suggests that the Cr present in these samples is mostly present in a metallic phase, and hence that the deposition method has successfully targeted the Pt surface sites. However, it is also apparent that the 200 °C sample is more oxidised than those annealed at 750 °C and 900 °C. Thus it appears that the Cr may be initially present as oxidised adatoms on the Pt surface, becoming more incorporated into a metallic phase as the annealing temperature is increased.

This view of the catalyst structure is confirmed by the XRD patterns shown in Fig. 3. The XRD pattern obtained for the 0.66 Cr/Pt/C sample annealed at 200 °C prepared by the $Cr(Bz)_2$ route (Fig. 3), corresponds to poorly crystalline fcc Pt particles, as indicated in the figure by the overlap of the peaks in the diffraction pattern with the vertical lines from the reference material. The 750 °C annealed sample (in Fig. 3) was also found to be mainly composed of a poorly crystalline cubic phase close in crystallographic parameters to Pt/C, but a second poorly crystalline cubic phase, similar in crystallographic parameters to cubic Pt_3Cr was also found, as indicated by the vertical bars in the figure. Subtle differences between the observed phase and the reference pattern may indicate the presence of a solid solution and/or stoichiometric differences. In addition, a trace amount of Cr_2O_3 was also present.

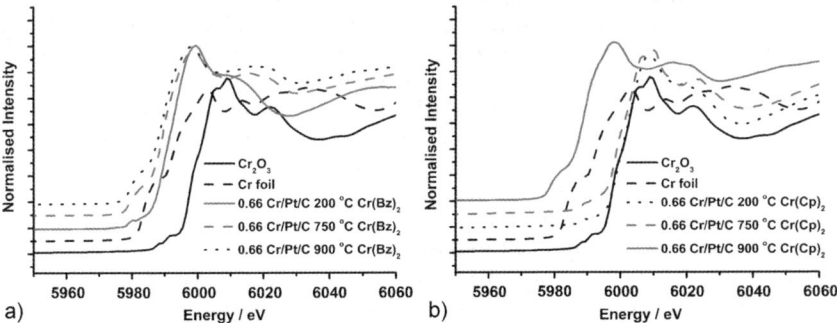

Fig. 2 XANES spectra at the Cr K edge for 0.66 Cr/Pt/C prepared by the (a) $Cr(Bz)_2$ and the (b) $Cr(Cp)_2$ routes annealed at 200 °C, 750 °C, and 900 °C. Spectra were collected in an atmosphere of $H_2(g)$.

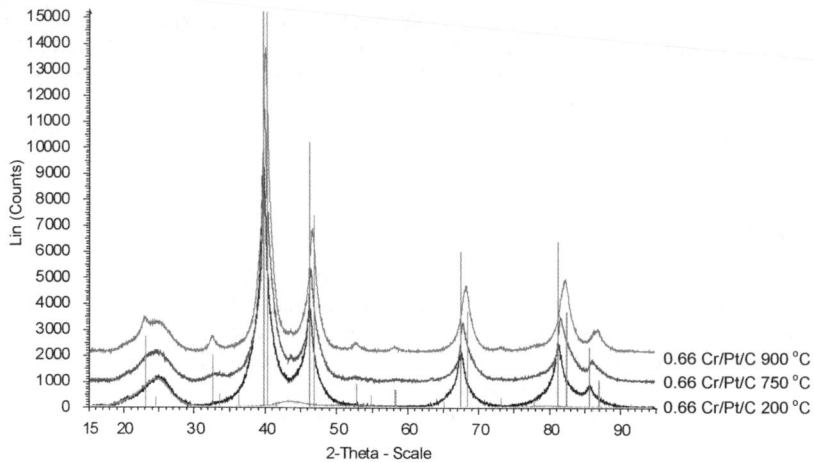

Fig. 3 XRD patterns of 0.66 Cr/Pt/C Cr(Bz)$_2$ as a function of annealing temperature. JCPDS reference data are included for comparison.

The 900 °C annealed sample was found to be composed of one phase, which is similar in crystallographic parameters to a cubic Pt$_3$Cr phase.

The same trend of gradual, temperature dependent formation of the alloy phase is not observed for the Cr/Pt/C catalysts prepared by the Cr(Cp)$_2$ route. The edge positions of the 200 °C and 750 °C annealed samples are both 5997 eV, which is the same energy observed for Cr$_2$O$_3$. The 750 °C sample has a near identical XANES spectrum to that obtained for Cr$_2$O$_3$, whilst the 900 °C sample has a XANES spectrum very similar to that of the Cr(Bz)$_2$ prepared Cr/Pt/C samples. No evidence of a crystalline Cr$_2$O$_3$ phase was found following any of the heat treatments in the XRD patterns, whilst only the cubic Pt phase was evident in the XRD pattern of the 200 °C sample and a cubic PtCr phase (intermediate between Pt and Pt$_3$Cr) was observed in the patterns of the 750 °C and 900 °C samples (not shown). This suggests that initially the Cr is present as an amorphous oxide around the Pt, which on heating to 750 °C becomes Cr$_2$O$_3$, and at 900 °C the Cr is driven into the Pt particle to form the alloy phase.

The k^2 weighted EXAFS data and associated Fourier transforms and the fitting parameters determined for the Pt$_3$Cr/C reference catalysts and 0.66 Cr/Pt/C catalysts are detailed below in Fig. 4 through to 6 and Tables 1 through to 3.

The 40 wt% Pt$_3$Cr catalyst was prepared using a route that has previously been shown to yield a well alloyed catalyst, as evidenced by XRD, and the EXAFS fitting parameters are in agreement with this assumption. The differences in the EXAFS and fitting parameters between the 40 wt% and 20 wt% Pt$_3$Cr/C reference catalysts are largely accounted for by the difference in particle diameter; similar Cr–Pt and Cr–Cr distances were obtained and larger coordination numbers were found for the 40 wt% catalyst with larger particles. The small Cr–O contribution found in the fitting of the 20 wt% catalyst is also associated with the smaller particle size; smaller particles are more readily oxidised and the relative contribution of atoms at the surface to the EXAFS is greater for smaller particles.

The EXAFS data of 0.66 Cr/Pt/C prepared by the Cr(Cp)$_2$ route show no signs of alloy phase formation until the catalyst is annealed at 900 °C. The EXAFS data for the 200 °C and 750 °C annealed catalysts are broadly similar, with the majority of the oscillatory amplitude in the low k space region, suggesting that the Cr is mostly in an environment surrounded by low Z neighbours. Heating to 750 °C results in oscillations over the full k-range, but the EXAFS fitting parameters only include Cr–Cr, and Cr–O interactions. Annealing at 750 °C appears to form Cr$_2$O$_3$, in agreement with the XANES analysis. Cr$_2$O$_3$ is the most stable oxide of chromium

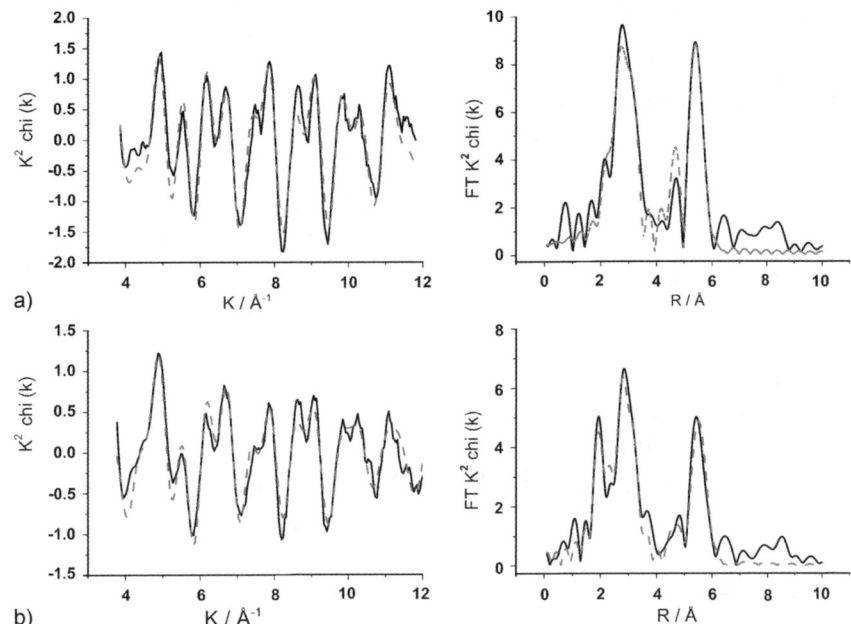

Fig. 4 (left) k^2 weighted experimental data and fit along with (right) the Fourier transform for (a) 40 wt% Pt$_3$Cr/C and (b) 20 wt% Pt$_3$Cr/C. Data (black line) and fit (dashed line).

and is readily formed above 250 °C.[26] As there was no evidence of Cr$_2$O$_3$ using XRD, it is likely that this is finely dispersed over the catalyst surface rather than existing as the bulk oxide. By 900 °C an alloy phase similar to the Pt$_3$Cr/C reference catalysts has been formed, as is shown by the presence of Cr–Pt contributions at around 2.7 Å and 4.7 Å and similar multiple scattering contributions.

The interpretation of the EXAFS for the 0.66 Cr/Pt/C catalyst series prepared by the Cr(Cp)$_2$ route cannot be extended to that prepared using the Cr(Bz)$_2$ precursor. In this instance, the 200 °C annealed sample is in general agreement with the previous analysis, although there is a significant reduction in the Cr–O coordination number, from 3.3 ± 0.2 to 2.1 ± 0.2. This agrees well with the XANES analysis, which suggested that the Cr species were present in a less oxidised state than for the equivalent catalyst prepared using the Cr(Cp)$_2$ route. After annealing to 750 °C, the fitting parameters confirm that an alloy has been formed as Cr–Pt shells are once again present around 2.7 Å and 4.7 Å. Additional heat treatment to 900 °C resulted in only a small change in the structural parameters derived from the EXAFS data.

Cr K edge EXAFS were also collected for the 0.167 Cr/Pt/C Cr(Bz)$_2$ catalysts annealed at 200 °C, 750 °C, and 900 °C. The spectra and fitting parameters (not shown) were very similar to those of the corresponding 0.66 Cr/Pt/C Cr(Bz)$_2$ catalysts, with only slight variations in the coordination numbers which were within the errors associated with the fitting.

The fitting parameters of the 0.66 Cr/Pt/C Cr(Bz)$_2$ 750 °C and 900 °C catalysts and 0.66 Cr/Pt/C Cr(Cp)$_2$ 900 °C are in general agreement with each other and those obtained for the 20 wt% Pt$_3$Cr/C reference catalyst. The 0.66 monolayer Cr/Pt/C catalysts also have approximately a 3 : 1 Pt : Cr atomic ratio, thus if annealing results in the formation of the Pt$_3$Cr cubic phase, similar values would be predicted. Although similar fitting parameters are generated, this does not confirm that the catalysts possess the same physical structure. As EXAFS is an averaging technique similar materials can possess different structures but still give rise to the same average local structure.

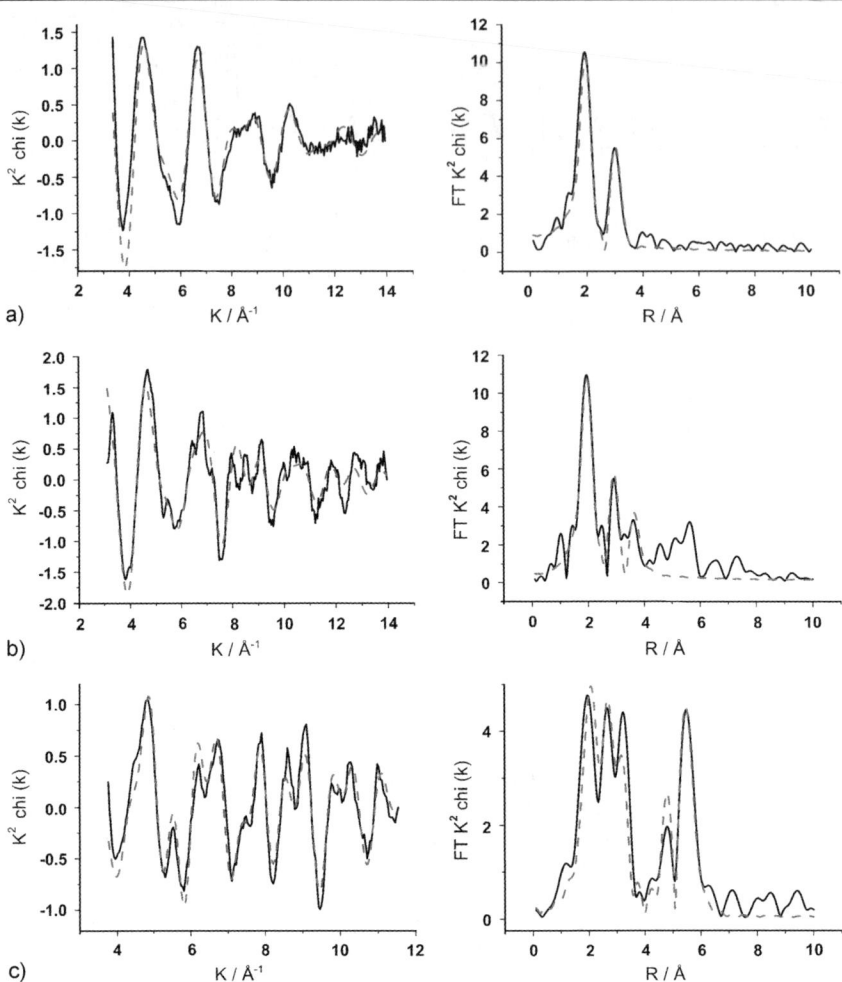

Fig. 5 (left) k^2 weighted experimental data and fit along with (right) the Fourier transform for 0.66 Cr/Pt/C Cr(Cp)$_2$ annealed at (a) 200 °C, (b) 750 °C , and (c) 900 °C. Data (black line) and fit (dashed line).

Additional clues regarding the structure of the nanoparticles may be inferred from the presence of Cr–O neighbours. For example, the fit obtained for the 900 °C 0.66 Cr/Pt/C Cr(Bz)$_2$ catalyst had 1.5 ± 0.1 Cr–O and 3.6 ± 0.3 Cr–Pt neighbours in the first coordination shell. The relatively large first shell Cr–O coordination number suggests that either there is a disproportionately large amount of Cr at the surface of the catalyst or that the sample contains separate Cr oxides. Close examination of a TEM EDX line profile showed that Cr was only present in regions of the catalyst where Pt was present, *i.e.* that Cr was not deposited on the carbon. Thus, we interpret the EXAFS as indicating that there is some segregation of Cr towards the surface of the particle.

EXAFS were also collected at the Pt L$_3$ edge for the 0.66 Cr/Pt/C Cr(Cp)$_2$ and Cr(Bz)$_2$ catalysts annealed at 750 °C and 900 °C, a 0.167 Cr/Pt/C Cr(Bz)$_2$ catalyst annealed at 900 °C, the 20 wt% Pt/C substrate catalyst and the 20 wt% and 40 wt% Pt$_3$Cr/C reference catalysts. (The spectra and fits are not shown for reasons of brevity.) The structural models indicated by the Pt L$_3$ EXAFS agree well with those derived from the Cr K edge EXAFS for the Cr containing catalysts. In particular, no

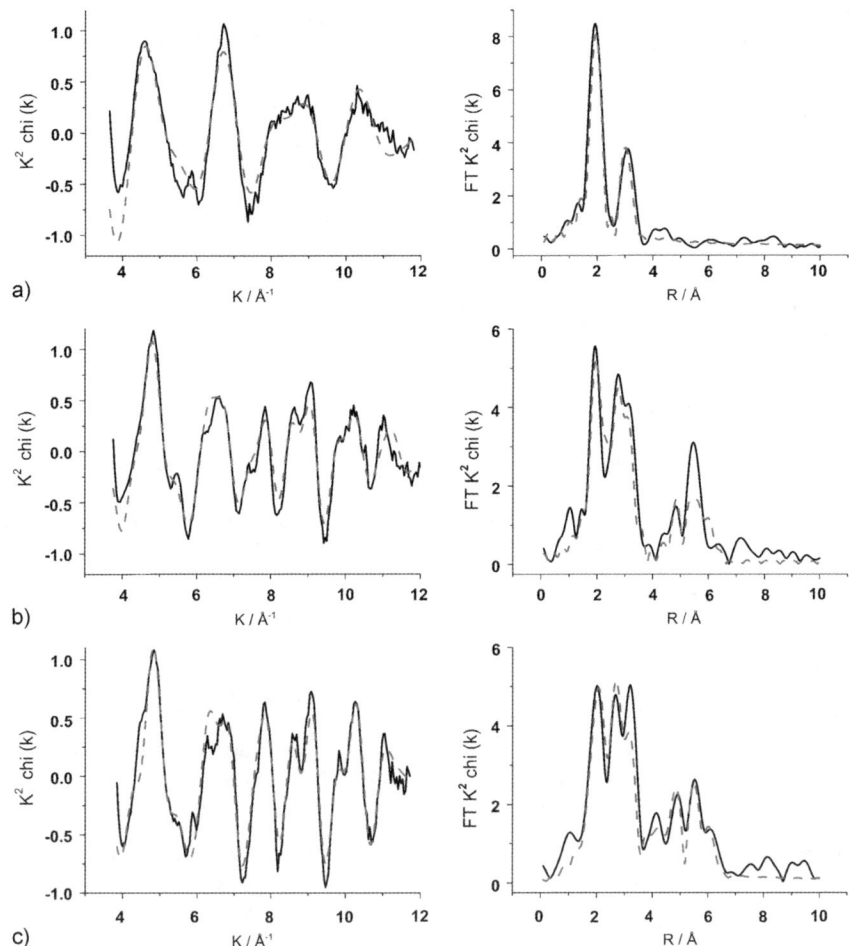

Fig. 6 (left) k^2 weighted experimental data and fit along with (right) the Fourier transform for 0.66 Cr/Pt/C Cr(Bz)$_2$ annealed at (a) 200 °C, (b) 750 °C , and (c) 900 °C. Data (black line) and fit (dashed line).

Pt–Cr shells were required in the fit of the EXAFS for the 0.66 Cr/Pt/C Cr(Cp)$_2$ catalyst annealed at 750 °C, but were required in the fits of this catalyst annealed at 900 °C (N(Cr–Pt) = 0.9 ± 0.2, R = 2.72 ± 0.02 Å, $2\sigma^2$ = 0.011 ± 0.002 Å^2) and the 0.66 Cr(Bz)$_2$ catalyst annealed at both temperatures, with similar fitting parameters.

As discussed in the introduction, one of the proposed explanations for the enhanced ORR activity of alloy catalysts is a contraction in the Pt–Pt distance. The first shell Pt–Pt distances obtained from analysis of the Pt L$_3$ EXAFS are summarised in Table 4.

The effect of Cr modification on the ORR activity of the Pt/C substrate catalyst may be assessed by examination of the Tafel plots presented in Fig. 7. With the exception of the Cr(Cp)$_2$ 750 °C catalyst, the modified catalysts all exhibit enhanced ORR activity (greater current at a given potential) than the parent Pt/C catalyst, with the greatest activity observed for the Cr(Cp)$_2$ 900 °C catalyst. As discussed above, the Cr(Cp)$_2$ modified catalyst annealed at 750 °C does not correspond to the formation of a well mixed PtCr particle, whilst both the Cr(Bz)$_2$ and Cr(Cp)$_2$ modified catalysts annealed at 900 °C have structures, as evidenced by both the XRD and EXAFS data, that correspond well to the Pt$_3$Cr alloy phase. Thus, there is clear

Table 1 Structural parameters for Pt₃Cr/C catalysts corresponding to the fits shown in Fig. 5. EXAFS data. Un1 refers to unit 1 and designates mixed occupancy of the shell

Catalyst	Shell	N	$R/\text{Å}$	$2\sigma^2/\text{Å}^2$	E_f/eV	R_{exafs} (%)
	Fitting parameters					
40 wt % Pt₃Cr/C in air	Cr–Pt (un1)	6.9 ± 0.6	2.719 ± 0.007	0.011 ± 0.001	−2.5 ± 0.8	36.1
	Cr–Pt	9.2 ± 2.4	4.76 ± 0.02	0.010 ± 0.002		
	Cr–Pt–Cr (un1)	6.9 ± 0.6	5.48 ± 0.01	0.019 ± 0.002		
20 wt % Pt₃Cr/C in air	Cr–O	1.2 ± 0.1	1.99 ± 0.01	0.005 ± 0.002	−3.4 ± 0.7	30.8
	Cr–Pt (un1)	3.9 ± 0.2	2.720 ± 0.005	0.010 ± 0.006		
	Cr–Cr	1.9 ± 0.7	4.79 ± 0.03	0.012 ± 0.006		
	Cr–Pt–Pt (un1)	3.9 ± 0.2	5.50 ± 0.01	0.017 ± 0.002		

Table 2 Structural parameters for 0.66 Cr/Pt/C Cr(Cp)₂ corresponding to the fits shown in Fig. 6. Un1 refers to unit 1 and designates mixed occupancy of the shell

Annealing temp.	Shell	N	$R/\text{Å}$	$2\sigma^2/\text{Å}^2$	E_f/eV	R_{exafs} (%)
	0.66 Cr/Pt/C Cr(Cp)₂ Fitting parameters					
200 °C	Cr–O	3.3 ± 0.2	2.00 ± 0.01	0.007 ± 0.002	−5.1 ± 1.0	38.5
	Cr–Cr	2.0 ± 0.3	3.00 ± 0.01	0.012 ± 0.003		
750 °C	Cr–O	3.6 ± 0.2	2.00 ± 0.01	0.009 ± 0.002	−3.3 ± 0.9	42.5
	Cr–Cr	1.6 ± 0.3	2.93 ± 0.02	0.010 ± 0.003		
	Cr–Cr	1.5 ± 0.6	3.69 ± 0.03	0.004 ± 0.003		
900 °C	Cr–O	1.6 ± 0.2	2.02 ± 0.01	0.009 ± 0.003	−6.2 ± 1.0	34.4
	Cr–Pt (un1)	3.3 ± 0.3	2.714 ± 0.008	0.012 ± 0.001		
	Cr–Pt	5.4 ± 1.4	4.77 ± 0.02	0.010 ± 0.002		
	Cr–Pt–Cr (un1)	3.3 ± 0.3	5.49 ± 0.02	0.020 ± 0.003		

evidence for a correlation between catalytic activity and PtCr alloy formation, and in particular the formation of the Pt₃Cr phase.

The importance of the formation of the Pt₃Cr alloy phase in determining the ORR activity of the catalyst was further investigated by comparing the specific activities of the 0.66 and 0.167 Cr/Pt/C Cr(Bz)₂ 900 °C catalysts and those obtained for the Pt₃Cr reference catalysts. The Tafel plots are presented in Fig. 8.

Comparison of the specific activities of ORR catalysts using this RDE method are usually made by comparing the values at 0.9 V,[1] as the mass transport is well defined at this potential and the reaction proceeds at the rate determined by the kinetics of the reaction, *i.e.* under kinetic control. The maximum ORR activity, 19.2 μA cm⁻² Pt, is observed for the 40 wt% Pt₃Cr/C catalyst, compared to 5.6 μA cm⁻² Pt for the 20 wt% Pt/C catalyst. This, approximately 3.5 fold, enhancement is in excellent agreement with previously published work[6] and is typical of catalysts consisting of Pt₃Cr alloy particles. Slightly smaller, but near identical, specific activities were obtained for the 20 wt% Pt₃Cr/C and 0.66 Cr/Pt/C Cr(Bz)₂ 900 °C catalysts, in support of the conclusion that the structure of the catalyst particles are very similar. The slight improvement obtained with the higher loaded Pt₃Cr/C catalyst is most likely to be related to the slightly larger particle size of this catalyst, which has previously been shown to result in enhanced ORR activity, in ref. 1, 27 and 28 and

Table 3 Structural parameters for 0.66 Cr/Pt/C Cr(Bz)$_2$ corresponding to the fits shown in Fig. 7. Un1 refers to unit 1 and designates mixed occupancy of the shell

Annealing temp.	Shell	N	$R/\text{Å}$	$2\sigma^2/\text{Å}^2$	E_f/eV	R_{exafs} (%)
	0.66 Cr/Pt/C Cr(Bz)$_2$ Fitting parameters					
200 °C	Cr–O	2.1 ± 0.2	1.99 ± 0.02	0.003 ± 0.001	−1.2 ± 2.0	36.7
	Cr–Cr	1.3 ± 0.3	2.97 ± 0.02	0.011 ± 0.004		
750 °C	Cr–O	1.7 ± 0.1	2.00 ± 0.01	0.008 ± 0.002	−4.8 ± 1.7	32.6
	Cr–Pt (un1)	3.1 ± 0.2	2.716 ± 0.006	0.012 ± 0.001		
	Cr–Pt	3.7 ± 1.0	4.77 ± 0.02	0.012 ± 0.003		
	Cr–Pt–Pt (un1)	3.1 ± 0.2	5.56 ± 0.02	0.016 ± 0.003		
900 °C	Cr–O	1.3 ± 0.2	2.02 ± 0.02	0.009 ± 0.003	−3.5 ± 1.5	38.5
	Cr–Pt (un1)	3.4 ± 0.3	2.718 ± 0.007	0.011 ± 0.001		
	Cr–Pt	1.4 ± 1.0	3.95 ± 0.02	0.014 ± 0.009		
	Cr–Pt	3.7 ± 1.1	4.77 ± 0.01	0.008 ± 0.003		
	Cr–Pt–Pt (un1)	3.4 ± 0.3	5.56 ± 0.01	0.013 ± 0.004		

Table 4 Pt–Pt first shell coordination distances obtained from fitting of the Pt L$_3$ EXAFS data

Catalyst	$R(\text{Pt–Pt})/\text{Å}$
20 wt% Pt/C	2.758 ± 0.004
20 wt% Pt$_3$Cr/C	2.733 ± 0.003
40 wt% Pt$_3$Cr/C	2.735 ± 0.004
0.66 Cr/Pt/C Cr(Cp)$_2$ 750 °C	2.757 ± 0.004
0.66 Cr/Pt/C Cr(Cp)$_2$ 900 °C	2.749 ± 0.004
0.66 Cr/Pt/C Cr(Bz)$_2$ 750 °C	2.748 ± 0.004
0.66 Cr/Pt/C Cr(Bz)$_2$ 900 °C	2.740 ± 0.004

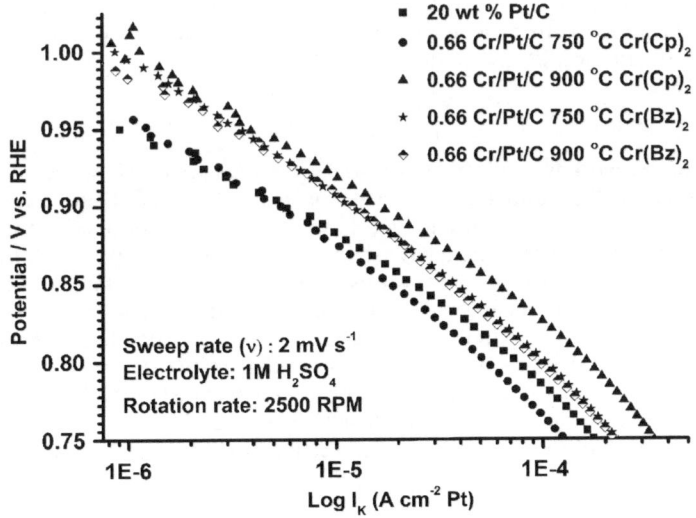

Fig. 7 Specific activities for the ORR at room temperature at the 20 wt% Pt/C and 0.66 Cr/Pt/C modified catalysts supported on glassy-carbon rotating disc electrodes. Data were obtained in oxygen saturated 1 mol dm^{-3} H$_2$SO$_4$ and are shown from the forward (negative going) sweep which was from 1.0 V to 0.05 V *vs.* RHE at 2 mV s^{-1} with a rotation rate of 2500 rpm.

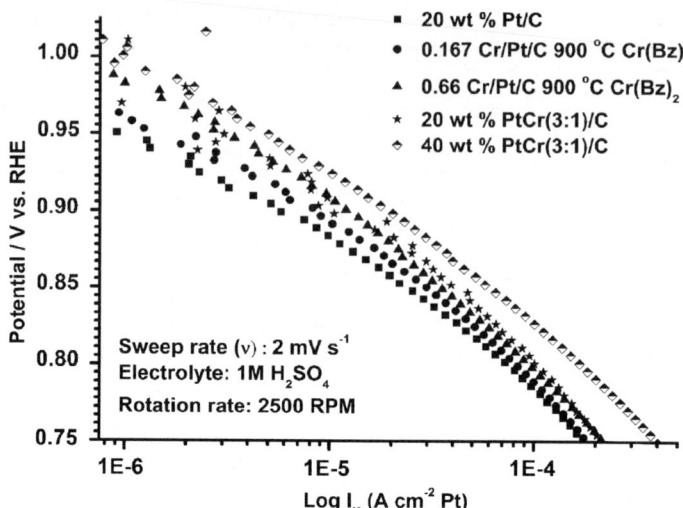

Fig. 8 Specific activities for the ORR at room temperature at the 20 wt% Pt/C 0.167 and 0.66 Cr/Pt/C Cr(Bz)$_2$, and Pt$_3$Cr/C catalysts supported on glassy-carbon rotating disc electrodes. Data were obtained in oxygen saturated 1 mol dm^{-3} H$_2$SO$_4$ and are shown from the forward (negative going) sweep which was from 1.0 V to 0.05 V *vs.* RHE at 2 mV s^{-1} with a rotation rate of 2500 rpm.

references therein, as no discernable difference in the structure of particles, *i.e.* Pt–Pt distance or the identity of the alloy phase, was found.

The effect of the Pt : Cr ratio can be seen by comparing the specific activities of the 0.167 and 0.66 Cr/Pt/C Cr(Bz)$_2$ 900 °C catalysts, which are 8.1 μA cm^{-2} Pt and 13.2 μA cm^{-2} Pt, respectively. The Cr atoms were shown to be in very similar coordination environments in the nanoparticles that make up these two catalysts, as indicated by the Cr K edge EXAFS data. However, no Pt–Cr shells were required in fitting the EXAFS of the 0.167 Cr/Pt/C catalyst, indicating that any alloy phase formation was restricted to the surface of the catalyst particle. This result supports the conclusion that the maximum activity is associated with the formation of nanoparticles that consist of the Pt$_3$Cr alloy phase throughout their structure.

4. Conclusions

The aim of this work was to prepare well-defined model systems in which the surface composition of the nanoalloy particles was controlled and then to relate the structural properties of the resulting catalysts to their performance towards the ORR. The results reported herein have shown that the Pt$_3$Cr alloy phase is needed to provide the optimal enhancement towards the reduction of oxygen. If the formation of the alloy was restricted to the surface of the catalyst nanoparticles, only a small enhancement towards the ORR was observed. When the Pt$_3$M phase is present, the specific activity of the catalyst is 2 to 3.5 times that of the standard 20 wt% Pt/C catalyst. The formation of the Pt$_3$Cr alloy phase is associated with a reduction of Pt–Pt bond distance, from 2.76 Å for the 20 wt% Pt/C to 2.73 Å for the 40 wt% Pt$_3$Cr/C, and it is this reduction in the Pt–Pt distance that gives the best indication of ORR activity.

Acknowledgements

This work was supported by the EPSRC through the provision of an Industrial CASE studentship for P. P. W., the Open University through the provision of a studentship for Y. Q., and Johnson Matthey. The assistance of Fred Mosselmans, Bob Bilsborrow, Steven Fiddy, and Chris Corrigan at the SRS is gratefully acknowledged.

References

1 H. A. Gasteiger, S. S. Kocha, B. Sompalli and F. T. Wagner, *Appl. Catal., B*, 2005, **56**, 9.
2 H. A. Gasteiger, W. Gu, R. Makharia, M. F. Mathias and B. Sompalli, in *Handbook of Fuel Cells—Fundamentals Technology and Applications*, ed. W. Vielstich, A. Lamm and, H. A. Gasteiger, Wiley, Chichester, UK, 2003, vol. 3, ch. 46, p. 593.
3 V. Jalan and E. J. Taylor, *J. Electrochem. Soc.*, 1983, **130**, 2299.
4 J. T. Glass, G. L. Cahen, G. E. Stoner and E. J. Taylor, *J. Electrochem. Soc.*, 1987, **134**, 58.
5 M. T. Paffett, J. G. Beery and S. Gottesfeld, *J. Electrochem. Soc.*, 1988, **135**, 1431.
6 S. Mukerjee and S. Srinivasan, *J. Electroanal. Chem.*, 1993, **357**, 201.
7 S. Mukerjee, S. Srinivasan, M. P. Soriaga and J. McBreen, *J. Electrochem. Soc.*, 1995, **142**, 1409.
8 A. N. Mansour, J. W. Cook and D. E. Sayers, *J. Phys. Chem.*, 1984, **88**, 2330.
9 V. R. Stamenkovic, B. S. Moon, K. J. J. Mayrhofer, P. N. Ross and N. M. Markovic, *J. Am. Chem. Soc.*, 2006, **128**, 8813.
10 V. R. Stamenkovic, B. S. Mun, K. J. J. Mayrhofer, P. N. Ross, N. M. Markovic, J. Rossmeisl, J. Greeley and J. K. Norskov, *Angew. Chem., Int. Ed.*, 2006, **45**, 2897.
11 V. R. Stamenkovic, B. S. Mun, K. J. J. Mayrhofer, P. N. Ross and N. M. Markovic, *J. Am. Chem. Soc.*, 2006, **128**, 8813.
12 V. R. Stamenkovic, B. Fowler, B. S. Mun, G. F. Wang, P. N. Ross, C. A. Lucas and N. M. Markovic, *Science*, 2007, **315**, 493.
13 T. Toda, H. Igarashi, H. Uchida and M. Watanabe, *J. Electrochem. Soc.*, 1999, **146**, 3750.
14 T. Toda, H. Igarashi and M. Watanabe, *J. Electroanal. Chem.*, 1999, **460**, 258.
15 E. M. Crabb, M. K. Ravikumar, D. Thompsett, M. Hurford, A. Rose and A. E. Russell, *Phys. Chem. Chem. Phys.*, 2004, **6**, 1792.
16 E. M. Crabb, M. K. Ravikumar, Y. Qian, A. E. Russell, S. Maniguet, J. Yao, D. Thompsett, M. Hurford and S. C. Ball, *Electrochem. Solid-State Lett.*, 2002, **5**, A5.
17 E. M. Crabb and M. K. Ravikumar, *Electrochim. Acta*, 2001, **46**, 1033.
18 E. M. Crabb, R. Marshall and D. Thompsett, *J. Electrochem. Soc.*, 2000, **147**, 4440.
19 E. M. Crabb and R. Marshall, *Appl. Catal., A*, 2001, **217**, 41.
20 N. Binstead, EXCURV98, CLRC, Daresbury Laboratory, Computer Program.
21 S. J. Gurman, N. Binstead and I. Ross, *J. Phys. C: Solid State Phys.*, 1984, **17**, 143.
22 J. J. Rehr and R. C. Albers, *Phys. Rev. B*, 1990, **41**, 8139.
23 J. J. Rehr, R. C. Albers and S. I. Zabinsky, *Phys. Rev. Lett.*, 1992, **69**, 3397.
24 T. J. Schmidt, H. A. Gasteiger, G. D. Stab, P. M. Urban, D. M. Kolb and R. J. Behm, *J. Electrochem. Soc.*, 1998, **145**, 2354.
25 A. Panterlouris, H. Modrow and M. Pantelouris, *Chem. Phys.*, 2004, **300**, 13.
26 A. Moen and D. G. Nicholson, *Chem. Mater.*, 1997, **9**, 1241.
27 K. Kinoshita, *Electrochemical Oxygen Technology*, Wiley, New York, 1992.
28 D. Thompsett, in *Handbook of Fuel Cells—Fundamentals Technology and Applications*, ed. W. Vielstich, A. Lamm and H. A. Gasteiger, Wiley, Chichester, UK, 2003, vol. 3, ch. 37, p. 467.

Structure–performance relationships of Rh and RhPd alloy supported catalysts using combined EDE/DRIFTS/MS

Andy J. Dent,[b] John Evans,[ab] Steven G. Fiddy,[c] Bhrat Jyoti,[a] Mark A. Newton[d] and Moniek Tromp*[a]

Received 25th April 2007, Accepted 29th June 2007
First published as an Advance Article on the web 14th September 2007
DOI: 10.1039/b706294j

Energy dispersive extended X-ray absorption fine structure spectroscopy (ED-XAFS), diffuse reflectance infrared Fourier transform spectroscopy (DRIFTS) and mass spectrometry (MS), have been combined for the structure–function study of Rh and RhPd supported catalysts for the reduction of NO by CO. The combined results show that although alloying of Rh with Pd prevents the dissociative oxidation of the Rh by NO, it does not prevent the extensive disruptive oxidation of Rh by CO. The influence of oxidative disruption by molecular CO in such systems may therefore be far more pervasive and catalytically important than has been previously observed. The overall metal particle size observed in the RhPd alloy system during the CO/NO reaction is significantly larger than for the Rh-only system for the entire temperature range employed. The catalytically active sites, however, are likely to be similar, with the overall activity of the alloy system to be reduced due to inactive RhPd alloy nanoparticles.

Introduction

Palladium, rhodium and platinum are core components of modern three-way catalysts for automobile emission control. The interaction of any of these components, or the combination of components, with reactive gases is poorly understood. Energy dispersive extended X-ray absorption fine structure spectroscopy (ED-XAFS), diffuse reflectance infrared Fourier transform spectroscopy (DRIFTS) and mass spectrometry (MS), have been combined for the structure–function study of Rh supported catalysts.[1,2] The combined set-up provides complementary structural and electronic information on the catalytic system, while determining its performance simultaneously. With reactions involving CO and NO, primary reactants in automobile exhausts, the active adsorbates act as probe molecule for surface species.

For the single metal Rh supported system we have recently shown that the nature of the rhodium centres is highly dependent on the reactant gases (CO, NO, H_2 and O_2) and their pressure, as well as the temperature of the reaction.[1,3–6] The metallic rhodium particles rapidly (dissociatively) oxidise under oxidising conditions and

[a] University of Southampton, School of Chemistry, Southampton, UK SO17 1BJ. E-mail: M.Tromp@soton.ac.uk; Fax: +44 (0)2380 593781; Tel: +44 (0)2380 594165
[b] Diamond Light Source, Chilton, Didcot, UK OX11 0QX
[c] CCLRC Daresbury Laboratory, Warrington, UK WA4 4AD
[d] European Synchrotron Radiation Facility, 38042, Grenoble, France

recluster under reducing conditions. At higher temperatures, Rh metal aggregation becomes more favoured while increasing the catalytic activity in the CO/NO redox reaction.[5] The oxidative disruption of Rh particles by CO has been known for almost fifty years.[7] Though this process may be rapid,[8] evidence to date suggests the capacity of molecular CO to induce corrosion of Rh nanoparticles is limited to highly dispersed particles, *i.e.* particles displaying a Rh–Rh coordination number below 5 in EXAFS ($\sim$13 Rh atoms per particle).[8,9] We have recently shown that NO can be more destructive toward supported Rh particles than CO, completely oxidising much larger Rh particles (Rh–Rh coordination number 7–8, *i.e.* atomicity $\sim$35–75 Rh atoms per particle) in a few seconds in a plug flow micro reactor.[10,11] The rapid oxidation of Rh nanoparticles by easily dissociable oxidants, *e.g.* NO, may explain the observed variations in behaviour. Such facile oxidation of Rh can be effectively curtailed through alloying with a small amount of Pd (1 wt% Pd with 4 wt% Rh).[12] Palladium is used in combination with rhodium especially for the effective oxidation of CO and hydrocarbons. The bimetallic RhPd system exhibits a better durability under high temperature oxidising conditions and a lower light-off temperature during NO reduction by H_2, however, with a considerably lower selectivity to N_2O than pure Rh and similar to Pd/Al_2O_3.[12] The overall alloy particle size increases in comparison to the single metallic ones. An increase in surface enrichment in Pd is proposed to prevent the oxidation of Rh.[12]

Here, the structure–performance relationship between a single metal Rh supported systems is compared to the RhPd alloy system for the temperature controlled CO/NO reaction. Energy dispersive extended X-ray absorption fine structure spectroscopy (ED-XAFS), diffuse reflectance infrared Fourier transform spectroscopy (DRIFTS) and mass spectrometry (MS), have been combined for the structure-function study of Rh and RhPd supported catalysts.

Experimental

Rh catalysts were made *via* aqueous impregnation of $RhCl_3 \cdot 3H_2O$ to γ-Al_2O_3 (Degussa, Aluminium Oxide C). RhPd catalysts were synthesised *via* co-impregnation using $RhCl_3 \cdot 3H_2O$ and $PdCl_2$ in acidified (HCl) water (pH 1.3). Samples were dried overnight before calcination for 6 h at 673 K and sieved to yield a particle diameter range of 90–120 μm. Finally the samples were reduced for 5 h at 573 K under flowing H_2. Samples were loaded in a previously described reaction cell,[1] allowing synchronous EDE and DRIFTS experiments, while monitoring reactants and products *via* mass spectrometry. Once loaded, the samples were purged with He, and re-reduced in H_2 at 300 K. The catalysts are subsequently cooled down to reaction temperature in H_2 after which the system is purged with He. A reactive 2.5% CO/2.5% NO/95% He feedstock (50 ml min^{-1}) was then introduced synchronously with the start of the DRIFTS/EDE/MS experiment. Experiments were performed as a function of temperature, 10 K min^{-1} up to 573 K.

EDE measurements were carried out at ID24 at the ESRF, using an asymmetrically cut Si[111] monochromator in the Laue configuration,[13] allowing simultaneous sampling of Rh and Pd K edges.[12] EXAFS detection was performed *via* a phosphor masked, Peltier-cooled, Princeton charge-couple-device (CCD) camera. DRIFTS measurements were performed using a Digilab FTIR7000 spectrometer with a high sensitivity, linearised MCT detector (1200–2300 cm^{-1}; 4 cm^{-1} resolution). A Pfeiffer Omnistar quadrupole mass spectrometer was employed. DRIFTS and EDE spectra were collected at a rate of 0.167 Hz corresponding to *ca.* 1 spectrum K^{-1}. Energy calibration was achieved using Rh and Pd foils. Background subtraction used the program PAXAS, with EXAFS analysis using EXCURV98.

XPS measurements were performed at the National Centre for Electron Spectroscopy and Surface analysis (NCESS), at Daresbury Laboratory, United Kingdom. 'Fresh' samples, *i.e.* reduced in H_2 but exposed to air after, were analysed using monochromated Al Kα X-ray radiation and sample charge compensation.

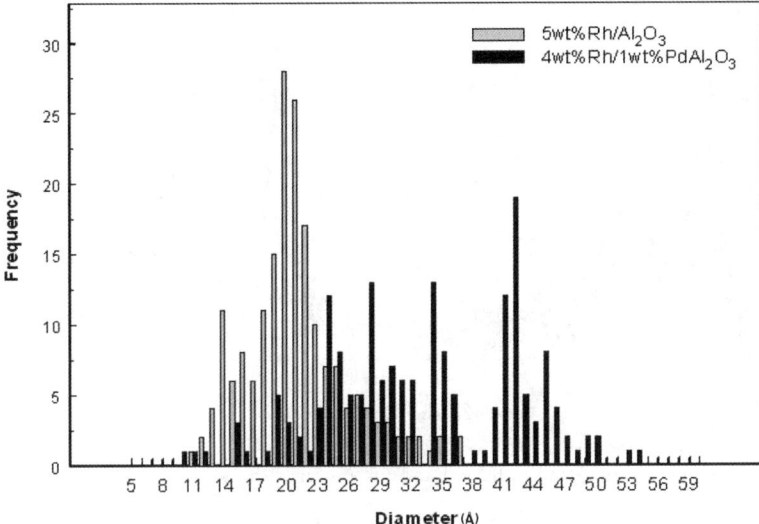

Fig. 1 Particle size distribution for a 5 wt% Rh/γ-Al$_2$O$_3$ and a 4 wt% Rh/1 wt% Pd/γ-Al$_2$O$_3$ sample.

A 150 mm radius Scienta hemispherical analyser operating with pass energy of 150 eV and a 0.8 mm entrance slit yielded a nominal resolution of 0.35 eV. Powdered samples were mounted with double sided carbon onto a stub and measured at a base pressure of *ca.* 5×10^{-9} mbar. TEM measurements were recorded in brightfield mode and were carried out on a JEOL FX 2000 microscope.

Results

Fig. 1 displays the particle size distribution derived from TEM for a 5 wt% Rh/ γ-Al$_2$O$_3$ and a 4 wt% Rh/1 wt% Pd/γ-Al$_2$O$_3$ sample. The corresponding statistical data is given in Table 1.

A Rh particle size distribution between 11 and 32 Å is observed for the 5 wt% Rh/ γ-Al$_2$O$_3$ sample, with an average particle size of 21.5 Å.[4,14] The Rh-only system has recently been shown to be readily oxidised in air (second time scale).[4] The supported particles as observed with TEM are thus oxidic rather than metallic. The particle size distribution derived from TEM as displayed in Fig. 1 and Table 1 are therefore misleading, since it represents oxidic Rh particles on support with the equivalent Rh metallic particles to be significantly smaller, *i.e.* an average particle size of $\sim$11 Å can be estimated assuming Rh$_2$O$_3$ particles on support in TEM. The 4 wt% Rh/1 wt% Pd system exhibits a larger overall particle size distribution, with an average particle size of 32.5 Å and standard deviation of 16.3, three times larger than that of

Table 1 Particle size distributions for 5 wt% Rh/γ-Al$_2$O$_3$ and 4 wt% Rh/1 wt% Pd/γ-Al$_2$O$_3$ as determined with TEM

Sample	5 wt% Rh/γ-Al$_2$O$_3$	4 wt% Rh/1 wt% Pd/γ-Al$_2$O$_3$
Points	189	183
Average particle size diameter/Å	21.5	32.5
Standard deviation	5.1	16.3
Average number of atoms in particle	225.4	989
Average number of surface atoms	117	416
Dispersion	0.52	0.43

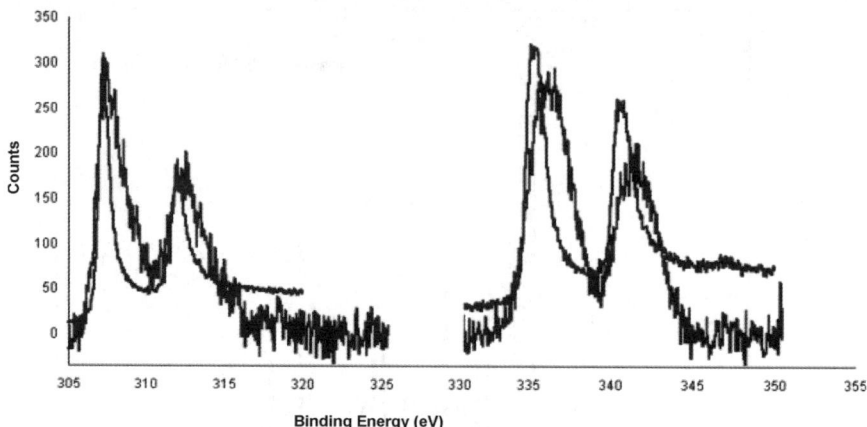

Fig. 2 Rh (a) and Pd (b) 3d XPS spectra obtained for the fresh 4 wt% Rh/1 wt% Pd/γ-Al$_2$O$_3$ sample (grey lines), compared to the Rh and Pd foil (black lines).

the Rh only sample. Looking at the distributions in more detail, the mixed Rh/Pd sample displays a bimodal distribution, with the predominant particle size regime ranges *ca.* 23–30 Å and 40–47 Å. Several reasons for the bimodal distribution can be given; (i) phase separation, *i.e.* the metal particles form in separated clusters with smaller Rh metal particles and larger Pd particles, (ii) alloying of the two metal components has occurred, where 'Rh-like' nucleation sites yield smaller particles, and 'Pd-like' sites yield the larger particles observed.

In Fig. 2 the Rh and Pd 3d$_{5/2}$ and 3d$_{3/2}$ XPS bands of the 4 wt% Rh/1 wt% Pd/γ-Al$_2$O$_3$ are given, in comparison to the respective Rh and Pd foil. Table 2 summarises the relevant XPS data for the alloy system and compares it to the analogous single metal systems. Whereas the Rh only supported samples have shown a significant upward energy shift of the Rh 3d$_{5/2}$ and 3d$_{3/2}$ XPS bands relative to the Rh(0) foil,[14] with the band positions more resembling Rh$_2$O$_3$, for the mixed Rh/Pd metal system as shown here no significant shift in the Rh XPS bands is observed. The Rh binding energy (BE) of *ca.* 307 eV correlates closely to that of the BE of bulk metallic Rh, indicating that introduction of Pd to the catalyst make-up has effectively insulated the Rh component against any facile oxidation as seen with the elemental 5 wt% Rh/γ-Al$_2$O$_3$ analogue system. At the same time, the Pd 3d$_{5/2}$ and 3d$_{3/2}$ XPS bands show an upward shift in BE compared to the Pd bulk reference. This indicates that there is a significant charge transfer from Pd, the Pd component being oxidised under the UHV conditions of TEM and more resembling PdO (Table 2).[15]

In addition, the Rh/Pd ratio as observed with XPS is plotted as a function of Rh metal loading for a series of mixed metal systems (with a total metal (Rh and Pd) loading of 5 wt%), alongside with the theoretically expected ratios (Fig. 3). Analysis of this Fig. 3 shows that the catalyst systems do not appear to be homogeneous from an XPS point of view, with lower amounts of Rh and/or higher amounts of Pd

Table 2 XPS Rh 3d$_{5/2}$ and Pd$_{5/2}$ binding energies

Sample	Rh 3d$_{5/2}$ BE/eV	Pd 3d$_{5/2}$ BE/eV	Rh(3d$_{5/2}$)/ Pd(3d$_{5/2}$) exp.	Rh(3d$_{5/2}$)/ Pd(3d$_{5/2}$) theory
Rh foil	307.2			
Rh$_2$O$_3$[14]	308.5			
5 wt% Rh/γ-Al$_2$O$_3$	308.9			
4 wt% Rh/1 wt% Pd/γ-Al$_2$O$_3$	307.4	336.2	0.975	4.0
Pd foil		335.2		
PdO[15]		335.7–337.0		

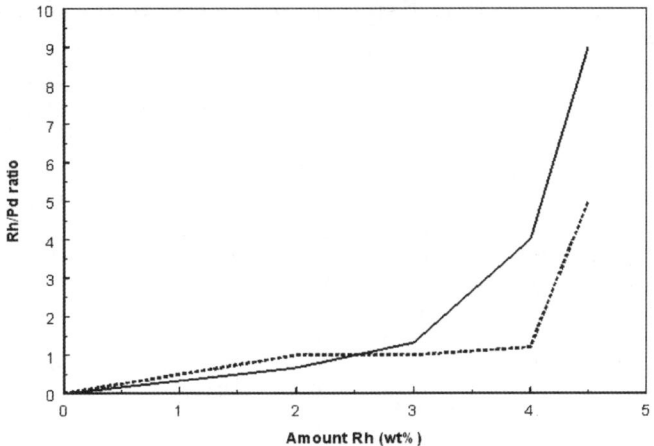

Fig. 3 Rh/Pd ratio as a function of Rh wt%, for a total metal loading of 5 wt%; theoretical (black solid line) and experimentally as determined with XPS (black dotted line).

observed for all samples from *ca.* 2.5 wt% Rh metal loading. For the 4 wt% Rh/1 wt% Pd system almost 4 times as much Pd is observed than theoretically expected. These XPS results suggests that the Pd is present on the surface of alloyed Rh/Pd metal particles, surface segregation, and as such protecting the Rh from the extensive oxidation as observed for the Rh only system.

The samples, *i.e.* 5 wt% Rh/γ-Al$_2$O$_3$ and 4 wt% Rh/1 wt% Pd/γ-Al$_2$O$_3$, have been characterised with Rh K-edge XAFS. The XANES and EXAFS (k^3-weighted) results of the 'fresh' samples, *i.e.* pre-reduced in H$_2$, stored in air and measured under He at room temperature, as well as the *in situ* (re)reduced samples are presented in Fig. 4, with the EXAFS analysis results given in Table 3.

Inspection of the XANES regions (Fig. 4a) shows that incorporation of 1 wt% Pd results in a metallic Rh signature, in contrast to the oxidic features observed for the 5 wt% Rh system. In addition, a significant difference in the EXAFS intensity (Fig. 4b) between the two systems is observed. Structural analysis of the EXAFS data (Table 3) reveals an extensive oxidation of the Rh metal for the 5 wt% Rh/γ-Al$_2$O$_3$ sample in its 'fresh' state, with a 1st shell RhO coordination number of 3, while large metallic particles with a RhM (M = Rh or Pd) coordination number of *ca.* 8 for the 4 wt% Rh/1 wt% Pd/γ-Al$_2$O$_3$ sample are observed. After *in situ* reduction, both systems are fully reduced with coordination numbers of 7.0 and 9.7 for 5 wt% Rh/γ-Al$_2$O$_3$ and 4 wt% Rh/1 wt% Pd/γ-Al$_2$O$_3$, respectively.

The CO/NO reaction is monitored with a combination of EDE, DRIFTS and MS. Fig. 5 shows the *in situ* Rh K edge EDE data taken for 5 wt% Rh/γ-Al$_2$O$_3$ and 4

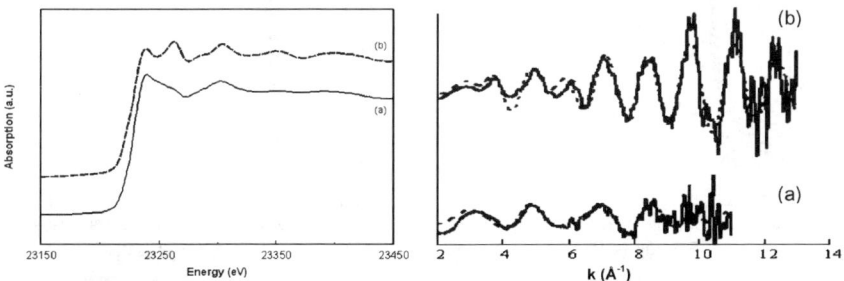

Fig. 4 Rh K edge XANES and k^3-weighted EXAFS data experiment (solid line) and fit (dotted line) for (a) a 5 wt% Rh/γ-Al$_2$O$_3$ and (b) 4 wt% Rh/1 wt% Pd/γ-Al$_2$O$_3$ sample.

Table 3 Rh K-edge EXAFS results for fresh and *in situ* reduced samples, measured in He at room temperature

Sample	Scatterer(s)	CN	$R/\text{Å}$	$2\sigma^2/\text{Å}^{-2}$	Ef/eV	R value (%)	k-range/ Å^{-1}
5 wt% RhCl/γ-Al₂O₃-*fresh*	Rh	2.2 (±0.2)	2.66 (±0.03)	0.012 (±0.002)	−2.4	57	2–11
	O	3.0 (±0.3)	1.97 (±0.01)	0.005 (±0.001)	−2.4		
5 wt% RhCl/γ-Al₂O₃-*reduced*	Rh	6.8 (±0.6)	2.67 (±0.02)	0.011 (±0.001)	−5.2	34	2–12.5
4 wt% Rh/1 wt% Pd/γ-Al₂O₃-*fresh*	Rh/Pd	7.7 (±0.6)	2.67 (±0.02)	0.012 (±0.002)	2.2	56	2–13
4 wt% Rh/1wt% Pd/γ-Al₂O₃-*reduced*	Rh/Pd	9.7 (±0.6)	2.68 (±0.02)	0.011 (±0.002)	2.4	50	2–13

wt% Rh/1 wt% Pd/γ-Al₂O₃, respectively, with increasing temperatures up to 573 K under the reactive mixture of 5% NO/5% CO/90% He. Prior to the catalytic testing, the samples have been reduced *in situ*, yielding fully reduced Rh for both systems (EXAFS results in Table 3).

The 5 wt% Rh/γ-Al₂O₃ system evidences an oxidic structure when exposed to the catalytic mixture at low temperatures. However, the 4 wt% Rh/1 wt% Pd/γ-Al₂O₃ alloy system shows an overall metallic Rh XANES signature under the same experimental conditions. Inspection of the XANES regions of the catalytic systems at higher temperatures reveals that there is no evidence for the formation of oxidic Rh for either system. EXAFS demonstrates that upon introduction of the reactive feedstock, an immediate and significant decrease in particle size for both systems is observed (Fig. 6). The 5 wt% Rh system indexes a coordination value of ∼3 up to 350 K, which decreases steadily to a value of ∼2.5 at ∼430 K. Above this temperature the value steadily returns to 3, and it is only above 550 K that the CN is seen to increase to a value of *ca.* ∼5.5.[1] The Rh coordination numbers for the alloy system are higher relative to the Rh-only system and remain constant, within error, between a value of 6 and 6.5 throughout the temperature range employed.

The Pd K edge EDE data, as simultaneously obtained with the Rh K edge EDE data, are presented in Fig. 7. Upon reduction in H₂ the Pd is fully reduced (starting from the oxidic Pd, Fig. 1 and Table 2, for the fresh sample in TEM). Although the

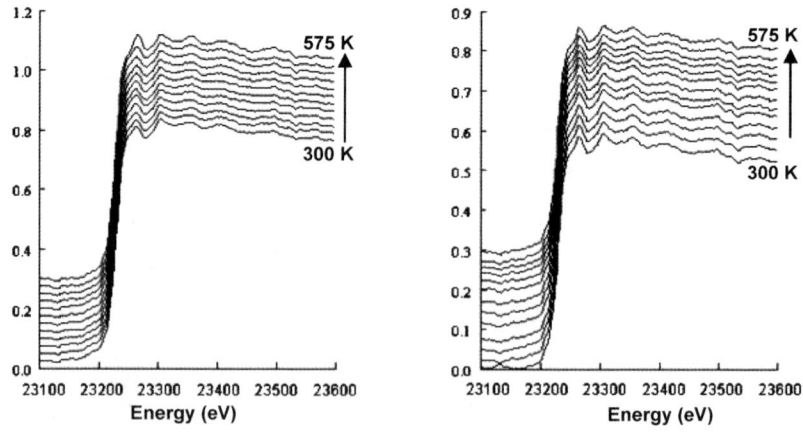

Fig. 5 Normalised Rh K EDE XANES spectra during NO reduction by CO for (a) a 5 wt% Rh/γ-Al₂O₃ and (b) 4 wt% Rh/1 wt% Pd//γ-Al₂O₃ sample as a function of temperature. One spectrum every 25 K is shown.

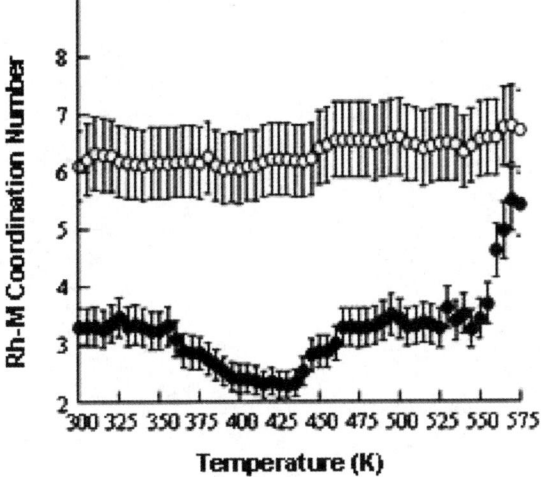

Fig. 6 Rh–M (M = Rh/Pd) coordination numbers during NO reduction by CO over 5 wt% Rh/γ-Al₂O₃ (filled circles) and 4 wt% Rh/1 wt% Pd/γ-Al₂O₃ (open circles).

data is not of sufficient quality to allow EXAFS analyses, the Pd K edge XANES and initial EXAFS features are under reaction conditions existing in a overall metallic environment, rather than an oxide or surrounded by other small scatterers such as carbon or nitrogen. This metallic signature is observed throughout the temperature dependent experiment. These results are consistent with previous studies, where similar temperature dependent experiments using the NO reduction by H_2 reaction observed a metallic Pd structure existing throughout the temperature regime.[12]

Fig. 8 shows the synchronously obtained DRIFTS data for the 5 wt% Rh/γ-Al₂O₃ and 4 wt% Rh/1 wt% Pd/γ-Al₂O₃ under investigation. It is clear that similar IR active species are formed on both of the systems, despite the different EDE spectra displayed by both. However, clear differences in overall concentrations of species are apparent, with significantly less species formed on the alloy system. The changes in DRIFTS intensities for selected vibrations are shown in Fig. 9: note that the DRIFTS intensities for the 5 wt% Rh/γ-Al₂O₃ are about five times larger than that for the 4 wt% Rh/1 wt% Pd/γ-Al₂O₃ system.

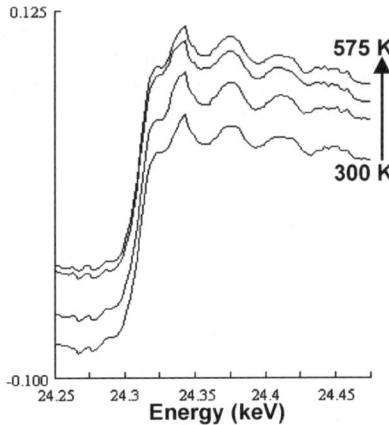

Fig. 7 Pd K EDE XANES spectra during NO reduction by CO for a 4 wt% Rh/1 wt% Pd/γ-Al₂O₃ sample as a function of temperature. Spectra taken at 300, 375, 475 and 575 K are given.

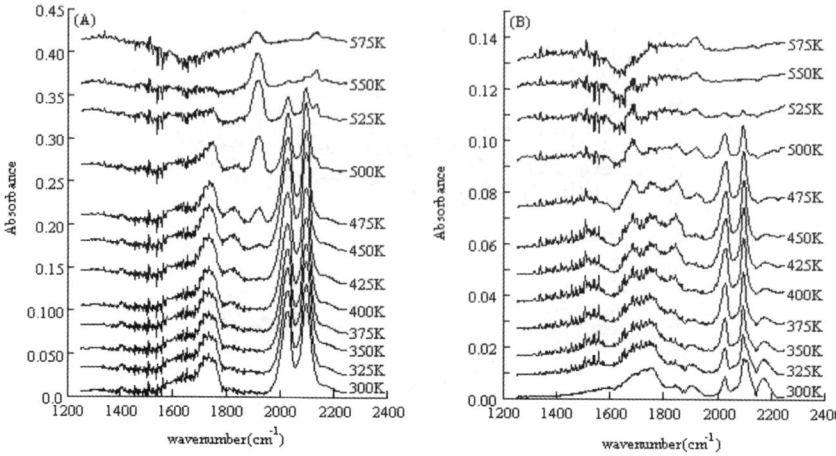

Fig. 8 DRIFTS spectra (a) 5 wt% Rh/γ-Al$_2$O$_3$ and (b) 4 wt% Rh/1 wt% Pd/γ-Al$_2$O$_3$ as a function of temperature. One spectrum every 25 K is shown, starting from 300 K (bottom spectrum).

Most of the IR absorptions observed are associated with Rh present in an oxidation state higher than 0, as opposed to be being dominated by stretches of CO or NO absorbed upon metallic Rh: The mononuclear Rh^I(CO)$_2$ with ν_{sym} ~ 2100 cm^{-1} and ν_{asym} ~ 2025 cm^{-1},[16] dominates the spectrum below ~ 525 K. The difference in intensities of the two peaks associated with the Rh(CO)$_2$ species at lower temperatures, with an additional peak present at ~ 2175 cm^{-1}, suggest that another species apart from the *geminal* dicarbonyl entity is present on the alloy system. This is most likely due to CO or NO bound to Pd affecting the overall DRIFTS spectra. Literature has several alternative assignments for this additional broad peak. CO bound to Pd, more specifically a Pd^{2+}, is suggested.[17,18] However, based on the metallic Pd signature observed in Pd K edge XANES this seems unlikely. Rh(NCO) and Pd(NCO) species have also been assigned to 2175 cm^{-1},[19] and cannot be excluded under these reaction conditions. Freund *et al.*[20] put forward an alternative explanation, being adsorption of CO at low coordinated Pd edge and corner sites. Since the peak is mainly observed for the alloy systems, the last explanation seems the most plausible.

The *geminal* dicarbonyl peaks are rapidly formed from 300 K and remain present up to about ~ 450 K, after which they gradually decrease and completely disappear at ~ 550 K for both systems. Note that the low temperature behaviour of this Rh(CO)$_2$ asymmetric vibration, as observed in the previous study, ref. 1, resulting in a maximum around 450 K, was not reproduced, while the high temperature end is identical. The Pd–CO vibration disappears with increasing temperatures, *i.e.* at ~ 550 K the temperature is probably too high for CO to exist as a stable adsorbate on the supporting Pd metal particle.

The bands at 1840 and 1745 cm^{-1} can be assigned to the symmetric and asymmetric stretches of the Rh(NO)$_2$ species, respectively.[21] Additionally, the peak at 1745 cm^{-1} could originate from a monodisperse Rh(NO)$^-$ species.[22,23] The higher intensity of the 1750 cm^{-1} compared to the 1830 cm^{-1} peak with their different temperature behaviour (Fig. 9) suggests multiple species are indeed present on both systems. These NO related vibrations have a maximum ~ 470 K after which they decrease rapidly and are unobservable from ~ 525 K. The peak at ~ 1920 cm^{-1}, formed for temperatures of ~ 450 K and above for both systems, is due to the Rh(NO)$^+$ species[24] (determined to most likely be a highly transient form of

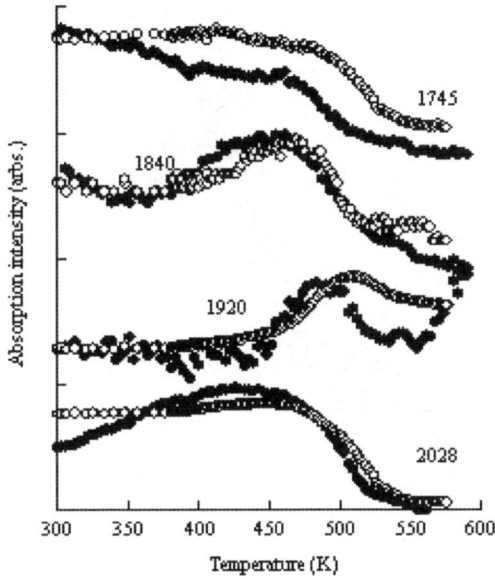

Fig. 9 Variations in DRIFTS intensities for vibrations as indicated for 5 wt% Rh/γ-Al$_2$O$_3$ (open circles) and 4 wt% Rh/1 wt% Pd/γ-Al$_2$O$_3$ (filled circles).

Rh(NO)$_2$). Moreover, the vibration has a maximum intensity at $\sim$500 K and is the only significant species remaining for temperatures above 525 K.

In addition to the peaks observed for both systems, the alloy system reveals a weak but separated peak at $\sim$1680 cm^{-1} in the entire temperature range, which is likely to be associated with NO bound to Rh metal.[25,26] This species that could very well be present on the Rh-only system as well, is now visible only as a shoulder on the high intensity symmetric Rh(NO)$_2$ vibration. Moreover, although not present in significant proportions, the presence of a Rh(CO)(NO) species forming at *ca.* 2100 and 1755 cm^{-1} (for ν(CO) and ν(NO), respectively) under the more intense peaks in these regions cannot be disregarded.[27] Literature suggests different Pd^{0}NO and Pd$^{\delta+}$ vibrations to be present in the wavenumber range 1570–1750 cm^{-1} and 1760–1815 cm^{-1},[18,28] and PdCO (Pd^0) and Pd$_2$CO type species around 2090 cm^{-1} and 1900–1980 cm^{-1}.[17,18] The presence of these species cannot be excluded based on the results shown, however, the vibrations observed in the alloy system now ascribed to Rh–CO and Rh–NO type species show similar temperature dependence suggesting these vibrations to be similar in both systems.

The mass spectrometry results obtained are represented in Fig. 10 by the percentage NO ($m/z = 30$) in the feedstock and the ratio of m/z 44/22, as a function of temperature. Ionisation of N$_2$O in the mass spectrometer leads to more fragmentation of that ion than for CO$_2$. As such, the probability of generating a double charged molecular ion (mass 22) is considerably reduced for N$_2$O compared to CO$_2$. The ratio of mass 44/22 therefore provides a means to measure the net CO$_2$/N$_2$O selectivity obtained during NO reduction; an increased 44/22 ratio means increased selectivity toward N$_2$O. It is clear from Fig. 8 that both systems display a very similar catalytic behaviour: the 5 wt% Rh/γ-Al$_2$O$_3$ has a slightly lower light off temperature, both $\sim$470 K, with a higher overall conversion from $\sim$500 K with a maximum of $\sim$50% *versus* 35% for the RhPd catalysts at 573 K. The N$_2$O selectivity for both catalysts is almost identical, with a maximum around 525 K, *i.e.* the temperature where significant catalysis starts to take place. The selectivity to N$_2$O decreases, and thus the selectivity to N$_2$ and CO$_2$ increases, with higher overall conversions of NO (at higher temperatures).

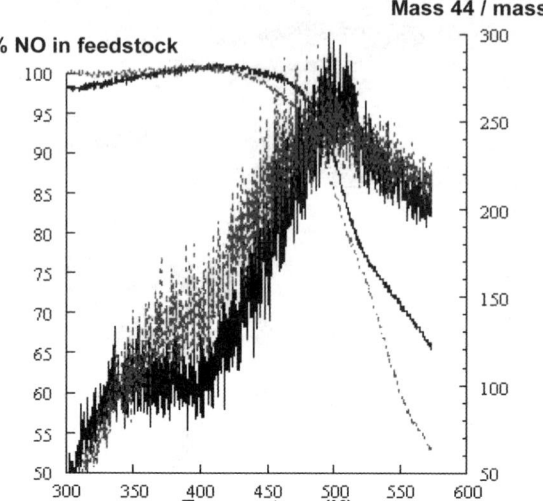

Fig. 10 NO percentage in the feedstock shown alongside the ratio of mass 44/mass 22 during NO reduction by CO as a function of temperature over 5 wt% Rh/γ-Al$_2$O$_3$ (grey-dotted) and 4 wt% Rh/1 wt% Pd/γ-Al$_2$O$_3$ (black).

Discussion

The mass spectrometry and DRIFTS results obtained for the 4 wt% Rh/1 wt% Pd/γ-Al$_2$O$_3$ and elemental 5 wt% Rh/γ-Al$_2$O$_3$ systems show the catalysts to display strikingly similar behaviour in terms of selectivity and activity during the reduction of NO by CO, with very similar types of species formed on the metal surface. The structural data extracted from synchronously obtained EDE however presents evidence that on a structural level, *e.g.* metal phase changes and particle morphology, the catalysts systems are very different.

The characterisation by TEM, XPS and XAFS shows that the fresh samples, *i.e.* reduced in H$_2$ and subsequently stored in air and measured under either vacuum or He at room temperature, have a very different susceptibility to oxygen or air. The Rh-only system is readily oxidised in air as demonstrated before.[4] The 5 wt% Rh TEM data is thus representative of the (partly) oxidic Rh particles sizes instead of metallic Rh particles. The addition of 1 wt% Pd protects the Rh component from oxidation and bigger particles, now overall metallic, with a larger size range distribution are obtained. XPS and XAFS indicate that the mixed metal system forms alloys, with XPS suggesting a surface enrichment of oxidic Pd (for the pre-reduced samples under the UHV conditions of TEM). *In situ* reduction in 5% H$_2$/He fully reduces both metals in both systems with Rh–M (M = Rh and/or Pd) coordination numbers of ~7 and ~10 for the 5 wt% Rh/γ-Al$_2$O$_3$ and 4 wt% Rh/1 wt% Pd/γ-Al$_2$O$_3$, respectively, resembling metal particle sizes of ~55 and ~200 to 1000 atoms (in fcc configuration), *i.e.* diameters of ~10 and 20–34 Å.

The collective, complementary results for both catalysts indicate that the Rh and RhPd particles rapidly fragment upon addition of the CO/NO feedstock at 300 K, a significant overall reduction in metal coordination number and thus particle sizes is observed. This disruption of the initial nanoparticles occurs for both catalytic systems into what is most likely a mixture of discrete mononuclear Rh(CO)$_2$ units and small Rh clusters. The fragmentation is considerably more extensive for the Rh-only system compared to the Rh/Pd alloy, which explains why the small amount of oxidic Rh is not picked up with XAFS in the alloy case; large amounts of metallic Rh (in Rh or RhPd particles) remain present showing an overall reduced form in XANES. These results are unexpected when considering the alloy system, based upon the previous knowledge of these bimetallic systems.[12] The 4 wt% Rh/1 wt% Pd system has been demonstrated to effectively insulate the Rh component against any facile NO oxidation, remaining Rh–Rh/Pd coordination number around 10 for the entire temperature range applied.[12]

The insulation of the Rh component in the alloy catalyst was explained and evidenced by the segregation of Pd to the surface of the particles, preventing the oxidation of Rh as well as the dissociation of NO, using H_2 as reductant.[12] Although Rh is known to dissociate CO, the oxidative disruption is generally regarded to proceed as a molecular adsorption. Contrary to dissociative adsorption, the segregated Pd at the surface of the particles should not significantly affect their capacity to adsorb CO (or NO). Analysis of all the available data therefore suggests that molecularly adsorbed CO is able to strip Rh with significant efficiency, under the current conditions employed, even from the PdRh alloy nanoparticles although in much lower amounts. This process occurs to a significant degree until the alloy nanoparticles are 'Pd-rich' enough to resist the disruption, *i.e.* at a Rh–Rh/Pd coordination number of 6 corresponding to metal particles of *on average* 19 metal atoms.

The following model for stabilisation of these M_{19} (here M = Rh or Pd) particles can be envisaged. In Fig. 11 the 'spherical' fcc configuration of 19 metal atoms is displayed. In the M_{19} particle an (abc) layered structure in fcc configuration can be identified. One of the corners of the metallic 'sphere' is likely to be embedded in the catalyst support. Because of the lower bond strength (lower heat of atomisation) of Pd compared to Rh, the Pd is likely to coordinate the lower coordinate sites, *i.e.* the other five 4-coordinate 'corner' sites in the M_{19} structure. All Rh atoms in these sized alloy particles are thus present in highly coordinated sites, making it very difficult, if not impossible, for any molecule to strip further Rh atoms out of these particles. This gives a Rh/Pd atom/atom ratio of $\sim 3 : 1$. The overall Rh/Pd atom/atom ratio of the alloy sample is 4 : 1. The outstanding amount of Rh was initially stripped from the alloy particles as observed in DRIFTS, and forms reduced monometallic Rh particles at higher temperatures as for the Rh only system (*vide infra*). Rearrangement of the metal alloy particles to form these highly stable alloy particles can take place under the reaction conditions employed (here already at 300 K) due to the exothermic nature of the reactions taking place. The results also suggest that the Pd

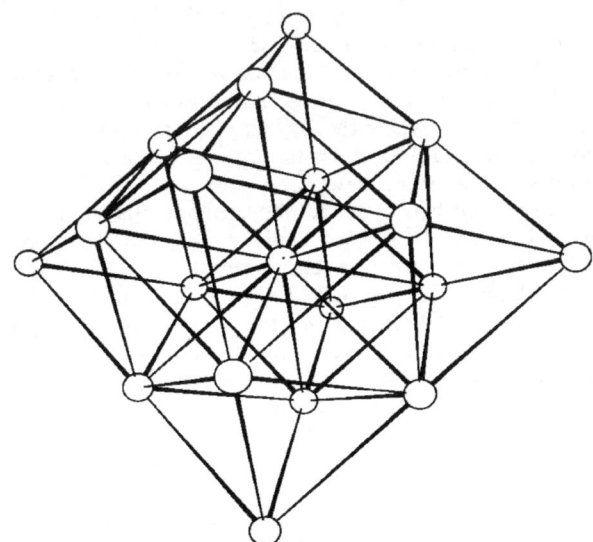

Fig. 11 M_{19} cluster in fcc configuration.

component is not significantly affected by the aforementioned process, as Pd remains in an overall metallic state throughout the experiments performed.

The 5 wt% $Rh/\gamma-Al_2O_3$ displays an immediate reduction in RhRh coordination from 7 to ~ 3 upon addition of the CO/NO feedstock at 300 K to 350 K, after which it decreases to a value of $ca.$ 2.5 at 450 K. It is at this temperature that the $Rh(CO)_2$ and $Rh(NO)_2$ vibrations show a maximum intensity. From 450 K, the RhRh coordination increases, with a significant decrease in $Rh(CO)_2$ and $Rh(NO)_2$ vibrations, while the $Rh(NO)^+$ species is formed and the catalyst starts to become active in the CO/NO conversion. For temperatures between ~ 470 K and 530 K the RhRh coordination is stable at ~ 3, and $Rh(CO)_2$ and $Rh(NO)_2$ disappear while the $Rh(NO)^+$ species go through a maximum. At even higher temperatures all CO and NO adsorbates disappear from the surface, leaving only the $Rh(NO)^+$ present in small amounts, the metal coordination number increases, forming reduced Rh particles, and the catalyst becomes considerably more active.

The average coordination number for the alloy system is seen to remain constant, within error, between a value of 6 and 6.5 throughout the temperature conditions employed while very similar surface species with similar temperature behaviour are formed, in smaller quantities and at overall slightly lower temperatures. From about 530 K a significant increase in catalytic activity from the 5 wt% compared to the 4 wt% Rh/1 wt% Pd system is observed. For the alloy system no adsorbate species can be identified at these high temperatures while for the 5 wt% Rh system $Rh(NO)^+$ species remain present. Since the majority of the catalytic activity is only observed after adsorbate species are removed from the metal surface, and the metal particles aggregate, the species observed in DRIFTS might very well all be spectator species.

The argument proposed above to explain the particle fragmentation also serves to explain the similar catalytic behaviour displayed by the elemental Rh and alloyed RhPd catalysts in terms of activity and selectivity. As large amounts of the Rh, for the 5 wt% Rh system the majority of the Rh metal, is stripped from its particulate environment, it is reasonable to suggest that it is held on the surface as a relatively unreactive organometallic surface species; the species will in essence behave similarly on both catalytic systems. That the Rh–Rh/Pd coordination number does not return to values resembling 'fully' metallic RhPd alloy nanoparticles (*i.e.* Rh–Rh/Pd = ~ 10) while most of the surface species are removed also suggests that re-alloying of the Rh with Pd does not occur, essentially resulting in a phase separated system (of Rh and RhPd particles). The Rh-only metal particles formed on the alloy system probably have an identical activity and selectivity to the Rh nanoparticles on the Rh-only system, *i.e.* the active sites are identical. However, the RhPd alloy particles are less reactive or not reactive at all and thereby reduce the overall activity of the system. It is worthy of note that these observations are in contrast to similar studies using NO reduction by H_2,[12] where the alloyed nanoparticles do not fragment and the bigger alloy nanoparticles significantly promote the catalysis occurring in comparison to the single Rh metal system.

It should be emphasised here that the above hypothesis should be approached with caution as XAFS is a macroscopic technique. For example, the postulation of a phase separated system may not be accurate due to the averaging nature of the EXAFS technique. Additionally, the EXAFS backscattering does not allow for the distinction between Rh and Pd; therefore the overall disruptive process may be wholly, or indeed partly, due to particle size effects. Also, if the Pd was an active component, then subsequent correlations made between the two systems would not be probed by analysing the Rh K edge alone.

Conclusions

In summary, the complementary data obtained during the temperature controlled NO reduction by CO highlight the extensive, disruptive oxidation of Rh as well as RhPd alloy nanoparticles, the capacity of which exceeds measurements made in

This journal is © The Royal Society of Chemistry 2008

previous studies. For the two structurally very different catalytic systems studied, *i.e.* 5 wt% Rh and 4 wt% Rh/1 wt% Pd, the catalytically active sites are likely to be similar, with the overall activity of the latter reduced due to inactive RhPd alloy nanoparticles in which the Rh is only present in highly coordinated sites with the Pd occupying the potentially catalytically active low coordination sites.

That these unexpectedly corrosive processes were previously unknown, or at the very least underestimated, will have a profound impact on the core understanding of their catalytic influence. This in turn will directly affect technological applications such as catalyst design like the three-way automotive catalyst for which the NO/CO reaction is a paramount consideration.

Acknowledgements

The authors would like to thank the ESRF for a Long Term proposal and the EPSRC for funding this research (M. A. N., B. J. (GR/60744/01) and M. T.). We also wish to thank the staff of the ESRF and of ID24 in particular for the provision of the facilities. The following people are gratefully acknowledged for their assistance with the experiments and their helpful discussions: Dr Graham Beamson and Dr Danny Law at NCESS for XPS, Dr P. J. F. Harris at the University of Reading for TEM.

References

1 M. A. Newton, B. Jyoti, A. J. Dent, S. G. Fiddy and J. Evans, *Chem. Commun.*, 2004, 2382.
2 M. A. Newton, A. J. Dent, S. G. Fiddy, B. Jyoti and J. Evans, *Catal. Today*, 2007, **126**, 64.
3 M. A. Newton, S. G. Fiddy, G. Guilera, B. Jyoti and J. Evans, *Chem. Commun.*, 2005, 118.
4 M. A. Newton, A. J. Dent, S. Diaz-Moreno, S. G. Fiddy, B. Jyoti and J. Evans, *Chem.–Eur. J.*, 2006, **12**, 1975.
5 A. J. Dent, J. Evans, S. G. Fiddy, B. Jyoti, M. A. Newton and M. Tromp, *Angew. Chem., Int. Ed.*, 2007, **46**, 5365.
6 M. A. Newton, A. J. Dent, S. G. Fiddy, B. Jyoti and J. Evans, *Phys. Chem. Chem. Phys.*, 2007, **9**, 256.
7 A. C. Yang and C. W. Garland, *J. Phys. Chem.*, 1957, **61**, 1504.
8 A. Suzuki, Y. Inada, A. Yamaguchi, T. Chihara, M. Yuasa, M. Nomura and Y. Iwasawa, *Angew. Chem., Int. Ed.*, 2003, **42**, 4795.
9 H. F. T. van' t Blik, J. B. A. D. van Zon, T. Huizinga, J. C. Vis, D. C. Koningsberger and R. Prins, *J. Phys. Chem.*, 1983, **87**, 2264.
10 T. Campbell, A. J. Dent, S. Diaz-Moreno, J. Evans, S. G. Fiddy, M. A. Newton and S. Turin, *Chem. Commun.*, 2002, 304.
11 M. A. Newton, A. J. Dent, S. Diaz-Moreno, S. G. Fiddy and J. Evans, *Angew. Chem., Int. Ed.*, 2002, **41**, 2587.
12 M. A. Newton, B. Jyoti, A. J. Dent, S. Diaz-Moreno, S. G. Fiddy and J. Evans, *ChemPhysChem*, 2004, **5**, 1056.
13 M. Hagelstein, C. Ferrero, U. Hatje, T. Ressler and W. Metz, *J. Synchrotron Radiat.*, 1995, **2**, 174.
14 A. J. Dent, S. Diaz-Moreno, J. Evans, S. G. Fiddy, B. Jyoti, M. A. Newton and M. Tromp, to be submitted.
15 (a) *e.g.* E. Lundgren, G. Kresse, C. Klein, M. Borg, J. N. Andersen, D. De Santis, Y. Gauthier, C. Konvicka, M. Schmid and P. Varga, *Phys. Rev. Lett.*, 2002, **88**, 246103; (b) Th. Pillo, R. Zimmermann, P. Steiner and S. Hufner, *J. Phys.: Condens. Matter*, 1997, **9**, 3987.
16 V. P. Zhdanov and B. Kasemo, *Surf. Sci. Rep.*, 1997, **29**, 31.
17 D. K. Paul and S. D. Worley, *J. Phys. Chem.*, 1990, **94**, 8956.
18 A. M. Venezia, L. F. Liotta and G. Deganello, *Langmuir*, 1999, **15**, 1176.
19 F. Solymosi and T. Bansagi, *J. Catal.*, 2001, **202**, 205.
20 T. Lear, R. Marshall, J. A. Lopez-Sanchez, S. D. Jackson, T. M. Klapotke, M. Baumer, G. Ruppenrechter, H.-J. Freund and D. Lennon, *J. Chem. Phys.*, 2005, **123**, 174706.
21 J. Liang, H. P. Wang and L. D. Spicer, *J. Phys. Chem.*, 1985, **89**, 5840.
22 M. A. Newton, D. G. Burnaby, A. J. Dent, S. Diaz-Moreno, J. Evans, S. G. Fiddy, T. Neisius, S. Pascarelli and S. Turin, *J. Phys. Chem. A*, 2001, **105**, 5965.

23 M. A. Newton, D. G. Burnaby, A. J. Dent, S. Diaz-Moreno, J. Evans, S. G. Fiddy, T. Neisius and S. Turin, *J. Phys. Chem. B*, 2002, **106**, 4214.

24 H. Arai and H. Tominaga, *J. Catal.*, 1976, **43**, 131.

25 T. W. Root, G. B. Fisher and L. D. Schmidt, *J. Chem. Phys.*, 1986, **85**, 4879.

26 M. A. Newton, A. J. Dent, S. G. Fiddy, B. Jyoti and J. Evans, *J. Mater. Sci.*, 2007, **42**, 3288.

27 B. E. Hayden, A. King, M. A. Newton and N. Yoshikawa, *J. Mol. Catal. A: Chem.*, 2001, **167**, 33.

28 M. E. Alikhani, L. Krim and L. Manceron, *J. Phys. Chem. A*, 2001, **105**, 7817.

Synthesis, characterization, electronic structure and catalytic performance of bimetallic and trimetallic nanoparticles containing tin

John Meurig Thomas,[ab] Richard D. Adams,[b] Erin M. Boswell,[b] Burjor Captain,[b] Henrik Grönbeck[c] and Robert Raja[d]

Received 23rd April 2007, Accepted 21st May 2007
First published as an Advance Article on the web 28th September 2007
DOI: 10.1039/b706151j

When anchored on a high-area, siliceous supports, nanoparticle catalysts, consisting of two or three different metals, but totaling no more than twenty atoms in all, exhibit exceptional activities and selectivities in solvent-free, one-step hydrogenation reactions at low temperatures (<420 K) and much lower pressures (*e.g.* 30 bar) than those required in current industrial manufacture. The two selective hydrogenations illustrated here are the conversion of (a) cyclododecatriene (CDT) to cyclododecene (CD) and (b) dimethyl terephthalate (DMT) to cyclohexane dimethanol (CHDM); each of these products is extensively used in the polymer industry. All our mixed-metal nanoparticles are derived from an appropriately chosen parent (precursor) mixed-metal carbonyl having phenyl-containing tin ligands, *e.g.* $Ru_4(\mu_4\text{-}SnPh)_2(CO)_{12}$. Various techniques are used to characterize the denuded, anchored cluster catalysts; and it is expected that aberration-corrected high-resolution electron microscopy (and other techniques, which are outlined) will be invaluable in such characterization. Density functional theory has provided important insights into the structures and electronic properties of our catalysts and their precursors.

I. Introduction

Tin has been shown to be a useful modifier of platinum group metals for applications to a variety of heterogeneous catalytic processes.[1–3] There is some evidence that tin helps to anchor metallic nanoparticles onto nanoporous supports in a highly dispersed and uniform manner.[3] In recent studies, it has been shown that triphenyl-stannane, Ph_3SnH, is an excellent reagent for introducing variable numbers of phenyltin containing ligands into polynuclear metal carbonyl complexes.[4–7] Significantly, it has been shown that these tin-containing multimetallic complexes can be precursors to multimetallic nanoparticles that exhibit exceptionally good catalytic activities when anchored on suitable silica supports.[3]

For several reasons, the bimetallic and trimetallic entities that we describe in this report cannot be designated as nanoalloys—in the sense normally associated with

[a] Department of Materials Science, University of Cambridge, Cambridge, UK CB2 3QZ
[b] Department of Chemistry and Biochemistry, University of South Carolina, Columbia SC 29208, USA
[c] Department of Applied Physics and Competence Center for Catalysis, Chalmers University of Technology, SE-412 96, Göteborg, Sweden
[d] School of Chemistry, University of Southampton, Highfield, Southampton, UK SO17 1BJ

that term.[8,9] First, the total number of atoms in the nanoparticle bi- and trimetallics prepared by us, is very much smaller (close to a factor of a hundred less) than the number present in the typical nanoalloys used by Yacaman *et al.*,[9,10] and others.[11] In recent structural determinations[10] of Au–Pd nanoalloys, for example, the nanoparticles may consist of several hundred atoms, typically clusters consisting of 561 or 923 atoms, whereas in our work the total number generally is less than 50 and in most cases, they may be as small as that of the molecular precursor, 6–10 metal atoms.[12–15] Whereas in conventional nanoalloys the number of atoms is large enough to form energy bands, that are similar to those of bulk metals and alloys, in our nanoparticles the energy levels are fewer and less closely spaced. Moreover, because our nanoparticles are anchored to a high-area oxide support (invariably mesoporous silica), well-defined, ionic–covalent bonds form between the metal atoms of the nanoparticle and the oxygens (which are, in turn, attached to silicon) of the support. Such bonds may be directly identified by EXAFS measurements combined with an amalgam of computational procedures (*e.g.* DFT/MM/MD).[16,17]

Second, since the small number of atoms bind firmly to the underlying support, it may be more appropriate to regard the nanoparticle as a molecular entity with its activity and selectivity being more like that of multinuclear single-site catalysts, such as that found in many metallo-enzymes, *e.g.* hydrogenases, than of a bulk metal or alloy.

Third, the notion (and reality) of such structural infractions as twins and coherent intergrowths—as is seen by Yacaman *et al.* in a 923-atom nanoalloy of AuPd[10]—is meaningless in our molecular bimetallic nanoparticles. In the nanoalloys of Yacaman *et al.*[10,18] and others,[19] one may discern directly, by aberration-corrected electronic microscopy, thin bands of hexagonal close-packed and face-centered cubic packed sheets. In a typical molecular nanoparticle of the kind that we have studied (also by aberration-corrected electron microscopy,[15] it is directly established (in line with theoretical predictions[20]) that a single bimetallic cluster of $Ru_{10}Pt_2$ does indeed possess molecular character; and that when six or more such clusters coalesce into larger entities containing *ca.* 200 atoms they adopt the regular crystalline, and faceted, state of a bulk metal (see Fig. 2 to 4 of ref. 15).

There are other differences, which we shall elaborate on below. For example, our bi- and trimetallic species are derived from precursor molecules that can be denuded (chemically) to expose only the mixed-metal constituent atoms. The nanoalloys that almost all others prepare are generally formed by co-precipitation, or "incipient wetness" methods that rely on sequential addition of separate solutions such as $HAuCl_4$ and $Pd\ Cl_2$,[10] (to form Au–Pd nanoalloys). Moreover, such nanoalloys (like the nanometals now in extensive use[21]), are utilized in their "passivated" form, which means that they have surfactant species such as poly(vinyl pyrrolidone) (PVP), or poly-diallyl dimethyl ammonium chloride (PDDA), the molecular weight of which may be as large as 450 000 Da, covering their surfaces.[22] Our nanoparticles are free from such surfactants or any other stabilizing molecular entity.

II. Precursor species and the preparation of bimetallic and trimetallic nanoparticle catalysts

In association with a colleague, B. F. G. Johnson, one of us (J. M. T.) began[12,23,24] a decade or so ago to synthesize bimetallic nanoparticles such as Pd_6Ru_6 and $Ru_{16}Cu_4$ from carbonylate precursors such as $[Pd_6Ru_6(CO)_{24}]^{2-}$ and $[Cu_4Ru_{16}C_2(CO)_{32}Cl_2]^{2-}$ which could be freed from their ligands (after immobilization on mesoporous silica, see ref. 12–14 for details) by gentle heating in oxygen or dry air. Such a preparation could be followed by joint, *in situ*, use of FTIR (to track the elimination of the carbonyl groups) and a combination of EXAFS and XRD.[25–27] The EXAFS data of the two components of the bimetallic were unified with estimates from DFT, (as well as MD and molecular mechanics) to arrive at the equilibrium structure of the supported catalyst, as previously described.[16]

To ensure that no bimetallic or trimetallic nanoparticle catalysts were laid down on the exterior wall of the mesoporous silica, pretreatment with diphenyldichlorosilane was used, so as to convert all pendant Si–OH groups at that surface into non-reactive diphenylsilane derivatives. In this way all the nanoparticles were grafted on to the interior surfaces of the silica—see ref. 12 and 14.

Herein we focus almost exclusively on the bi- and trimetallic nanoparticles derived from a range of well-defined precursors such as those shown in Fig. 1. All these precursor structures were characterized by X-ray crystallography, as described in a series of papers by two of us (R. D. A. and B. C.).[3e,28–31]

The procedure used to convert these Ru–Sn precursors into active nanoparticle catalysts is as follows: a slurry of the cluster precursor dissolved in CH_2Cl_2 solvent is shaken with the siliceous support and evaporated to dryness. The siliceous support with anchored cluster is then decarbonylated by calcination in vacuum at *ca.* 200 °C for 2 h, prior to use in the catalytic reactions.

As no synchrotron beam-time is available to us prior to submission of this paper, we plan to convey the results of the EXAFS measurements (carried out so as to determine their atomic structure) at the Discussion Meeting in September, '07. We have, however, examined some of the nanoparticles using high angle annular dark

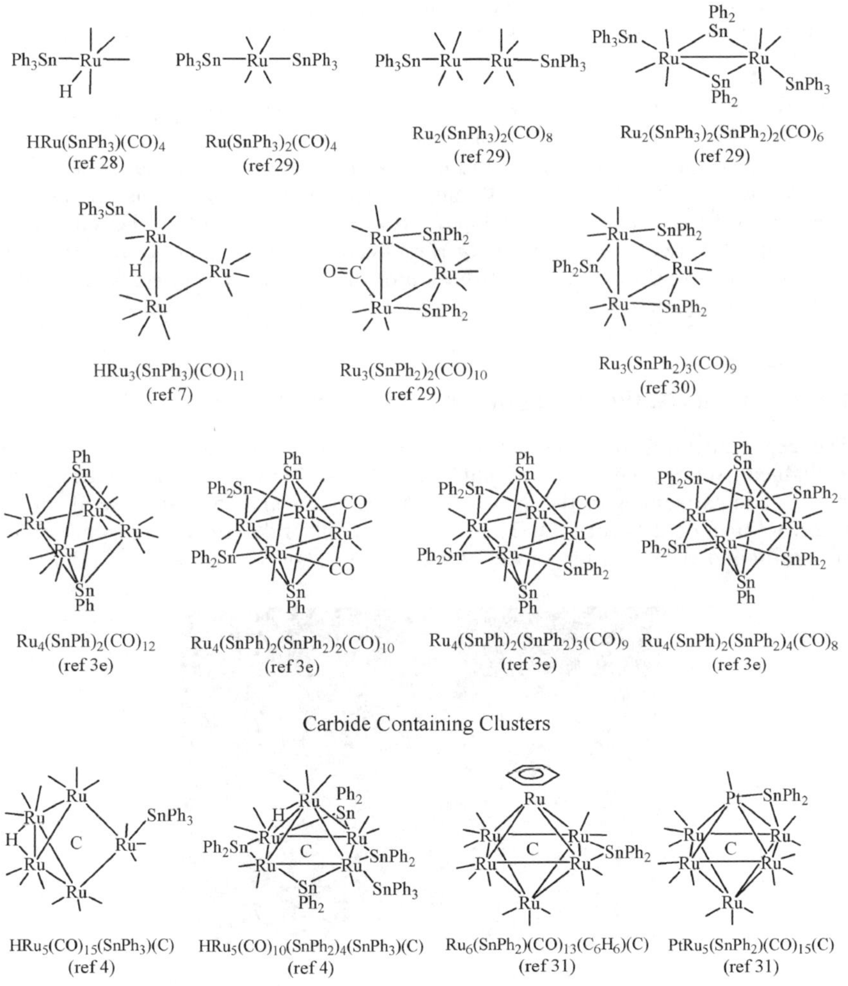

Carbide Containing Clusters

Fig. 1 A large collection of potential Ru_xSn_y molecular cluster precursors.

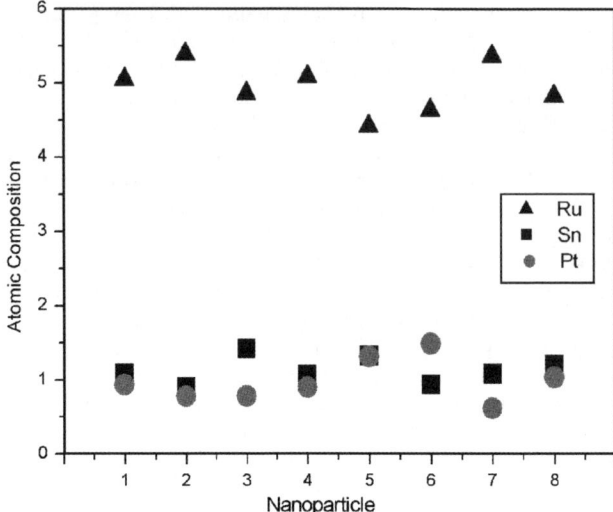

Fig. 2 A plot showing the uniformity of the composition of several nanoparticles of the Ru$_5$PtSn catalyst on Davison silica 38 Å mesopore obtained by electron-induced X-ray emission.[3d]

field (HAADF) electron microscopy with the STEM technique as previously described.[3d,23] The results confirm that the nanoparticles remain intact; but in some of the preparations a distinctly non-uniform distribution of the individual nanoparticles was produced. With Ru$_5$Pt–Sn, electron-induced X-ray emission spectroscopy yielded the results shown in Fig. 2, from which it is seen that the integrity of the multi-nuclear core of the precursor is retained. A tomographic study (see ref. 32 and 33) reveals that the spatial distribution of this trimetallic catalyst is reasonably uniform, see Fig. 3.

III. Characterization using density functional theory

Few experimental methods exist to determine structures of gas phase clusters and available methods such as ion mobility[34] and trapped ion electron diffraction[35] still require comparisons with theoretical results. In particular, methods based on the Density Functional Theory (DFT) have been very helpful in providing an insight on structural and electronic properties of small atomic clusters.

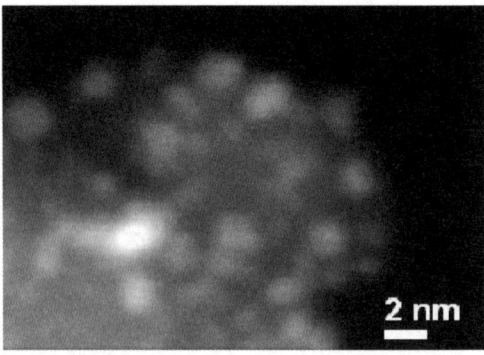

Fig. 3 A high angle annular dark field (HAADF) image of Ru$_5$PtSn nanoclusters on Davison silica 38 Å mesopore.[3d]

Here, DFT has been used to explore Ru_5PtSn and Ru_4Sn_2 together with their precursors in the gas-phase and supported on an α-cristobalite surface. The details of the calculations have been described previously.[36] In brief, DFT was applied with the gradient corrected exchange-correlation (xc) functional according to Perdew, Burke and Ernzerhof (PBE).[37] The one-electron Kohn–Sham orbitals were expanded in a localized basis set (double numerical with polarization functions).[38] Pseudopotentials[39] were applied to describe the interaction between the valence electrons and the cores of Ru, Pt and Sn.

The performance of the computational approach for the systems under consideration was checked by calculations for Ru_2. The bond length, binding energy and vibrational frequency were calculated to be 2.29 Å, 2.38 eV and 330 cm^{-1}, respectively. These results are in fair agreement with the available experimental data of the binding energy and the vibrational frequency, namely 2.0 ± 0.2 eV and 347 cm^{-1}, respectively.[40]

The optimized structures for the gas-phase systems are reported in Fig. 4. For the precursor carbonylates, only the experimentally determined isomers were considered, whereas relaxations of several initial configurations were performed for $PtRu_5CSn$, $PtRu_5Sn$ and Ru_4Sn_2.

The structure of the carbonylates has been experimentally determined *via* X-ray diffraction measurements of the crystalline solid.[3e,31] In the $PtRu_5(CO)_{15}$-

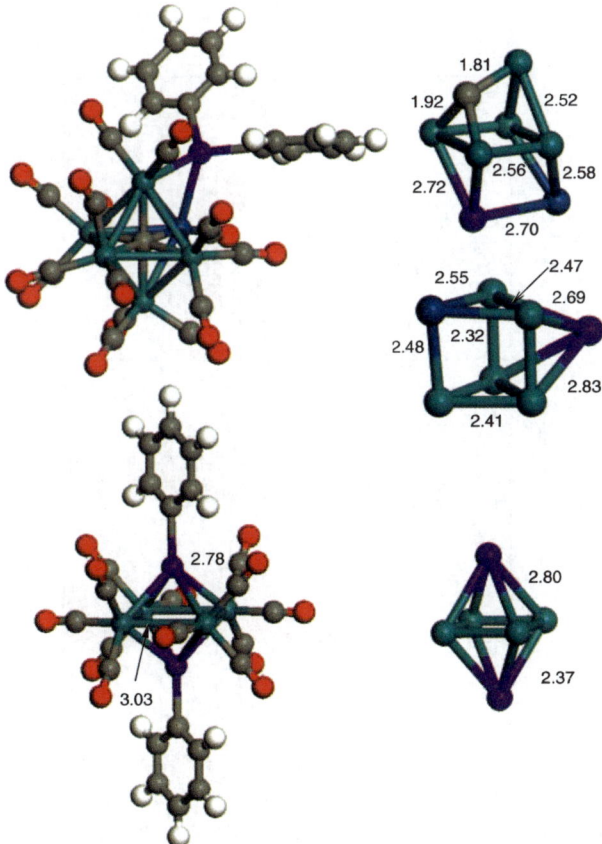

Fig. 4 Structural models of the ground state geometries of $PtRu_5(CO)_{15}(\mu\text{-}SnPh_2)(\mu_6\text{-}C)$, $PtRu_5CSn$, $PtRu_5Sn$, $Ru_4(\mu_4\text{-}SnPh)_2(CO)_{12}$ and Ru_4Sn_2. Selected inter-atomic distances are given in Å. Color code: Ru: dark green, Pt: blue, Sn: purple, C: grey, O: red and H: white.

(μ-SnPh$_2$)(μ$_6$-C) compound, the five ruthenium atoms and single Pt atom form an octahedron with an interstitial carbon atom. The CO molecules are attached atop the metal atoms and the SnPh$_2$ group in a bridge configuration. The geometrical relaxation results only in minor distortions with respect to the experimental structure. In the theoretically computed structure, the Ru–Ru distances are, on average, 2.95 Å, which is slightly longer than the experimental value of 2.90 Å. A similar trend is present for Ru$_4$(μ$_4$-SnPh)$_2$(CO)$_{12}$. Experimentally, the Ru–Ru and Ru–Sn bond lengths are determined to be 3.03 Å and 2.78 Å, respectively. The slight overestimation of the inter-atomic distances can be attributed to the applied xc-functional.

Removal of the ligands introduces large structural relaxations in the PtRu$_5$CSn core. In the energetically favored isomer, the carbon atom is bound to the surface of the cluster in a three-fold position. As it is uncertain whether the C atom is retained within the cluster upon oxygen treatment, we have also considered the case when the carbon atom is removed from the nanoparticle. The stable structure can be described as a trigonal prism with Sn as a cap. In contrast to PtRu$_5$CSn, the relaxations of Ru$_4$Sn$_2$ are less pronounced. The octahedral symmetry is preserved with a central ruthenium square capped by the two Sn atoms. However, the Ru–Ru distance in the central square is shorter in the cluster without ligands. The Sn–Ru bonds are somewhat elongated.

The electronic density of states (EDOS) for the gas-phase clusters are reported in Fig. 5. The projections of the Kohn–Sham orbitals on the atomic metal states (Ru, Pt and Sn) are shown as shaded curves. The carbonylated precursors have large

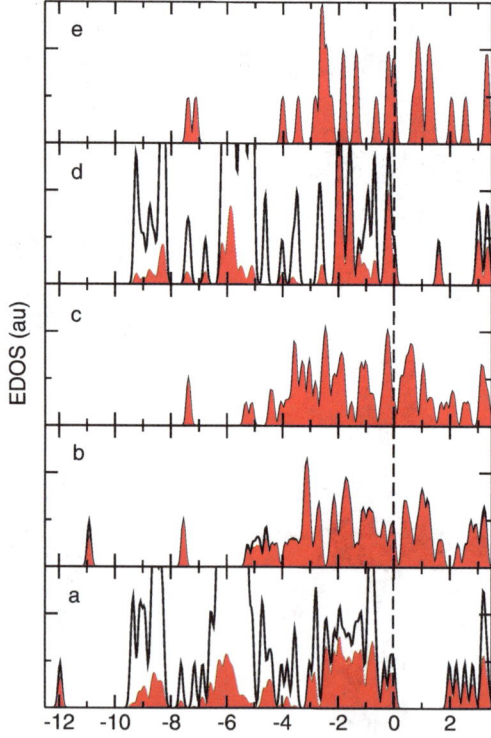

Fig. 5 Electronic density of states for (a) PtRu$_5$(CO)$_{15}$(μ-SnPh$_2$)(μ$_6$-C) (b) PtRu$_5$CSn, (c) PtRu$_5$Sn (d) Ru$_4$(μ$_4$-SnPh)$_2$(CO)$_{12}$ and (e) Ru$_4$Sn$_2$. The EDOS is obtained by a 0.15 eV Gaussian broadening of the one-electron Kohn–Sham energies. The shaded curves correspond to projection on metal (Pt, Ru and Sn) states. The EDOS is reported with respect to the HOMO level.

HOMO–LUMO separations of 2 eV, and 1.6 eV for $Ru_5(CO)_{15}(\mu\text{-}SnPh_2)(\mu_6\text{-}C)$ and $Ru_4(\mu_4\text{-}SnPh)_2(CO)_{12}$, respectively. The LUMO for Ru_4Sn_2, which has mainly Ru(4d) character, is a single state between the HOMO and a manifold of unoccupied states at 3 eV above the HOMO.

The Ru_5PtCSn cluster has a small HOMO–LUMO separation of 0.3 eV. The ground state is calculated to be in a triplet state, but this configuration is very close in energy to an electronic singlet configuration. From the projections onto the metal states, we conclude that carbon and ruthenium states are strongly hybridized. In the EDOS, such states appear at 5 eV, and 11 eV below the HOMO level, respectively. The state at about 7.5 eV below the HOMO level is of Sn(5s) character. A small HOMO–LUMO separation is also present for Ru_5PtSn. In fact, the electronic structure for Ru_5PtCSn and Ru_5PtSn show clear similarities. Ru_4Sn_2 has a slightly larger HOMO–LUMO separation than Ru_5PtCSn, namely 0.6 eV. The larger separation is manifested in a preference for the singlet state, which is preferred by 0.2 eV with respect to the triplet state. Just as for Ru_5PtCSn, the states at 7.5 eV below the HOMO level correspond to Sn(5s).

In this work, we have used a thin layer of the (001) facet of α-cristoballite as a model substrate for the anchored state of the clusters (see ref. 36 for details). The surface was initially covered with hydroxyl groups. However, the interaction between the cluster and the oxygen atoms was found to be strong enough to drive desorption of H_2. In particular, the release of H_2 was found to be exothermic by 2.7 eV and 1.8 eV for Ru_5PtCSn and Ru_4Sn_2, respectively. The stable adsorption configurations are shown in Fig. 6. In the case of Ru_5PtCSn, both top and side views are reported. Both clusters are predicted to be linked to the surface via Ru–O bonds, with bond distances of about 2.0 Å. Configurations with Sn bonded to the oxygen atoms are higher in energy. For Ru_4Sn_2, the lowest isomer with an Sn–O contact was found to be about 0.4 eV above the ground state.

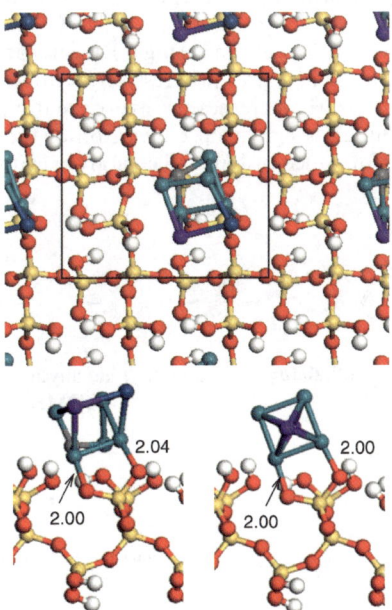

Fig. 6 Upper part: top view of $PtRu_5CSn$ supported on α-cristoballite. The surface cell is indicated. Lower part: side view of $PtRu_5CSn$ and Ru_4Sn_2 anchored to the surface. Selected inter-atomic distances are given in Å.

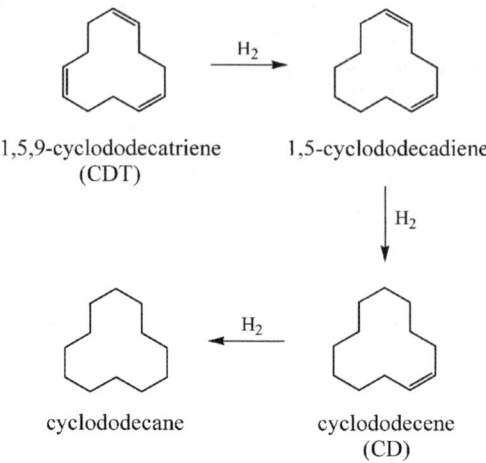

1,5,9-cyclododecatriene
(CDT)

1,5-cyclododecadiene

cyclododecane

cyclododecene
(CD)

Scheme 1

IV. Catalytic performance

Two commercially important selective hydrogenations were chosen as appropriate reactions to test the performance of our Ru–Sn-based bimetallic nanoparticle catalysts, Scheme 1 and Scheme 2.

The first of these reactions (Scheme 1) yields cyclododecene (CD) which is used extensively (as described elsewhere[3e]) to produce many industrially significant products that include laurolactam (the stepping stone to both nylon-12 and nylon 612) ketones, lactones and polymer intermediates. The second (Scheme 2) yields CHDM, an extremely important linker molecule that is deemed superior to ethylene glycol in the polymer industry.

Fig. 7 compares the performance of Ru_4Sn_2, Ru_3Sn_3, Ru_4Sn_4, and Ru_4Sn_6 with Ru_6Sn and Ru_5SnPt for the hydrogenation of CDT to CD (for a kinetic plot for the hydrogenation of 1,5,9-cyclododecatriene using a Ru_4Sn_6 catalyst see Fig. 8). As can be seen the selectivity for CD increases progressively with the increase in the Sn/Ru ratio. The addition of Pt clearly increases the hydrogenation activity, but decreases the selectivity for CD due to a significant conversion of the CD to cyclododecane. Fig. 9 compares the performance of bimetallic catalysts with a trimetallic Ru_5PtSn

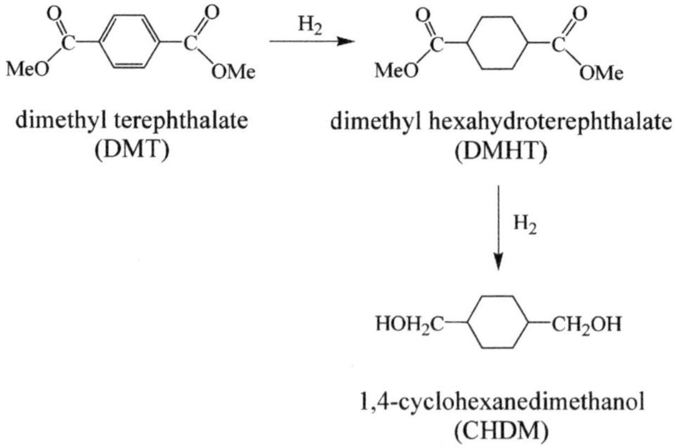

dimethyl terephthalate
(DMT)

dimethyl hexahydroterephthalate
(DMHT)

1,4-cyclohexanedimethanol
(CHDM)

Scheme 2

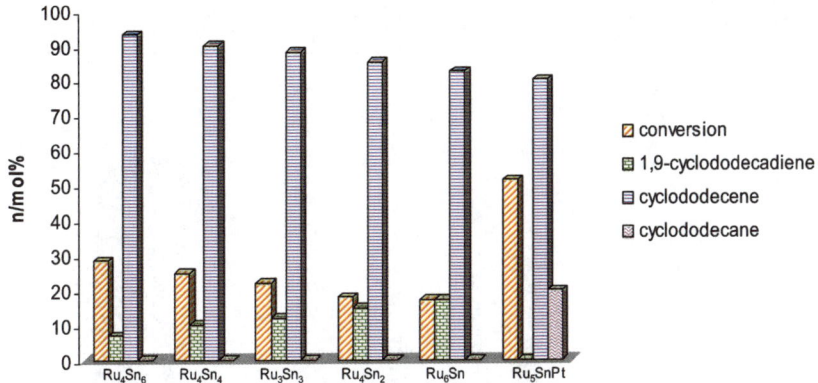

Fig. 7 Comparison of catalytic performance of Ru_4Sn_6, Ru_4Sn_4, Ru_3Sn_3, Ru_4Sn_2 and Ru_5SnPt with the previously reported (chlorine-containing) Ru_6Sn.[3c] Reaction conditions: substrate $\cong 50$ g, catalyst $\cong 25$ mg (cluster anchored on mesopore $\sim 2\%$ metal loading), H_2 pressure $\cong 30$ bar, $T = 373$ K, $t = 8$ h.

catalyst for the DMT to CHDM conversion.[3d] The molecular structures of the Ru_4Sn_6 and Ru_5PtSn cluster precursors are shown in Fig. 10 and 11, respectively.

It is noteworthy that these bimetallic nanoparticle catalysts are very efficient in selectively converting both the CDT and DMT to the desired products. Turnover frequencies are exceptionally high, and exceed those of nanometallic (small particle) platinum, ruthenium and palladium for the same catalytic conversions.[3d] And it is demonstrably the case that fine-tuning, involving the whole sweep of Ru_nSn_m (and $Ru_nSn_mPt_o$) bi- and trimetallic nanoparticles, for example an Ru_5Sn_5 cluster complex as shown in Fig. 12, are likely to yield yet further, superior hydrogenation catalysts for these and other reactions. It is also important to note that activities and selectivities have not yet been optimized with respect to operating temperatures.

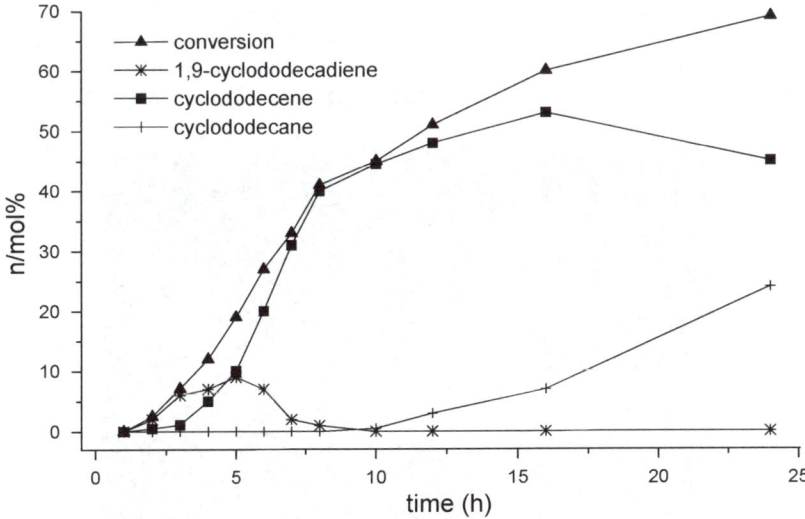

Fig. 8 Kinetic plot for the hydrogenation of 1,5,9-cyclododecatriene using Ru_4Sn_6 at 393 K. At 393 K, and under 30 bar H_2 the solvent-free conversion of 1,5,9-cyclododecatriene to cyclododecene proceeds smoothly, and for the first 10 h almost exclusively, in the presence of the Ru_4Sn_6 bimetallic nanoparticle catalyst supported on mesoporous silica.

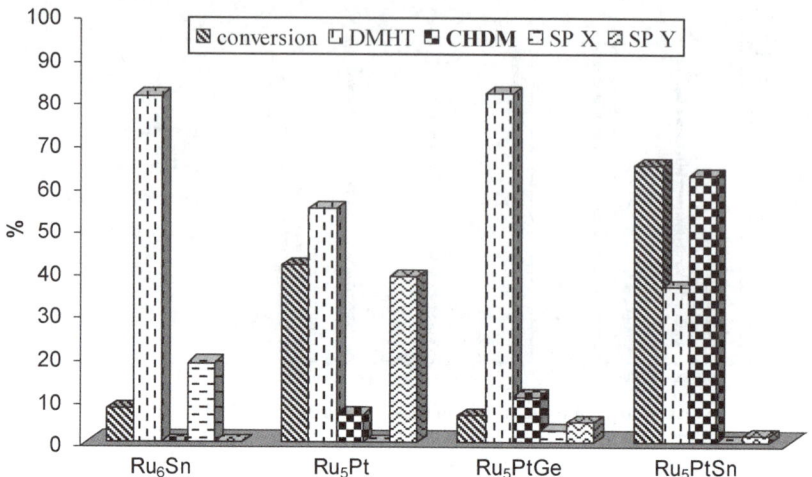

Fig. 9 Comparison of activities for some metallic nanoparticle catalysts for the hydrogenation of dimethyl terephthalate. Reaction conditions: $t = 24$ h, $T = 373$ K, 20 bar H_2 pressure. SP X and SP Y are undesirable side products of the hydrogenation reaction.

V. Discussion

(i) First we deal with *catalytic performance*, where we emphasize that both the reactions have been catalyzed under environmentally favourable, solvent-free conditions, at very modest temperatures and at quite low pressures of hydrogen (30 bar). In this context, it is worth emphasizing that the commercial production of CHDM from DMT currently involves two catalytic reactors, one to hydrogenate the benzene ring, the other, using a second catalyst, to convert the terminal ester groups to primary alcohols. (One of these reactors alone requires a pressure of hydrogen of 400 bar).

The second noteworthy feature is that the trimetallic nanoparticles are much more active than the bimetallic ones in the reactions we have studied to date. Third, we note from the DFT calculations that, rather unexpectedly, the lowest energy structure for all the anchored, naked clusters studied by us to date is one that involves a preferential Ru–O–Si link to the cristobalite (silica) surface. On general chemical grounds, one would have expected the Sn–O–Si link to be the primary

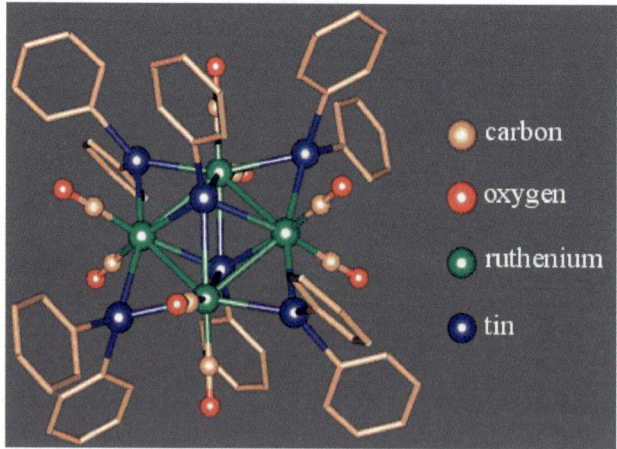

Fig. 10 The molecular structure of $Ru_4(\mu_4\text{-}SnPh)_2(\mu\text{-}SnPh_2)_4(CO)_8$.

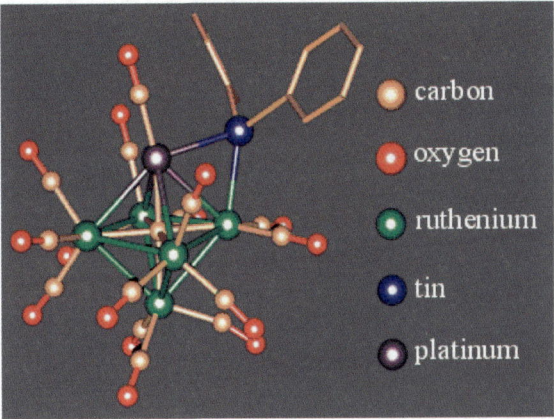

Fig. 11 The molecular structure of PtRu$_5$(μ-SnPh$_2$)(CO)$_{15}$(μ_6-C).

mode of attachment to the mesoporous support. We note, however, that one of the configurations that has a slightly elevated energy in the case of the anchored Ru$_4$Sn$_2$ does have both a Ru–O and also a Sn–O bond (to the silicon)—see Fig. 6. Experiments, now in progress, involving ^{119}Sn Mössbauer spectroscopy[41] and EXAFS measurements[42,43] should resolve this issue unambiguously.

We also plan to undertake some laser-ablation mass-spectrometric studies of those anchored nanoparticle catalysts (like Ru$_5$PtSn) derived from precursors that contain μ_6-C atoms in the parent cluster. One cannot be absolutely certain that, in our gentle oxidation of the precursor anchored to the silica, this μ_6-C (carbide) is also driven off (like the CO and Ph groups, which have a readily recognizable spectroscopic signature). The DFT calculations on Ru$_5$PtSn and on Ru$_5$PtSnC do not show vast differences, so it is probably not of great importance whether or not the carbidic carbon is retained in the anchored catalyst.

(ii) The precise *atomic structure* of the anchored nanoparticle catalysts will be better understood once the results of the above-mentioned Mössbauer and XAFS studies are known. But it is conceivable that it will be even better understood once the full deployment of recent advances in high resolution electron microscopy are made. It is important to recall that conventional HRTEM images are affected[44] by

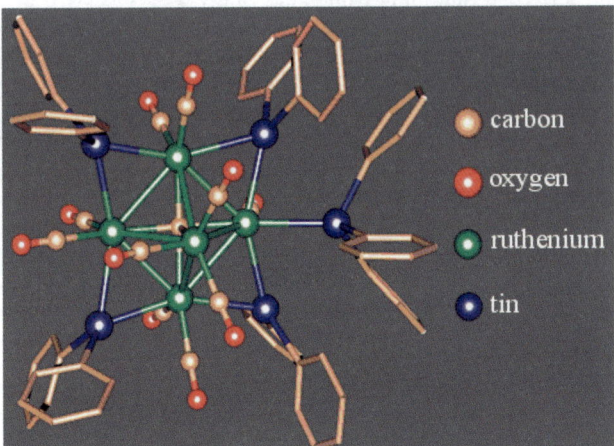

Fig. 12 The molecular structure of HRu$_5$(μ-SnPh$_2$)$_4$(SnPh$_3$)(CO)$_7$(μ_5-C). Hydride ligand not shown.

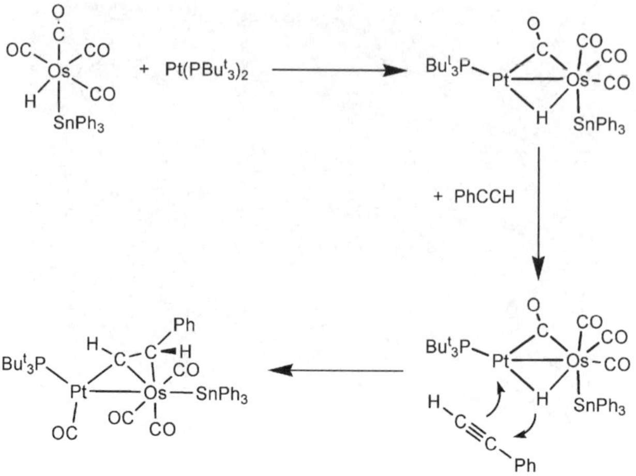

Scheme 3

electron microscope objective lens aberrations, limited spatial resolution and complicated interpretations, so that the visibility of nanoparticles are influenced by strong phase contrast from the underlying support. Thanks to recent major advances in hardware (aberration correction[45,46]) and computation (exit wave-function restoration[47,48]), it is now possible to examine minute supported nanoparticles in unprecedented detail and resolution. Such advances, applied to our samples should prove particularly enlightening.

(iii) We also want to comment briefly on the synergy found in catalysis using bimetallic nanoparticles, a topic initiated by Sinfelt.[49] It is well-known[14,28,50] that Ru and Pt in intimate structural contact surpass in activity the catalytic nature of nanoparticle Ru and Pt separately: one metal activates hydrogen, the other the olefinic bond. More needs to be learned in practice and in theory about insertion reactions at metal–hydrogen bonds, rather along the lines recently reported by two of us (R. D. A. & B. C.) in connection with the bimetallic cooperativity involved in the insertion of an alkyne into a metal Os–H bond.[51] In particular, it was shown (Scheme 3) that the addition of an electronically unsaturated platinum containing fragment promotes an important reaction both by activating a selected Os–H bond and then facilitating the addition and insertion of a selected alkyne reagent PhC_2H into that bond, in a tin containing trimetallic complex.

(iv) Bimetallic catalysts are also used for certain types of *oxidation* reactions. For example, it has been shown that the addition of iron to platinum produces a superior catalyst for the preferential oxidation (PROX) of CO by oxygen in the presence of very high concentrations of hydrogen.[52,53] In very recent studies, it has been shown that the platinum–iron cluster complexes $PtFe_2(COD)(CO)_8$ and $Pt_5Fe_2(COD)_2(CO)_{12}$, COD = 1,5-cyclooctadiene, are good precursors to highly dispersed PtFe nanoparticles that are among the best catalysts yet reported for CO PROX in hydrogen.[54]

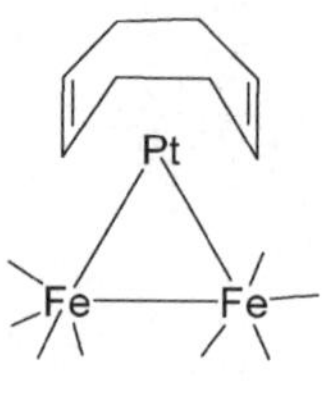

1

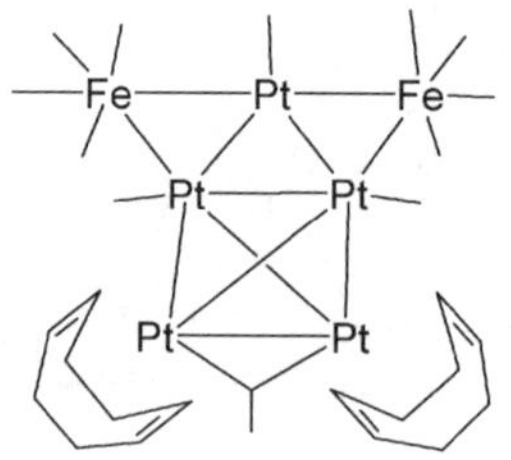

2

The Pt$_5$Fe$_2$/SiO$_2$ sample was more active than Pt/SiO$_2$ for PROX with a selectivity of approximately 92% at 50 °C. The deactivation of the Pt$_5$Fe$_2$/SiO$_2$ catalyst with time was substantially slower than the Pt/SiO$_2$, indicating that the highly reducing environment under the PROX conditions helps to maintain the properties of the active Pt–Fe bimetallic sites.

(v) Recent studies have shown that nanoparticles of gold are also excellent oxidation catalysts.[55] The recent work of Hutchings indicates that bimetallic catalysts, in which gold is combined with palladium and bismuth, exhibit good activities and selectivities for the aerial oxidation of primary alcohols to aldehydes[56] and cyclic olefins to epoxides,[57] respectively, in solvent free environments, *e.g.* eqn (1) and (2).

$$\text{RCH}_2\text{OH} \xrightarrow{\text{Au/Pd}} \text{R–C}\overset{H}{\underset{O}{\diagdown}} \qquad (1)$$

$$\text{(cyclohexene)} \xrightarrow{\underset{\text{Au/Bi}}{O_2}} \text{(cyclohexene oxide)} \qquad (2)$$

Tin is also known to be a catalyst for certain types of oxidation reactions in some of its higher oxidation states.[58,59] We are inclined to believe that tin will also prove to be a useful modifier for these new bimetallic oxidation catalysts.

Interest in catalysis by heterometallic nanoparticles created from molecular cluster complexes continues to grow.[14,60] We feel that there are many benefits, such as improved selectivity, that can be derived by the incorporation of tin into nanoparticles for use in catalytic hydrogenation and oxidation reactions. Molecules, such as shown in Fig. 1, may be precursors for synthesizing them efficiently.

References

1 (*a*) R. Burch, *J. Catal.*, 1981, **71**, 348; (*b*) R. Burch and L. C. Garla, *J. Catal.*, 1981, **71**, 360; (*c*) R. Srinivasan and B. H. Davis, *Platinum Met. Rev.*, 1992, **36**, 151; (*d*) T. Fujikawa, F. H. Ribeiro and G. A. Somorjai, *J. Catal.*, 1998, **178**, 58; (*e*) Y.-K. Park, F. H. Ribeiro and G. A. Somorjai, *J. Catal.*, 1998, **178**, 66.

2 (*a*) G. W. Huber, J. W. Shabaker and J. A. Dumesic, *Science*, 2003, **300**, 2075; (*b*) J. W. Shabaker, D. A. Simonetti, R. D. Cortright and J. A. Dumesic, *J. Catal.*, 2005, **231**, 67; (*c*) M. Guidotti, V. Dal Aanto, A. Gallo, E. Gianotti, G. Peli, R. Psaro and L. Sordelli, *Catal. Lett.*, 2006, **112**, 89; (*d*) R. D. Cortright, J. M. Hill and J. A. Dumesic, *Catal. Today*, 2000, **55**, 213.

3 (*a*) S. Hermans, R. Raja, J. M. Thomas, B. F. G. Johnson, G. Sankar and D. Gleeson, *Angew. Chem., Int. Ed.*, 2001, **40**, 1211; (*b*) B. F. G. Johnson, S. A. Raynor, D. B. Brown, D. S. Shephard, T. Mashmeyer, J. M. Thomas, S. Hermans, R. Raja and G. Sankar, *J. Mol. Catal. A: Chem.*, 2002, **182–183**, 89; (*c*) S. Hermans and B. F. G. Johnson, *Chem. Commun.*, 2000, 1955; (*d*) A. B. Hungria, R. Raja, R. D. Adams, B. Captain, J. M. Thomas, P. A. Midgley, V. Golvenko and B. F. G. Johnson, *Angew. Chem., Int. Ed.*, 2006, **45**, 4782; (*e*) R. D. Adams, E. M. Boswell, B. Captain, A. B. Hungria, P. A. Midgley, R. Raja and J. M. Thomas, *Angew. Chem., Int. Ed.*, 2007, in press.

4 (*a*) R. D. Adams, B. Captain, W. Fu and M. D. Smith, *Inorg. Chem.*, 2002, **41**, 5593; (*b*) R. D. Adams, B. Captain, W. Fu and M. D. Smith, *Inorg. Chem.*, 2002, **41**, 2302.

5 R. D. Adams, B. Captain, J. L. Smith, Jr, M. B. Hall, C. L. Beddie and C. E. Webster, *Inorg. Chem.*, 2004, **43**, 7576.

6 R. D. Adams, B. Captain and L. Zhu, *Organometallics*, 2006, **25**, 2049.

7 K. Burgess, C. Guerin, B. F. G. Johnson and J. Lewis, *J. Organomet. Chem.*, 1985, **295**, C3.

8 R. L. Ferrando, J. Jellinek and R. L. Johnson, *Chem. Rev.*, 2007, in press.

9 S. J. Mejia-Rosales, C. Fernandez-Navarro, E. Perez-Tijerina, J. M. Montejano-Carrizalles and M. Jose-Yacaman, *J. Phys. Chem. B*, 2006, **110**, 12884.

10 S. J. Mejia-Rosales, C. Fernandez-Navarro, E. Perez-Tijerina, D. A. Bloom, L. F. Allard and M. Jose-Yacaman, *J. Phys. Chem. B*, 2007, **111**, 1256.

11 J. L. Fernandez, D. A. Walsh and A. J. Bard, *J. Am. Chem. Soc.*, 2005, **127**, 357.

12   D. S. Shephard, T. Maschmeyer, B. F. G. Johnson, J. M. Thomas, G. Sankar, D. Ozkaya, W. Z. Zhou, R. D. Oldroyd and R. G. Bell, *Angew. Chem., Int. Ed. Engl.*, 1997, **36**, 2242.
13   S. Hermans, R. Raja, J. M. Thomas, B. F. G. Johnson, J. M. Thomas, G. Sankar and D. Gleeson, *Angew. Chem., Int. Ed.*, 2001, **40**, 1211.
14   J. M. Thomas, B. F. G. Johnson, R. Raja, G. Sankar and P. A. Midgley, *Acc. Chem. Res.*, 2003, **36**, 20.
15   E. P. W. Ward, I. Arslan, P. A. Midgley, A. Bleloch and J. M. Thomas, *Chem. Commun.*, 2005, 5805.
16   S. T. Bromley, G. Sankar, C. R. A. Catlow, T. Maschmeyer, B. F. G. Jenkins and J. M. Thomas, *Chem. Phys. Lett.*, 2001, **340**, 524.
17   J. M. Thomas, R. Raja, B. F. G. Johnson, T. J. O'Connell, G. Sankar and T. Khimyak, *Chem. Commun.*, 2003, 1126, see also *Science*, 2003, **300**, 867.
18   M. Jose-Yacaman, E. Perez-Tijerina and S. M. Rosales, *J. Mater. Chem.*, 2007, **17**, 1.
19   R. L. Johnson, *Philos. Trans. R. Soc. London, Ser. A*, 1998, **356**, 211.
20   C. L. Cleveland, U. Landman, T. G. Schaaff, M. N. Shafigallin, P. W. Stephens and R. L. Whetten, *Phys. Rev. Lett.*, 1997, **79**, 1873.
21   (*a*) J. Zhu, V. F. Puntes, I. Kiricsi, C. X. Miao, J. W. Ager, A. P. Alivisatos and G. A. Somarjai, *Langmuir*, 2003, **19**, 4396; (*b*) R. M. Rioux, H. Song, J. D. Hoefelmeyer, P. Yang and G. A. Somorjai, *J. Phys. Chem. B*, 2005, **109**, 2192.
22   J. L. Elechiguerra, L. Larious-Lopez and M. Jose-Yacaman, *Appl. Phys. A*, 2006, **84**, 11.
23   D. Ozkaya, W. Zhou, J. M. Thomas, P. A. Midgley, V. J. Keast and S. Hermans, *Catal. Lett.*, 1999, **60**, 113.
24   W. Zhou, J. M. Thomas, D. S. Shepherd, B. F. G. Johnson, D. Ozkaya, T. Maschmeyer, R. G. Bell and Q. F. Ge, *Science*, 1998, **280**, 705.
25   J. W. Couves, J. M. Thomas, D. Waller, R. H. Jones, A. J. Dent, G. E. Derbyshire and G. N. Greaves, *Nature*, 1994, **354**, 465.
26   J. M. Thomas, *Angew. Chem., Int. Ed.*, 1999, **38**, 3588.
27   J. M. Thomas and R. Raja, *J. Organomet. Chem.*, 2004, **689**, 4110.
28   R. D. Adams, B. Captain, E. Trufan and L. Zhu, *J. Am. Chem. Soc.*, 2007, **129**, 7545.
29   R. D. Adams, B. Captain and E. Trufan, unpublished results.
30   R. D. Adams, B. Captain and E. Trufan, *J. Am. Chem. Soc.*, 2007, in press.
31   R. D. Adams, B. Captain and W. Fu, *J. Organomet. Chem.*, 2003, **671**, 158–165.
32   P. A. Midgley, M. Weyland, J. M. Thomas and B. F. G. Johnson, *Chem. Commun.*, 2001, 207.
33   P. A. Midgley, E. W. P. Ward, A. B. Hungria and J. M. Thomas, *Chem. Soc. Rev.*, 2007, **36**, 1477.
34   G. von Helden, M. T. Hsu, P. R. Kemper and M. T. Bowers, *J. Chem. Phys.*, 1991, **95**, 3835.
35   S. Kruckeberg, D. Schooss, M. Maier-Borst and J. H. Parks, *Phys. Rev. Lett.*, 2000, **85**, 4494.
36   H. Grönbeck and J. M. Thomas, *Chem. Phys. Lett.*, 2007, **443**, 524.
37   J. P. Perdew, K. Burke and M. Ernzerhof, *Phys. Rev. Lett.*, 1996, **77**, 3865.
38   B. Delley, *J. Chem. Phys.*, 1990, **92**, 508.
39   B. Delley, *Phys. Rev. B*, 2002, **66**, 155125.
40   H. Wang, Y. Liu, H. Haouari, R. Craig, J. R. Lombardi and D. M. Linsday, *J. Chem. Phys.*, 1997, **106**, 6534.
41   R. D. Adams, B. Captain and R. Herber, in progress.
42   J. M. Thomas, R. Raja, B. F. G. Johnson, T. J. O'Connell, G. Sankar and T. Khimyak, *Chem. Commun.*, 2003, 1126.
43   J. M. Thomas and G. Sankar, *Acc. Chem. Res.*, 2001, **34**, 571.
44   J. M. Thomas, O. Terasaki, P. L. Gai, W. Zhou and J. M. Gonzalez-Calbert, *Acc. Chem. Res.*, 2001, **34**, 583.
45   M. Haider, S. Uhlemann, E. Schwan, H. Rose, B. Kabuis and K. Urban, *Nature*, 1998, **392**, 768.
46   J. M. Thomas and W. Zhou, *ChemPhysChem*, 2003, **4**, 927.
47   A. I. Kirkland and R. R. Meyer, *Microsc. Microanal.*, 2004, **10**, 401.
48   L. C. Gontrand, L. Y. Chang, C. J. D. Hetterington, A. I. Kirkland, D. Ozkaya and R. E. Dunin-Borkowski, *Angew. Chem., Int. Ed.*, 2007, in press.
49   J. H. Sinfelt, *Bimetallic Catalysis: Discoveries, Concepts and Applications*, Wiley, New York, 1983.
50   R. D. Adams, *J. Organomet. Chem.*, 2000, **600**, 1.
51   R. D. Adams, B. Captain and L. Zhu, *J. Am. Chem. Soc.*, 2006, **118**, 13672.
52   O. Korotkikh and R. Farrauto, *Catal. Today*, 2000, **62**, 249.
53   M. Kotobuki, A. Watanabe, H. Uchida, H. Yamashita and M. Watanabe, *J. Catal.*, 2005, **236**, 262.

54 A. Siani, B. Captain, O. S. Alexeev, E. Stafyla, A. B. Hungria, P. A. Midgley, J. M. Thomas, R. D. Adams and M. D. Amiridis, *Langmuir*, 2006, **22**, 5160.
55 M. Haruta, *Nature*, 2005, **473**, 1098.
56 D. I. Enache, J. K. Edwards, P. Landon, B. Solsona-Espriu, A. F. Carley, A. A. Herzing, M. Watanabe, C. J. Kiely, D. W. Knight and G. J. Hutchings, *Science*, 2006, **311**, 362.
57 M. D. Hughes, Y.-J. Xu, P. Jenkins, P. McMorn, P. Landon, D. I. Enache, A. F. Carley, G. A. Attard, G. J. Hutching, F. King, E. H. Stitt, P. Johnston, K. Griggin and C. J. Kiely, *Nature*, 2005, **437**, 1132.
58 A. Corma, L. T. Nemeth, M. Renz and S. Valencia, *Nature*, 2001, **412**, 423.
59 M. Boronat, P. Concepcion, A. Corma and M. Renz, *Catal. Today*, 2007, **121**, 39.
60 (*a*) O. S. Alexeev and B. C. Gates, *Ind. Eng. Chem. Res.*, 2003, **42**, 1571; (*b*) P. Braunstein and J. Rose, in *Metal Clusters in Chemistry*, ed. P. Braunstein, L. A. Oro and P. R. Raithby, Wiley-VCH, Weinheim, 1999, vol. 2, ch. 2.2, p. 616.

General Discussion

Dr Ellis opened the discussion of Professor Hutching's paper: Given that both palladium and gold are active for hydrogen peroxide synthesis in their own right, yet the alloy is more active than either, what do you think is the nature of the active sites in your catalysts?

What can modelling of these particles tell us about the active sites?

Professor Hutchings responded: Clearly the addition of Au to Pd nanocrystals dramatically enhances the activity and selectivity of the catalyst for hydrogen peroxide synthesis. Although both Au and Pd are active, Au is much less reactive than Pd. There can be two effects leading to this enhancement:

(a) site isolation in which the less reactive Au, disrupts the surface structure of the more active Pd, maintaining the sites for hydrogenation of molecular oxygen, but eliminating the non-selective sites for hydrogen peroxide hydrogenation and hydrogen combustion, since different surface structures are likely to be required for these reactions.

(b) an electronic effect whereby Au electronically promotes the activity of Pd through the formation of a dilute alloy.

The effect appears to be related to Au–Pd alloys and a few experiments we have carried out with other systems give very inferior catalysts, hence on balance it is unlikely to be just site isolation. As the carbon-supported catalysts shows the most active and selective performance and this catalyst has the highest concentration of small particles (see Fig. 3 in the paper) we consider that small particles (*ca.* 2–3 nm) that contain mainly Pd with a few percent of Au are quite possibly the most active structures in the catalyst. It is quite possible that modelling will be able to provide valuable insights into the activation of molecular oxygen and hydrogen on the surface of metallic nanoparticles, and we look forward to such studies being completed.

Professor Johnston said: To what extent is calcination essential in making the Pd come to the surface? What sort of segregation is observed (if any) if the particles are heated under a reducing environment? Calculations that we have recently performed (using semi-empirical potentials, but also confirmed at the DFT level)—albeit for much smaller clusters—indicate an energetic preference for Au to segregate to the surface.[1]

1   L. O. Paz-Borbon, R. L. Johnston, G. Barcaro and A. Fortunelli, submitted for publication.

Professor Hutchings replied: We consider the calcination in air at elevated temperatures to be essential to induce this segregation to form the core shell structure. We have recently studied this process in great detail[1] for the Au–Pd/Al$_2$O$_3$ catalyst. When the material is dried at *ca.* 100 °C the core–shell structure is not pronounced and the structure is that of a homogeneous alloy; the core shell gradually develops on heating at higher temperatures in air becoming fully established at 400 °C. As the structure forms so the activity of the catalyst decreases but the stability is enhanced, particularly when calcined at 400 °C. The environment that is used for the heat treatment is very important, since when an Au–Pd/Al$_2$O$_3$ material that has been calcined at 400 °C (and at this stage has a Pd-rich shell and an Au-rich core) is subsequently heated in hydrogen at the same or a higher temperature the structure inverts to form a new core–shell structure with an Au-rich surface and a Pd-rich core.[1] However we do not consider that this effect can be induced under the reaction conditions as the temperature is too low (2 °C), the oxygen is in excess with respect to the hydrogen and the catalyst activity is maintained on reuse. For Au–Pd/TiO$_2$ catalysts, heating the calcined catalyst in H$_2$ at 500 °C leads to a reduction of the Pd^{2+} to Pd^0 but inversion of the core–shell structure is not observed, at least by XPS measurements;[2] this contrasts with our observations for the Au–Pd/Al$_2$O$_3$ catalysts[1] and this further shows the importance of the support structure for these catalysts.

1 A. A. Herzing, A. F. Carley, J. K. Edwards, G. J. Hutchings, and C. J. Kiely Micro-structural Development and Catalytic Performance of Au–Pd Nanoparticles on Al$_2$O$_3$ Supports: The Effect of Heat Treatment, Temperature and Atmosphere, submitted for publication.
2 J. K. Edwards, B. Solsona, P. Landon, A. F. Carley, A. Herzing, C. J. Kiely and G. J. Hutchings, *J. Catal.*, 2005, **236**, 69.

Professor Bond commented: The incredibly powerful methods now available examining the structure of individual nanoalloy particles provides an opportunity to carry forward our understanding of catalytic reaction mechanisms in a way previously unknown. It is however necessary to enter one or two caveats. First, the number of particles actually resolved is quite small, and one cannot be certain that they are representative of the whole. Second, as I have said in my comments on Professor Kiely's paper, it is also helpful to know the location of those particles within the support grain, as when the pores are liquid-filled as they will be during reaction, only those close to the surface may be able to contribute. Third, the detailed structure of supported metal catalysts can depend critically on the chemistry being performed in the preparation, and this is notoriously hard to control and reproduce. The availability of these powerful and refined methods of characterisation requires a subtle analysis of the factors in the preparation that determine the ultimate structure, and selecting the conditions of the preparation with the greatest care. For example, the rate at which a sample is dried can determine the location of the alloy particles within the pore structure.[1]

There is a further difficulty to be faced. It cannot be absolutely certain that the techniques of characterisation accurately reflect the structure that pertains under reaction conditions. It is possible, for example, that the core–shell structure develops (when it does) on exposure to the air before the sample is placed under vacuum; the exothermic adsorption of oxygen on Pd atoms may well account for their appearance on the surface, this process of 'extractive chemisorption' having been established long ago for strongly adsorbing molecules. The role of the carbon support seems to be to prevent the ordering that the other systems experience; topotactic and interfacial energy factors may be important.

1 G. C. Bond, *Heterogeneous Catalysis, Principles and Applications*, 2nd edn., Oxford University Press, Oxford, 1987, p. 79.

Professor Hutchings replied: We agree with the caveats raised concerning micro-scopy as a characterization tool for catalysts, albeit a very powerful tool. For this reason we advocate the use of a whole range of techniques to characterize if this is possible, since by using complementary techniques one can build up a more detailed understanding of the active site. In this case we have also used XPS which is a powerful surface sensitive technique that averages the data over many catalyst particles and thereby analyses considerably more particles than that accessed by electron microscopy. XPS can play a role in determining whether the metal particles are in the pores or on the surface, since the escape depth of the photoelectrons means that components deeper in the pores are not detected. We have demonstrated this in a recent study of Au/Fe$_2$O$_3$ where particle size and catalyst activity varied with calcination temperature.[1,2] The core shell structure is observed in Au–Pd catalysts calcined in air and is stable. It is likely that 'extractive chemisorption' does occur, but during this calcination step and not during the transfer to the vacuum system.

Uncalcined catalysts do not show any core–shell structure and these are of course exposed to the air.

1   G. J. Hutchings, M. S. Hall, A. F. Carley, P. Landon, B. E. Solsona, C. J. Kiely, A. Herzing, M. Mckee, J. A. Moulijn, A. Overweg, J. C. Fierro-Gonzalez, J. Guzman and B. C. Gates, *J. Catal.*, 2006, **242**, 71.
2   P. Landon, J. Ferguson, B. E. Solsona, T. Garcia, S. Al-Sayari, A. F. Carley, A. Herzing, C. J. Kiely, M. Makkee, J. A. Moulijn, A. Overweg, S. E. Golunski and G. J. Hutchings, *J. Mater. Chem.* 2006, **16**, 199.

Professor Sir Thomas commented: The highest pressures that 'environmental' HREM experts (like Pratibra Gai) have been able to operate with so far is *circa* 300 Torr, which, although very helpful, is far removed from the pressures used for certain oxidations and selective hydrogenations (*circa* 30 bar). Useful results have nevertheless been retrieved using such techniques as Gai and colleagues have developed.

Professor Hutchings answered: This point is raised in relation to what is the nature of the active species during the catalysis reaction, and this, of course, is a key question. At present we have reported the analysis of the fresh catalyst. The detailed work we carry out involving aberration corrected STEM-XEDS mapping cannot yet be carried out *in-situ* within the environmental microscope developed by Gai and co-workers. However, we calcine the catalysts at temperatures up to 400 °C for 3 h, and then use them at 2 °C in a very dilute reaction mixture for a few minutes. Hence we consider that the structures formed at the high temperature should be resilient under these very mild reaction conditions. However, we agree that *in-situ* EM character-ization of nanoparticle composition is a highly desirable goal.

Professor Kiely asked: In relation to Professor Sir Thomas' question about the applicability of environmental TEM experiments of the type pioneered by Pratibha Gai to our supported AuPd systems, I would like to add three pertinent comments. Firstly, our reported analysis of the composition of individual particles relies upon being able to perform high spatial resolution STEM-XEDS analysis using an aberration corrected probe. To date, no instrument has yet been constructed that combines aberration correction technology and an environmental cell chamber, although I believe one has been designed and will shortly to be delivered to Professor Gai at the University of York. Secondly, although the imaging performance of the environmental microscope will not be severely compromised by the presence of a gas atmosphere around the sample, the spatial resolution of the STEM-XEDS mapping technique will be degraded. This is inevitable because any incident high energy electrons that have been scattered by ambient gas molecules prior to hitting the sample, are capable of generating X-rays from areas that are remote from the volume of sample being analyzed under the footprint of the focused probe. Thirdly, the supported AuPd system for the direct production of hydrogen peroxide would be a non-ideal choice for *in-situ* TEM studies, because it involves solid catalysts, gaseous reactants and liquid products.

Professor Sir Thomas answered: All the points that you raise are valid, especially the final one, which reminds us that well-nigh all *in situ* methods so far introduced[1] to study catalysts in the act of turnover are inapplicable when one has to contend with gaseous reactants leading to liquid products from a solid catalyst. One must design a procedure, *e.g.* involving incoming X-ray and IR photons together with a mass-spectrometric or gas-chromatographic means for a fuller picture of the state of the AuPd surface.

1   J. M. Thomas, *Chem.–Eur. J.*, 1997, **3**, 1557.

Professor Henry asked: What is important in *in situ* experiments is to have the right coverage? We have observed, by *in situ* HRTEM, gold nanoparticles in various gas environments (O_2, H_2) it turns out that in the mbar pressure range the shape of the particles changes reversibly with the nature of the gas from a Wulff shape (truncated octahedron) to a more or less spherical shape.

Professor Sir Thomas responded: Your comment is interesting and valid. My colleagues and I have reported [1–3] on the ease with which nanoparticles of Au, Pt, Pd become spherical (on a graphite surface) when heated at modest temperatures—well below their bulk melting points—in gaseous O_2.

1   J. M. Thomas, E. L. Evans and J. O. Williams, *Proc. R. Soc. London, A*, 1972, **331**, 417.
2   J. M. Thomas and P. L. Gai, *Adv. Catal.*, 2004, **48**, 171.
3   P. P. Edwards and J. M. Thomas, *Angew. Chem., Int. Ed*, 2007, **46**, 2.

Professor Johnson commented: Two points:

(1) Have you considered preparing well-defined AuPd clusters and de-ligating them to produce well-defined nanoparticles of known size and composition?

(2) What other systems have you examined; I wonder particularly about AgRu, or CoRu—my suggestion being based simply on the ability of Co to produce O_2 and Ru systems to activate and hold hydrogen.

Professor Hutchings answered: Your suggestion of using well-defined clusters would be a useful way forward, especially if they can have relatively large numbers of atoms (*i.e.* hundreds). At present we have not investigated this approach.

We have evaluated the use of Au–Pt, Au–Ru and Au–Rh catalysts and these are all very inferior to the Au–Pd catalysts.[1] Of these the Au–Pt catalysts were more active than the corresponding Au catalyst, whereas the Au–Ru and Au–Rh catalysts showed much lower activity. Hence, for Au the addition of Ru is not beneficial. We have not examined the Co–Ru system and this could, of course, be potentially much better.

1   G. Li, J. K. Edwards, A. F. Carley and G. J. Hutchings, *Catal. Today*, 2007, **122**, 361.

Dr Theobald commented: Why is there a tendency to form core–shell alloy particles on oxide supports but not on carbon?

Is the high activity of the carbon supported alloy due to the random alloy rather than the core–shell structure or is the carbon support getting involved?

Professor Hutchings responded: Concerning the first point about the tendency to form core–shell structures, this a very interesting point. To date we have observed core–shell structures on TiO_2-, Al_2O_3- and Fe_2O_3-supported Au–Pd nanocrystals when these materials are calcined in air at 400 °C. In a very recent detailed study[1] of the Au–Pd/Al_2O_3 catalyst we have studied the formation of the core–shell structure as a function of calcination temperature. When the material is dried at *ca.* 100 °C the core–shell structure is not pronounced and the structure is that of a homogeneous alloy; the core shell gradually develops on heating at higher temperatures in air becoming fully established at 400 °C. For the carbon-supported material the material clearly remains as a homogeneous alloy. In all cases following calcination the palladium is present as Pd^{2+} and Pd^0 whereas the gold is present as Au^0, and so we don't consider that it is preferential oxidation on the oxidic supports that is causing this effect. Rather, it is probably the nature of the surface species on the support, *i.e.* degree of hydroxylation, that exerts a controlling influence.

We do not consider that the carbon is playing an active role in the hydrogen peroxide synthesis reaction, but as shown in the paper (Fig. 7) the support does play a key role in the subsequent hydrogenation reaction in which hydrogen peroxide is reacted with water. The origin of the higher activity of the carbon-supported Au–Pd nanoparticles is due to the higher concentration of small nanoparticles (2–7 nm) when compared with the oxide-supported catalysts (see Fig. 3 in the paper).

1   A. A. Herzing, A. F. Carley, J. K. Edwards, G. J. Hutchings, and C. J. Kiely, Micro-structural Development and Catalytic Performance of Au–Pd Nanoparticles on Al₂O₃ Supports: The Effect of Heat Treatment, Temperature and Atmosphere, submitted for publication.

Professor Henry said: About the fact the core shell structure occurs on TiO_2 and Al_2O_3 supports and not on amorphous carbon, are the metal particles in epitaxy on the crystalline support?

Professor Hutchings responded: We have not observed any epitaxial relationships between the crystalline support and the Au, Au–Pd or Pd nanoparticles.

Dr Russell said: Do you notice any disruption to/alteration of the carbon support in the area around the metal particles?

Professor Hutchings replied: Within experimental error we do not see any reaction of the carbon support, but we will perform more detailed electron microscopy studies to determine if more subtle effects are playing a role.

Professor Bond asked: It should also be possible to make bimetallic Pd–Au catalysts by deposition–precipitation, since hydrolysis of the $PdCl_4^{2-}$ ion by base leads to the precipitation of hydrous $Pd(OH)_2$ after three of the four Pd–Cl bonds have been hydrolysed. It should not be too hard to find conditions under which both the Pd and the Au are deposited together.

If it is desired to remove the product from the neighbourhood of the catalyst as quickly as possible, it might be worth considering another reactor configuration; for example, a pressurised trickle-column reactor might afford at least a dilute solution of the peroxide when operating at low conversion.

Professor Hutchings answered: The suggestion of using deposition–precipitation as a method of preparation is well made and we will certainly investigate this. The initial stage of work has focused on catalyst discovery using the relatively simple co-impregnation method. We have started to use an alternative method involving a sol–gel method which produces much smaller nanoparticles that the co-impregnation method and our initial studies with this method shows that the catalysts can be as active but utilise much lower concentrations of Au–Pd (typically 1 wt%).

At present we have only used a stirred batch autoclave for our experiments. The suggestion of using a continuous flow reactor is very valid and we will be investigating the use of such reactors in the next stage of the work.

Professor Sermon communicated: I imagine that the maximum activity you see for 2.5wt%Au–2.5wt%Pd/C (55) compared to 5wt%Au/C (1) would be even larger on a turnover frequency (TOF) or atom basis (bearing in mind the fact that you have almost twice as many Pd atoms as Au atoms at a particular loading). You mention that no halide promoter is present, but is there XPS evidence of precursor Cl^- having left the active surface (rather than the C) completely?

Professor Hutchings communicated in reply: The improved dispersion of the Au–Pd nanoparticles on the carbon (see Fig. 3 in the paper) gives rise to the enhanced activity and this is related to the increased surface area of Au–Pd available for reaction. The optimal ratio for Au–Pd varies with the support being Pd : Au = 2 : 1 on a molar basis for Au–Pd/TiO₂ and Pd : Au = 1 : 2.6 on a molar basis for Au–Pd/

Al_2O_3.[1,2] You are correct that there is residual chloride present on the catalyst, and this indeed may be playing an *in-situ* promotional role. The key point we are making with respect to promoters is that we do not add additional halide. Typically Pd catalysts for the direct synthesis reaction have to have substantial quantities of bromide added to ensure the stability of the hydrogen peroxide under reaction conditions.[3] This added bromide has to be removed from the product before it can be used. Our Au–Pd catalysts do not require the addition of bromide, and indeed we find added bromide to be deleterious.[4]

1 J. K. Edwards, B. Solsona, P. Landon, A. F. Carley, A. Herzing, M. Watanabe, C. J. Kiely and G. J. Hutchings, *J. Mater. Chem.*, 2005, **15**, 4595.
2 J. K. Edwards, B. Solsona, P. Landon, A. F. Carley, A. Herzing, C. J. Kiely and G. J. Hutchings, *J. Catal.* 2005, **236**, 69.
3 V. R. Choudhary and C. Samanta, *J. Catal.*, 2006, **238**, 28: V. R. Choudhary, C. Samanta and P. Jana, *Appl. Catal., A*, 2007, **317**, 234.
4 J. K. Edwards, A. Thomas, A. F. Carley, A. A. Herzing, C. J. Kiely and G. J. Hutchings, Au–Pd supported nanocrystals as catalysts for the direct synthesis of hydrogen peroxide from H_2 and O_2, *Green Chem.*, 2008, DOI: 10.1039/b714553p

Professor Sermon communicated: Have you prepared your Pd–Au nanoalloy catalysts using non-halide precursor? I was just wondering (although your catalysts are free from halide parameter) whether any Cl^- could remain from $HAuCl_4$ or $PdCl_2$ associated with the active site? It would be interesting to compare the activities of Au/C, PdAu/C and Pd/C on a metal atom basis or a TOF basis.

Professor Hutchings communicated in reply: At present we have only used halide-containing precursors for Au–Pd catalysts, however 5wt%Pd/TiO_2 has been prepared using $Pd(NO_3)_2$ and $PdCl_2$ precursors and the activity for both was the same within experimental error. Indeed there is residual chloride in the calcined active Au–Pd catalysts and this could be playing a role in promoting the stability of the product, since this is the role of halide additives in the direct synthesis reaction. The calcined catalysts can be used in water/methanol several times (see Fig. 1 in the paper) but we have yet to determine the fate of the halide that remains on the catalyst as a function of use. The use of a halide-free precursor would indeed be interesting but as yet we have not investigated the use of starting materials other than $HAuCl_4$.

Professor Hutchings opened the discussion of Dr Caps' paper: The observation that the oxidation of CO is enhanced when hydrogen is present is a very significant finding. Haruta and co-workers[1] have shown that hydrogen aids the activation of dioxygen for propene epoxidation. Your finding might suggest that your catalysts could be active for selective oxidation, since catalysts that can activate H_2 and O_2 are known to be selective for H_2O_2 synthesis[2] and alcohol oxidation.[3] However, it might be that a peroxy initiator[4] might be required. You could investigate the possibility using gas/solid or gas/liquid/solid systems that could use the small amount of catalyst available by this preparation method.

1 A. K. Sinha, S. Sedan, S. Tsokota and M. Haruta, *Angew. Chem.*, 2004, **43**, 1546.
2 J. K. Edwards, B. Solsona, P. Landon, A. F. Carley, A. Herzing, C. J. Kiely and G. J. Hutchings, *J. Catal.*, 2005, **236**, 69.
3 D. I. Enache, J. K. Edwards, P. Landon, B. Solsona-Espriu, A. F. Carley, A. A. Herzing, M. Watanabe, C. J. Kiely, D. W. Knight and G. J. Hutchings, *Science*, 2006, **311**, 362.
4 A. K. Sinha, S. Seelan, S. Tsubota and M. Haruta, *Angew. Chem., Int. Ed.*, 2004, **43**, 1546.

Dr Caps answered: We believe indeed that catalysts that can activate both H_2 and O_2 have a potentially wide range of applications in oxidation catalysis. This is apparently the case for the catalysts presented in this study prepared by laser vaporisation, but it is also the case for many gold supported-catalysts prepared by chemical or even metallurgical routes.[1] It actually seems to be a general feature of gold catalysts. Your suggestion of applying this potential of gold catalysts to other reactions is thus very interesting. We will certainly investigate the catalytic behaviour of these materials in the aerobic epoxidation of alkenes that we have started to study.[2,3] Although the samples obtained by laser vaporisation have low gold loadings, the fact that similar gold size distributions are achieved on all supports will allow us to carry out systematic studies of the support effect in these reactions.

1 M. Lomello-Tafin, A. Ait Chaou, F. Morfin, V. Caps, J.-L. Rousset, *Chem. Commun.*, 2005, 388.
2 P. Lignier, F. Morfin, L. Piccolo, J.-L. Rousset, V. Caps, *Catal. Today*, 2007, **122**, 284
3 P. Lignier, F. Morfin, S. Mangematin, L. Massin, J.-L. Rousset, V. Caps, *Chem. Commun.*, 2007, 186.

Professor Sir Thomas commented: Your electrostatic quadrupolar selector that you described at the end of your talk is impressive. But the smallest particle size that you seem to be able to reach by this route is 2.3 ± 0.3 nm. This means that your nanoalloy particles contain from 900 to 1200 atoms. This is beyond the cluster limit that I have talked about earlier. Can this technique be pushed even further so that you could reach say 0.5 to 0.8 nm diameter clusters?

Dr Caps replied: The electrostatic quadrupole that we use will not allow us to select particles with diameters less than 1.5 nm, which would require the use of very low voltages. However, lower diameters could potentially be achieved by tuning the composition of a vaporised alloy rod made of immiscible elements. For example, by vaporising a bimetallic rod, such as $Au_{25}Fe_{75}$, and selecting the 1.5 nm AuFe particles, we could, upon exposure to air, obtain separated 0.9 nm Au clusters and FeO_x particles. The gold particle size could be further lowered by lowering the concentration of gold in the vaporised rod.

Professor Bond remarked: The result that I found most surprising was the relative inactivity of the AuFe cluster for CO oxidation, since Fe oxides are known to be effective supports for Au for this reaction. I noted that you showed the classic diagram of the variation of edge, corner and planar atoms (*viz.* the variation coordination number types) with particle size. Now these calculations were performed only on perfect crystal forms, which with catalysts made by chemical means are statistically improbable. Real particles will typically have rough surfaces caused by incomplete outer layers. These calculations must therefore be used with great care.

Dr Caps answered: The low activity of the clusters produced from the vaporisation of an $Au_{25}Fe_{75}$ alloy is indeed surprising. However, these systems have not undergone any thermal treatment other than heating under reaction conditions up to 300 °C. The effective Au/Fe_2O_3 catalysts that you refer to have been calcined to 400 °C, a temperature higher than that required to activate Au/TiO_2 or even Au/Al_2O_3 materials ($\sim$250–300 °C) for an PROX reaction. We might thus not have the adequate Au/FeO_x interaction to achieve high CO oxidation/PROX activity. Furthermore, one must consider the way in which these Au/FeO_x systems are formed: exposure to air apparently causes iron atoms to be expelled from the AuFe particle, due to the low miscibility of these two elements (*cf.* binary phase diagram) and the strong affinity of iron for oxygen. Non stoichiometric, amorphous iron oxides are formed (XRD, XPS) and the potentially mobile gold atoms might well diffuse on the alumina support and migrate away from the iron nanooxides.

We agree that the occurrence of perfect crystal forms in chemically- or even physically-made catalysts is improbable. Our catalysts are prepared using low-energy cluster beam deposition, a physical method in which the metal particles are softly landing on any kind of support with no fragmentation observed. Although these systems are initially believed to be closer to model catalysts (as isolated Pd particles directly deposited on a C/Cu microscopy grid and separated by each other by tens of nanometers suffer from no interaction, at the nanoscale level, with a powder support or with another metal particle), transfer to air and subsequent exposure to reaction conditions have proven to dramatically affect the geometry of the particles. The diagram showing the frequency of atoms in planes, edges and corners as a function of the particle size was merely an illustration of the utility of size-selection for the study of the growth mechanism of carbon nanotubes over catalytic metal particles for field emission applications, as it shows the theoretical drastic changes in the particle morphology even for a very narrow size range (1.2 to 4.4 nm).

Professor El-Shall asked: The catalytic activities of your catalysts for CO oxidation are not high in spite of the very small 2 nm particle size. The activities of Au/TiO_2 nanoparticles prepared by the deposition–precipitation method are much higher since 100% CO conversion occurs at very low temperatures much below 0 °C. What is the reason for the low activity? How is the catalyst-support interaction characterized in your systems?

Dr Caps replied: The catalytic activity and especially the temperature at which a given conversion is achieved depends on the reactions conditions used (concentration of each reactant gas, total flow rate, space velocity...). Under our conditions, our AuTi systems are indeed 7–8 times less active than the reference catalyst of the World Gold Council for CO oxidation[1] despite a slightly lower average gold particle size (2–3 vs 3.7 nm). Several factors might contribute to the superiority of the reference catalyst prepared by deposition–precipitation:

—the use of the peculiar P25 support which is a mixture of anatase (80%) and rutile (20%) titania. P25 has indeed proven superior to all other titania supports considered for chemical preparations. It has been proposed that the anatase/rutile grain junctions might stabilize highly active gold particles.[2] In our case, we have used a Hombikat T100 titania support from Sachtleben Chemie with similar surface area (60 $m^2 g^{-1}$) but 100% anatase crystallites.

—the different metal–support interaction: the interaction between a physically deposited metal particle or a chemically-grafted metal precursor and the support might be significantly different even after subsequent heat treatment.

—a chemical promotion by the sodium cations present in the solid (~ 500 ppm) arising from the base (NaOH) used during the preparation. One of the advantages of the catalysts prepared by our physical method is precisely the absence of heteroatoms. These "clean" catalysts allow us to probe the actual activity of supported gold nanoparticles with no interference from other elements.

1   C. Rossignol, S. Arrii, F. Morfin, L. Piccolo, V. Caps, J.-L. Rousset, *J. Catal.*, 2005, **230**, 476.
2   M. Haruta, *Gold Bull.*, 2004, **37**, 27.

Professor Johnson asked: In your description you refer to surface atoms on your nano-catalysts. Given the apparent disparity of results using catalysts prepared by different synthetic routes, it is possible that even for these very small systems, surface irregularities are significant *i.e.* more-or-less steps, kinks *etc*?

Dr Caps replied: As noted by Prof. Bond (see later), there must indeed be a gap between the theoretical models of metal particles and the morphology of the particles that are actually involved in catalytic reactions. Surface irregularities might indeed be significant even for these very small systems.

Professor Sermon asked: I was interested in your results for H_2-promotion of your CO oxidation. I was reminded that with nanoalloys on supports we still do not know their temperature during catalysed reactions. In an exothermic reaction this could be higher than the support as a result of the reaction. Is the addition of H_2 changing the surface chemistry alone, raising the temperature of the nanoalloy surface or is it doing both?

Dr Caps replied: One way in which hydrogen could raise the temperature of the gold nanoparticles would be by reacting with oxygen to form water, thereby releasing heat. However, results in our group show that the CO oxidation rate is enhanced before hydrogen conversion is detected.[1] Furthermore, the fraction of water that could be formed without being detected could account for less than 1% of the CO oxidation rate promotion by H_2, as shown by our latest results on Au/Al_2O_3[2] and based on the study of the water concentration effect over supported gold catalysts.[3] We believe that the temperature effect is minimal, although it cannot be fully excluded, and that the addition of H_2 is mostly changing the surface chemistry, the nature of the reaction intermediates and the way by which O_2 is activated.

1 C. Rossignol, S. Arrii, F. Morfin, L. Piccolo, V. Caps, J.-L. Rousset, *J. Catal.*, 2005, **230**, 476–483.
2 E. Quinet, F. Morfin, F. Diehl, P. Avenier, V. Caps, J.-L. Rouset, Hydrogen Effect on the Preferential Oxidation of Carbon Monoxide over Alumina-supported Gold Nanoparticles, *Appl. Catal. B*, submitted.
3 M. Daté, M. Okumura, S. Tsubota, M. Haruta, *Angew. Chem., Int. Ed.* 2004, **43**, 2129–2132.

Professor Evans opened the discussion of Professor Sermon's paper: Previously we have reported time resolved XAFS studies on the formation of Pt–Ge particles from metallorganic precursors. During a temperature ramp under hydrogen, and using a mesoporous silica as a support, reduction of platinum occurred before there was any reaction of the germanium.[1,2] At about 500 K, the platinum particle size reduced as new Pt–Ge particles were formed. So this pattern supports your two stage view of Pt–Sn alloy formation. What structures can be observed after catalysis once the carbonaceous layer is formed?

1 S. G. Fiddy, M. A. Newton, A. J. Dent, G. Salvini, J. M. Corker, S. Turin, T. Campbell and J. Evans, *Chem. Commun.*, 1999, 851.
2 S. G. Fiddy, M. A. Newton, T. Campbell, A. J. Dent, I. Harvey, G. Salvini, S. Turin and J. Evans, *Phys. Chem. Chem. Phys.*, 2002, **4**, 827.

Professor Sermon replied: We believe (as you did in your Pt–Ge system) that there is evidence of partial nanoalloy formation (*e.g.* from the TPR peak at $\sim$480K). We now have work in hand on HRTEM characterisation of such nanocatalysts[1] and hope to clarify this in due course.

1 F. Besenbacher, J. V. Lauritsen and S. Wendt, *Nano Today*, 2007, **2**(4), 30.

Professor Bond remarked: It is quite clear that the PtAu/C system behaves very differently from the $PtSn/Al_2O_3$ system, but it is not obvious whether this is due to the presence of the modifying metal or to the change of support. Thus in Fig. 5, 6 and 7 the Pt/Al_2O_3 and $PtSn/Al_2O_3$ afford very similar results for both reactants, and in particular the 2MP/3MP ratio is always close to two. This shows that the n-hexane first undergoes dehydrocyclisation to MCP, which is then hydrogenolysed to give the same product ratio as when MCP is the reactant. This ratio does not vary with either time-on-stream, temperature or Sn content (at least up to the 50% point), which would suggest that the active centre is unaffected by these changes. It is therefore not easy to see what useful information is given about the surface composition.

What is most surprising about the results for the PtAu/C system is the very high 2MP/3MP ratios that are often observed. Can you conceive of any mechanism for the hydrogenolysis of MCP that would so favour the breaking of the C–C bond nearer to the point of substitution?

Professor Sermon responded: It seemed to us that 2MP/3MP ratios starting with either MCP or n-hexane were interesting for fingerprinting nanoalloy surfaces. Others have used a similar approach with hydrogenative ring opening (HRO) of *trans*-methyl-ethyl-cyclopropane (MECP)[1] where there was a maximum 2MP/3MP ratio at Pd : Pt = 1 : 1, and isomerization of n-hexane over Pd–Pt.[2] You are right to point out that the 2MP/3MP ratio is higher in Fig. 6 and 7 for PtAu than PtSn. 2MP/3MP > 2 have been seen[3] for n-hexane and methylcyclopentane fingerprinting of alumina-supported Pd–Re nanoalloy particle surfaces (despite n-hexane conversion dropping with time on stream at 513 K). Thus they found 2MP/3MP ≈ 3 from n-hexane (see footnote to their Table 2) and for MCP the 2MP/3MP was up to 4.5 at 513 K. Furthermore 2MP/3MP values of ≈3–4 were seen even for EuroPt-1 6.3%Pt/silica at 496 K.[4] We believe that in the PtAu catalysts there is 3MP depression (*via* fractal or surface atom configuration mechanisms) rather than 2MP elevation. If 2MP/3MP does not vary with nanoalloy composition to 50% then that in itself is telling us something about the surface composition.

1 N. Gyorffy, L. Toth, M. Bartok, J. Ocsko, U. Wild, R. Schlogl, D. Teschner and Z. Paal, *J. Mol. Catal.*, 2005, **238A**, 102.
2 A. Barrera, J. A. Montoya, M. Viniegra, J. Navarrete, G. Espinosa, A. Vargas, P. de Angel and G. Perez, *Appl. Catal.*, 2005, **290A**, 97–109.
3 W. Juszczyk and Z. Karpinski, *Appl. Catal.*, 2001, **206A**, 67.
4 Y. Zhuang and A. Frennet, *Appl. Catal.*, 1999, **177A**, 205–217.

Professor Hutchings asked: There are likely to be glucose residues on the surface of the catalysts following preparation. They are possibly at very low concentration and will be severely oxidised/reduced; however, these residues could be active as catalysts for the reactions studied. There are several literature examples of surface carbon species being active as catalysts. Can you comment on this?

Professor Sermon responded: That is an interesting point. Of course (i) the catalyst is pre-calcined at 673K for 2h and then pre-reduced at 673K for 2h and (ii) the MCP and n-hexane rapidly cover the nanoalloy surfaces with carbonaceous deposits. However, we will explore this point further.

Dr Baletto asked: How can the structure details influence the catalytic properties of the nanoalloy? To fill the gap between experiment and numerical approaches, what does the theoretical work have to say to the experimental community?

Professor Bond answered: The comparison of catalytic activity with theoretical calculations might be made more fruitful if the calculations were extended to the mode and strength of binding of potential reactants, especially of small molecules such as hydrogen. They could then be compared with measurements of the 'incestuous' reactions of hydrogen, *i.e. para*-hydrogen conversion and hydrogen–deuterium exchange. These reactions have passed out of fashion and are generally forgotten, but they are useful when it is necessary to keep the system simple: here one has only one reactant to worry about.

Professor Jellinek commented: There is a view shared by quite a few that catalysis can be understood in terms of binding energies, preferred binding sites or arrangements, and the barrier heights that separate these sites and arrangements from the approaching reactants. Information on these is usually sought through static (*i.e.*

fixed nuclear configuration) electronic structure computations. I do not belong to the group that shares this exclusive view. Whereas the mentioned data are important (and the more accurate they are the better), catalysis is a dynamical/kinetic phenomenon, which involves breaking and making chemical bonds. This bond rupture and formation takes place as a consequence of coupling between different degrees of freedom and the energy flow between them. The mechanisms underlying this energy flow (these mechanisms can involve transformations of the catalyst as well) and the nature of the parameters they depend on (these are the parameters one can use to improve catalytic performance) can be identified and understood only through dynamical/kinetic numerical simulations or experiments. For the simulations to be feasible on the desired time scales one has to use computationally efficient, yet accurate, semiempirical potentials. Catalysis is another field that underscores the importance of developing such potentials.

Professor Evans opened the discussion of Dr Russell's paper: Both of your organometallic precursors are air sensitive but will probably decompose in different ways on your electrode materials. The chromocene is likely to lose the first ring very easily, but may retain the second under substantially more forcing conditions. May this contribute to the higher temperature required for it to alloy with platinum? Do you have any information on this activation process?

Dr Russell replied: Yes, both organometallic precursors are quite air sensitive with the chromocene being the most sensitive. Our data suggest that the chromocene $(Cr(Cp)_2)$ precursor decomposes on the Pt/C catalyst indiscriminately, leaving Cr_2O_3 on the catalyst which is located close enough to the Pt/C that upon heating to 900 °C manages to mix well enough to form some of the Pt_3Cr phase, but not to the same extent as the $Cr(Bz)_2$ modified catalyst as evidenced by the larger Pt–Pt distance (2.755 Å as compared to 2.740 Å). We have further plans to study the details of the mechanism of the controlled surface modification reaction in more detail, but other than noting that when ruthenocene was used to modify Pt/C in a separate study that cyclopentane was found in the exhaust gas by mass spectrometry we have no further information at this time.

Professor Johnston remarked: Is the greater incorporation of Cr observed using $(Bz)_2Cr$ due to the fact that it is Cr^0 (rather than Cr^{2+} as in $(Cp)_2Cr$)?
Have you tried any other sources of Cr^0?

Dr Russell answered: We believe the greater incorporation probably may mostly be attributed to the greater stability (reduced air sensitivity) of the $Cr(Bz)_2$ compared to the $Cr(Cp)_2$, such that the controlled surface modification reaction stands a chance of working as intended. The increased stability is related to the fact that it is a Cr^0 compound. No, we have not tried other sources of Cr^0.

Dr Schofield asked: What are the particle sizes in the catalysts described and how far does the Cr penetrate into the particles at 900 °C? What is the driving force for mixing?

Dr Russell replied: I will answer the question in 3 parts:
(1) As can be seen from the profiles obtained by analysis of the TEM images shown in Fig. 1, the typical particle sizes are between 3 and 10 nm in diameter.
(2) From the analysis of the EXAFS data we conclude that the when the surface alloy is formed (as in the case at 750 °C) that the alloy must represent at least the top

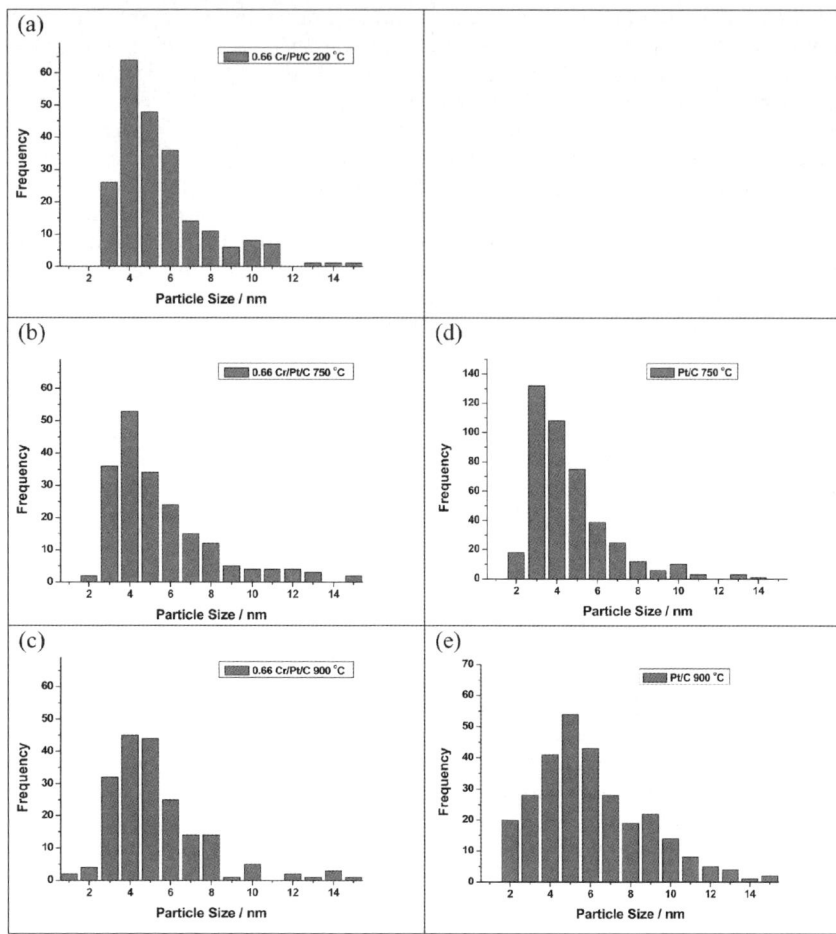

Fig. 1 Particle size distributions obtained by analysis of the TEM images for 0.66 Cr/Pt/C annealed at (a) 200 °C, (b) 750 °C. (c) 900 °C and the unmodified Pt/C annealed at (d) 750 and (e) 900 °C.

2 to 3 layers of the particles, as a significant contraction of the Pt–Pt distance is observed, from 2.758 Å for unmodified Pt/C to 2.748 Å for the 750 °C treated sample. Further heat treatment results in additional mixing of the layers and a further contraction of the Pt–Pt distance to 2.740 Å. (See table 4 in the manuscript).

(3) The driving force for mixing is the formation of the more stable Pt_3Cr phase. As noted by Ruban, Skriver, and Norskov[1] if a Cr layer is deposited onto a Pt substrate, the surface segregation energies are such that Pt will be strongly segregated to the surface, *i.e.* Cr will tend to go sub-surface, and thus the formation of an alloy (mixed) phase is strongly favored.

1 V. Ruban, H. L. Skriver, and J. K. Norskov, *Phys. Rev. B*, 1999, **59**, 15990.

Professor Henry commented: Why is it necessary to have the bulk Pt_3Cr alloy to have a good activity, rather than a surface alloy?

Dr Russell answered: The 'bulk' Pt_3Cr phase is required so that the Pt–Pt bond distance is reduced and the electronic nature of the Pt is modified. The surface alloy can be viewed in the extreme as a Cr decorated Pt nanoparticle. Cr is not simply acting as a promoter in the electrocatalysis, so a 'bulk' alloy is needed to realise the full improvement in activity.

Dr Tromp asked: The results and discussion section describes and discusses the XANES data obtained, as shown in Fig. 2. First of all I have a problem with the general description of the XANES data. The description of the origin of pre-edge features in the Cr K edge as described in paragraph two is not a good and general one. Pre-edge features originate from and are indicative of the geometry, *i.e.* hybridisation of different orbitals (partially) allowing dipole forbidden transitions (in this case the 1s to 4p transition is the dipole allowed one). Mixing of p with s and/or d orbitals in certain geometries will make these partially allowed and thus visible in the Cr K edge XANES). CrO_3 indeed displays a sharp pre-edge and its assignment is correct, but in this study other samples with many other pre-edge features (with a different origin) are observed. Moreover, an edge position shift of 1.6 eV per unit change in valency is far from a general rule, and likely only valid in the specific reference 25 of your paper.[1] Also this reference recognises that shifts are not necessarily easy (*i.e.* directly) to address. We have recently published a paper on Cr compounds which shows that the edge shift is very dependent on ligand type and geometry, and that positive as well as negative edge shifts can take place, sometimes without changing the (official) valence state of the metal.[2]

In this same study we have seen a maximum edge shift of 8–10 eV for samples with very different ligand systems and valence states. It seems therefore very unlikely that an edge shift of 20 eV can be real (studying oxidic to metallic samples) as observed in Fig. 2. Similar studies in the literature show a maximum of 5–6 eV for the same reference materials. Can you comment on this? Is there perhaps a problem with the experiment?

Since there seems to be a problem with the energy calibration, this should be noted and the changes in XANES and subsequent assignments (paragraph 3 in 'Results and Discussion' on page 5, and paragraph 2 on page 7) should be amended. It is probably better to describe the pre-edge features observed and make assignments and draw conclusions based on that? (The pre-edge features are very distinct and indicative, and are in this case more reliable).

1   A. Panterlouris, H. Modrow and M. Pantelouris, *Chem. Phys.*, 2004, **300**, 13.
2   M. Tromp, J. Moulin, G. Reid and J. Evans, *AIP Conf. Proc.*, 2007, **882**, 669.

Dr Russell replied: We thank Dr Tromp for her detailed comments, and yes the potential shifts observed in our data are too large to be accounted for by a simple change in oxidation state. Unfortunately we were not able to collect the data using a third ionisation chamber and metal reference foil, so it is not possible to align the spectra properly and, perhaps, we have erred by pointing out the edge shifts. However, our conclusions are supported by the differences in the pre-edge features seen in the various spectra as pointed out by Dr Tromp and justified by the detailed analysis of the EXAFS data.

Professor Ferrando opened the discussion of Dr Tromp's paper: Are there any calculations (DFT for example) supporting your proposed PdRh structure of size 19?

Dr Tromp responded: No, we have not yet performed any calculations on this M19 cluster. The cluster is proposed based on the combined results of the different techniques as presented here, in comparison to the results obtained for a mono-metallic Rh system. It is a proposal for the average alloy structure present, which can explain all characterisation results as well as the catalytic ones. Theoretical studies are currently being pursued.

Dr Schofield asked: In the discussion of XPS in the paper, you suggest you have a fourfold enhancement of Pd on the surface of the particles. Half of the particles in question have a particle size of < 3 nm. Previous experience of XPS suggests that, because the penetration depth of the beam is around 3 nm, particles this small will be detected in their entirety, rather than only the surface. On this basis, how accurate is XPS in establishing core–shell structure in particles which are so small?

Dr Tromp replied: XPS studies have been performed *ex situ* (vacuum) on the starting catalytic materials, *i.e.* after pretreatment, prereduction, and subsequent exposure to air. TEM on exactly the same materials shows that the average RhPd particle size is about 3 nm, about 50% of the particles is bigger (with an overall bimodal distribution). The electron mean free path for these XPS experiments is ~2–3 nm. The XPS probes a significantly smaller amount of Rh than Pd, as can been seen. The Rh signal is thus more attenuated than the Pd. This can be due to a smaller amount of Rh *vs* Pd, but the Rh and Pd XAS absorption edge steps indicate that ~4 wt% Rh and ~1 wt% Pd are present in the bulk sample, as expected. The reduced amount of Rh probed with XPS can thus only be due to surface enrichment of the Pd. We acknowledge that exact quantification of these results is difficult, because of the surface sensitive nature of the technique and indeed the possibility that smaller particles contribute differently to the XPS signal than larger particles (bulk *vs* surface). Therefore, no further quantitative assignments towards the amount and type of surface *vs* bulk atoms in the alloy particles is done here. EXAFS results on the starting systems as described in ref. 1 support the Pd surface enrichment.

1   M. A. Newton, B. Jyoti, A. J. Dent, S. Diaz-Moreno, S. G. Fiddy and J. Evans, *ChemPhysChem*, 2004, **5**, 1056.

Professor Sermon communicated: Does the range of particle sizes formed matter? Is there a way of narrowing the range of particle sizes that you see?

Dr Tromp communicated in reply: The particle size range does matter as can be seen in this paper from the comparison between the monometallic and the alloy system. The alloy particle size distribution can be decreased by more controlled synthesis methods like different calcination and reduction procedures or the deposition of well-defined multimetallic (*e.g.* carbonyl) clusters on the support. However, once catalytic reactions are performed with highly oxidising reactants and at high temperatures there is no control on the particle size and distribution any more.

Professor Evans opened the discussion of Professor Sir Thomas' paper: In the past, all our attempts to support low valent clusters often were thwarted by side reactions with the oxide surface,[1] or to direct oxidative grafting on the surface,[2] although this was slowed by use of a face bridging (phosphinidene) ligand.[3] Some organometallics will react with clean oxide surfaces at temperatures below ambient, *e.g.* [RhCl(CO)$_2$]$_2$ on TiO$_2${110}[4] and then convert to metallic particles in minutes at ambient conditions.[5] What is it in your systems that seems to prevent such problems?

1   S. C. Brown, J. Evans, *Chem. Commun.*, 1978, 1063.
2   V. D. Alexiev, N. Binsted, J. Evans, G. N. Greaves, R. J. Price, *Chem. Commun.*, 1987, 395
3   V. D. Alexiev, N. Binsted, S. L. Cook, J. Evans, R. J. Price, N. J. Clayden, C. M. Dobson, D. J. Smith, G. N. Greaves, *J. Chem. Soc., Dalton Trans.*, 1988, 2649
4   J. Evans, B. E. Hayden, F. Mosselmans, A. Murray, *Surf. Sci.*, 1994, **301**, 61.
5   M. A. Newton, R. J. Bennett, R. D. Smith, M. Bowker, J. Evans, *Chem. Commun.*, 2000, 1677.

Professor Sir Thomas answered: Your work has been very important in the context of our studies. But there are some significant differences between your methods of anchoring clusters and ours. Up until very recently we used carbonylate precursors[1] to anchor the cluster on to the high-area silica. We have never used phosphinidene ligands nor TiO_2 as a support. But here, in addition to readily decomposable CO ligands we also used SnO_2 groups. All the indicators are that, certainly the COs (and probably the SnO_2 moieties) are driven away on thermal treatment *in vacuo* or N_2 during the course of preparing the naked cluster.

1   J. M. Thomas, B. F. G. Johnson, R. Raja, G. Sarkar and P. A. Midgley, *Acc. Chem. Res.*, 2003, **36**, 20.
2   T. Maschmeyer, in *Turning Points in Solid-State, Materials and Surface Science*, ed. K. D. M. Harris and P. P. Edwards, Royal Society of Chemistry, Cambridge, 2007.

Professor Bond remarked: From the results that you have presented, it is difficult to know what importance to assign to the presence of tin in your clusters, because you do not show any results for a tin-free cluster; the reaction characteristics you see may well therefore be due simply to ruthenium atoms.

I note that in the reaction of the cyclododecatriene the reaction starts to become slower at about 50% conversion and to become non-selective. Your catalytic system therefore falls into the same category as that of other reductions of multiply unsaturated molecules, namely, high selectivity is found only while enough of the diene or triene is still around to prevent the re-adsorption and reaction of the monoene. Examples that spring to mind are (i) fat hardening, where Ni catalysts can give high selectivity to oleic acid, and (ii) the Pd-catalysed hydrogenation of ethyne to ethene. In this last case and in some others it is possible by the use of selective poisons to stop the reaction at the monoene stage, but it does not appear to happen with your Ru–Sn clusters.

It is very easy to understand why these reactions have shown such good selectivity; it is simply because the more points of potential attachment to the surface that you have, the stronger will be the adsorption, and thus the more difficult it is for the monoene to compete for the surface. No doubt that is also the case here, and I expect it could be confirmed by appropriate calculations on the interaction of the various molecules with the clusters. It might even be possible to compare the calculated interaction energies with experimentally measured heats of adsorption.

Professor Sir Thomas answered: In so far as your first paragraph is concerned it is simply not true to say that we have no comparative results for a tin-free cluster. The *Angew. Chem.* (2007) paper referred to in our paper clearly shows that a Pt–Ru cluster behaves very differently from a trimetallic one (consisting of Pt–Ru–Sn). This is seen in our paper also in regard to the dimethyl terephthalate results.

Your suggestions in the second and third paragraphs are most helpful. You could well be right that it is partly because of the excess of cyclododecatriene that we achieve such good selectivities for the monoene—as in the oleic acid case that you quote. But I have less faith in the calorimetric measurements that you suggest. The trouble with sensitive calorimeters (like the one used by King *et al.*) is that there is great uncertainty as to which species is adsorbed, so one cannot relate the enthalpy measurements in a logical way to the various adsorbed species.

Professor Pastor asked: I would like to come back to the electron holography images you showed and ask you about some details on the capabilities of this technique. What would be the spatial resolution and sensitivity of this method? Would it be appropriate for mapping magnetic domains and domain walls in nanowires? What are the characteristic measuring times? Could it be used to follow the relaxation dynamics of ensembles of interacting nanoparticles?

Professor Sir Thomas answered: The technique of electron nanoholography which I outlined at the end of my presentation is a very new one. From recent work—see ref. 1 and 2—we know enough to be able to answer almost all of your questions in the affirmative. It is certainly capable of mapping magnetic domains and domain walls in nanowires and in colloidal-dimension "necklaces".[2] The spatial resolution is high; better than a few nanometres under ideal conditions. And, although the times of measurement are not small—minutes rather than seconds—it is quite conceivable that relaxation dynamics of ensembles of interacting nanoparticles are already manageable.[3] My colleague Dr Dunin-Borkowski, who has just moved from Cambridge to the Technical University of Denmark, is actively pursuing the kind of questions that interest you.

1 R. E. Dunin-Borkowski, M. R. McCartney and D. T. Smith, *Electron Holography of Nanostructured Materials*, ed. H. S. Naliva, American Scientific Publishers, Stevenson Ranch, CA, 2004, vol. 3, p. 41.
2 J. M. Thomas, E. T. Simpson, T. Kasama and R. E. Dunin-Borkowski, 2007, submitted.
3 A. Tondmura, *The Quantum World Unveiled by Electron Waves*, World Scientific, Singapore, 1998.

Professor Hutchings remarked: Concerning the origin of the high selectivity to the partial hydrogenation product, I would like to add that origin of the effect could be due to preferential adsorption. Have you considered competitive experiments in which the diene is co-fed with the alkene to determine if the diene is preferentially reacted?

Professor Sir Thomas answered: This is an interesting and sensible suggestion. We have indeed considered doing such an experiment. But it is easier to describe than to carry it out. The reason for this is that the reaction takes place under quite high pressure (*ca* 20 bar); and although, on "per gram" terms, the surface area of the active clusters is relatively high, the small amounts of catalyst that we use have low (total) surface areas. We are, however, aiming to do MRI experiments (such as those described by Gladden *et al.*[1]), in the near future to shed light on the precise causes of the remarkable selectivities that we observe.

1   L. F. Gladden, M. D. Mantle and A. J. Sederman, *Adv. Catal.*, 2006, **50**, 2.

Professor Meyer commented: I merely wanted to point out that if your cluster had been anchored to the support through the tin atoms, then you would likely lose the selectivity in the reaction that you desire. Therefore, although it was surprising that the cluster is anchored through Ru, it is also necessary for this to be a successful catalyst. One could however conceive of other metals which may be even better than Ru as anchors (maybe Fe?).

Professor Sir Thomas replied: It is premature, in my opinion, to argue that Fe might be better than Ru as an anchor. Moreover, it is not yet incontrovertibly established, although our calculations suggest otherwise, that the Ru–O linkage is the key bond for anchoring the cluster. More experimental and computational work is needed.

Professor Henry addressed Professor Sir Thomas and commented: During this session we have discussed in detail the size of clusters, their shape (facets, edges), chemical composition (bulk and surface) and their effect on the associated physical and chemical properties, assuming that these parameters are fixed. However for catalysis all these parameters can evolve dynamically during the reaction, then it becomes necessary to have an *in situ* dynamic approach. This problem has been for a long time difficult to solve until now, in recent years, several new techniques have opened the way to solve it. These new techniques are for example; Environmental High Resolution TEM, High-Pressure (compared to UHV) STM and AFM, X-ray diffraction....

It appears for example by using E-HRTEM under a pressure of 2 to 10 mbar that gold particles (around 4 nm) supported on TiO_2 or carbon change quickly and reversibly their shape between vacuum (truncated octahedron = Wulff shape), hydrogen (truncated octahedron) and oxygen (rounded).[1] The presence of gases can also change the surface composition of bimetallic particles and under a high pressure of oxygen noble metal particles can become oxidized.

1 S. Giorgio, S. Sao Joao, S. Nitsche, D. Chaudeson, G. Sitja and C. R. Henry, *Ultramicroscopy*, 2006, **106**, 503.

Professor Sir Thomas responded: Your points are well taken. The dynamics of morphological changes of nanoparticles are a reality; and are not easy to study under truly *in situ* conditions (*e.g.* 1 atm pressure or more of reactant gas). Whilst I greatly applaud and admire the work on E-HRTEM pioneered by Prof. P. L. Gai,[1,2] it must be admitted that the pressures that one can cope with (as she herself has often stated) are still well below those that are encountered in practical situations.

Your reference to *in situ* STM studies is highly significant, and the comparative studies of CO oxidation in O_2 are very revealing. Hendriksen *et al.*[3] have shown, using this technique that a Pd catalyst is, in fact PdO and that the kinetics follows the Mars–van Knevelen mechanism (of sacrificial oxidation). This is a far cry from the UHV-oriented studies on Pd, where the Langmuir–Hinshelwood kinetics apply and the pressure dependence of the reaction rate is very different from that of the "realistic" (1 atm pressure O_2) situation.

I agree with Professor Henry, one must also take into consideration the reversible change of shape of nanoparticles with temperature and pressure.

So far as clusters are concerned, it is quite possible that, if they are anchored securely to the support, they may not undergo structural changes anything like as easily as nanoparticles (where anchoring is less important) do.

1 P. L. Gai and E. D. Boyes, *Electron Microscopy in Heterogeneous Catalysis*, Institute of Physics, Bristol, 2003.
2 J. M. Thomas and P. L. Gai, *Adv. Catal.*, 2004, **48**, 171.
3 B. L. M. Hendricksen, S. C. Bobaru and J. M. W. Frenken, *Top. Catal.*, 2006, **36**, 43.

Professor Bond said: Those who are working at the interface between materials and solid state science on the one hand and chemical reactivity on the other sometimes express surprise that the connection between the two is less clear than they might have been hoped, and disappointment that knowledge of the former is not very helpful in understanding the latter. This is especially so when the aspect of chemical reactivity that is being examined is catalytic activity. I believe that the difficulties are largely due to a lack of sufficient attention to the catalytic phenomenon, that is to say, too little time and effort being devoted to careful study of the reaction and its mechanism. We tend to ask questions and expect answers without seeking information that is readily to hand, quickly and cheaply available, but essential for establishing the desired correlations.

I am continually troubled by papers that I read where inordinate time and money is expended on characterising some catalysts (sometimes very poor ones), while the catalytic component of the paper is short and quite inadequate. It is as if for a piece of work taking a week the catalysis is started after tea on the Friday afternoon. We need first of all to define what we mean by 'activity'. The assessment of a catalyst is often reported as a conversion obtained under a single set of experimental conditions; this leads to statements such as 'catalyst A is ten times more active than catalyst B'. If I read this in a paper I am refereeing, I write a critical report, because such a statement has very little value. This is because typically the relative rates depend on the temperature and reactant concentrations used, so a quite different hierarchy of 'activities' can result on changing these conditions, and all of them are equally valid. We showed many years ago that the activity patterns exhibited by Ni–Cu alloys for ethyne hydrogenation assumed quite different forms when the temperature was altered, because there were different activation energies.[1] As the bare minimum it is necessary to report activation energy (E_{app}), pre-exponential factor (ln A), and orders of reaction; these can indicate even if they cannot establish whether activity differences are caused by changes in the active site concentration or in the electronic character, and to what extent the reactants are chemisorbed. With some further work, the true activation energy and the heats of adsorption of the reactants can be extracted from the temperature dependence of the kinetics.[2] In particular it is inadequate to follow the common practice of those working in the field of environmental catalysis of simply recording a conversion vs temperature plot and a T50. There is also a widespread lack of awareness of when the reaction is under diffusion control. Some kind of analysis of the reaction mechanism is essential before trying to discuss correlations with solid state composition and structure. An attempt was made in 1964[3] to define the smallest amount of information needed before we can say we know the probable mechanism; for convenience this has been repeated recently.[4,5]

To me it is quite incomprehensible why so little care is taken to discover even the barest outlines of the test reaction. It may be because practitioners are more conversant and comfortable with physical methods, the exercise of which seems to demonstrate a kind of academic virility, and less happy paddling in the murky waters of catalysis. Or perhaps it is felt useful to retain unsolved problems, since they provide the excuse for performing further work.

1   G. C. Bond and R. S. Mann, *J. Chem. Soc.*, 1959, 3566.
2   G. C. Bond, F. Rosa C. and E. L. Short, *Appl. Catal., A*, 2007, **329**, 46.
3   G. C. Bond and P. B. Wells, *Adv. Catal.*, 1964, **15**, 91.
4   G. C. Bond, *Metal-Catalysed Reactions of Hydrocarbons*, Springer, New York, 2005, prologue and ch. 5.
5   G. C. Bond, C. Louis and D. T. Thompson, *Catalysis by Gold*, Imperial College Press, London, 2006, ch. 1.

Professor Sir Thomas replied: Whilst I can agree, and even endorse, some of the remarks and philosophical asides made by Professor Bond, I do not think that the situation is anything like as complicated as he makes it out to be. Rather than labour the point here, I would refer him and others to Section 1.3 (*Definition of Catalytic Activity*) of my book (with W. J. Thomas).[1] There, *inter alia*, we state:

"...where A' is a temperature-independent pre-exponential factor and E' is the apparent activation energy of the catalytic reaction E' cannot be expected to be the true activation energy, even if the catalyst structure remains unchanged with varying temperature, because the concentration of reactant at the catalyst surface will, in general, be temperature-dependent. For this and other reasons......it is best not to define catalytic activity in terms of activation energy. Far more convenient is the use of the concept *of turnover frequency* or *turnover number*. The turnover frequency (TOF) is simply the number of times, n, that the overall catalytic reaction in question takes place per catalytic site per unit time for a fixed set of reaction conditions (temperature, pressure or concentration, reactant ratio, extent of reaction). In words,

TOF = (number of molecules of a given product)/((number of active sites) × (time))

or TOF = $(1/S)(dn/dt)$

where S is the number of active sites.

In heterogeneous catalysis, it is sometimes difficult to determine the number of active sites. For such situations, S is often replaced by the total, readily measurable, area A of the exposed catalyst. Clearly, $(dn/dt)/A$ sets a lower limit to the TOF."

This is the approach that my colleagues and I have taken for the past 20 years or so; and, on this basis, sensible comparisons can be made between the catalytic performance of a range of (similar) catalysts.[2]

1 J. M. Thomas and W. J. Thomas, *Practise of Heterogeneous Catalysis*, Wiley-VCH, Weinheim, 1997, p. 26.
2 J. M. Thomas, R. Raja and D. W. Lewis, *Angew. Chem., Int. Ed.*, 2006, **44**, 6456.

Professor Bond replied: I think we are all perfectly familiar with the idea of turnover frequency (TOF), but we do not all share Sir John's affection for it. As I believe he himself appreciates, its worth is limited by the accuracy with which we know the size of the active ensemble, and their number in the working catalyst. Although the SSITKA method allows *in situ* estimates to be made, it leads to numbers that are in some cases very small and not precisely measurable.[1] I have no wish to complicate the situation unnecessarily, but values of TOF at best provide only a snapshot of activities under one set of conditions, and reveal nothing about variables having scientific as well as practical importance, such as lifetime or the effects of temperature and reactant concentrations. There is a great deal more to the understanding of the causes of catalytic activity than just making "sensible comparisons" based on values of TOF. As Einstein said, "We must make things as simple as possible—but not simpler".

1 J. G. Goodwin Jr., Soo Kim and W. D. Rhodes, in *Specialist Periodical Reports - Catalysis*, ed. J. J. Spivey and G. W. Roberts, Royal Society of Chemistry, Cambridge, 2004, vol. 17, p. 320.

Dr Russell said: The same problem exists in electrocatalysis. For example many people report 'better' oxygen reduction catalysts, but the catalyst they are comparing to is dreadful. Often this is a result of poor storage of the commercial catalyst. This is where characterisation is so important. So you know what you are comparing to.

Energy dispersive X-ray spectroscopy of bimetallic nanoparticles in an aberration corrected scanning transmission electron microscope†

Andrew A. Herzing,[a] Masashi Watanabe,[ac] Jennifer K. Edwards,[b] Marco Conte,[b] Zi-Rong Tang,[b] Graham J. Hutchings[b] and Christopher J. Kiely*[a]

Received 25th April 2007, Accepted 8th June 2007
First published as an Advance Article on the web 13th September 2007
DOI: 10.1039/b706293c

The technique of X-ray energy dispersive spectroscopy (XEDS) spectrum imaging in a dedicated scanning transmission electron microscope (STEM) is discussed in relation to its applicability to bimetallic nanoparticles. It is shown that the recent availability of aberration corrected microscopes and multivariate statistical analysis (MSA) techniques has allowed us to overcome many of the intrinsic limitations previously encountered when attempting STEM-XEDS spectrum imaging on nanoscopic volumes of material. We demonstrate through a variety of applications to Au–Ag and Au–Pd bimetallic nanoparticle systems, that STEM-XEDS can provide invaluable high spatial resolution compositional information on (i) alloy homogeneity and phase segregation effects within individual nanoparticles, (ii) particle size–alloy composition correlations, (iii) the detection of trace amounts of alloying element and (iv) metal component distribution in extremely highly dispersed catalyst systems.

Introduction

One of the great advantages of electron microscopes is the ability to couple high resolution imaging with spectroscopic techniques such as X-ray energy spectroscopy (XEDS), electron energy loss spectroscopy (EELS) and Auger electron spectroscopy (AES). A scanning transmission electron microscope (STEM) equipped with an XEDS detector is now commonly referred to as an analytical electron microscope (AEM).

The incident high energy electrons generated in the electron gun of the microscope can ionize atoms within the sample by causing the ejection of an inner shell electron. A subsequent relaxation process can occur within the excited atom whereby the ejected electron is replaced with an electron from a higher energy level which loses a portion of its energy by the emission of an X-ray photon.[1] The emitted X-ray

[a] *Center for Advanced Materials and Nanotechnology, Lehigh University, 5 East Packer Avenue, Bethlehem PA 18015-3195, USA. E-mail: chk5@lehigh.edu*
[b] *School of Chemistry, Cardiff University, Main Building, Park Place, Cardiff, UK CF10 3AT*
[c] *National Center for Electron Microscopy, Lawrence Berkeley National Laboratory, Berkeley CA 94720-8250, USA*

† The HTML version of this article has been enhanced with additional colour images.

photons from the sample possess energies that are characteristic of specific electronic transitions within the target atoms. Hence the X-rays emitted from the specimen can be collected *via* a solid-state detector and recorded to form an X-ray energy dispersive spectrum (EDS), which contains a wealth of chemical information about the sample. The energies of the X-ray emission peaks can be used as a fingerprint of the identity of the various elements present within the sample. Furthermore the areas under the X-ray emission peaks can be analysed either by the Cliff–Lorimer *k*-factor method[2] or the Watanabe–Williams ζ-factor method[3] to provide a quantitative measure of sample composition. The use of XEDS in the AEM has become widespread because of the high spatial resolution of the chemical analysis technique (*i.e.* approaching the dimensions of the incident probe for ultra-thin TEM samples).

While electron imaging techniques such as phase contrast (lattice) or annular dark field (ADF) imaging have been used extensively to characterize nanoparticulate and supported catalyst materials, only a few instances of the application of STEM-XEDS analysis technique to such systems can be found in the literature.[4–8] The underlying reason for this derives from the fact that X-ray generation within individual nanoparticles is a rather inefficient process at such high spatial resolution. Furthermore, the X-ray detection systems used in the AEM only collect a small fraction of the X-rays emitted from the nanoscopic entity, rendering the STEM-XEDS chemical mapping of nanoparticles to be relatively impractical. In this contribution we explain how the recent introduction of aberration corrected electron microscopes has drastically improved this situation and has made STEM-XEDS spectrum imaging of nanoparticles a practical reality.

It is now very well established that the use of microscopes with intense aberration corrected electron probes with sub-nanometer dimensions can improve the spatial resolution of high angle annular dark field (HAADF) images of nanoparticles to sub-nanometer values.[9–11] Furthermore, it is now possible to perform STEM-EELS chemical analysis on materials with atomic-column-by-atomic-column precision using aberration corrected probes.[12] However there are relatively few aberration corrected STEM's in existence which are equipped with XEDS systems. We have recently demonstrated[13,14] that aberration corrected STEM-XEDS spectrum imaging coupled with multivariate statistical analysis is an extremely powerful technique for studying impurity segregation to grain boundaries,[13] nanoprecipitates in steel,[15] and for identifying segregation effects in bimetallic catalyst nanoparticles.[16]

In this contribution we describe the application of the STEM-XEDS technique in point analysis, line profile and spectrum image modes as applied to Au–Ag core-shell alloy particles. The practical limitations of performing XEDS analysis on bimetallic nanoalloy particles in a conventional STEM are then explained. Methods of overcoming these practical limitations by using a combination of aberration corrected STEM spectrum imaging and multivariate statistical analysis (MSA) are also described. Finally we present four case studies on Au-based bimetallic supported catalyst systems using XEDS analysis in an aberration corrected AEM. In each case we demonstrate a new facet of nanoalloy chemical analysis that is facilitated by the application of the C_s corrected STEM-XEDS technique. These are namely (i) the visualization of core-shell morphology development, (ii) the determination of particle size–nanoalloy composition relationships, (ii) the detection of trace alloying elements and contaminants in bimetallic nanoalloys and (iv) mapping the spatial distribution of extremely highly dispersed nanoalloys on oxide supports.

Experimental

Two distinct types of bimetallic nanoparticles were examined in this study. Ligand-stabilized Au–Ag core-shell particles were prepared by a two phase colloidal synthesis method.[17] Samples for scanning transmission electron microscopy (STEM) examination were prepared by depositing a single drop of the dilute colloidal solution onto a continuous carbon film supported by a 300 mesh copper TEM grid and then simply allowing the toluene solvent to evaporate.

Au–Pd/Al$_2$O$_3$, Au–Pd/C and AuPd/CeO$_2$ supported catalyst samples were prepared by impregnation of the support *via* an incipient wetness method using aqueous solutions of PdCl$_2$ (Johnson Matthey) and/or HAuCl$_4 \cdot$ 3H$_2$O (Johnson Matthey).[18] Samples of the supported bimetallic alloys were prepared for STEM examination by dispersing the catalyst powder in high purity ethanol, then allowing a drop of the suspension to evaporate onto a holey carbon film supported by a 300 mesh copper TEM grid.

Samples were subjected to chemical microanalysis and high angle annular dark-field (HAADF) imaging in a VG systems HB603 STEM operating at 300 kV. This instrument was fitted with a Nion C_s aberration corrector and an Oxford Instruments INCA TEM 300 system for X-ray energy dispersive (XEDS) analysis. The effective electron probe size used in STEM-XEDS spectrum imaging experiments was 0.2 nm and dwell times between 200 and 400 ms per pixel were typically employed. Multivariate statistical analysis (MSA) was employed in some instances to reduce the effect of random noise and aid quantitative analysis of these XEDS spectrum images.[3]

The STEM-XEDS technique—point analyses, linescan profiles and spectrum images

The most common way of performing XEDS is in *point analysis* mode, whereby an X-ray spectrum from 0–20 kV is acquired using a stationary probe from a specific point on the specimen. For example, Fig. 1 shows a high angle annular dark field (HAADF) image of a pair of 40 nm diameter core-shell Au–Ag ligand protected nanoparticles. In order to prove the existence of a core–shell morphology one can simply compare point spectra obtained from the center and very edge of the particle, as shown in the pair of spectra situated on the left hand side of Fig. 1. It is clear from comparing the relative intensities of the Au M$_\alpha$ (2.120 eV) and Ag L$_\alpha$ (2.983 eV) peaks at the edge and centre, that the Au content is significantly depleted at the particle edge, suggesting the existence of an Ag-rich alloy at the surface of the nanoparticle.

Another analytical approach which gives more information on the Au–Ag elemental distribution is to take whole series of XEDS spectra from successive points as the probe is tracked across the diameter of the core–shell nanoparticle. The *linescan profiles* of the Au L$_\alpha$ (8.492 eV) and Ag L$_\alpha$ (2.983 eV) signal intensity can then be plotted as a function of probe position across two orthogonal directions of the core–shell particle as shown on the right hand side of Fig. 1. In this case 40 point spectra have been acquired at 1 nm intervals across the diameter of the particle using

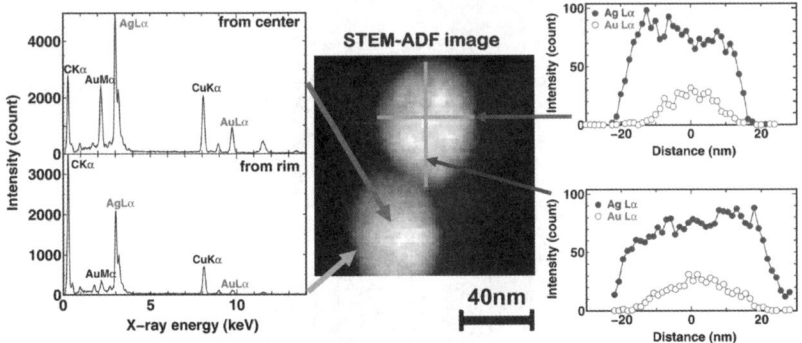

Fig. 1 (Left) Point XEDS spectra taken from the centre and edge of the AuAg nanoparticle imaged (centre) in ADF mode. (Right) The corresponding Ag L$_\alpha$ and Au L$_\alpha$ linescan profiles taken across the diameter of the same AuAg nanoparticle.

a 1.5 nm full-width-tenth-maximum (FWTM) electron probe. Comparison of the Ag and Au linescan profiles clearly shows a significant depletion of Au from the first 5 nm surface layer of the nanoparticle. The shape of the Ag and Au profiles is consistent with the thickness variation expected as one traverses the diameter of a spherical nanoparticle.

A particularly powerful technique that combines the imaging and analytical capabilities of the AEM is known as *spectrum imaging* (SI).[19,20] The first step in the SI process is to collect an HAADF image of the area of interest. This image is then divided into an array of pixels with a user defined resolution; usually 128 × 128. The STEM probe is then scanned over the same area of the sample that was used to form the image, but this time it is stopped for a pre-determined time interval, or dwell time, at each pixel. During this time interval (usually a few 100–200 ms), an entire X-ray spectrum from 0–20 keV is acquired. The probe then proceeds to the next pixel and collects another complete spectrum and this process is repeated until a spectrum has been acquired from all of the pixels. The acquisition of one spectrum image can be quite time consuming. For example, using a typical dwell time of 200 ms and a pixel resolution of 128 × 128, the acquisition process takes approximately one hour. The resulting array of spectra is known as a spectrum image data cube (Fig. 2) and it is defined by the coordinates of the image pixels (x and y) and the energy channels of the spectra (E). Obviously, this data cube will contain an immense amount of data. For example, for an SI with a resolution of 128 × 128, the data cube will contain 16 384 individual spectra. If the energy dispersion is set at 20 eV channel^{-1}, each spectra will contain 1024 channels from 0–20 keV and thus a total of 16.8 million data points are contained in the data cube! Fortunately, fully integrated detector and software packages, such as Oxford Instrument's INCA microanalysis system,[21] have been developed to aid the electron microscopist in manipulating this large amount of data.

Fig. 3 shows a high angle annular dark field image of the same Au–Ag core-shell particles that were analyzed in Fig. 1. Also shown are the Ag L_α (2.983 keV) and Au L_α(8.492 keV) X-ray maps subsequently extracted from the spectrum image data cube. These maps show in an intuitively obvious way that the particles have core–shell morphologies because the Ag X-ray signal has a spatial distribution that closely mimics that of the HAADF image whereas the Au X-ray signal originates from a much more spatially compact area located at the center of the nanoparticles. A further advantage of the SI technique is that the information contained in the data cube can be interrogated in a number of different ways. For instance, point analyses

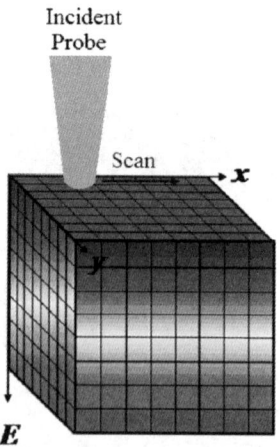

Incident
Probe

Scan

Fig. 2 Schematic diagram illustrating the spatial dimensions (x, y) and energy dimension (E) of a spectrum image data cube.

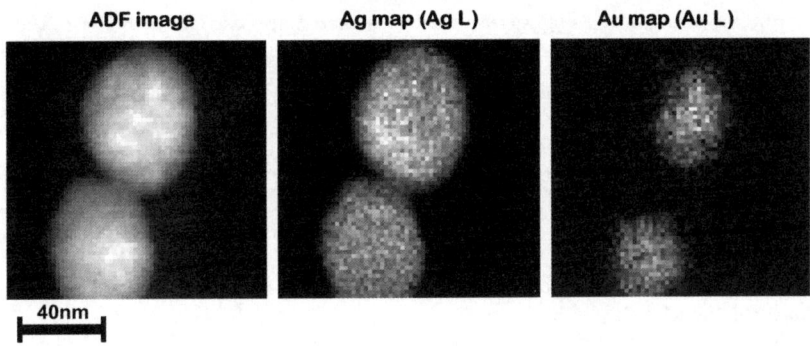

| ADF image | Ag map (Ag L) | Au map (Au L) |

40nm

Fig. 3 ADF image (left) and the corresponding Ag L$_\alpha$ (centre) and Au L$_\alpha$ (right) XEDS maps of the same pair of AuAg nanoparticles as shown in Fig. 1.

for any individual pixel, or any combination of pixels, can later be extracted from the stored SI data cube and analyzed. Furthermore, linescan profiles from any vertical or horizontal line of pixels can also be subsequently extracted from the SI data cube. Indeed, if a particle exhibited sufficient spherical symmetry one could in principle perform radial binning of the pixels from the center of the particle outwards to improve the counting statistics of the diametric line profile.[22]

Limitations of STEM-XEDS analysis on nanoparticulate systems

The high spatial resolution of chemical analysis that is possible in the AEM comes at the expense of the amount of X-ray signal that is generated and collected, both of which are severely limited by the very nature of the instrument.[23,24] The signal generation is limited due to the extremely small volume in which the nanometer scale STEM probe interacts with the necessarily very thin sample. Techniques such as SEM and electron probe microanalysis (EPMA) use a broad, intense beam to analyze relatively large, bulk samples and the resulting interaction volume of the electron beam with the samples is of the order of 1 μm^3. This limits the spatial resolution of SEM-based XEDS, but it does allow a large number of X-rays to be generated. However, the smaller probe and much thinner samples utilized in a typical TEM increases the spatial resolution of the technique, but also shrinks the interaction volume by several orders of magnitude to about 10^{-5} μm^3 and thus severely decreases the number of X-rays generated. This limitation becomes even more pronounced in the AEM, which typically utilizes a 1 nm probe to analyze a sample that is approximately 10 nm thick, producing an interaction volume of only 10^{-8} μm^3.

To further exacerbate the situation, the actual process of collecting the relatively low X-ray signal generated in the AEM is itself highly inefficient. Again, the origin of this inefficiency is directly related to the conditions used to achieve a high spatial resolution, one of which is a short focal length for the objective lens which is determined by its distance from the sample. Practically, this means that the gap in the objective lens pole piece has to be relatively small and thus the physical size of the X-ray detector introduced near to the sample is severely limited. Typically, a compromise is reached between the detector size and the pole–piece gap which allows the detector to subtend a collection solid-angle of approximately 0.3 sr. The characteristic X-ray distribution is approximately spherically symmetric with X-rays being emitted equally in all directions from the point of interaction. Thus, the limited angular size of the X-ray detector means that only 0.01–1.00% of all the emitted X-rays are ever collected, since only those X-rays which directly enter the detector contribute to the spectrum.

Potential advantages of using an aberration corrected microscope for STEM-XEDS spectrum imaging

From the standpoint of collecting X-ray spectrum images from nanoscopic entities in the AEM, the inefficiency in X-ray signal generation and collection is a severe limitation and must be overcome if one is to produce statistically relevant data from nanoparticles that are smaller in dimension than those shown in Fig. 1 and 3. One potential way of doing this would be to increase the dwell time of the probe, which would increase the number of X-rays generated simply by prolonging the acquisition time. However, this is to be avoided if at all possible because the increased collection time will exacerbate the deleterious contributions of specimen drift and beam damage of the sample.

Alternatively, increasing the current in the electron probe would also increase the number of X-rays generated by injecting more ionizing electrons into the sample. Once again, the ability of the microscopist to do this is limited by the physical constraints necessary to achieve a high spatial resolution. The probe current can be increased by using a larger probe-forming aperture and thus allowing a greater number of the electrons emitted from the source to contribute to the beam. However, using a larger aperture size, which defines the probe convergence angle (β), increases the effect of spherical aberration (C_s) which is proportional to β^3. Decreasing β by using a smaller aperture will improve the image resolution by decreasing the effects of C_s but will also simultaneously severely limit the probe current. Therefore, a suitable value of β is usually chosen which represents a compromise between probe size and current, which are typically 1.5 nm and 1 nA, respectively in the AEM.

This trade-off can be overcome to some extent in an aberration (C_s) corrected instrument, because, in the absence of C_s, a larger aperture can be used without degrading the resolution. In this case, the maximum aperture size is now dictated by higher-order aberrations which do not become significant until a much higher β is reached. Thus, for the Lehigh C_s-corrected VG HB603, simulations have shown that a 1.1 nm (FWTM) probe in the corrected state contains 6.0 nA of current while in the uncorrected state it is limited to 0.5 nA.[13] This represents a 12-fold increase in probe current without a reduction in resolution! Conversely Fig. 4 shows that maintaining a constant probe current of 0.5 nA, the probe size in the corrected state

(a)　　　　　　　　　　　**(b)**

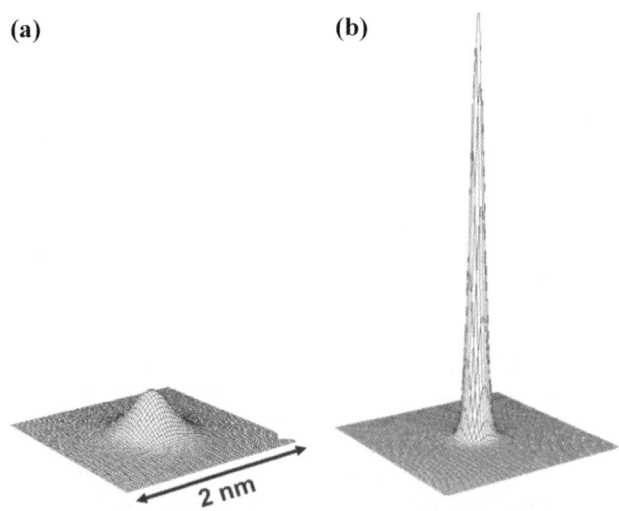

Fig. 4 Simulations of the intensity distribution at constant current (0.5 nA) for (a) the uncorrected electron probe (1.1 nm FWTM) and (b) the fully optimized corrected probe (0.35 FWTM) for the Lehigh VG HB603 STEM.

is 0.35 nm FWTM compared to 1.1 nm FWTM in the uncorrected state, a 68% reduction in probe diameter without a loss of current!

The ability to increase the probe current without degrading the instrument's spatial resolution for imaging and analysis decreases the minimum detection limit for chemical analysis by energy dispersive X-ray (XEDS) spectroscopy from 2–3 atoms in a thin metal foil to 1–2 atoms.[13] The Lehigh VG HB603 STEM had previously held the world record with respect to STEM-XEDS spatial resolution and detectability,[24] so the addition of the aberration corrector represents a significant improvement to this already impressive instrument.

Multivariate statistical analysis of STEM-XEDS spectrum images

The inherently poor generation and collection efficiency of X-rays in the AEM is a major limiting factor in the application of spectrum imaging at the nanoscale. Aberration-correction, at least in part, significantly improves the data collected *via* this technique by partially overcoming the trade-off between probe current and resolution. However, it is also possible to further improve the collected SI data set *via* post-acquisition processing techniques. One particularly powerful method of doing this is known as multivariate statistical analysis (MSA). This is a set of processing techniques which analyzes the spectrum image data cube as a whole and identifies the various components within it which vary independently.[25,26]

The data cube is analyzed as a matrix consisting of rows defined by the images (*i.e.* the energy slices) and columns defined by the pixel intensities. If a relatively simple transformation is performed whereby the original data matrix (X) is multiplied by its transpose (X^t), the result is known as the variance–covariance matrix (Y) of the data cube. The next step in the MSA processing is to calculate the eigenvalues and eigenvectors of this variance–covariance matrix. Each eigenvector defines an individual component (or image) within the data matrix and its associated eigenvalue is proportional to that components contribution to the total variance of the data set. Now, it is simply a matter of the microscopist deciding which of these components to use to reconstruct the data set and which to discard as 'noise'.[27]

Practically speaking, this is performed by arranging the assigned eigenvalues in descending order and the result is known as a scree plot, named after the similarly shaped pile-up (or scree) which occurs at the foot of mountains after an avalanche or landslide. The scree plot (shown schematically in Fig. 5) quickly decays as it follows the first few eigenvalues with the largest magnitudes and it soon gives way to a linear region defined by the components with lower eigenvalues. The onset of linearity in the plot indicates that the remaining, low eigenvalue components are not

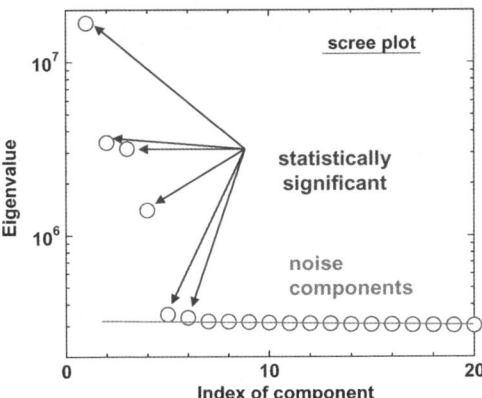

Fig. 5 Schematic diagram of a typical scree plot derived from the multivariate statistical analysis (MSA) of a STEM-XEDS spectrum.

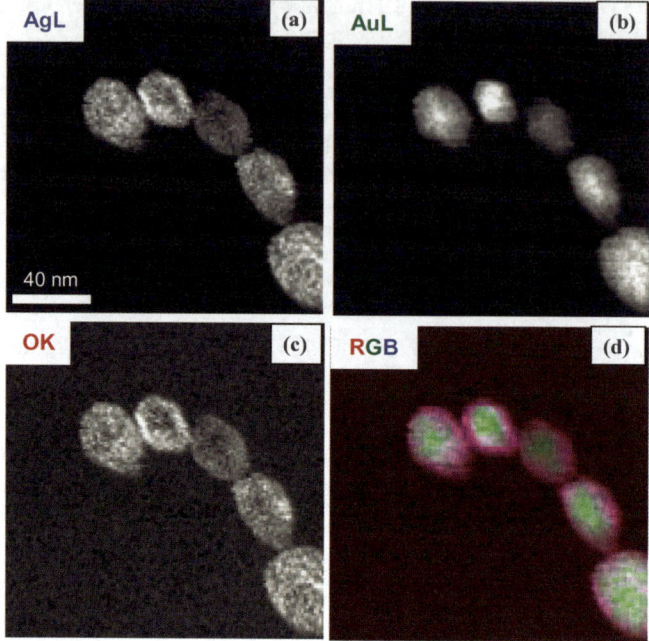

Fig. 6 The (a) Au L_α, (b) Pd L_α and (c) O K XEDS maps obtained from a string of AuAg core–shell nanoparticles. (d) The reconstructed RGB overlay image of the maps shown in (a), (b) and (c) [Red – O; Green – Au: Blue – Pd]

independently varying and cannot be distinguished from the background noise within the spectrum. These latter components can then be readily identified and removed from the data set, which is then reconstructed with only those components that are distinguishable from the noise. However, before these lower eigenvalues are definitely discarded, it is also possible for the microscopist to inspect the actual image that results from each component. This procedure rules out the arbitrary removal of important spectral data.

Fig. 6 shows a montage of SI data for a string of Au–Ag core–shell nanoparticles. Reconstructed components corresponding to three of the highest eigenvalues for this data cube are shown in Fig. 6(a), (b) and (c) and represent the MSA processed Au L_α (8.492 keV), Ag L_α (2.983 keV) and O K (0.525 keV) X-ray maps, respectively. Fig. 6(d) presents a red–green–blue (RGB) overlay map of the Au, Ag and O chemical distribution which is an appealing and informative alternative way of presenting the data. The Au-rich core and Ag-rich shell morphology is again apparent. The equilibrium Au–Ag phase diagram[28] suggests that there should be complete mutual solid solubility of the two components over the entire Au–Ag binary alloy composition range. The surface energies of Ag and Au are reported to be 1.3 J m^{-2} and 1.5 J m^{-2}, respectively.[29] Hence we might expect there to be a slight tendency for Ag to migrate to the surface of the nanoparticle. However, Fig. 6(c) shows the O K component of the data set (Fig. 6(c)), which has a strong spatial correlation to the Ag signal. It is apparent that there is some driving force for the Ag to segregate to the particle surface in order to satisfy its desire to interact with atmospheric oxygen to form silver oxide.

Application 1: development of core–shell morphologies within AuPd/Al$_2$O$_3$ catalysts

The ability to detect and visualize core–shell morphologies in bimetallic clusters has proven to be of considerable importance in the study of supported Au–Pd catalysts.

These materials exhibit considerable potential for the direct production of H_2O_2 from molecular H_2 and O_2 under mild reaction conditions.[30,31] We have previously reported that AuPd nanoparticles supported on Fe_2O_3, TiO_2 and Al_2O_3 exhibit very distinct Pd-rich shell/Au-rich core morphologies in catalysts that have been calcined in air at 400 °C.[18,32–34] This is again somewhat counter-intuitive because Pd surface segregation is not predicted by the thermodynamics of the Au–Pd system. Instead, the equilibrium phase diagram, predicts that an FCC solid solution of Au and Pd is formed over the entire compositional range with special ordering compounds existing at the Au_3Pd and $AuPd_3$ compositions.[28] Furthermore, the surface energies of Au and Pd are 1.50 J m^{-2} and 2.05 J m^{-2}, respectively,[29] suggesting that if anything were to migrate to the nanoparticle surface it should be the Au component. In order to investigate the development of the Au–Pd nanoparticle morphology more closely we have used STEM-XEDS spectrum image analysis on a systematic series of Au–Pd/Al_2O_3 samples that have been subjected to different heat treatments: namely (i) dried at 120 °C, but left uncalcined, (ii) calcined at 200 °C, (iii) calcined at 400 °C and (iv) calcined at 400 °C then subsequently reduced at 500 °C in H_2. The typical AuPd nanoparticle morphologies generated by these samples are shown in the matrix of Au M_α, Pd L_α and the RGB color overlay maps presented in Fig. 7. It is clear that the uncalcined sample (Fig. 7, row 1) shows homogeneous AuPd nanoalloy particles, whereas the samples calcined at 200 °C (Fig. 7, row 2) and 400 °C (Fig. 7, row 3) show the progressive development of a Pd-rich shell and Au-rich core morphology. This characteristic core–shell morphology is presumably brought about by the preferential formation of Pd–O bonds at the alloy surface since

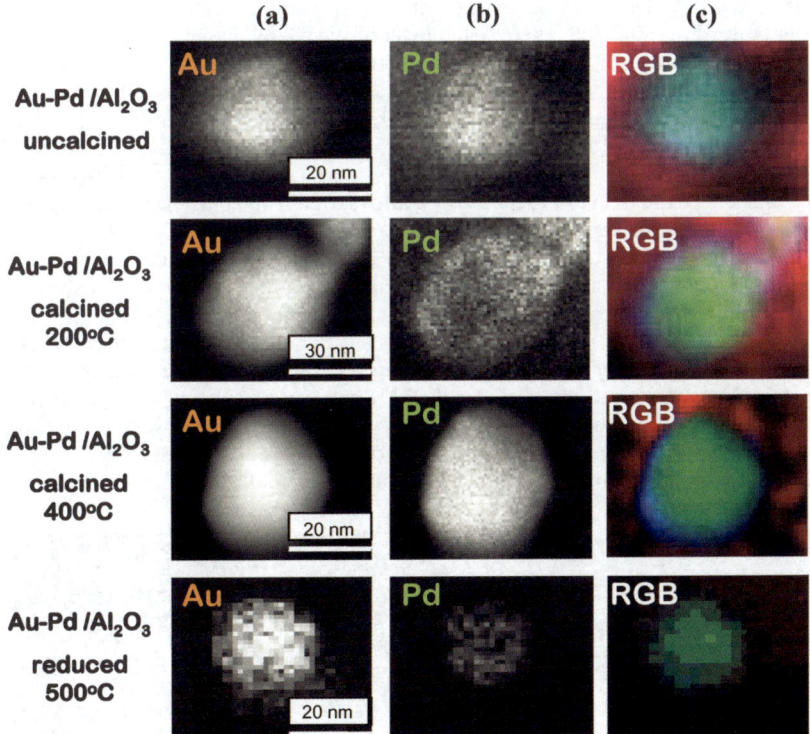

Fig. 7 Montage of (a) Au L_α XEDS map (b) Pd L_α XEDS map and (c) RGB overlays [Green – Au: Blue – Pd] for a series of AuPd/Al_2O_3 samples subjected to different thermal treatments. Row 1 – dried at 120 °C: Row 2 – calcined at 200 °C: Row 3 – calcined at 400°C: Row 4 – calcined at 400 °C then reduced at 500 °C in H_2.

in this 200–400 °C temperature range palladium oxidizes more readily than gold. Interestingly, it is the 400 °C sample which is subsequently reduced in H_2 at 500 °C, then the Pd-rich shell/Au-rich core morphology becomes unfavorable and the nanoalloy particles either revert to homogeneous mixtures or form an *inverted* core–shell with gold enrichment on the surface and Pd in the interior. An example of the latter inverted core shell morphology shown in Fig. 7 (row d). The application of the STEM-XEDS technique in a C_s corrected AEM, in conjunction with MSA analysis of the SI data cube, is clearly providing a new level of insight into core–shell morphology development in nanoalloy particles.

Application 2: particle size/alloy composition correlations in AuPd/C catalysts

Another aspect where the STEM-XEDS spectrum imaging technique can provide additional characterization insight relates to the fact that one invariably encounters a size distribution of supported metal particles in real catalyst systems. For example, Fig. 8(a) shows an HAADF image of supported metal particles in a 2.5 wt% Au–2.5 wt% Pd/activated carbon sample (after calcination at 400 °C) in which a tri-modal particle size distribution is apparent. The smallest particles fell in the range of 2–5 nm range, intermediate size particles were approximately 10–40 nm in size, while an occasional particle or two was observed that exceeded 100 nm in size. The corresponding Pd L_α and Au M_α STEM-XEDS maps presented in Fig. 8(b) and (c), respectively, show that the composition of the metal nanoparticles is size dependent. The intermediate size particles are clearly visible in both the Au M_α and Pd L_α XEDS maps indicating that they are AuPd alloys, whilst the smaller (2–5 nm) particles are only clearly visible in the Pd L_α map.

To investigate the size dependent composition of these particles further, a much higher collection time (400 ms) for each pixel of the X-ray map was employed. While significant electron beam damage was inflicted on the sample by this technique, it was necessary in order to improve the counting statistics of the data. Fig. 9 shows a matrix of ADF images, Au M_α and Pd L_α XEDS maps and RGB overlay for typical particles in the small (3 nm), intermediate (15 nm) and large (50 nm) size ranges. By comparing the relative intensities of the Au M_2 and Pd L_α maps, it is qualitatively clear that the particles are in fact all AuPd alloys, but as the particle size increases the Pd-to-Au ratio decreases. The RGB (Pd – Blue, Au – Green, C – Red) overlay images also reflect this trend since they show a gradual change of color from blue to green with increasing particle diameter. Quantitative analyses of the data show that the small, intermediate and large particles to have compositions of 97 : 3, 65 : 35 and 5 : 95 expressed in wt% Pd : wt% Au, respectively. The intermediate size particles are the closest to the nominal Au : Pd (1 : 1) composition, whereas the smallest particles are in fact a very dilute alloy of Au in Pd. Such a direct visual representation of particle size/composition correlations is now allowing us to assess how various modifications to the catalyst preparation procedure simultaneously affect both the size and composition distribution of the resultant nanoalloy particles.

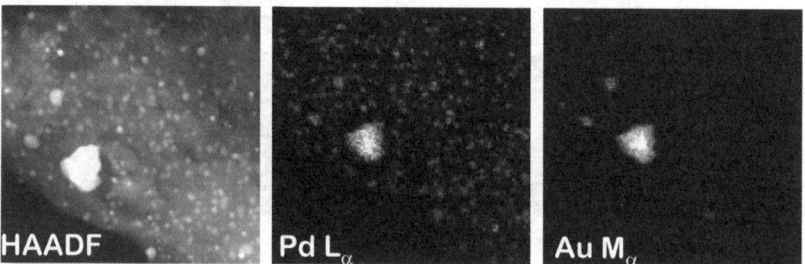

Fig. 8 HAADF image (left) and the corresponding Ag L_α (centre) and Au M_α (right) XEDS maps of a AuPd/activated carbon catalyst.

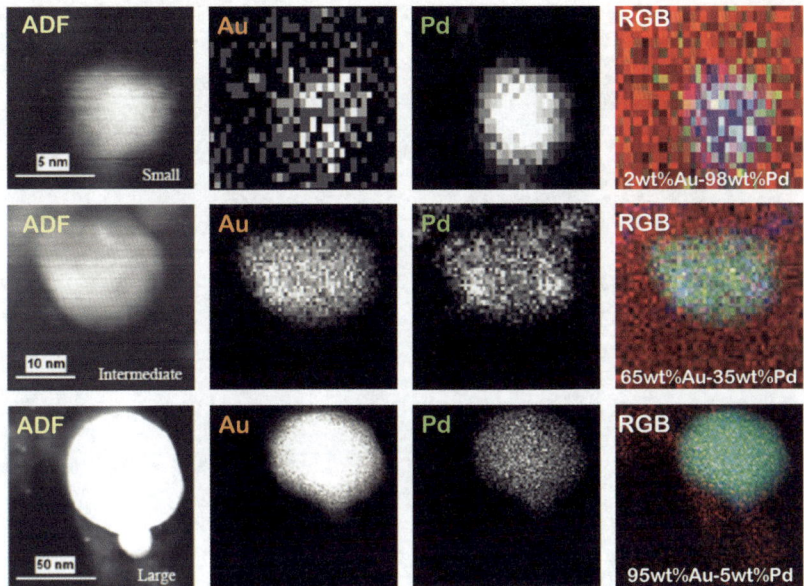

Fig. 9 Montage of (a) ADF image, (b) Au L_α XEDS map, (c) Pd L_α XEDS map and (d) RGB overlays [Red – C: Green – Au: Blue – Pd] from small (row 1), intermediate (row 2) and large (row 3) particles found in a AuPd/C catalyst.

The RGB color overlay maps shown in Fig. 9 also clearly demonstrate that the intermediate scale particles are in fact homogeneous random AuPd alloys rather than core–shell nanoparticles. This is in direct contrast to similar size AuPd particles supported on oxides such as Fe_2O_3,[32,34] TiO_2,[33] Al_2O_3[18] where MSA analysis revealed a Pd-rich shell surrounding a Au-rich core after calcination at 400 °C. The question arises as to why core–shell morphologies form spontaneously on the oxide supports but not on activated carbon. This may be due to the oxidation efficacy of the support surface, since carbon is a reducing support. As noted earlier, the formation of Pd-rich shell structures may be due to the surface of the nanocrystals being oxidized and PdO_x, therefore, forms on the surface. This process is hindered by the reducing nature of the carbon support. Indeed, carbon is used in the extraction of Au^{3+} salts during the commercial production of gold and during this process the Au^{3+} is reduced to metallic gold.[35] This process may also occur during the preparation of the Au–Pd alloys used in this study and this may stabilize the random alloys rather than core–shell structures.

Application 3: the detection of minor alloying elements and contaminant species in Au–Pd/C catalysts

Combining STEM-XEDS spectrum imaging with MSA processing can also be beneficial for detecting relatively minor amounts of alloying elements and contaminants in catalyst samples. This point can be demonstrated by examining data obtained from a bimetallic Au–Pd (95 : 5) catalyst supported on activated carbon in which the catalyst contained a *total* metal loading of only 1 wt%. Hence, detecting the small amount of the secondary alloying element in this catalyst represents a significant challenge for STEM-XEDS analysis. The degree of difficulty can be illustrated by considering the example of a hemispherical metal particle that is 3 nm in size, which would contain approximately 450 atoms. If it were assumed that

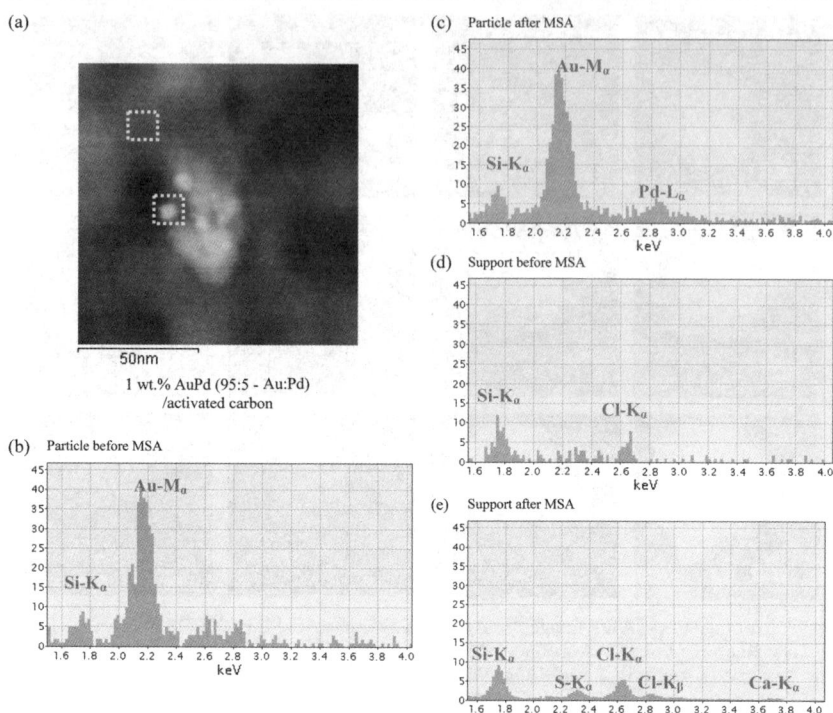

Fig. 10 (a) ADF image of a AuPd (95 : 5)/C catalyst. Summed (3 × 3) matrix of (b) unprocessed and (c) MSA processed XEDS spectra obtained from the region marked which encompasses a metal nanoparticle. Summed (3 × 3) matrix of (d) unprocessed and (e) MSA processed XEDS spectra obtained from the support only region.

complete alloying of the Au and Pd occurred, this metal particle would contain only around 20 Pd atoms.

When the Au–Pd catalyst was examined, discrete 3–5 nm particles were observed by HAADF imaging. (Fig. 10(a)). A STEM-XEDS spectrum image was then acquired and the unprocessed spectra were extracted from a 3 × 3 square of (*i.e.* 9) nine pixels on one of the metal particles (Fig. 10(b)) as well as from a 3 × 3 square just on the support itself (Fig. 10(d)). The unprocessed sum spectrum from the metal particle consisted of the Au–M_α peak (2.120 keV) and the Si K internal fluorescence peak (1.739 keV) from the detector. A slight perturbation is apparent near the expected energy of the Pd–L_α peak (2.863 keV), but the signal is not strong enough to be definitively assigned to Pd. The unprocessed sum spectra from the support (Fig. 10(d)) revealed the probable presence of chlorine and, again, the silicon internal fluorescence peak. In order to improve the statistical quality of the STEM-XEDS data, MSA processing was then carried out on the entire SI data cube and the same two sets of 3 × 3 spectra were then extracted and summed. The considerable improvement to the particle sum spectrum after MSA spectrum (Fig. 10(c)) definitively shows the presence of Pd and confirms that the nanoparticle is in fact a nanoalloy. Furthermore, in addition to the Cl–K_α peak (2.621 keV), the Cl–K_β peak (2.815 keV) is now also resolved in the MSA processed sum spectra from the support material (Fig. 10(e)). The chlorine signal, which seems to be associated with both the particle and the support, is likely to have originated from the use of $PdCl_2$ and $HAuCl_4 \cdot 3H_2O$ precursors in the catalyst preparation procedure. The summed spectrum after MSA processing from the support ((Fig. 10(e)) also clearly shows the presence of sulfur and calcium impurities in the activated carbon which were not expected.

Therefore, the combination of aberration-corrected STEM-XEDS and MSA reconstruction in favorable circumstances can allow the relatively unambiguous identification of very minor alloying and contaminant elements within catalysts samples at an exceptionally high spatial resolution.

Application 4; spectrum imaging highly dispersed metallic components in AuPd/CeO$_2$ catalysts

Another situation sometimes encountered in supported metal catalysts is where the metal component is so finely dispersed on an oxide support (*e.g.* atomic dispersions, clusters of 2–3 atoms, or sub-nm clusters) that it is difficult to image the active mono- or bi-metallic component using techniques such as lattice imaging or ADF imaging. In the case where there is a large atomic mass (*z*) difference between the metal overlayer and the support material, high angle annular dark field (HAADF) imaging in an aberration corrected STEM can be used to great effect in imaging individual atoms and sub-nm metal clusters. For example, there have been reports for the La/Al$_2$O$_3$,[36] Pt/Al$_2$O$_3$,[10] PtRu/Al$_2$O$_3$,[11] Au/C,[37,38] and Au/Fe$_2$O$_3$[39–41] systems where C_s corrected HAADF imaging has allowed direct visualization of even the most highly dispersed metallic components. There are instances however, where the cations in the support have comparable *z* values to those of metal overlayer atoms, where even C_s corrected HAADF cannot conclusively image the highly dispersed species. An important example of this is the AuPd/CeO$_2$ system[42] where the metal dispersions are so fine and *z*-values are so close (*i.e.* Ce = 58, Pd = 46, Au = 79) that HAADF imaging is not effective. Techniques such as atomic absorption spectroscopy (AAS) and X-ray photoelectron spectroscopy (XPS) can prove the presence of surface AuPd, but give no information on the spatial distribution of these species. STEM-XEDS spectrum imaging can however, give useful information on this aspect of overlayer dispersion as demonstrated in Fig. 11. The material shown consists of a 2.5 wt% Au–2.5 wt% Pd mixture impregnated onto hollow spherical agglomerates of

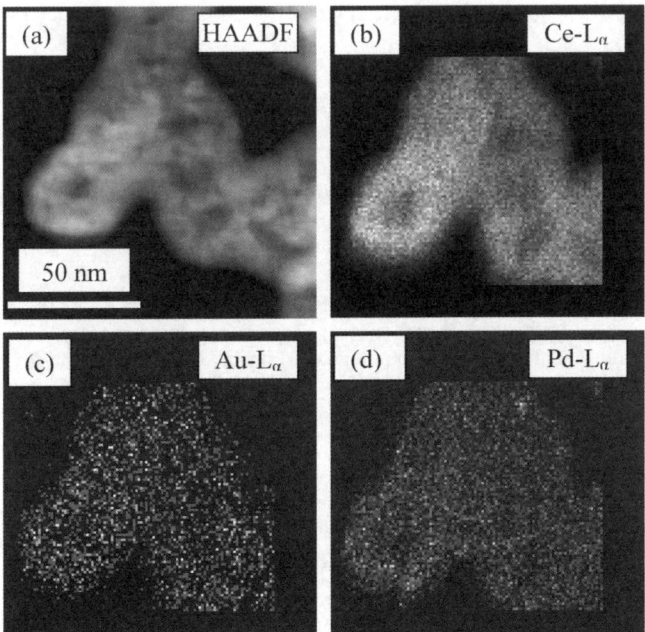

Fig. 11 (a) ADF image, (b) Ce L$_\alpha$ XEDS map, (c) Au L$_\alpha$ XEDS map and (d) Pd L$_\alpha$XEDS map from a AuPd/CeO$_2$ catalyst.

nanocrystalline (~ 5 nm) CeO_2 prepared using a novel supercritical CO_2 preparation route.[42] The HAADF image in Fig. 11(a) shows no evidence of the Au or Pd. Phase contrast lattice images also showed no evidence of discrete metallic clusters. However the Au L_α (8.492 keV) and Pd L_α (2.863 keV) XEDS maps shown in Fig. 11(c) and (d), respectively, demonstrate that the two metallic components were intimately mixed. Furthermore, comparison of the spatial distribution of Au and Pd signals with that of the Ce L_α (4.838 keV) XEDS map presented in Fig. 11(b) shows that both metallic components are uniformly dispersed over the entire nanocrystalline CeO_2 support sphere.

Conclusions

STEM-XEDS spectrum imaging in an aberration corrected analytical electron microscope combined with multivariate statistical analysis has been shown to be a suitable way of overcoming the low X-ray generation and poor detection efficiency that has previously been a limiting factor when applying the XEDS method to individual nanoparticles. Using ligand stabilized AuAg nanoparticles and supported AuPd catalysts as examples, we have shown the STEM XEDS spectrum imaging technique can be applied in novel ways to characterize a variety of important features in nanoalloy systems. These include (i) alloy homogeneity and phase segregation effects within individual nanoparticles, (ii) particle size—alloy composition correlations, (iii) the detection of trace amounts of alloying element and (iv) metal component distribution in extremely highly dispersed catalyst systems.

Acknowledgements

This work formed part of the EU AURICAT project (Contract HPRN-CT-2002-00174) and the EPSRC/Johnson Matthey funded ATHENA project and we thank them for funding this research. We also thank the World Gold Council (through the GROW scheme) and Cardiff University (AA Reed studentship) for providing support for J. K. E. C. J. K. and A. H. would also like to acknowledge the generous support of NSF Materials Research Science and Engineering Center (NSF DMR-0079996).

References

1   D. B. Williams and C. B. Carter, *Transmission Electron Microscopy*, Plenum Press, New York, 1996, ch. 4, p. 51.
2   G. W. Lorimer, in *Quantitative Electron Microscopy*, ed. J. N. Chapman and A. J. Craven, Proceedings of the 25th Scottish Universities Summer School in Physics, 1983, ch. 8, p. 305.
3   M. Watanabe and D. B. Williams, *J. Microsc.*, 2006, **221**, 89.
4   C. E. Lyman, *J. Mol. Catal.*, 1983, **20**, 357.
5   C. E. Lyman, H. G. Stenger and J. R. Michael, *Ultramicroscopy*, 1987, **22**, 129.
6   R. E. Lakis, C. E. Lyman and H. G. Stenger, *J. Catal.*, 1995, **154**, 261.
7   R. Prestvik, B. Totdal, C. E. Lyman and A. Holmen, *J. Catal.*, 1998, **176**.
8   J. Liu, *J. Electron Microsc.*, 2004, **54**, 251.
9   S. Wang, A. Y. Borisevich, S. N. Rashkeev, M. V. Glazoff, K. Sohlberg, S. J. Pennycook and S. T. Pantelides, *Science*, 2004, **3**, 143.
10 K. Sohlberg, S. Rashkeev, A. Y. Borisevich, S. J. Pennycook and S. T. Pantelides, *ChemPhysChem*, 2004, **5**, 1893.
11 A. Y. Borisevich, S. Wang, S. N. Rashkeev, S. T. Pantelides, K. Sohlberg and S. J. Pennycook, *Microsc. Microanal.*, 2006, **10**(S2), 460.
12 M. Varela, S. D. Findlay, A. R. Lupini, H. M. Christen, A. Y. Borisevich, N. Dellby, O. L. Krivanek, P. D. Nellist, M. P. Oxley, L. J. Allen and S. J. Pennycook, *Phys. Rev. Lett.*, 2004, **92**, 095502–1.
13 M. Watanabe, D. W. Ackland, A. Burrows, C. J. Kiely, D. B. Williams, O. L. Krivanek, N. Dellby, M. F. Murfitt and Z. Szilagyi, *Microsc. Microanal.*, 2006, **12**, 515.
14 M. Watanabe, D. W. Ackland, A. Burrows, C. J. Kiely, D. B. Williams, M. Kanno and R. Hynes, *Microsc. Microanal.*, 2005, **11**(S2), 2132.

15 M. Watanabe, D. B. Williams and M. G. Burke, *J. Mater. Sci.*, 2006, **41**, 4512.
16 D. I. Enache, J. K. Edwards, P. Landon, B. Solsana-Espriu, A. F. Carley, A. Herzing, M. Watanabe, C. J. Kiely, D. W. Knight and G. J. Hutchings, *Science*, 2006, **311**, 362–365.
17 B. Rodriguez-Gonzalez, A. Sanchez-Iglesias and L. M. Liz-Marzin, *Faraday Discuss.*, 2004, **125**, 133.
18 B. E. Solsana, J. K. Edwards, P. Landon, A. F. Carley, A. A. Herzing, C. J. Kiely and G. J. Hutchings, *Chem. Mater.*, 2006, **18**, 2689–2695.
19 C. Jeanguillaume and C. Colliex, *Ultramicroscopy*, 1989, **28**, 252–257.
20 J. A. Hunt and D. B. Williams, *Ultramicroscopy*, 1991, **38**, 47–73.
21 Oxford Instrument's INCA website: http://www.X-raymicroanalysis.com/pages/main/-inca.htm.
22 N. Braidy and J. A. Botton, *Microsc. Microanal.*, 2006, **12**(S2), 654.
23 J. J. Friel and C. E. Lyman, *Microsc. Microanal.*, 2006, **12**, 2.
24 C. E. Lyman, J. I. Goldstein, D. B. Williams, D. W. Ackland, S. Von Harrach, A. W. Nicholls and P. J. Statham, *J. Microsc.*, 1994, **176**, 85.
25 P. Trebbia and N. Bonnet, *Ultramicroscopy*, 1990, **34**, 165.
26 P. Trebbia and C. Mory, *Ultramicroscopy*, 1990, **34**, 179.
27 N. Bonnet, *J. Microsc.*, 1998, **190**, 2.
28 H. Okamoto and T. B. Massalski, *Phase Diagrams of Binary Gold Alloys*, ASM International, 1987.
29 F. R. de Boer, R. Boom, W. C. R. Mattens, A. R. Miedema and A. K. Niessen, *Cohesion in Metals*, North Holland, Amsterdam, 1988.
30 P. Landon, P. J. Collier, A. J. Papworth, C. J. Kiely and G. J. Hutchings, *Chem. Commun.*, 2002, **18**, 2058.
31 P. Landon, P. J. Collier, A. F. Carley, D. Chadwick, A. J. Papworth, A. Burrows, C. J. Kiely and G. J. Hutchings, *Phys. Chem. Chem. Phys.*, 2003, **5**, 1917.
32 B. E. Solsana, J. Edwards, P. Landon, A. F. Carley, A. Herzing, C. J. Kiely and G. J. Hutchings, *J. Mater. Chem.*, 2005, **15**, 4595.
33 B. E. Solsana, J. Edwards, P. Landon, A. F. Carley, A. Herzing, C. J. Kiely and G. J. Hutchings, *J. Catal.*, 2005, **236**, 69.
34 J. Edwards, P. Landon, A. F. Carley, A. A. Herzing, M. Watanabe, C. J. Kiely and G. J. Hutchings, *J. Mater. Res.*, 2007, **22**, 831.
35 J. K. Edwards, A. F. Carley, A. A. Herzing, C. J. Kiely and G. J. Hutchings, *Faraday Discuss.*, 2007, this volume.
36 Oak Ridge National Lab STEM Group Website: http://stem.ornl.gov/highlights/cat.html.
37 A. Bleloch, L. M. Brown, R. M. Brydson, A. J. Craven, P. J. Goodhew and C. J. Kiely, *Microsc. Microanal.*, 2002, **8**(2), 470.
38 P. E. Batson, N. Dellby and O. L. Krivanek, *Nature*, 2002, **418**, 617.
39 A. R. Lupini, A. G. Franceschetti, S. T. Pantelides, S. Dai, B. Chen, W. Yan, S. H. Overbury and S. J. Pennycook, *Microsc. Microanal.*, 2004, **10**(S2), 462.
40 M. Varela, A. R. Lupini, K. van Benthem, A. Y. Borisevich, M. F. Chisholm, N. Shibata, E. Abe and S. J. Pennycook, *Annu. Rev. Mater. Rev.*, 2005, **35**, 539.
41 A. A. Herzing, *PhD Thesis*, Lehigh University, Bethlehem, PA, USA, 2007.
42 Z.-R. Tang, J. Edwards, D. I. Enache, J. K. Bartley, S. H. Taylor, A. F. Carley, A. A. Herzing, C. J. Kiely and G. J. Hutchings, *J. Catal.*, 2007, **249**, 208.

Highly size-controlled synthesis of Au/Pd nanoparticles by inert-gas condensation

E. Pérez-Tijerina,*[a] M. Gracia Pinilla,[a] S. Mejía-Rosales,[a] U. Ortiz-Méndez,[b] A. Torres[b] and M. José-Yacamán[c]

Received 18th April 2007, Accepted 11th May 2007
First published as an Advance Article on the web 15th October 2007
DOI: 10.1039/b705913m

Gold/Palladium nanoparticles were fabricated by inert-gas condensation on a sputtering reactor. With this method, by controlling both the atmosphere on the condensation chamber and the magnetron power, it was possible to produce nanoparticles with a high degree of monodispersity in size. The structure and size of the Au/Pd nanoparticles were determined by mass spectroscopy, and confirmed by atomic force microscopy and electron transmission microscopy measurements. The chemical composition was analyzed by X-ray microanalysis. From these measurements we confirmed that with the sputtering technique we are able to produce particles of 1, 3, and 5 nm on size, depending on the choice of the synthesis conditions. From TEM measurements made both in the regular HREM, as well as in STEM-HAADF mode, we found that the particles are icosahedral in shape, and the micrographs show no evidence of a core-shell structure, in contrast to what is observed in the case of nanoparticles prepared by chemical synthesis.

I. Introduction

Nanoparticles containing two or more metals have unique catalytic, optical, and electronic properties.[1-6] For instance, the strong catalytic activity of many of the bimetallic nanoparticles produced by different methods is at least partially due to their high surface-to-volume ratio, although there are other factors that need to be considered, such as geometry, structure, and the distribution of the atomic species inside of the particle and into its surface. These properties are also strongly dependent on the size of the particles; in order to use efficiently these structures in the development of applications at industrial scales, the chosen synthesis method must allow a fair control of the mean size of the particle, with small deviations around this size. Of the different methods developed to produce metal nanoparticles, those based on chemical procedures are the most commonly used,[7-10] mainly because of their relative simplicity and economy. The bimetallic nanoparticles produced by chemical methods present in many cases a core-shell structure, as determined by transmission electron microscopy methods.[11-12] Apparently there is

[a] Laboratorio de Nanociencias y Nanotecnología, Facultad de Ciencias Físico-Matemáticas, Universidad Autónoma de Nuevo León, San Nicolás de los Garza, Nuevo León 66450, México
[b] Facultad de Ingeniería Mecánica y Eléctrica, Universidad Autónoma de Nuevo León, San Nicolás de los Garza, Nuevo León 66450, México
[c] Texas Materials Institute and Chemical Engineering Department, The University of Texas at Austin, Austin TX 78712 USA

no unique parameter that determines which of the atomic species will form the core while the other species stay close to or at the surface of the particle. The elemental distribution into the particle depends not only on the specifics of the synthesis method, but may also depend on the relative concentration of the atomic species.

Bimetallic Au/Pd has been one of the most studied systems because of its catalytic properties. This alloy has been used in many reactions, such as CO oxidation, hydrogenation of hydrocarbons, and synthesis of vinyl-acetate, among others.[13-15] Au/Pd nanoparticles produced by chemical methods are likely to have a core-shell structure and an icosahedral or cuboctahedral geometry; however, they may also form truncated octahedra, decahedra, or even more elaborate shapes. This wide variety of geometries, sizes, relative concentrations, and elemental distributions, reinforces the relevance of having a reliable procedure to create particles of specific sizes and shapes, and to analyze the particles with an adequate degree of accuracy by the use of experimental and theoretical tools.

In the present work, we describe the synthesis of Au/Pd nanoparticles using a sputtering system, in the presence of a coolant inert gas. The resulting nanoparticles were analyzed by mass spectroscopy, electron microscopy, atomic force microscopy, and X-ray microanalysis. The experimental measurements are contrasted with simulated TEM and STEM images in order to verify the geometry and the distribution of the two metals. We present the results of the different experimental analysis of the particles, and these results are discussed and compared with simulated electron microscopy images. Finally, we summarize the main conclusions of our discussion.

II. Experimental

Bimetallic Au/Pd nanoparticles were produced with a sputtering system Nanogen 50 (Nanoparticles source) from Mantis Deposition Ltd.,[16] using the Inert Gas Condensation method (ICG).[17-19] Fig. 1 shows a schematic diagram of the experimental setup. In the ICG process, a supersaturated vapor of metal atoms is originated by sputtering an AuPd alloy target in an inert gas atmosphere of Ar and He. The AuPd target was prepared by Aci Alloys, with a 99.99% purity and a 50 : 50 Au/Pd composition.

The Nanogen 50 system was kept at low temperature by a coolant mixture, and before the nanoparticles deposition the system pressure was set at 1×10^{-9} Torr. The size and production rate of the nanoparticles were controlled through the variation of (i) gas flow (Ar and He), (ii) partial pressure ($1-2 \times 10^{-1}$ Torr), (iii) magnetron power (that works on the range of 32–130 W), and (iv) zone condensation length (that can be varied from 50 mm to 130 mm). The rate of nucleation of the nanoparticles can be modified by the variation of Ar gas flow, as well as by changing the partial pressure. The argon flow produces a larger erosion of the metal target, generating a supersaturated metal vapour. When helium is introduced in the chamber, there are collisions between the nanoparticles and the He molecules and,

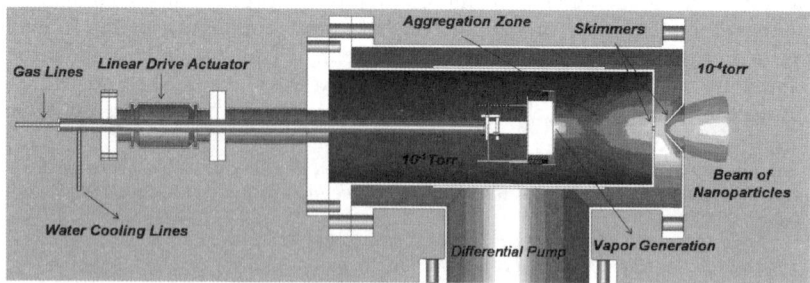

Fig. 1 Schematic diagram of the experimental setup.

when the helium flow gets increased, the number of collisions increases as well, and thus the mean free path of the nanoparticles is reduced, decreasing their size. The relation between the size of the nanoparticles and the magnetron power is monotonically increasing, until it reaches a saturation regime where an increment in the power decreases slightly the mean size of the nanoparticles. The increasing of the condensation zone length, by the manipulation of the linear drive actuator, allows the production of larger particles, due to the increase of the residence time of the nanoparticles on the aggregation zone. These parameters were optimized to produce particles of three different sizes: 1.1, 3.0 and 5.0 nm. The nanoparticles were deposited onto quartz substrates and copper grids/holey carbon; the latter choice was taken for the purposes of analysis in a JEOL 2010 TEM, and in a TECNAI 20 Field emission gun TEM. The depositions were made in a few minutes, such that the density of the nanoparticles deposited on the substrates was not too high. In order to retain the structural and morphological properties, the energy of cluster impact[20,21] was controlled; the energy of acceleration was kept at 0.1 eV atom^{-1}, assuring a soft landing of the particles on the substrate.

The high resolution TEM images were obtained at the optimum defocus. High annular dark field was used to obtain the size distribution of the nanoparticles. HRTEM images were calculated using the SimulaTem software developed at the UNAM in Mexico. The STEM simulated images were generated assuming that in a STEM image, every atom generates a signal of intensity approximately proportional to $Z^{1.7}$, Z being the atomic number.

III. Results and discussion

The nominal average size of the particles was calculated *in situ* from the evaporation conditions. We used the mass spectrometer to select the size of the particles. By controlling the evaporation conditions we were able to choose the production of three different sizes of particles: 1.1, 3, and 5 nanometers. HAADF images of the particles are shown in Fig. 2 along with plots that show the frequency distributions of the sizes, as measured by HAADF. As can be noted in the figure, the dispersion σ is very small (Fig. 2a). When the HAADF measurements are compared with the estimates made with the mass spectrometer, it is found that the real dispersion in size is even smaller than the one calculated by the Nanogen system. Such a comparison can be seen in Fig. 3.

The particle size distribution was also verified using atomic force microscopy. The results are shown in Fig. 4. On the AFM images we indicate with numerated lines the paths along which the height profiles were taken. These profiles are shown on the right side of the figure, six for the 5 nm particles, four for the 1.1 nm particles. From the profiles it was possible to verify the mean size, measured by the maximum heights on the profiles. On the AFM measurements, the size of the particles appear somewhat larger than on the HAADF measurements, but this apparent discrepancy may be easily understood if it is considered that the Z coordinate of the AFM tip is calibrated at only one height, while the relationship between the measured Z height and the actual Z height is not linear. Nevertheless, it can be noted that the size distribution in the 1 nm particles is very narrow, while the 5 nm particles show a slightly broader but still small distribution. From these observations we can conclude that, for potential applications where the deviation on the size of the particles is a critical factor, this production method is significantly better than the chemical synthesis methods, which in most cases do not produce narrow distributions.

We studied the crystalline structure and shape of the nanoparticles using HREM. A typical HREM micrograph of the Au/Pd 5 nm nanoparticles is shown in Fig. 5a. The apparent shape of the particle that appears in this figure is consistent with what we would expect in particles with a five-fold symmetry, such as the icosahedron or the decahedron. Very extensive results have been published on particles with these

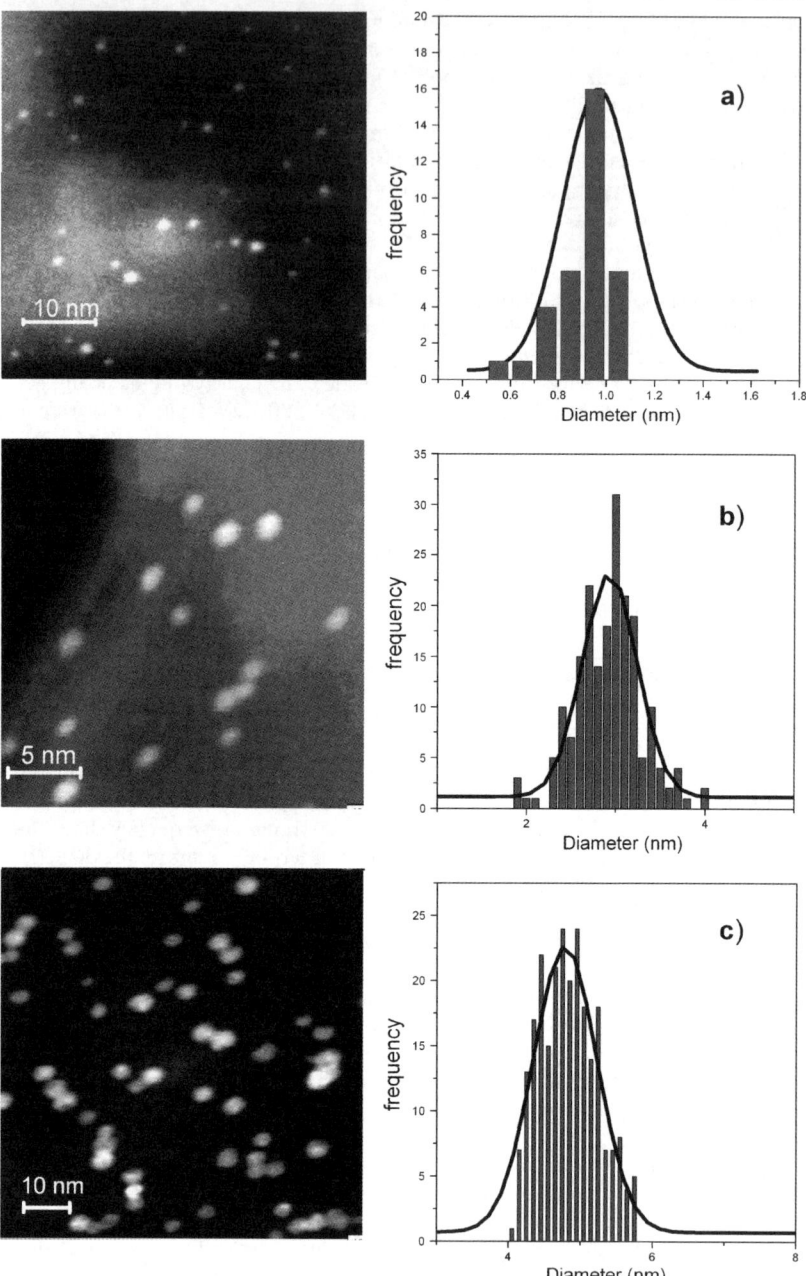

Fig. 2 HAADF images and frequency histograms of nanoparticles of size (a) 1.1 nm (σ = 0.44 nm), (b) 3 nm (σ = 0.3 nm), and (c) 5 nm (σ = 0.07 nm).

kinds of geometry.[22] From the FFT pattern, shown in Fig. 5c, it can be inferred that the particle in Fig. 5a corresponds to an icosahedron showing a two fold orientation. The simulated TEM image for a 3871-atom icosahedron (Fig. 5b), and its corresponding FFT (Fig. 5d), are also consistent with this interpretation. The atomistic model for an icosahedral particle with this orientation, of approximately the same

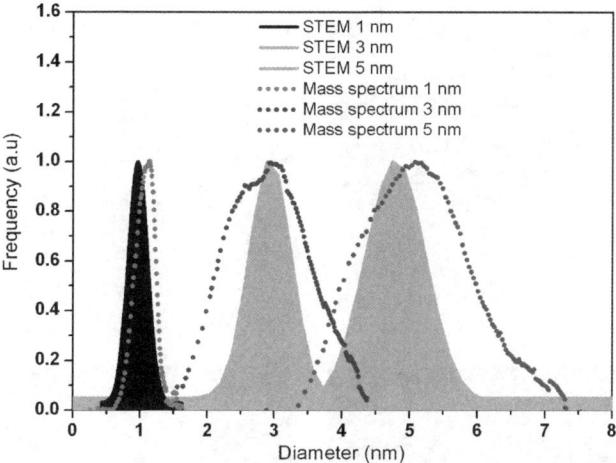

Fig. 3 Mass spectrometer profiles (line, before the selection by mass) and HAADF (filled area, after the selection by mass) of the size distribution.

size as the real particle, is shown in Fig. 5e. For some particles, their orientation with respect to the electron beam made it difficult to verify clearly the structure, but nevertheless we can conclude that the most common particle structure in our

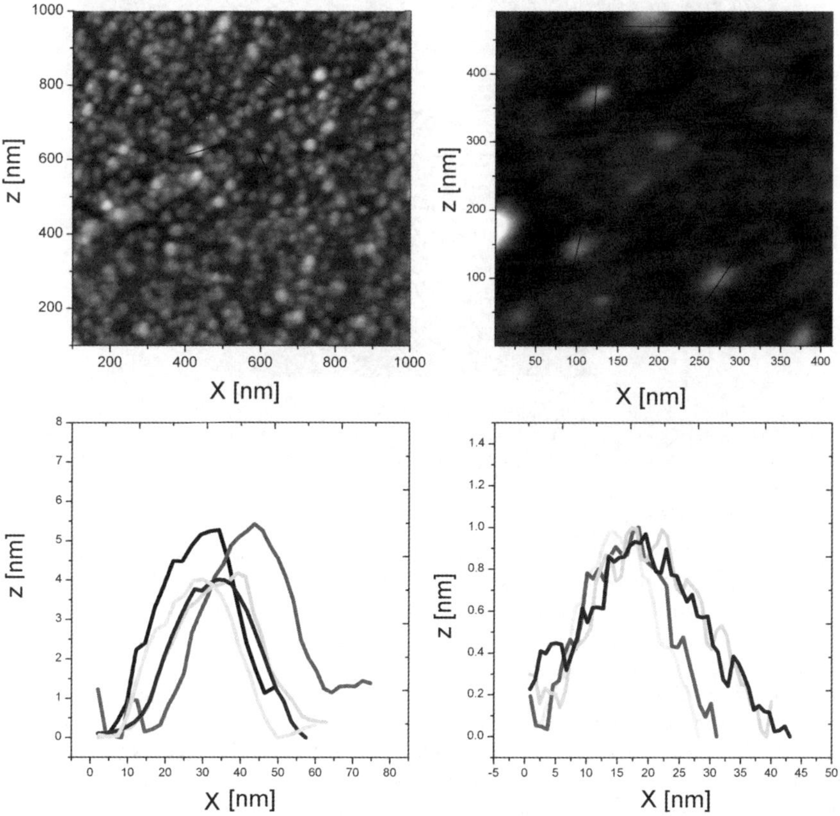

Fig. 4 Atomic force microscopy images of the samples of 1.1 nm and 5 nm, and Z height AFM profiles.

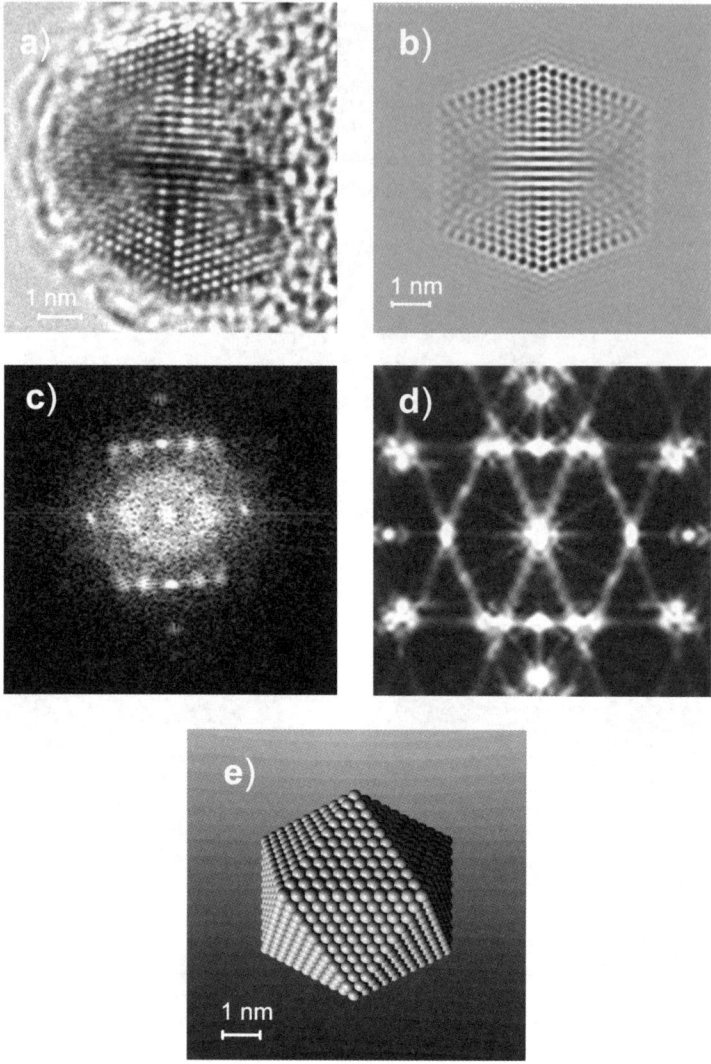

Fig. 5 (a) HRTEM micrograph of an icosahedral AuPd nanoparticle, (b) HRTEM simulation model of the nanoparticle, (c) experimental FFT pattern, (d) FFT pattern of simulation model, and (e) atomistic model of icosahedral nanoparticle.

depositions was the icosahedral; however, we also observed decahedral particles with less frequency. In a previous work,[12] we have reported that nanoparticles grown by the polyol method are mostly cuboctahedral in shape with an FCC crystalline structure. This is in sharp contrast with our present case, where the particles have Ic and Dh structures. This suggests that our nanoparticles are produced under non-equilibrium conditions due to the fast cooling of the metal plasma and to the nucleation rate of the nanoparticles.

It has been shown by García *et al.*[23] that the contrast in HAADF is very sensitive to the core-shell structure. The core-shell structure will be clearly revealed in HAADF images, since the image intensity I is proportional to $Z^{1.7}$, Z being the atomic number. This non-linear dependency must produce a high contrast between regions with a large concentration of Au atoms, and those regions where there is a majority of Pd atoms. In order to compare our HAADF images with what we would expect to find in core-shell structures, we calculated a series of HAADF images for particles with different orientations and elemental distributions. Some of these images are shown in Fig. 6. We focused our attention on particles with a palladium core (Fig. 6a), and particles with a gold core (Fig. 6b). The intensity profiles for the different cases are observed in Fig. 6c. By comparing the real HAADF-STEM images (such as the one shown in Fig. 7) with the simulated images shown in Fig. 6, we conclude that no core-shell structure is formed in the particles produced by the sputtering reactor, but the bimetallic particles are formed by an alloy which has an

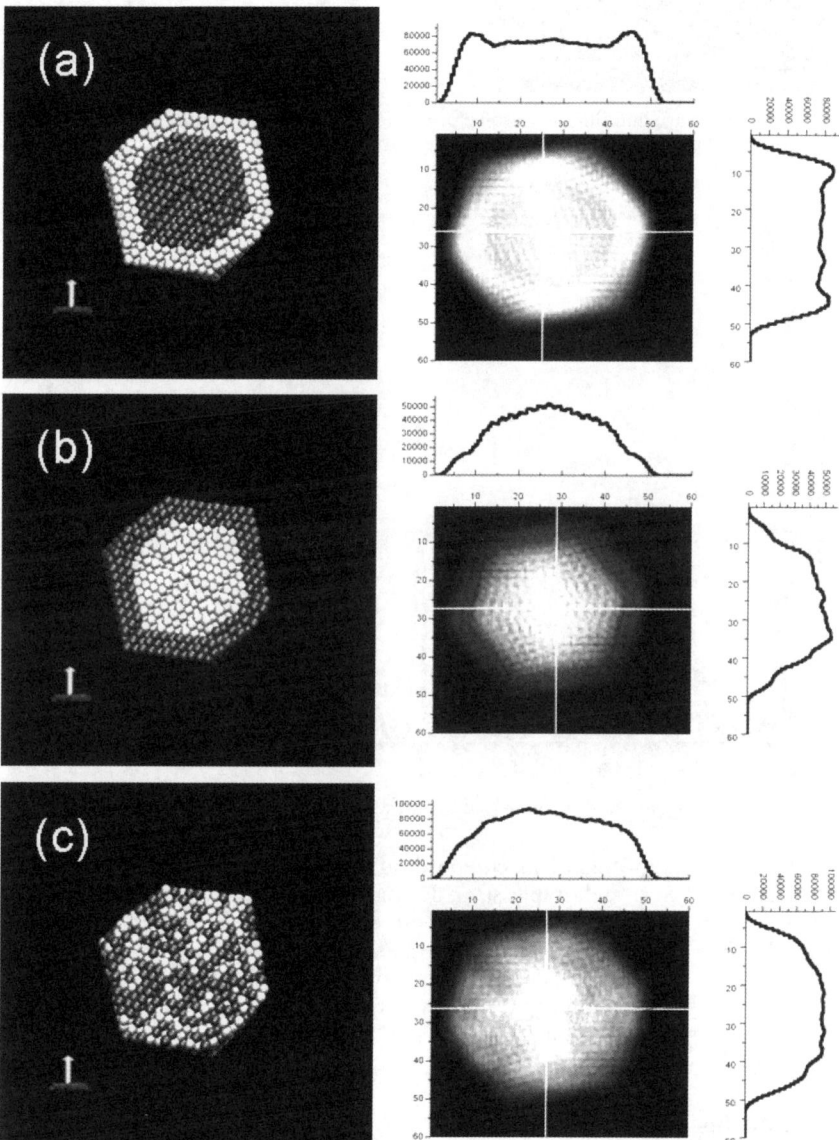

Fig. 6 Simulated STEM image of Au/Pd nanoparticle, along with intensity profiles. (a) Pd core, (b) Au core, (c) Random distribution of Au and Pd. The images at the left are cross-sections of the model nanoparticles, where Au atoms are represented in light tone, and Pd atoms are represented in a darker tone.

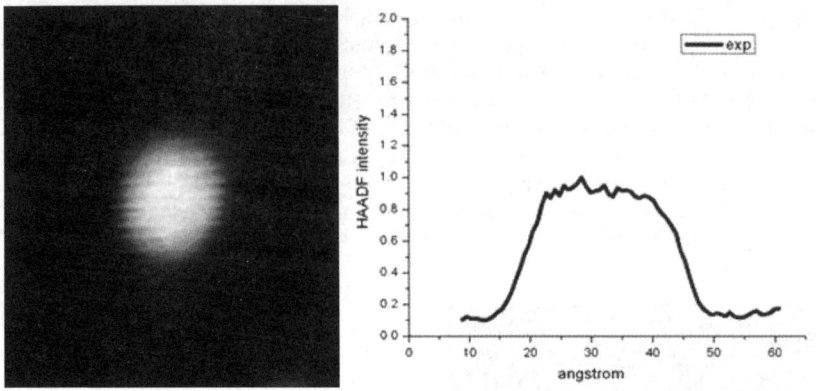

Fig. 7 STEM image and intensity profile of a Au/Pd nanoparticle synthesized by sputtering. The profile is characteristic of a homogeneous alloy.

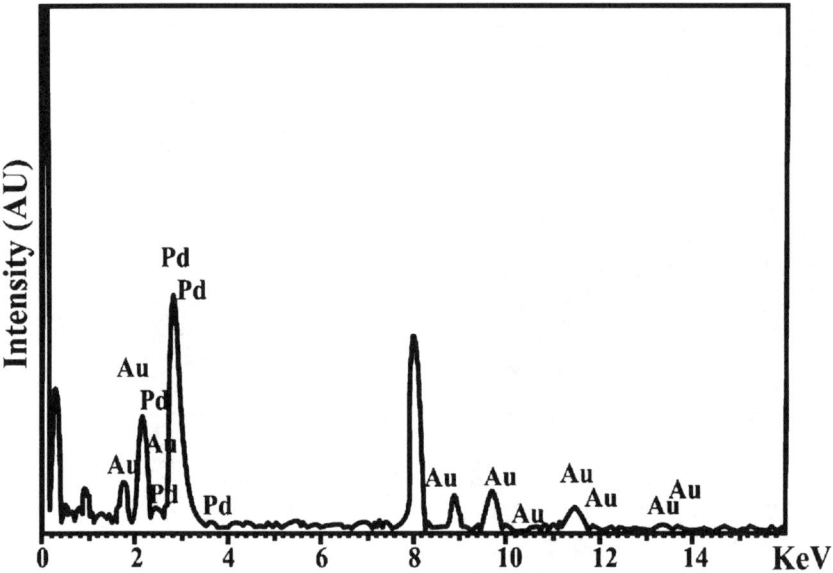

Fig. 8 X-Ray microanalysis profile of Au/Pd nanoparticles.

homogenous composition. This composition might be attributable to the particle getting trapped in a metastable state during its formation, and annealing of the samples might yield a core-shell structure such as those obtained by chemical methods. This annealing process will be the subject of a follow-on paper.

Finally, we analyzed the chemical composition of the nanoparticles by X-ray microanalysis. Fig. 8 shows the X-ray spectra, from where we obtained that the composition in the nanoparticles is Au: 18.8% and Pd: 81.2%, measured in atomic percentages.

IV. Conclusions

We synthesized Au/Pd bimetallic nanoparticles by an inert gas condensation process, in which the precise control of atmosphere conditions, condensation zone length, and selection by mass, allow a strict control of the size, composition and shape of the particles. With the conditions used in our experiments we were able to produce nanoparticles at three selected sizes, in the range of 1.1 to 5 nm. Images of the nanoparticles by HAADF-STEM show a narrow distribution on the size of the particles, and a high number density of particles in the substrate, with a low formation of aggregates. Even when the mass spectra of the nanoparticles suggest the possible overlapping between the size distributions for nanoparticles of 3 and 5 nm, these distributions are measured before filtering the nanoparticles, and after the filtering of the nanoparticles by their mass, the resulting distribution curves show no overlapping at all for the different selected mean sizes. We also performed AFM measurements, and the resulting small distributions for each deposition show a high correlation with the mass spectra and the HAADF-STEM results. The HRTEM micrographs and FFT patterns, and their comparison with simulated images, show that almost all of the produced nanoparticles have a five-fold symmetry (icosahedral geometry). When the HRTEM images are compared with simulated STEM images, we do not find evidence of the formation of a core-shell structure in the chemical composition of the nanoparticles, but our results suggest the formation of a homogeneous alloy of Au and Pd on the whole of the particles. These results may have a high relevance from the standpoint of functionality of the nanoparticles, since the catalytic activity, and the electronic and optical properties are strongly affected by the distribution of the chemical species on the most external regions of the particles. To study in depth the effect of the distribution of the two metals in the surface of the nanoparticle, it is necessary to measure chemical activity and to perform DFT calculations and atomistic simulations; a work in progress that will be the subject of a future paper.

Finally, we consider that our results demonstrate that the method of size selected inert gas condensation offers a useful alternative to control precisely not only the size but also the shape and the local composition of bimetallic nanoparticles.

Acknowledgements

This work was supported by the International Center for Nanotechnology and Advanced Materials of The University of Texas at Austin (ICNAM), the Council for Science and Technology of the State of Nuevo León, México, and the National Council for Science and Technology, México (CONACYT), grants (43772 & 207569) and NL-2004-C05-060.

References

1   M. Harada, K. Asakura and N. Toshima, *J. Phys. Chem.*, 1993, **97**, 5103–5114.
2   G. Schmid, H. West, J. o. Malm and C. Grenthe, *Chem.–Eur. J.*, 1996, **2**, 1099–1103.
3   P. Mulvaney, M. Giersig and A. Henglein, *J. Phys. Chem.*, 1993, **97**(24), 6334–6336.
4   G. Schmid, *Clusters and Colloids*, VCH, Weinheim, 1994.
5   M. Michaelis, A. Henglein and P. Mulvaney, *J. Phys. Chem.*, 1994, **98**(24), 6212–6215.
6   L. M. Liz-Marzan and A. P. Philipse, *J. Phys. Chem.*, 1995, **99**(41), 15120–15128.
7   T. Itakura, K. Torigoe and K. Esumi, *Langmuir*, 1995, **11**(10), 4129–4134.
8   M. J. Hostetler, C. J. Zhong, B. K. H. Yen, J. Anderegg, S. M. Gross, N. D. Evans, M. Porter and R. W. Murria, *J. Am. Chem. Soc.*, 1998, **120**(36), 9396–9397.
9   S. Link, Z. L. Wang and M. A. El-Sayed, *J. Phys. Chem. B*, 1999, **103**(18), 3529–3533.
10 R. G. Freeman, M. B. Hommer, K. C. Grabar, M. A. Jackson and M. J. Natan, *J. Phys. Chem.*, 1996, **100**(2), 718–724.
11 I. Srnova-Sloufova, F. Lednicky, A. Gemperle and J. Gemperlove, *Langmuir*, 2000, **16**(25), 9928–9935.
12 S. J. Mejia-Rosales, C. Fernández-Navarro, E. Pérez-Tijerina, J. M. Montejano-Carrizales and M. José-Yacaman, *J. Phys. Chem. B*, 2006, **110**(26), 12884–12889.
13 A. Jablonski, S. H. Overbury and G. A. Somorjai, *Surf. Sci.*, 1977, **65**(2), 578–592.
14 B. J. Wood and H. Wise, *Surf. Sci.*, 1975, **52**(1), 151–160.
15 C. W. Yi K Luo, T. Wei and D. W. Goodman, *J. Phys. Chem. B*, 2005, **109**(39), 18535–18540.
16 Mantis Deposition Ltd, Oxford, England. http://www.mantisdeposition.com.
17 K. Sattler, J. Mühlbach and E. Recknagel, *Phys. Rev. Lett.*, 1980, **45**(10), 821–824.

18 S. H. barker, S. C. Thorton, A. M. Keen, T. I. Preston, C. Norris, K. W. Edmonds and C. Binns, *Rev. Sci. Instrum.*, 1997, **68**(4), 1853–1857.
19 I. M. Goldby, B. Von Issendorff, L. Kuipers and R. E. Palmer, *Rev. Sci. Instrum.*, 1997, **68**(9), 3327–3334.
20 H. Haberland, Z. Insepov and M. Moseler, *Phys. Rev. B*, 1995, **51**, 11061–11067.
21 O. Rattunde, M. Moseler, A. Häfeler, J. Kraft, D. Rieser and H. Haberland, *J. Appl. Phys.*, 2001, **90**(7), 3226–3231.
22 J. A. Ascencio, C. Gutiérrez-Wing, M. E. Espinosa, M. Marín, S. Tehuacanero, C. Zorrilla and M. José-Yacamán, *Surf. Sci.*, 1998, **396**(1–3), 349–368.
23 D. García-Gutiérrez, C. Gutiérrez-Wing, M. Miki-Yoshida and M. José-Yacamán, *Appl. Phys. A*, 20043), 481–487.

Structures and optical properties of 4–5 nm bimetallic AgAu nanoparticles†

Z. Y. Li,*[a] J. P. Wilcoxon,[a] F. Yin,[a] Y. Chen,[a] R. E. Palmer[a] and R. L. Johnston[b]

Received 13th June 2007, Accepted 22nd June 2007
First published as an Advance Article on the web 17th September 2007
DOI: 10.1039/b708958a

Three types of bimetallic AgAu nanoparticles, with mean size of 4–5 nm, $Ag_{core}Au_{shell}$, $Au_{core}Ag_{shell}$ and alloyed AgAu, have been synthesized using an inverse micelle method. To image these small size nanoparticles, quantitative high angle annular dark field imaging using scanning transmission electron microscopy was successfully applied. Our results show that good control of nanoparticle size dispersion and composition modulation was achieved. Optical properties of the nanoparticles are correlated with direct internal structure analysis. The structural stability is discussed, based on thermodynamic considerations.

1. Introduction

Nanoparticles comprising two different metallic elements offer additional degrees of freedom for altering their physical properties by varying the atomic composition and atomic arrangement.[1] This could potentially enable a wide range of opportunities for discovering materials with novel physical properties. However, our present understanding of fundamental issues, such as the driving force for generating specific structures of bimetallic nanoparticles, as well as structure–property correlations, has been hindered by the lack of size control and knowledge of the precise chemical composition and atomic arrangement within a given bimetallic nanoparticle. The situation is most acute for nanoparticles less than a few nanometres in diameter. Currently it is still a significant challenge to synthesis and to characterize, with atomic precision, bimetallic nanoparticles with sizes smaller than about 5 nm.

Bimetallic AgAu nanoparticles have been chosen as a model system. This is probably one of the most studied bimetallic systems in the literature,[1–4] partly because both Ag and Au nanoparticles display distinctive optical plasmon absorbances in the visible range, with size and shape dependent properties. In the bulk, Ag and Au form alloys for all compositions with very little surface segregation due to their very similar lattice constants, 4.09 Å and 4.08 Å, respectively.[5] However, nanoparticles with either alloyed, or core–shell segregated structure can be synthesised through various methods.[6] As a consequence, the optical properties of these bimetallic particles can be tuned not only by varying their size and external morphology, but also by changing their composition and internal structure (chemical ordering).[2] In fact, optical absorption spectroscopy has been the most widely

[a] Nanoscale Physics Research Laboratory, School of Physics and Astronomy, University of Birmingham, Birmingham, UK B15 2TT. E-mail: ziyouli@nprl.ph.bham.ac.uk; Fax: +44-121-4147327; Tel: +44-121-4144593
[b] School of Chemistry, University of Birmingham, Birmingham, UK B15 2TT

† The HTML version of this article has been enhanced with colour images.

used technique for inferring the internal structure of the Ag–Au nanoparticles. Often, a single plasmon band has been used to indicate that the Ag–Au particles have an alloyed form, while the appearance of two distinct plasmon peaks has been used as a fingerprint for segregated bimetallic nanoparticles.[2,7–10] In practical terms, this approach is simple, but is only appropriate for Ag–Au nanoparticles with particle sizes between $\sim$10–100 nm. In the case of smaller core–shell particles, the situation is more complicated,[11] as the plasmon resonance absorbance energy depends not only on composition but also on nanostructure. While the outermost shell of the materials might dominate the interaction with the incident light for shell thicknesses larger than the frequency dependent absorbance length, the electron plasmon oscillations will also be damped by any interface between the metals, changing the shape of the absorbance curve.

The optical contributions from the core material may be effectively screened for large shell thicknesses. In any case, optical measurements provide number average information that may not reflect the heterogeneous nature of bimetallic nanoparticles. Therefore, it is extremely important to develop methods having a high spatially-resolved imaging capability for the internal structure to correlate nanostructure with optical properties. This will provide us with the knowledge of the actual structure of the as-synthesized nanoparticles and how changes in nanostructure over time affect the optical absorbance. High resolution imaging of individual nanoparticles will also provide the detailed inputs needed for accurate modelling to understand the underlying physical mechanisms that favour either mixing or segregation in bimetallic nanoparticles.

In this paper, we show that bimetallic Ag–Au nanoparticles of approximately 4–5 nm average diameter, with a very narrow size distribution, can be prepared through a chemical route. The internal structure is elucidated by a combination of a conventional electron transmission microscope (TEM) and high angle annular dark field (HAADF) imaging in a scanning transmission electron microscope (STEM). The plasmon absorptions of these nanoparticles in solution are examined and correlated with direct structural imaging of individual particles.

2. Nanoparticle synthesis

The bimetallic Ag–Au nanoparticles used in this study were prepared using an inverse micelle method, details of which have been described in previous publications.[6,12] The key aspect of this method, which distinguishes it from either liquid or gas atomic aggregation processes, is that the metal cluster growth is controlled by the micro-heterogeneous environment of the droplet-like inverse micelles. One advantage of this method is the inexpensive, ready availability of simple salts as atom sources for the growth process. The method has also produced nanoparticles with an extremely narrow size distribution as demonstrated by size-exclusion chromatography (SEC).[13]

Nanoalloys of Ag and Au, designated AgAu, were synthesized by co-reduction of 0.01 M $HAuCl_4$ and $AgBF_4$ in the presence of a cationic surfactant, tetraoctyl ammonium chloride, (0.1 M TOAC), which forms inverse micelles in toluene. After mixing of the individual gold and silver inverse micelle solutions, a reducing agent, $LiBH_4$ in tetrahydrofuran (THF), was added under rapid magnetic stirring. An equimolar amount of an alkyl thiol, typically dodecanethiol, $C_{12}H_{25}SH$, can be added to provide additional stabilization prior to reduction. To produce larger nanoparticles, with a size between 4–5 nm, the thiol is added after reduction. The reactions take place in an inert atmosphere glove box. A nearly instantaneous solution colour change from light yellow to dark orange–red occurs upon addition of the strong $LiBH_4$ reductant, with copious evolution of hydrogen bubbles. Stirring is continued for approximately an hour. AgAu nanoalloys, nearly free of the TOAC surfactant are obtained by precipitation using a 10-fold excess of dry methanol or acetone and redissolving the resulting precipitant in toluene or benzene.

The Ag$_{core}$/Au$_{shell}$ nanoparticles, designated Ag/Au, with the "/" corresponding to the interface between the metals and the first metal listed being the core, were synthesised by first growing a Ag seed nanocrystal followed by solution phase deposition of several generations of Au overlayers.[12] An organometallic Au atom source, gold triphenyl phosphine chloride in benzene is slowly injected into a solution of Ag seed nanoparticles with diameter $D = 2.6$ nm, and is reduced and deposited onto the surface of the Ag seeds. The preparation sequence can be reversed to produce inverted Au$_{core}$/Ag$_{shell}$ (Au/Ag) structures. In this case, a Ag organic compound dissolved in benzene, Ag(p-C$_6$H$_4$CO$_2$), is slowly injected and reduced in a solution containing Au seed nanoparticles. The Ag is deposited onto the surface of $D = 1.8$ nm Au seed particles in the presence of a stabilizing alkyl thiol molecule.

The freshly prepared samples were shown, by high resolution size-exclusion chromatography, to have a monodispersed size distribution with the average size increasing on each additional deposition step.[13] A single chromatographic elution peak for each generation with a narrow linewidth and homogeneous optical properties throughout the elution peak shows that the solution growth produces a series of discretely sized nanoparticles. The particles were passivated with C$_{12}$SH during the growth process to allow them to be precipitated using excess methanol, purified by acetone washing, and re-dispersed in hydrocarbons, toluene or octane, prior to imaging or absorbance studies. The samples were kept in dry nanoparticle film form at room temperature for storage. They were re-suspended in toluene just before optical measurements and electron microscopy studies.

3. Optical absorption spectroscopy

The absorption spectra of passivated nanoparticles were obtained *in situ* during size exclusion chromatography using an on-line photodiode array (PDA) with an adjustable bandwidth and wavelength range. Typical bandwidths for our system were either 2.4 nm or 4.8 nm when using a wavelength range of 290–795 nm. Fig. 1 shows three series of absorption spectra taken from Ag–Au binary nanoparticles: (a) AgAu alloy nanoparticles with varied composition, (b) Ag/Au nanoparticles with an increasing amount of Au, and (c) Au/Ag nanoparticles with an increasing amount of Ag. For the core/shell samples, the spectra from the respective seed samples are also included. The spectra have been normalized at the absorption maximum and shifted in the y-axis for clarity. The sizes obtained from TEM and SEC are also noted in Fig. 1.

It is clear that there is only one single distinctive peak in all spectra. In the alloyed particles, the absorption peak red shifts slightly as the Au contents increases. In the case of core/shell samples, whether the absorption maximum red or blue shifts depends on the outermost shell materials. For Ag/Au, the maximum red shifts from 460 for pure Ag to 524 nm for Ag : Au = 1 : 5. Au nanoparticles with diameter of 6 nm have an absorption maximum of around 523 nm in toluene[4] which is comparable to the Ag : Au = 1 : 5 sample, demonstrating how effectively the Au shell dominates the optical properties. In addition, the deposition of Au on the Ag seeds results in an unsymmetric shape of the absorbance peak. By comparing to a Au particle of similar size, the presence of Ag in the core reduces the peak width. On the other hand, thicker depositions of Ag in the shell of Au seed nanoparticles blue shift the maximum from 510 to 470 nm and decrease the linewidth (measured from half-width at half maximum to the red side of the peak) from 92 nm to 58 nm. Pure Ag nanoparticles of this size have even narrower and more symmetrical resonance peaks. In both Ag/Au and Au/Ag cases, the core optical properties can be dominated by the shell optical properties for compositions with core : shell > 1 : 2.

The origin of the intense absorption band in the UV-visible region has been attributed to the collective oscillation of the free conduction electrons induced by the incident light. These resonances are known as plasmons. It has been generally accepted that when nanoparticles are composed of more than one metal, both

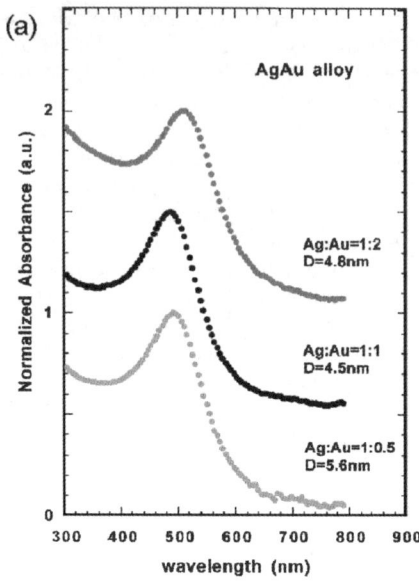

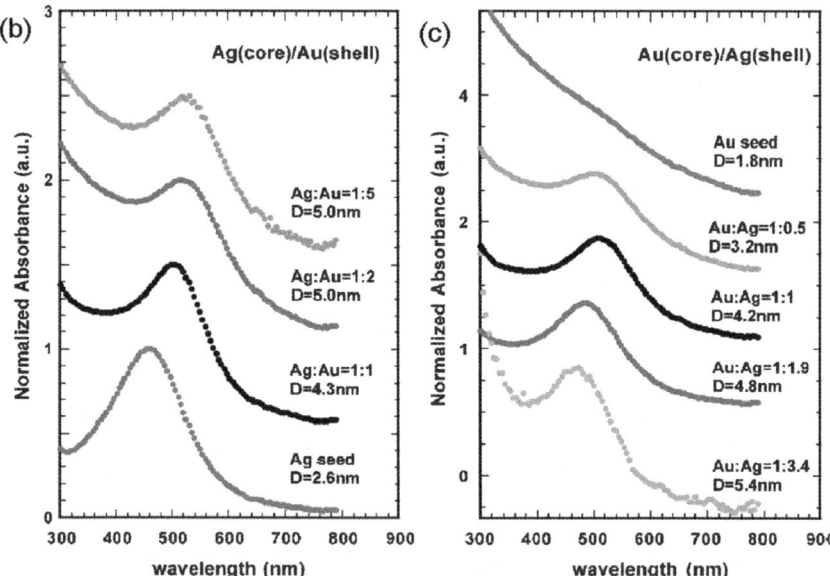

Fig. 1 UV-visible absorption spectra of three series of bimetallic Ag–Au nanoparticles with varied atomic composition: (a) AgAu alloyed nanoparticles, (b) Ag_{core}/Au_{shell} nanoparticles, and (c) Au_{core}/Ag_{shell} nanoparticles. For the core–shell samples, the spectra from the seed samples are also included. The spectra have been normalized at the absorption maximum and shifted in the y-axis for clarity.

composition and the degree of atomic segregation will determine the position and the width of the plasmon resonance. Although the results for AgAu nanoalloys presented in the literature are in reasonable agreement, the reported results and interpretation for core–shell nanoparticles are more equivocal, especially for sizes below 5 nm. In Fig. 2, we summarize the plasmon resonance positions, obtained

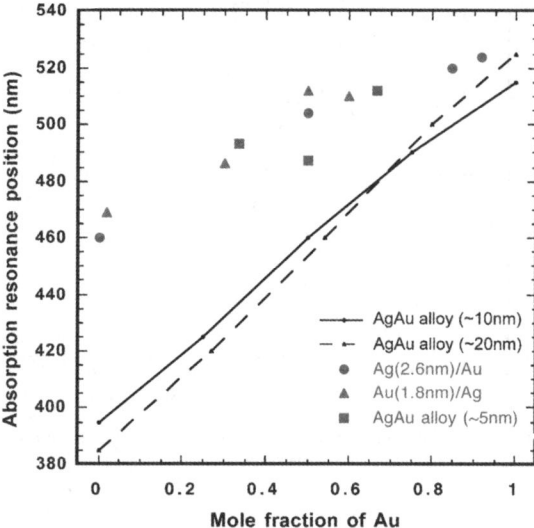

Fig. 2 Absorption maximum positions as a function of the mole fraction of Au in alloyed AgAu and core/shell structured, Ag/Au and Au/Au, nanoparticles. Data from alloyed AgAu particles with larger size are also plotted as a reference > ref. 9 and 14].

from our absorption spectra in Fig. 1, as a function of mole fraction of Au in the binary Ag–Au nanoparticles. Results in the literature for AgAu nanoalloys are also included,[9,14] where the plasmon peak was found to shift linearly between those for pure Ag and pure Au nanoparticles. The general trend of the red shift in plasmon resonance position with increasing Au content agrees well with the results in the literature for alloyed AgAu particles. However, our data show the resonance at a much higher wavelength than previously reported data for the same atomic composition for AgAu nanoalloys. In the case of pure Ag or Au nanoparticles, it has been demonstrated that, as the particle size decreases, the plasmon peak will red-shift significantly for Ag while there is a slightly blue shift for Au.[4] We, therefore, interpret the longer wavelength resonance as originating from effects of the smaller size of our nanoparticle samples. However, it should also be noted that, in the core/shell samples, the particle size and the shell thickness varies somewhat with atomic composition of the nanoparticles.

The linewidth of the absorption resonance is a measure of the damping of the electron oscillations responsible for the light absorbance. A narrower linewidth reflects decreased damping of the electron oscillations. It is found that there is considerably less damping for a AgAu alloyed particle (Fig. 1(a)) than in a pure 4–5 nm Au particle.[4] The linewidths of the 4–5 nm Au/Ag nanoparticles in Fig. 1(c), are all less than a pure Au nanoparticle of this size and even less damping occurs with increasing amounts of Ag. For Ag/Au nanoparticles with a 1 : 1 composition, we observe that decreasing the particle size blue shifts and somewhat narrows the linewidth (data not shown in this paper). This is significantly different from pure Ag nanoparticles which show a red shift and line broadening with decreased size, as well as pure Au nanoparticles which show a blue shift but complete loss of an absorption maximum for sizes less than 3 nm. Thus, creating a core/shell nanostructure can minimize the damping due to a decreasing nanoparticle size.

In Fig. 3, we display the absorption spectra for the same Ag : Au atomic composition of 1 : 1 for all three samples of similar total size for comparison. The largest damping of the electron oscillation set up by the incident electromagnetic field occurs when the core-shell structure is formed. There is a large red-shift in the

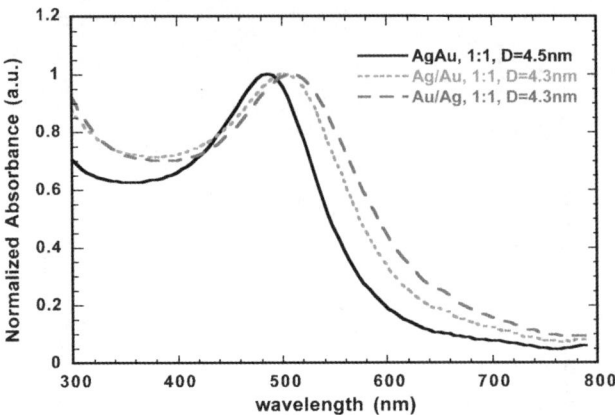

Fig. 3 Comparison of absorption spectra from bimetallic Ag–Au nanoparticles with the same atomic composition (1 : 1) but different nanostructures.

core–shell particles compared to the nanoalloys. The key observation here is that none of these spectra are as blue shifted or as narrow as a pure 4 nm Ag particle, nor are any linewidths as wide as for a pure Au particle. Therefore, the presence of Ag always leads to reduced attenuation of the electron resonance. It should be noted that these spectra were obtained shortly after synthesis. There is evidence that aging effects can lead to reduced differences between the nanostructures and their associated spectra by permitting interdiffusion and mixing of the atoms.

Although a general qualitative understanding of the plasmon resonance spectra has been reached, detailed quantitative modelling of the optical properties is still not available. To model the plasmon resonance, the following key parameters need to be considered: the frequency dependence of the dielectric function of metal materials, the dielectric constant of the medium in which the nanoparticles are embedded, and the size and shape of the particles. In the case of a nanoparticle with a core–shell geometry, contributions from the core material will be difficult to assess unless we know exactly the thickness of the shell materials and the sharpness of the interface between the two metals. As noted above, interdiffusion at the boundary of two materials will also need to be considered and this depends on sample age.

One current difficulty is that the structural data for nanoparticles, especially for those with sizes smaller than 5 nm, is very limited and not unambiguous. On the basis of molecular dynamics simulation, Shibata *et al.*[15] have suggested that the stability of Au/Ag core–shell nanoparticles depends on the particle size, with complete interdiffusion at room temperature for core sizes below 4.6 nm. However, our spectra from Fig. 3 indicate that nanostructural segregation is reflected in the spectral differences observed when the samples are less than 1 week old. Our earlier work also shows that the stability of Ag/Au core–shell nanoparticles of ~4 nm depends on the shell thickness. When the atomic composition ratio is Ag : Au = 1 : 2, the core–shell geometry is stable in toluene solution for a couple of years.[16]

4. Structural characterization

To understand the relationship of nanostructures to optical properties of binary Ag–Au nanoparticles, further structural characterization was undertaken using conventional transmission electron microscopy and scanning transmission electron microscopy. To prepare the specimen, a drop of the solution containing the particles was deposited onto a standard 3 mm diameter copper grid covered with a thin amorphous carbon (a-C) film under ambient conditions.

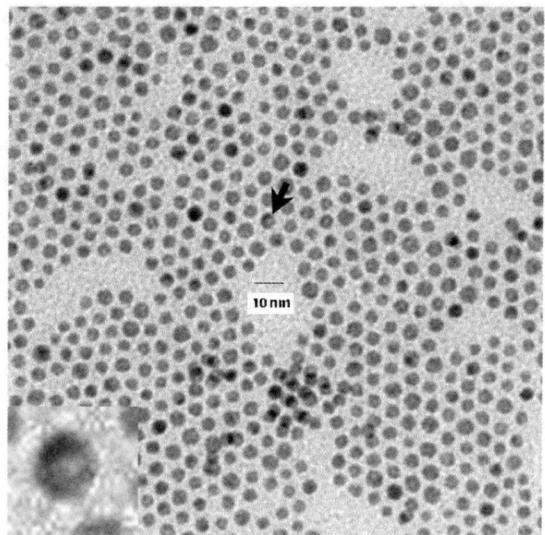

Fig. 4 A typical TEM image taken from Ag/Au nanoparticles with the atomic composition 1 : 1. The arrow shows one of the particles with contrast modulation indicating the core/shell structure. The image of this particle has been enlarged and shown as the insert.

Fig. 4 shows a representative overview of the Ag/Au (1 : 1) nanoparticles on the a-C film when the sample was 3 months old. Some particles in this image are oriented to reveal the lower contrast Ag core (see the insert for one of such particles indicated by an arrow). However, the structural information from the interference of the low-angle scattered electron waves in TEM is strongly intertwined with the experimental conditions, making quantitative interpretation non-trivial. This drawback can be overcome by collecting high-angle scattering originating from the interaction of a well-focused electron probe with the material in a STEM. Here an electron beam is focused down to a fine spot and is scanned across the specimen. The transmitted electrons are collected by a high angle annular dark field (HAADF) detector. In this set-up, the electron scattering is mostly incoherent and the contrast is proportional to nearly the square of the atomic number, Z.[17] Therefore the elemental information of the samples under investigation can be obtained directly from analysing image intensity contrast.[18–20] Fig. 5 displays two STEM images taken from the core–shell particles that have the same atomic composition 1 : 1 but reversed chemical ordering: (a) Au/Ag and (b) Ag/Au. The difference in intensity contrast between these two samples can be seen more clearly in line profiles, where a nearly flat intensity contrast at the centre of the projection of the Ag/Au particles is apparent in Fig. 5 (right panel).

To quantify the HAADF-STEM intensity profiles from the core–shell nanoparticles, simulations based on simple geometric considerations are performed: the particles are assumed to have spherically symmetrical structure, and the HAADF intensities are taken to depend exclusively on the height of the projected atomic column, with each atom contributing a Z-dependent intensity, see Fig. 6(a). Fig. 6(b) shows the simulation result for the line profile through the centre of a pure Au nanoparticle, without (dashed line) and with (solid line) taking account of the finite nature of the electron probe shape. For the core/shell nanoparticles, it is assumed that both the core and the shell are spherical and that the centre of mass of the core coincides with the centre of mass of the shell. The difference observed in Fig. 5 for Au/Ag and Ag/Au particles has been reproduced in Fig. 7, which can be attributed to the large difference in atomic number between Ag and Au ($Z = 47$ for Ag and 79 for Au). The close resemblance between the simulated profiles and our experimental

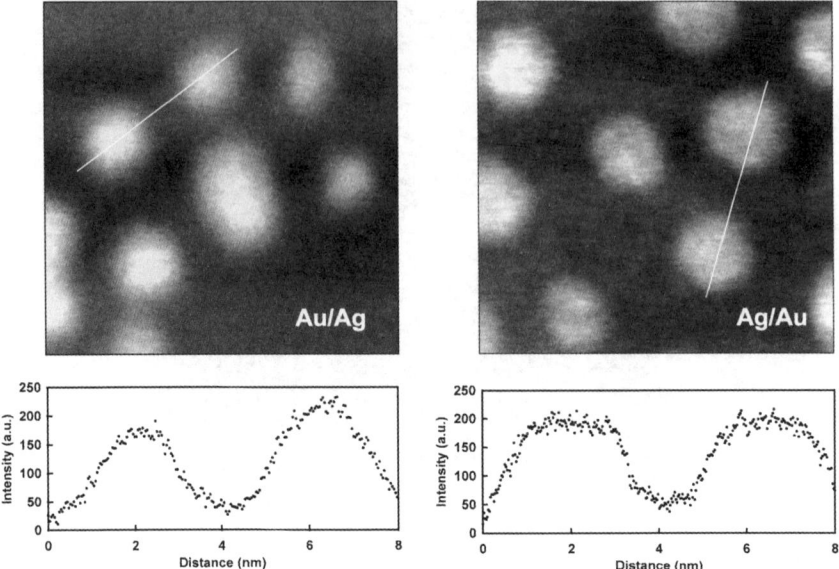

Fig. 5 HAADF-STEM images (13 nm × 13 nm) from core/shell nanoparticles, Au/Ag and Ag/Au, with atomic composition 1 : 1. The representative line profiles are shown under the corresponding images.

data suggests that the Ag/Au (1 : 1) particles maintained the core/shell structure as they were synthesised. From the measured core size and total size of the particles for Ag/Au and Au/Ag (1 : 1), together with the theoretical values of bond lengths (from the bulk metals) for Ag and Au, we can estimate that the shell thickness is just under 2 atomic layers (∼0.6 nm) and this is consistent with the experimental results in Fig. 5. Although from HAADF-STEM, we have not been able to identify unambiguously any sign of atomic intermixing at the interface, nevertheless we can be confident that the inner core is Ag rich and the overall geometry remains as core/shell. Our preliminary investigation using electron energy-loss spectroscopy (EELS)

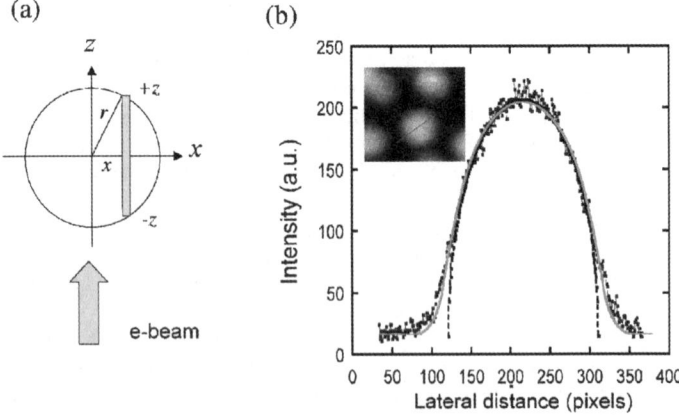

Fig. 6 (a) Schematic showing that a pure nanoparticle or a homogeneously alloyed nanoparticle is modelled by a sphere with radius of r. HAADF intensity is proportional to the projected atomic column depth along the electron beam. (b) Comparison of the experimental data from a pure Au nanoparticle with the simulated line profiles with and without taking into account of the finite size effect of the probe. The insert displays the STEM image from which the experimental line profile is taken.

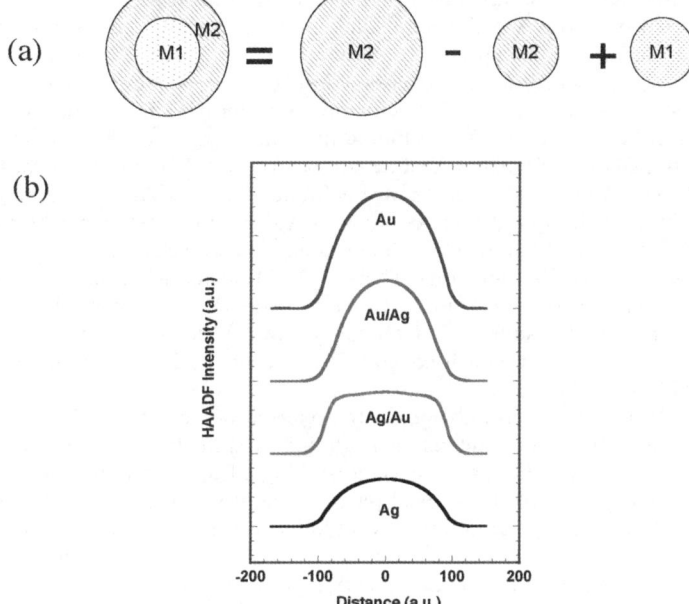

Fig. 7 (a) Schematic showing that a core/shell binary nanoparticle is modelled by a simply geometric consideration, where M1 and M2 denote material 1 and material 2, respectively. (b) Simulated HAADF intensity line profiles through the centres of the particles after taking into account the probe size effect. The atomic composition for Au/Ag and Ag/Au is 1 : 1.

with STEM on another set of Ag/Au samples indicates that the Ag extends into the outer Au shell.[21] So the distinct optical signature shown in Fig. 3 obtained from the core/shell particles compared to the nanoalloys of the similar size is due to the unique nanostructures for each case.

5. Thermodynamic stability

To understand the structure of core/shell nanoparticles (Ag/Au and Au/Ag), we here consider the factors controlling the thermodynamic stability of Ag–Au nanoparticles.[1] First, the metal–metal bond lengths in Ag and Au are very similar, so there is a negligible strain-induced driving force to induce segregation. The surface energy of Ag (78 meV A^{-2}) is less than that of Au (97 meV A^{-2}), which favours surface-enrichment by Ag. The cohesive energy of Au (3.81 eV/atom) is greater than that of Ag (2.95 eV atom^{-1}).[5] This, coupled with the weak exothermic enthalpies of formation of the bulk AgAu alloys, indicates that the metal–metal bond strengths are in the order Au–Au > Ag–Au > Ag–Ag, favouring core-enrichment by Au. These simple considerations are in agreement with the results of calculations by Johnston, Ferrando and co-workers,[22–24] based on the Gupta many-body empirical potential.[25–27] However, in the present case, "inverted" Ag/Au nanoparticles have been synthesised and their stability have been confirmed by HAADF-STEM. This suggests that this structure is at least kinetically stable. In the presence of passivating ligands (thiols in the present study) the stronger covalent binding of Au to S may act to stabilise structures with Au atoms at the surface of the nanoparticle. This also agrees with the distinct optical properties of each type of nanostructure studied. Chromatography confirms that Au–thiol bonds are stronger than Ag–thiol ones.[12]

Finally, as Au is more electronegative than Ag (Pauling electronegativities are 2.4 (Au) and 1.9 (Ag)), there should be some degree of electron transfer from Ag to Au atoms. The ionic contribution to the Au–Ag bonding favours Au–Ag mixing, as this increases the number of favourable Ag^+–Au^- interactions. It should be noted, however, that bulk Ag–Au alloys are solid solutions for all compositions, whereas charge transfer might be expected to favour ordered phases. Zhang and Fournier have recently modelled 55-atom icosahedral Ag–Au clusters using a pairwise additive Morse potential, coupled with an ionic term.[28] They calculated a 1% increase in cohesive energy after including the ionic term, with the stabilizing effect of charge transfer being greatest for 1 : 1 compositions. While the cluster surface was predicted to be Ag-rich, the ionic term was found to favour Ag/Au ordering on the surface. The contribution of ionic bonding in AuAg alloy clusters has been studied (for small clusters, with up to 20 atoms) by Bonačić-Koutecký and co-workers, who found, based on Density Functional Theory (DFT) calculations, that there is indeed a degree of electron transfer from Ag to Au,[29,30] and Ag–Au mixing. In contrast to the Gupta potential results,[25–27] a tendency towards low-coordinate Au surface atoms was also observed, which concurs with recent DFT calculations on Ag–Au clusters with up to 11 atoms.[31,32]

The question of the thermodynamically preferred degree of chemical ordering in Ag–Au clusters, therefore remains open, though it is likely to depend critically on cluster size and composition, as well as morphology. Rather than having completely core–shell segregated or fully alloyed particles, however, the competition between the various ordering–segregation effects could lead to the stabilisation of multi-shell onion-like structures: either alternating pure shells (*e.g.* Ag/Au/Ag) or with an intermediate interfacial alloy "shell" (*e.g.* Ag/AgAu/Au). Interestingly, spontaneous interfacial alloying and anomalous interfacial diffusion has been predicted by Ouyang *et al.* for Au/Ag particles, based on a size-dependent thermodynamic and kinetic model.[33]

6. Concluding remarks

We have studied the nanometre-sized core–shell and alloyed Ag–Au nanoparticles using both optical and structural tools. The optical absorption spectra reveal strong plasmon resonances and discernable differences in their maximum position and linewidth for nanoparticles with different internal structures. Conventional TEM confirms the narrow size distribution of the nanoparticles but without unambiguous information about their internal structure. On the other hand, we have shown that HAADF imaging using STEM has been able to identify the core–shell particles, even for nanoparticles with a shell thickness of only 1–2 atomic layers. Our experimental results indicate that it is possible to generate core/shell structured nanoparticles with either Au or Ag on the surface. The thermodynamical considerations suggest that Ag prefers either to be on the surface or to form interfacial alloys with Au, while Au prefers to segregate at the core. However, the outer-most Au layer may be stabilized by strong covalent bonding to the thiol ligands. Previous chromatographic studies comparing the binding of Au nanoparticles with Ag ones for alkyl thiols indicate that the Au–thiol bond is stronger than the Ag–thiol bond, which would stabilize the Ag/Au nanostructures as observed.

References

1 R. Ferrando, J. Jellinek and R. L. Johnston, *Chem. Rev.*, 2007, in press.
2 L. M. Liz-Marzan, *Langmuir*, 2006, **22**, 32.
3 P. Mulvaney, *Langmuir*, 1996, **12**, 788.
4 J. P. Wilcoxon, J. E. Martin and P. Provencio, *J. Chem. Phys.*, 2001, **115**, 998.
5 C. Kittel, *Introduction to Solid State Physics*, John Wiley & Sons Inc., New York, 1986, 6th edn.
6 J. P. Wilcoxon and B. L. Abrams, *Chem. Soc. Rev.*, 2006, **35**, 1162.
7 J. Sinzig, U. Radtke, M. Quinten and U. Kreibig, *Z. Phys. D: At. Mol. Clusters*, 1993, **26**, 242.
8 S. W. Han, Y. Kim and K. Kim, *J. Colloid Interface Sci.*, 1995, **99**, 15120.
9 S. Link, Z. L. Wang and M. A. El-Sayed, *J. Phys. Chem. B*, 1999, **103**, 3529.

10 J. H. Hodak, A. Henglein, M. Giersig and G. V. Hartlandet, *J. Phys. Chem. B*, 2000, **104**, 11708.
11 J. P. Wilcoxon, in *the Encyclopedia of Nanoscience and Nanotechnology*, ed. J. Swartz, C. Contescu and K. Putyera, Marcel Dekker Inc., 2004, p. 3177.
12 J. P. Wilcoxon and P. Provencio, *J. Am. Chem. Soc.*, 2004, **126**, 6402.
13 J. P. Wilcoxon, J. E. Martin and P. Provencio, *Langmuir*, 2000, **16**, 9912.
14 K.-S. Lee and M. A. El-Sayed, *J. Phys. Chem. B*, 2006, **110**, 19220.
15 T. Shibata, B. A. Bunker, Z. Zhang, D. Meisel, C. F. Vardeman II and J. D. Gezelter, *J. Am. Chem. Soc.*, 2002, **124**, 11989.
16 Z. Y. Li, J. Yuan, Y. Chen, R. E. Palmer and J. P. Wilcoxon, *Appl. Phys. Lett.*, 2005, **87**, 243103.
17 S. J. Pennycook, *Ultramicroscopy*, 1989, **30**, 58.
18 Z. Y. Li, J. Yuan, Y. Chen, R. E. Palmer and J. P. Wilcoxon, *Adv. Mater.*, 2005, **17**, 2885.
19 M. Di. Vece, N. P. Young, Z. Y. Li, Y. Chen and R. E. Palmer, *Small*, 2005, **2**, 1270.
20 S. J. Pennycook, B. Rafferty and P. D. Nellist, *Microsc. Microanal.*, 2000, **6**, 343.
21 R. Merrifield, Z. Y. Li, J. W. Wilcoxon, R. E. Palmer and A. L. Bleloch (to be published).
22 G. Rossi, R. Ferrando, A. Rapallo, A. Fortunelli, B. C. Curley, L. D. Lloyd and R. L. Johnston, *J. Chem. Phys.*, 2005, **122**, 194309.
23 B. C. Curley, G. Rossi, R. Ferrando and R. L. Johnston, *Eur. Phys. J. D*, 2007, **43**, 53.
24 F. Y. Chen, B. C. Curley, G. Rossi and R. L. Johnston, *J. Phys. Chem. C*, 2007, **111**, 9157.
25 R. P. Gupta, *Phys. Rev. B*, 1981, **23**, 6265.
26 V. Rosato, M. Guillopè and B. Legrand, *Philos. Mag. A*, 1989, **59**, 321.
27 F. Cleri and V. Rosato, *Phys. Rev. B*, 1993, **48**, 22.
28 M. Zhang and R. J. Fournier, *Mol. Struct. (THEOCHEM)*, 2006, **762**, 49.
29 V. Bonačić-Koutecký, J. Burda, R. Mitrić, M. Ge, G. Zampella and P. Fantucci, *J. Chem. Phys.*, 2002, **117**, 3120.
30 R. Mitrić, C. Bargel, J. Burda, V. Bonačić-Koutecký and P. Fantucci, *Eur. Phys. J. D*, 2003, **24**, 41.
31 G. F. Zhao and Z. Zheng, *J. Chem. Phys.*, 2006, **125**, 014303.
32 F. Y. Chen and R. L. Johnston, *Appl. Phys. Lett.*, 2007, **90**, 153123.
33 G. Ouyang, X. Tan, C. X. Wang and G. W. Yang, *Chem. Phys. Lett.*, 2006, **420**, 65.

Growth and structural properties of CuAg and CoPt bimetallic nanoparticles

Cyril Langlois,*[a] Damien Alloyeau,[ab] Yann Le Bouar,[b] Annick Loiseau,[b] Tetsuo Oikawa,[c] Christine Mottet[d] and Christian Ricolleau[a]

Received 18th April 2007, Accepted 26th June 2007
First published as an Advance Article on the web 21st September 2007
DOI: 10.1039/b705912b

Core/shell CuAg and alloyed CoPt have been synthesized using two vapor phase deposition techniques. For CuAg prepared by Thermal Evaporation (TE), the size and the morphology of the Cu cores are the key parameters to promote the formation of the core/shell arrangement. For CoPt synthesized by Pulsed Laser Deposition (PLD), the growth kinetics of nanoparticles, depending on the deposition rate, the substrate nature and the temperature, controls the nanoparticle morphology. The competition between the growth and the ordering kinetics governs the nanoparticle structure. By reducing the growth kinetics, as-grown $L1_0$ ordered nanoparticles are obtained according to the bulk phase diagram.

A. Introduction

Multi-component nanostructures have recently attracted much attention, from both the experimental and theoretical points of view, since it is possible to enhance or to obtain new physical and chemical properties that cannot be obtained in single component nanoparticles. When metals are combined in core/shell configurations or alloyed structures, one can observe dramatic changes in chemical,[1-4] magnetic[5,6] and optical properties[7-9] compared to those of the individual component. This is in addition to the size which directly influences the physico-chemical properties when the characteristic length of a physical property becomes similar to or larger than the particle size.

For both fundamental and technological purposes, new developments are strongly dependent on the ability to have accurate control on the growth of nanoparticles. It is essential to control chemical composition, crystallography, mean size, size distribution and morphology to tailor the physical and chemical properties of the nanoparticles. Up to now, bimetallic core/shell or alloyed nanostructures have been synthesized by several chemical and physical methods: colloids chemistry,[10,11] ion implantation,[12] as well as vapour phase deposition techniques like sputtering[13,14]

[a] Laboratory Materials and Quantum Phenomena, University Paris 7, 2 Place Jussieu, Paris, France. E-mail: Cyril.Langlois@paris7.jussieu.fr; Fax: +33 1 40 79 47 30; Tel: +33 1 40 79 58 10
[b] Laboratory for Microstructural Investigations—ONERA-CNRS, B.P., 92322 Châtillon France. E-mail: Damien.Alloyeau@onera.fr; Fax: +33 1 56 73 41 55; Tel: +33 6 62 64 12 22
[c] JEOL Ltd, 1-2 Musashino 3-Chome, Akishima, Tokyo 196-8558, Japan. E-mail: oikawa@jeol.co.jp; Fax: +81 42 546 8063; Tel: +81 42 542 2152
[d] CRMCN-CNRS, Campus de Luminy, Case 913 13288, Marseille Cedex 9, France. E-mail: mottet@crmcn.univ-mrs.fr; Fax: +33 4 91 41 89 16; Tel: +33 6 60 30 28 09

and electron-beam evaporation.[15,16] Among the vapour phase deposition techniques, multiple targets sequential pulsed laser deposition (PLD) has appeared to be one of the most flexible and promising growth techniques because of its capability to control the synthesis by varying parameters such as the laser energy, the pulse repetition rate and fluence (*i.e.* energy density), the target to substrate distance, *etc.* This technique has been widely and successfully used in the preparation of monometallic and semiconductor nanoparticles as well as thin oxide films with controlled composition due to its ability to reproduce very well the stoichiometry of the ablated target.[17]

In this paper, we investigate the growth and the structural properties of core/shell CuAg and alloyed CoPt nanoparticles prepared by thermal evaporation (TE) and pulsed laser deposition techniques, respectively. For the CuAg system, we report on the size effects governing the core/shell arrangement, through the study of morphological transitions of the Cu cores depending on their sizes. Concerning the CoPt nanoparticles, we present the influence of the substrate and the laser frequency of the PLD experiment on the growth and the structure of the nanoparticles. We will show how it is possible, by controlling the growth kinetics of nanoparticles, to control the structure and the morphology of the nanoparticles.

B. Experimental

CuAg and CoPt nanoparticles have been prepared on amorphous substrates using the same ultra-high vacuum (UHV) chamber either by classical thermal evaporation or pulsed laser deposition. During the evaporation, the residual pressure in the chamber is close to 10^{-8} mbar. The growth of the nanoparticles has been performed on two different substrates: a 10 nm thick amorphous carbon film (a-carbon) of a commercial 3 mm Cu grid or a 2 nm layer of amorphous alumina (a-Al$_2$O$_3$), deposited using PLD on top of the amorphous carbon. The deposition rate of each element is controlled by an *in situ* quartz crystal monitor, which indicates the nominal thickness of deposited materials on the quartz surface, in a continuous thin film approximation. As the metallic species do not wet the two amorphous substrates used here, nanoparticles are formed instead of continuous thin films and the nominal thickness of a sample obviously does not correspond to the particle thickness. During the deposition, the substrate is held at a fixed temperature T_s, between 25 °C and 750 °C. The heating system is a tungsten filament heated by the Joule effect, positioned just behind the TEM grid. The heat transfer occurs through radiation from the filament to the substrate and conduction from the substrate to the particles. At the end of the synthesis, samples were covered by a 3 nm thick a-Al$_2$O$_3$ layer deposited by PLD, in order to protect the nanoparticles from air oxidation. Nevertheless, the high vacuum conditions of the deposition chamber offer the possibility to anneal *in situ* the synthesized particles before the deposition of the protective amorphous alumina layer.

1. Thermal evaporation

The sources were Knudsen type cells containing high purity Cu and Ag pellets heated at 1040 °C and 970 °C, respectively. Copper was first evaporated to form pure Cu nanoparticles at the surface of the a-carbon substrate. In a second step, Ag was evaporated. Typically, deposition rates for Cu and Ag were around 0.25 nm per min. During the deposition of Cu and Ag, the substrate is heated at a temperature of 400 °C. This allows the diffusion of the two species on the surface of the substrate. The nanoparticles are annealed during half an hour at the same temperature.

2. Pulsed laser deposition

A typical target–substrate holder configuration is used to deposit separately the a-Al$_2$O$_3$, Co and Pt by PLD, using a KrF excimer laser at 248 nm. The pulse duration is 25 ns and the repetition rate is in the range of 1 to 15 Hz. The laser irradiation produces the plasma of the different species and vaporised atoms condense on the substrate placed at 5 cm from the target. The growth of the CoPt nanoparticles has been performed on amorphous carbon and amorphous alumina. CoPt nanoparticles are obtained by alternated irradiation of two pure metal targets allowing a composition controlled synthesis.[18,19] To ensure a good homogeneity of the particle composition, the nominal thickness of deposited materials at each successive step (corresponding to one series of laser pulses on Co and Pt targets) is set to 0.1 nm. The number of pulses on Co and Pt targets at each step was fixed in order to obtain a composition close to $Co_{50}Pt_{50}$. The laser energy can be chosen in the range of 150 to 250 mJ depending on the ablation threshold of the selected target. The deposition rate depends on the target nature, the energy and the frequency of the laser. We used energies of 190 mJ and 250 mJ for the cobalt and the platinum targets, respectively. These selected energies avoid the formation of micrometric droplets and minimizes the deposition rate which is, as we will show, a key parameter to control particle morphology. With this configuration, deposition rates for Co and Pt were, respectively 0.30 and 0.04 nm min^{-1}.

3. Characterisation tools

Structural characterisations of the nanoparticles were performed using several TEM techniques: conventional Bright Field imaging (BF), High Resolution imaging (HRTEM), Energy Filtered imaging (EFTEM) and electron tomography. The microscopes used for these studies were a JEOL JEM-2100F and a ZEISS LIBRA 200 FE field-emission TEM operating at 200 keV.

C. CuAg Core/shell nanoparticles

Success in obtaining bimetallic core/shell nanoparticles by a physical route mainly depends on: (1) the miscibility of the two metals, (2) the surface energies of those metals and (3) the interfacial energies between a substrate and the two metals. The CuAg system was chosen as a model because of silver and copper being immiscible elements (0.1% solid solution at 298 K). A lower surface energy for Ag (1210 mJ m^{-2} compared to 2130 mJ m^{-2} for Cu)[20] favours the silver surface segregation. CuAg core/shell nanoparticles are then expected to be formed on the amorphous carbon substrate, with Cu core and Ag shell. To improve the efficiency of the core/shell formation, a sequential procedure has been used during the thermal evaporation, with deposition of Cu firstly, and Ag in a second step.

1. 6@3 system

Nominal thicknesses of 6 nm and 3 nm for Cu and Ag, respectively were deposited and we obtained nanoparticles sizes around 25 nm (Fig. 1a). In order to study the growth of core/shell nanoparticles on the substrate, the samples (denoted 6@3 in the following) were fully characterised by transmission electron microscopy. The EFTEM technique was particularly efficient to identify the Ag shells and the Cu cores. For each element, a chemical map was obtained on which the black and white contrast corresponds to the localisation of the selected element (Fig. 1b and c). Gathering together the Cu and Ag chemical maps results in a color coded image as the one presented on Fig. 1d for the 6@3 system. Such chemical information cannot be obtained directly from the classical BF image (Fig. 1a) since it shows no contrast between the two elements.

Many different configurations regarding the silver localisation are noticed. Silver can form continuous shells around the Cu cores, as expected from the thermo-dynamical considerations discussed previously, but as well silver can grow either independently from the Cu cores, forming pure silver nanoparticles, or onto the Cu

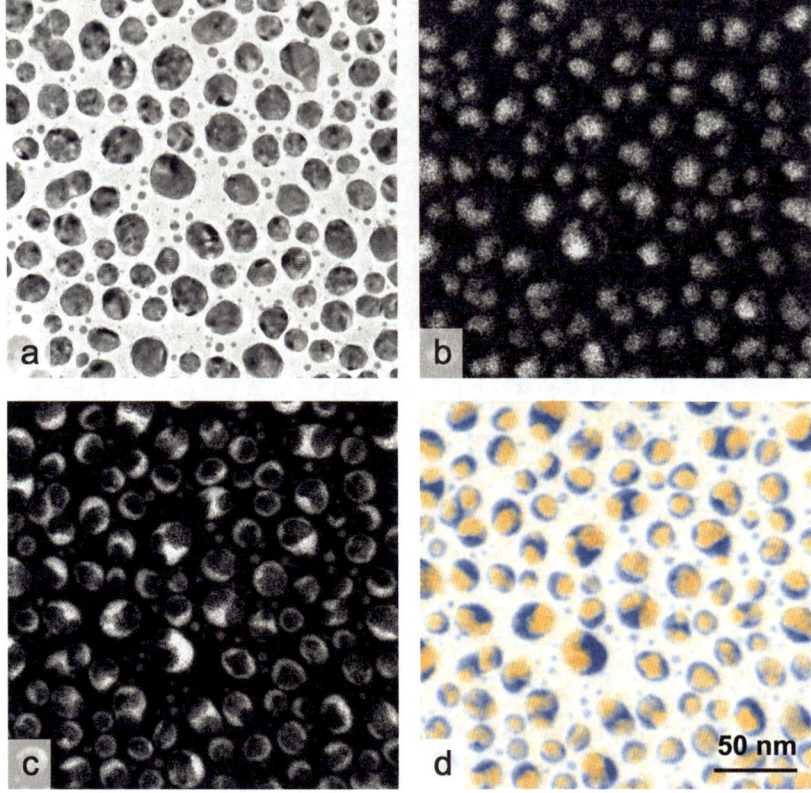

Fig. 1 (a) BF conventional TEM image of an area on the 6@3 sample, (b) Cu chemical map of the same area obtained by EFTEM, (c) Ag chemical map and (d) Colour-coded chemical map with Ag (blue) and Cu (yellow). Note the presence of pure Ag nanoparticles that have nucleated separately on the substrate.

cores, but only on one side of the Cu nanoparticles. The EFTEM images can hardly bring proof of a covering of the Cu cores by Ag because of the projection along the z axis perpendicular to the substrate to form the images. The complete capping of the Cu core by the Ag shell is evidenced by a well-known optical effect called 'Moirés patterns' that can be seen on HRTEM or even conventional TEM images (Fig. 2a). Large dark stripes superimposed on to the lattice fringes of the crystal arise when two crystals are overlapped. These patterns appear when interferences occur between two slightly different spatial frequencies (in orientation and/or spacing) in the reciprocal space. Here, the Moiré patterns originate from the epitaxial relationship between the core and the shell.

The crystallographic orientation is the same for the core and the shell, and this was noticed for all the observed nanoparticles. This can be seen on the power spectrum on Fig. 2b, on which Ag and Cu crystals are oriented along the same [110] zone axis of the Face Centred Cubic (FCC) structure, with the same radial orientation of the 200 reflections of the two crystals. The mirror corresponding to the twin plane is reported on the power spectrum. This is a striking example of the shell following exactly the underlying defect of the core.

On Fig. 3, showing a CuAg nanoparticle for which silver did not form a perfect shell around the core, the same epitaxial relationship is reported on the [100] zone axis given in the inset of Fig. 3b. Because of the difference in lattice parameters between the two materials, interfacial edge dislocations would be expected. Since the interface between the two crystals lies on a (220) plane, we carried out a Bragg

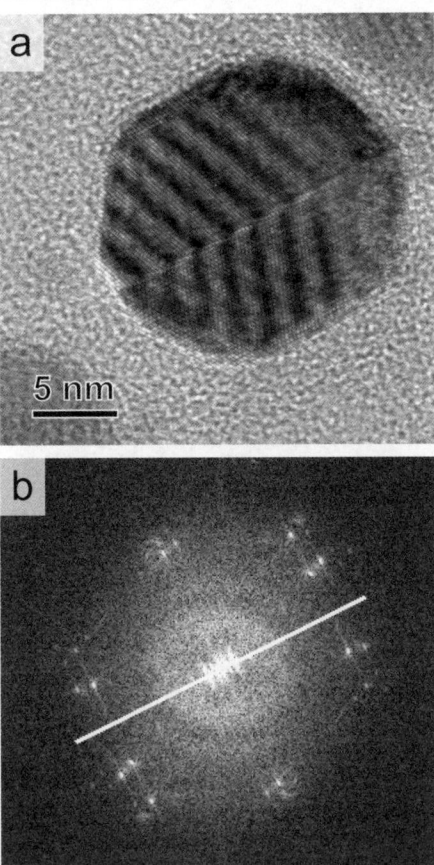

Fig. 2 (a) High resolution image of a CuAg nanoparticle of the 6@3 system, oriented along the [110] zone axis. Moiré patterns show that the silver shell caps completely the Cu core. (b) Power spectrum of the previous image that emphasize the epitaxial relationship between the core and the shell. The power spectrum shows a symmetry mirror (highlighted by a white line on the figure) coming from the twin boundary.

filtering of the image using the 220 type reflections of Cu and Ag. This allows to evidence the localisation of the excess (220) Cu planes composing the misfit dislocations. On Bragg-filtered images (Fig. 3b), the interfacial dislocations spacing is 1.1 nm, that is, around 8 $(220)_{Ag}$ interplane distances. When estimating the number of (220) planes between interfacial dislocations for an Ag film grown on a Cu film along the [220] direction from the simple formula:[21]

$$n = d_{(220)Ag}/(d_{(220)Cu} - d_{(220)Ag})$$

one would find $n = 8$. The rule of thumb used here is still valid at the nanoscale because we found the same dislocations periodicity on the image. This is a surprising situation since dislocations are usually attracted to free surfaces in order to decrease the elastic field around their core. Here, the epitaxial stress is important enough (a_{Ag} is 11% larger than a_{Cu}) to maintain six perfect dislocations inside approximately a $5 \times 5 \times 5$ nm^3 volume. The presence of interfacial dislocations reveals also that Ag and Cu have recovered their own lattice parameter. This has been verified by checking the lattice parameters on both sides of the nanoparticles. This is coherent with another simple model used for heteroepitaxial films, giving the critical thickness over which the interface is no more coherent and misfit dislocations appear.[22] This

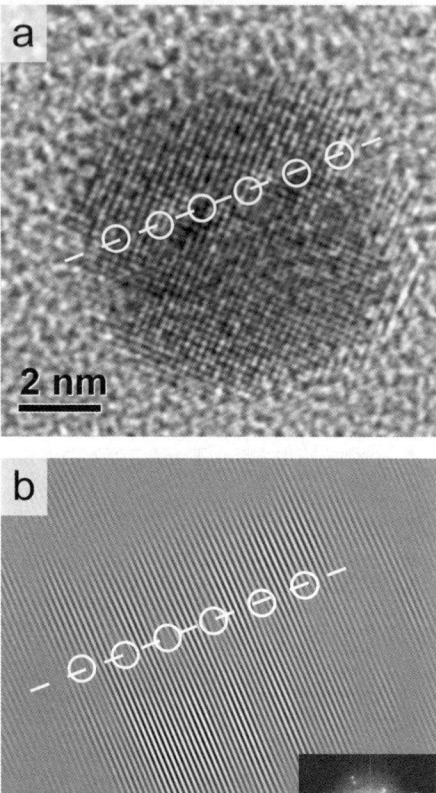

Fig. 3 (a) High resolution image of a CuAg nanoparticle with Ag located on the upper side and Cu on the lower side of the particle. The (220) interface plane is represented by a dotted line. The misfit dislocations are localized by white circles along the interface plane. (b) Bragg-filtered image using the 220 reflection that allows the excess Cu planes of the perfect misfit dislocations to be seen. The inset represents the power spectrum of micrograph (a) on which the two 220 reflections used for the filtering are circled.

critical thickness t_c is roughly given by:

$$t_c = b/\varepsilon$$

where b is the Burger's vector modulus (1/2 [110] for the FCC structures) and ε the misfit between the two lattice parameters defined by $\varepsilon = (a_{Ag} - a_{Cu})/a_{Cu}$. Applying this model leads to a critical thickness close to 2 nm, to be compared with the thickness of the Ag layer on Fig. 3a which is around 2.5 nm.

2. Cu morphological transitions

From Fig. 1d, it is worth noting that the core/shell nanoparticle sizes are the smallest (<12 nm) from the size distribution of the nanoparticles. Over this limit, most of the CuAg nanoparticles do not present the core/shell configuration. The size of the pre-existing Cu cores being unambiguously an important parameter in the formation of the core/shell arrangement, we turned to the structural characterisation of the Cu cores alone, especially under the identified size limit. Controlling the size of the Cu nanoparticles can be achieved by two different means. The first would be to lower the substrate temperature during the formation of the Cu core, decreasing the Cu diffusion on the surface. Cu atoms would then be fixed more efficiently on the nucleating sites, resulting in an increased number of particles per unit area, and subsequently smaller particle sizes. The second is to decrease the quantity of matter during deposition, keeping constant all the other parameters, especially the substrate temperature. The density of particles per square nanometer remains the same, and the mean nanoparticle size decreases because of the lack of atoms available for the growth. As for gold, Cu exhibits some morphological transitions between three different types: (1) an icosahedral shape constituted by 20 slightly distorted FCC tetrahedra gathered to form a 20 faces polyhedron; (2) a decahedral shape formed by 5 identical FCC sectors separated by imperfect twins, giving a 5-fold symmetry polyhedron; and (3) a rounded or faceted monocrystalline FCC structure. These transitions have been predicted using tight-binding semiempirical potentials for metal–metal interactions[23] and observed in a number of published papers, for clusters either free or supported on a substrate. We conducted HRTEM experiments on a sample with 2 nm of Cu deposited at 400 °C to check whether the synthesis method used here reproduces the same morphological transitions. The work by Koga *et al.*[24] was very helpful in identifying the different morphologies by interpreting the lattice fringe contrast on the HRTEM images. Although a good statistical sampling is difficult to realize, we obtained the following results, collected in Fig. 4.

Due to the relatively broad size distribution of the Cu nanoparticles, it is clear that the three types of morphologies and structures are present on the substrate, with a strong overlapping between icosahedral and decahedral domains. The transition from decahedra to monocrystalline FCC structures is sharper, with FCC Cu nanoparticles sizes starting from 8 nm. The size limit over which core/shell

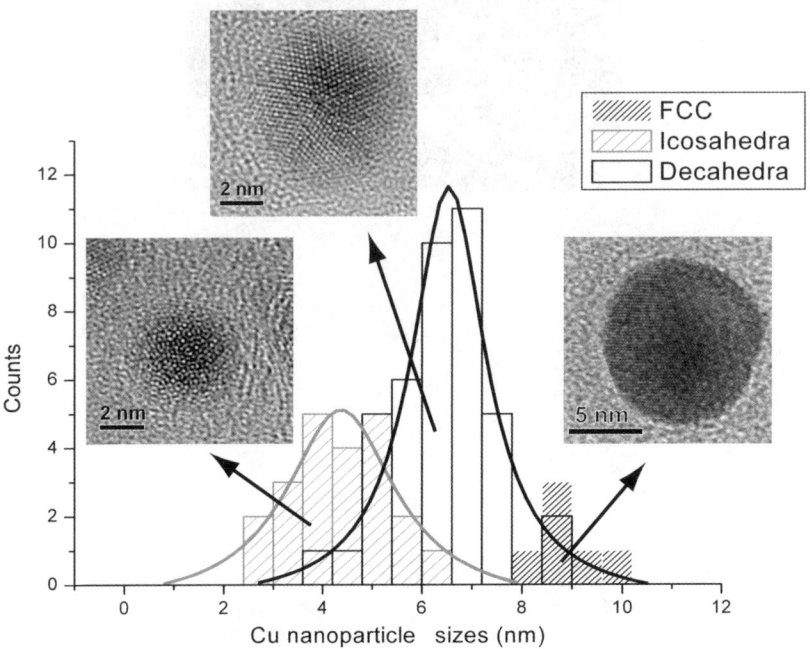

Fig. 4 Size distributions of the three different morphologies identified on a sample with 2 nm of Cu deposited at 400 °C. These distributions were obtained from conventional and high resolution TEM images. For each size range, a typical HRTEM image of a nanoparticle is shown.

arrangements are no more formed corresponds well to the decahedra-to-FCC structural transition.

3. Dependence with Cu core sizes

We carried out two other experiments to confirm that a Cu core size under 12 nm is the key parameter to obtain core/shell nanoparticles. The parameters were adjusted to obtain on one hand a Cu core size distribution centred under 12 nm, and on the other hand a distribution largely over 12 nm.

The first CuAg system has been synthesized with nominal thickness of 2 nm and 1 nm for, respectively Cu and Ag (denoted 2@1 in the following), keeping a substrate temperature of 400 °C and a constant ratio between the Cu and Ag nominal thicknesses. An EFTEM image corresponding to this system is presented on Fig. 5a. A majority of the nanoparticles present a core/shell configuration, with core sizes below 12 nm and shell thickness distribution centred on 2 nm (values ranging between 1 nm and 3.5 nm). The thickness of the Ag shell was determined by the width of the intensity profile on the EFTEM Ag chemical maps. On the same area of the microscopy grid corresponding to the 2@1 system, size counting on the EFTEM

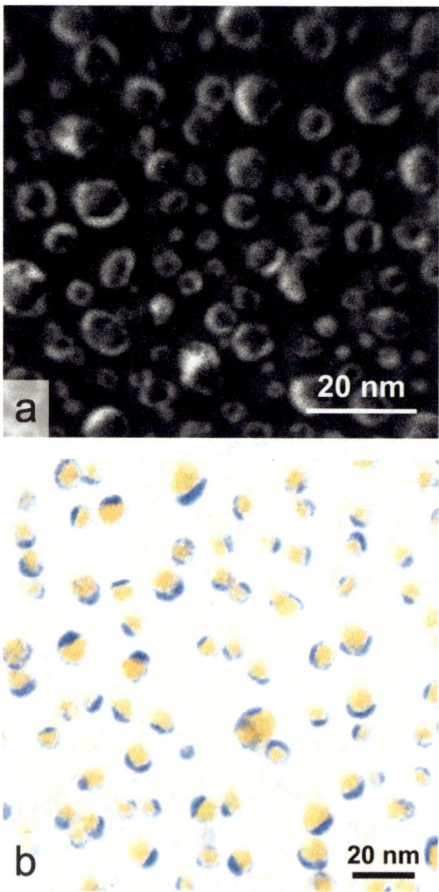

Fig. 5 (a) Ag chemical map obtained by EFTEM, corresponding to the 2@1 system deposited with a substrate temperature of 400 °C. (b) Colour-coded chemical map with Ag (blue) and Cu (yellow) obtained by EFTEM, corresponding to the 2@1 system deposited with a substrate temperature of 500 °C.

and TEM images as well as statistical measurements using a micro-Raman probe[13] have been carried out. The statistical approach confirms the local measurements.

The second experiment aimed at the formation on the substrate of Cu cores with sizes over 12 nm. This was achieved by depositing the same Cu thickness (2 nm), but with a substrate temperature of 500 °C (Fig. 5b). For Ag, no difference was noticed in the final microstructure between a deposition at a substrate temperature of 500 °C (as for Cu) and a deposition after cooling down to 400 °C. Almost no particles present a core/shell structure, which confirms the assumption of a Cu core size limit for the core/shell formation. The explanation must be linked to the way silver grows on the Cu nanoparticles. From crystal growth theory, it is well known that the growth rate is different depending on the Miller indices of the face exposed to incoming species. This remark must be related to the fact that for an icosahedron, all the faces present the same crystallographic orientation, that is (111) faces. For a decahedron, the two bases are also (111) faces, and depending on the growth along the 5-fold axis, the decahedron show (100) lateral faces. The situation is completely different for the monocrystalline FCC structure: the nanoparticles exhibit either no faceting, or faces typical from a cuboctahedron which show (111), (200) and (220) faces. For the two morphologies corresponding to sizes less than 10 nm, the growth of the Ag shell will be far more isotropic, and hence the thickness will be more homogeneous all around the Cu cores. This hypothesis is sustained by HRTEM observations on the 2@1 system deposited with a substrate temperature of 400 °C. Fig. 6 shows an example of a Cu decahedron with a complete Ag shell all over the particle. The Moiré patterns evidence the presence of Ag in an epitaxial relationship with the five subjacent Cu sectors. The [110] zone axis for the five different FCC sectors composing the decahedron are visible on the power spectrum presented on Fig. 6b.

Moreover, Moiré patterns stop before the edges of the particles, which means that only Ag is present on the edges, hence forming a continuous shell all around the particle. The shell thickness on this image (Fig. 6a) can then be estimated to about 2 nm. This is the limit predicted by the critical thickness model (Section C.1) over which the Ag shell is relaxed to its bulk lattice parameter.

D. CoPt bimetallic alloyed nanoparticles

This part focuses on the influence of the nature of the substrate (a-Al_2O_3 or a-carbon) and the synthesis parameters on the growth and the structure of CoPt nanoparticles. Bulk equiatomic CoPt exhibits a phase transition at 825 °C, between a tetragonal ordered phase ($L1_0$) at low temperature and a disordered FCC structure at high temperature.[25] The $L1_0$ ordered phase, due to the alternative atomic stacking of pure Co and pure Pt layers, has large magnetocrystalline anisotropy. By means of this property, important research efforts are devoted to study $L1_0$ ordered CoPt nanoparticles which are expected to be the future information storage media for extremely high density recording (EHDR).[26,27] The diffraction pattern of this structure presents superstructure reflections characterizing the ordered structure.

1. Influence of the substrate on nanoparticle growth

CoPt nanoparticle thin films with 2 nm nominal thickness have been prepared at a substrate temperature of 550 °C with a laser frequency of 5 Hz on a-carbon and a-Al_2O_3 substrates. Fig. 7a and b show TEM bright field images of the CoPt nanoparticles prepared on the a-carbon and the a-Al_2O_3 substrates, respectively. The mean size, the polydispersity (standard deviation divided by the mean size of the nanoparticles), the particle density and the coverage ratio for each sample are given in Table 1.

The coverage ratio, defined as the percentage of the covered surface of the substrate by the nanoparticles, has been determined from TEM bright field images

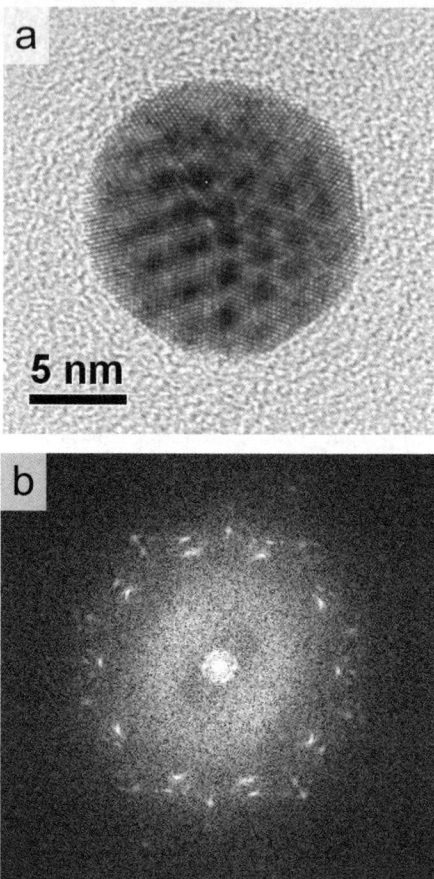

Fig. 6 (a) High resolution micrograph of a CuAg core/shell decahedron with Moiré patterns on each of the five FCC sectors. (b) Power spectrum of micrograph (a).

using Digital Micrograph software. Note that the contrast of particles smaller than 2 nm is too weak to consider them in the statistical characterisation.

The growth mechanisms involved in a PLD experiment are quite different from those observed in classical thermal evaporation techniques because the flux of atoms is discontinuous. These mechanisms can be described as follow: firstly, nucleation and size increase of the *nuclei* under high flux of atoms during the deposition time (*i.e.* several µs) and secondly, coalescence and Ostwald ripening between two pulses (*i.e.* several 0.1 s). The coalescence takes place by atomic rearrangement of two close enough nanoparticles, leading to the formation of atoms inter-diffusion bridges.[28] The Ostwald ripening implies the growth of large particles at the expense of the ones smaller than the critical nuclei.[29] In the case of PLD, due to the rapid decrease of supersaturation after a laser pulse, these two mechanisms occur alone over a long time and influence the nanoparticle morphology. As a consequence, in the late stage of the synthesis, new nucleation sites become available on the substrate between the biggest particles, making possible the growth of smaller CoPt nanoparticles. This explains the high polydispersity observed on the a-carbon substrate that shows a high number of very small particles in the range size from 2 nm to 4 nm (inset Fig. 7a). This phenomenon is less important in the case of the a-Al$_2$O$_3$ substrate (inset Fig. 7b). It suggests that the diffusion of Co and Pt on a-carbon substrate is higher than the ones on a-Al$_2$O$_3$.

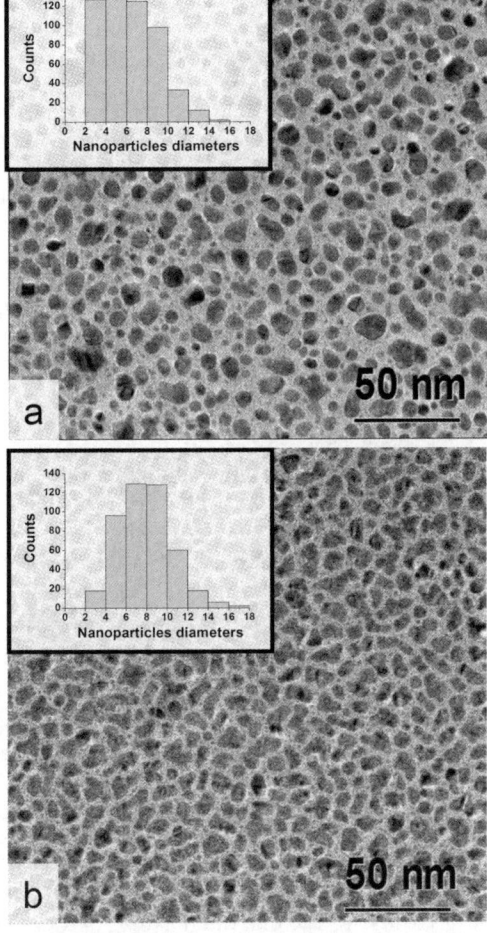

Fig. 7 CoPt FCC disordered nanoparticles synthesized by PLD at 550 °C with in the insets the size distribution of the particles. (a) Substrate: 10 nm thick amorphous carbon film on a commercial TEM grid. (b) Substrate: 2 nm thick amorphous alumina film deposited by PLD on the carbon film of a commercial TEM grid.

Since the nominal thickness is the same for both samples, by comparing the mean size and the coverage ratio of CoPt nanoparticles on a-carbon and on a-Al_2O_3, we can deduce that the nanoparticles are more flattened on a-Al_2O_3 substrate than on a-carbon substrate. This difference is closely related to the kinetic barriers involved during the nanoparticle formation. Kinetic energy of the atoms, given by the substrate temperature governs the ability to overcome the various diffusion barriers (metal–metal and metal–substrate upstepping and downstepping activation

Table 1 Mean size, polydispersity, particle density and coverage ratio of the samples prepared at 550 °C, with a laser frequency of 5 Hz on a-carbon and a-Al_2O_3 substrates

2 nm of CoPt on substrate	Mean size/nm	Polydispersity (%)	Particles density (10^2 part. μm^{-2})	Coverage ratio (%)
a-Carbon	6.3	45	110	42
a-Al_2O_3	7.9	30	90	50

energies).[30,31] At a given temperature, these diffusion barriers control the evolution of film morphology during the growth. From our experimental results, the two-dimensional growth on alumina may then be the signature either of strong interactions of CoPt with alumina (higher activation barrier for the 3-dimensional growth compared to carbon) or a lower effective substrate temperature in the case of alumina. To summarize, we show here that the kind of interactions between the substrate and the species deposited drive both the size distribution and the morphologies of the nanoparticles: the 3-dimensional growth of the nanoparticles is promoted on a-carbon whereas a-Al$_2$O$_3$ substrate minimizes the size dispersion. The following question remains: how can we know if the diffusion and the substrate–metal upstepping energy of the Co and Pt atoms are retained by a stronger a-Al$_2$O$_3$–metals interaction, or slowed down by the lower temperature of the alumina surface?

It is thus important to emphasise the thermal properties of the substrate, which are generally not taken into account enough in nanoparticle synthesis on either mono-crystalline or amorphous substrates. Contrary to carbon, alumina is a good thermal insulator. As a rough indication, the heat capacity, at room temperature, of the crystalline alumina (79.45 J mol^{-1} K^{-1}) is almost ten times higher than the one of graphite (8.58 J mol^{-1} K^{-1}).[32] This property of alumina is well known since in the electronic and optic industries, amorphous alumina thin films are promising materials for thin thermal insulator layers.[33] However, it is difficult to differentiate the respective role of the strong interactions of the deposited species with the substrate and low thermal conductivity of the a-Al$_2$O$_3$ layer, because they both have the same influence on the morphology of CoPt nanoparticles.

2. Thermal insulator properties of a-Al$_2$O$_3$ substrate

To demonstrate the strong influence of the thermal conductivity of the substrate, we have synthesized nanoparticles on a-Al$_2$O$_3$ with the same experimental conditions than the ones used for the synthesis of nanoparticles shown in Fig. 7b. However, we have increased the thickness of the a-Al$_2$O$_3$ substrate to 10 nm. As can be seen in Fig. 8, the particles are very close to the percolation threshold with irregular shapes elongated in the plane of the substrate. These flat morphologies compared to Fig. 7b indicate the tendency of the metals to wet the surface because the thermal energy brought to the system is not high enough to permit 3D growth mechanism. The coverage ratio of 80% is close to the one obtained on particles formed on a 2 nm thick a-Al$_2$O$_3$ layer heated at 400 °C. The particle morphologies studied as a

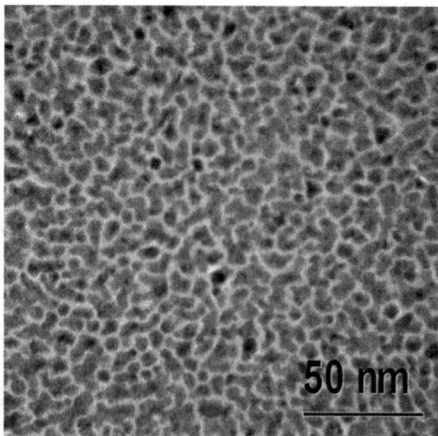

Fig. 8 CoPt nanoparticles, with a 2 nm nominal thickness, synthesized by PLD on a 10 nm thick amorphous alumina layer deposited by PLD on the carbon film of a commercial TEM grid heated at 550 °C.

function of the a-Al$_2$O$_3$ layer thickness clearly show the thermal influence of this substrate.

3. Thermal influence of the substrate on the structure of the nanoparticles

The structure of previous samples (T_s = 550 °C) was FCC disordered which is a non equilibrium phase compared to the bulk material phase diagram. This FCC structure appears using the PLD synthesis route because the growth kinetics of the particles is faster than the ordering kinetics. To obtain as-grown L1$_0$ ordered nanoparticles, the substrate has to be heated over 650 °C.[18] At T_s = 550 °C, the ordering mechanisms require a post-synthesis annealing. CoPt nanoparticles, prepared in the same conditions as the ones studied in part D1, were annealed over 2 h at 550 °C and were then covered by a thin a-Al$_2$O$_3$ layer after the annealing. Fig. 9a and b show the morphology of the nanoparticles after the annealing on the a-carbon and a-Al$_2$O$_3$ substrates, respectively. Since substrates are amorphous, the nanoparticles are randomly oriented on the surface and the diffraction patterns display a typical ring pattern characteristic of a powder-like diffraction. Because of the very weak intensity of the superstructure reflections (see inset of Fig. 9c), we used imaging plate detectors which have a very high sensitivity and a large dynamic range. Moreover, to highlight the recorded intensity we performed a rotational averaging intensity profile using Process Diffraction software. This intensity profile is drawn as a function of the modulus of the diffraction vector $g = 2\pi/d$ where d is the interplanar distance. Fig. 9c and d show, respectively averaging plots corresponding to the diffraction patterns

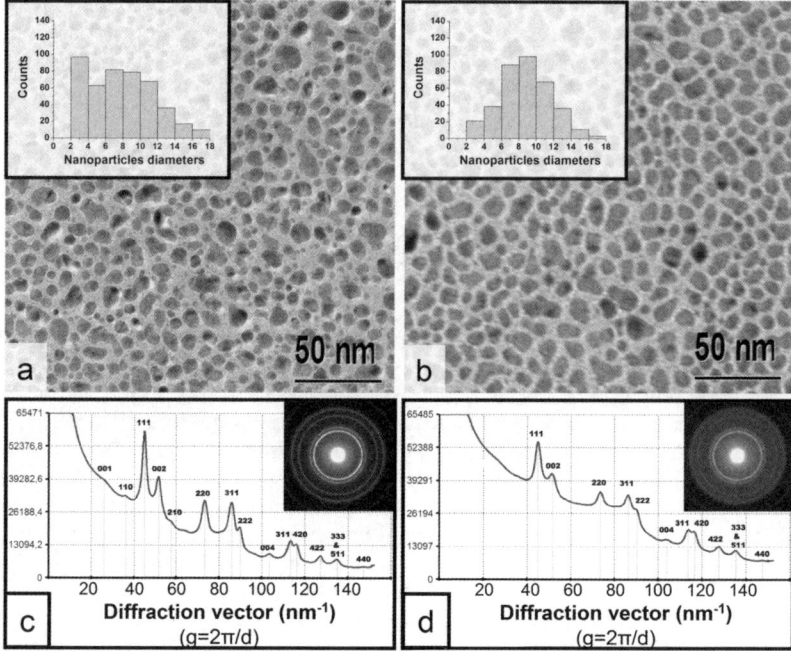

Fig. 9 CoPt nanoparticles deposited by PLD with a laser frequency of 5 Hz (30 min) on a substrate heated at 550 °C and then annealed in the deposition chamber at 550 °C during 2 h. In the inset: size distribution of the particles. (a) Substrate: 10 nm thick amorphous carbon film of a commercial TEM grid. (b) Substrate: 2 nm thick amorphous alumina film of deposited by PLD on the carbon film of a commercial TEM grid. Electron diffraction patterns of CoPt nanoparticles grown (30 min) and annealed (120 min) at 550 °C, and their corresponding rotational averaging profiles. (c) Ordered L1$_0$ nanoparticles on amorphous carbon film. (d) Disordered FCC nanoparticles on amorphous alumina film.

of each sample: CoPt/a-carbon and CoPt/a-Al$_2$O$_3$. All rings are indexed according to the FCC disordered or the L1$_0$ ordered structures of the CoPt system.

The diffraction pattern of the particles annealed over 2 h after their synthesis on a-carbon presents the 001, 110 and 211 superstructure reflections characteristic of the L1$_0$ ordered lattice. On the contrary, the diffraction pattern of particles grown and annealed on a-Al$_2$O$_3$ presents only the fundamental reflections of the FCC disordered structure. This result has been confirmed by HRTEM. Energy Dispersive X-ray (EDX) analyses show that particles are close to the equiatomic composition on both samples. The annealed particles mean size is smaller on a-carbon (7.8 nm) than the one on a-Al$_2$O$_3$ (8.8 nm). We can then rule out the possibilities that the different structural states observed on both samples are due to deviation of the composition of the CoPt particles or to size effects. During the annealing, the contact surface between the substrate and the particles is limited. Taking into account the particle size, the proportion of atom linked to the substrate is too low to consider a strong influence of the chemical interaction on the ordering of the particles. It still requires an assumption to explain this difference: the thermal insulator properties of amorphous alumina.

During the annealing, the increase of the mean particle size on a-carbon (20%) is twice as high as the one of the particles on a-alumina (10%). It shows the low diffusion of Co and Pt atoms on a-Al$_2$O$_3$ substrate. These alumina properties have been highlighted already by Takahashi et al.[34] who compared the morphology of FePt nanoparticles after an annealing of 1 h at 600 °C in MgF$_2$ and a-Al$_2$O$_3$ matrix. They showed that the covering alumina layer reduces the coalescence and Oswald ripening phenomena avoiding the size increase of particles. Then they explained by size effects the disordered structure of the particles in alumina compared with the ordered ones in MgF$_2$ but they did not consider the eventually different effective particle temperature. The goal of our structural studies is not to deny their interpretation but to emphasize the strong influence of the thermal properties of the substrate.

4. Influence of the laser frequency on the growth and the structure of the nanoparticles

The influence of the laser frequency in the PLD experiment can be seen by comparing the morphology and structure obtained with a laser frequency of 5 Hz (Fig. 7a) with the ones obtained with a lower frequency. We have synthesized a sample of 2 nm nominal thickness prepared on a-carbon substrate heated at 550 °C with a laser frequency of 1 Hz. Fig. 10a shows a TEM bright field image of this nanoparticle thin film. By decreasing the laser frequency from 5 to 1 Hz, we have increased the polydispersity from 45 to 66% and the mean size from 6.3 to 9.2 nm. With a 1 Hz laser frequency, the time between two pulses is five times longer than the time at 5 Hz. As a consequence, the histogram in the inset of Fig. 10a shows clearly a bimodal size distribution, the first population has a mean size of 3 nm whereas the second one has a mean size of 13 nm. It is the signature of the Ostwald ripening and coalescence mechanisms which occur over a longer time which influence the particle morphology. The synthesis of samples takes 30 min and 150 min, using a laser frequency of 5 Hz and 1 Hz, respectively. To dissociate the laser frequency effect from the synthesis time effect, we compared nanoparticles synthesized with a laser frequency of 5 Hz and annealed at 550 °C during 120 min (Fig. 9a), to ones obtained with a laser frequency of 1 Hz (Fig. 10a). The higher mean size and polydispersity of the latter clearly show the influence of the laser frequency. The Oswald ripening and coalescence phenomena during the deposition at 1 Hz (150 min) are more effective than those occuring during the synthesis at 5 Hz (30 min) and the post synthesis annealing (120 min).

The shape of the nanoparticles, fabricated with a 1 Hz laser frequency, has been determined by means of a 3D electron tomography experiment. The electron

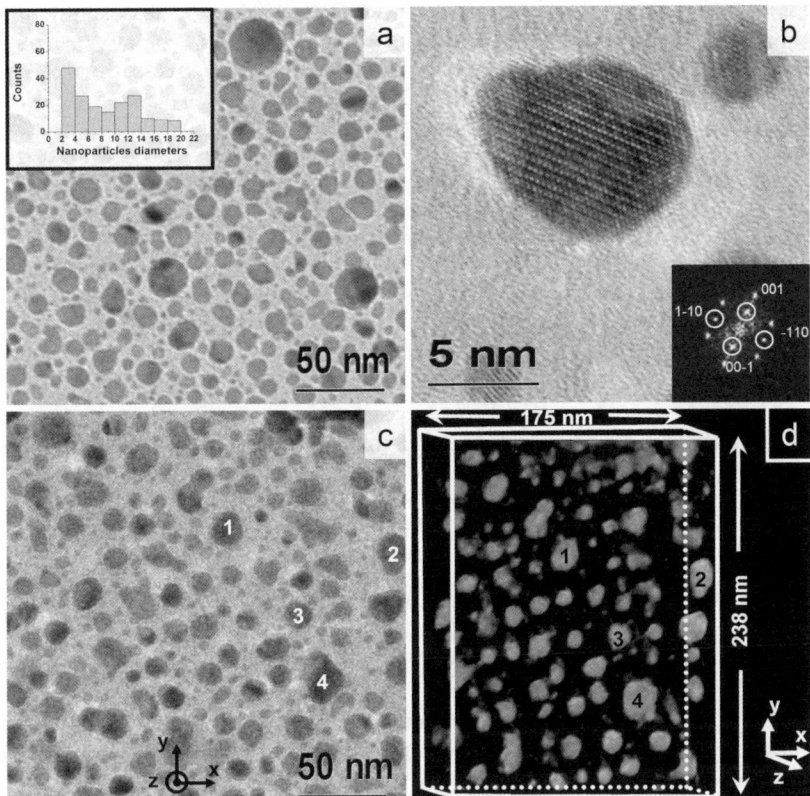

Fig. 10 Ordered L1$_0$ CoPt nanoparticles synthesized by PLD with a laser frequency of 1 Hz (2h30) on amorphous carbon film at a temperature of 550 °C. (a) BF image of the nanoparticles. (b) HRTEM image of a CoPt L1$_0$ ordered nanoparticle oriented along the [110] direction, with in the inset the power spectrum on the image where the superstructure reflections are encircled. (c) TEM images of the selected area for the tomography experiment, (d) 3D reconstruction of the selected area obtained by electron tomography. The correspondence between 2D images of the particles and the 3D volumes is emphasized by the numbers on these two images.

tomography technique allows the 3D morphology to be reconstructed from 2D observations of the nanoparticles. One or several series of projections of the nanoparticles have to be recorded in bright field imaging conditions. Such a series of projections is obtained by tilting the specimen in a range from −70° to 70° with respect to the electron beam and taking an image for each tilt angle. Once the series of projections is recorded, the reconstruction of the object is computed by using a back-projection algorithm.[35] A bright field image of the selected area is presented in Fig. 10c and its 3D representation (tomogram) is presented in Fig. 10d. The thickness of the particles has been measured by slicing the tomogram in the two perpendicular x and y directions (Fig. 10d). The thickness of the nanoparticles are measured directly on the sliced images, corresponding to the (x,z) or the (y,z) planes, by the size of the nanoparticles along the z direction. For 80% of the nanoparticles, the measured thickness is equal to the diameter in the plane of the substrate. It can be explained by the 3D growth mechanisms involved between two pulses (metal–metal and metal–substrate upstepping), discussed in part D1. The shape of the last 20% of the nanoparticles, which is larger than thick, results from coalescence mechanisms.

A low laser pulse frequency during the growth also influences the structure of the nanoparticles. With a laser frequency of 1 Hz and $T_s = 550$ °C, the $L1_0$ structure is stabilized without further annealing. This result has been observed from electron diffraction of the particles and Fig. 10b shows a HRTEM image of a single $L1_0$ ordered CoPt nanoparticle oriented along the [110] zone axis. In such images, the order is clearly evidenced by the succession of bright and dark fringes which can be related to the stacking of pure Co and Pt planes along the [001] direction as indicated by the power spectrum of the image given in the inset of Fig. 10b.

To conclude, when decreasing the laser frequency, we reduce the nanoparticle growth kinetics, which is then slower than the ordering one. In these conditions, the diffusion of the metallic atoms at 550 °C allows us to obtain as-grown nanoparticles with the $L1_0$ equilibrium structure according to the bulk phase diagram.

Conclusion

Growth and structural properties of immiscible CuAg and alloyed CoPt nanoparticles have been studied. In CuAg, prepared by TE, the Ag layer is completely relaxed when its thickness is equal to 2 nm. The mechanism involved in the stress relaxation has been demonstrated to be the same as the one occurring in thin film epitaxy. We have shown that there is a critical size for the Cu core, up to which the formation of the core/shell arrangement is promoted. This configuration is probably favoured by the underlying morphologies of the Cu cores which present icosahedral and decahedral configurations under a size smaller than 8 nm. Such shapes are mainly constituted of (111) faces implying that the growth rate of the Ag layer is the same on all these faces.

In CoPt, prepared by PLD, the morphology and the structure of the nanoparticles has been investigated as a function of the nature of the substrate and the laser frequency. The diffusion of Co and Pt atoms is favoured on a-carbon compared to amorphous alumina, or by using a lower frequency during the deposition process. These phenomena result from the competition between the kinetic energy of the atoms, given by the substrate temperature, and the kinetic barriers, depending on the metal/substrate interaction, which drive the formation of 3D morphologies. The thermal properties of the substrate can have a strong influence on the structure of the nanoparticles. Finally, we have shown that it is possible to synthesize as-grown ordered $L1_0$ CoPt nanoparticles by reducing the growth kinetics (*i.e.* by decreasing the laser frequency).

References

1 S. Sao-Joao, S. Giorgio, J. M. Penisson, C. Chapon, S. Bourgeois and C. Henry, *J. Phys. Chem. B*, 2005, **109**, 342.
2 B. Skarman, T. Nakayama, D. Grandjean, R. E. Benfield, E. Olsson, K. Niihara and L. R. Wallenberg, *Chem. Mater.*, 2002, **14**, 3686.
3 N. Toshima, *Pure Appl. Chem.*, 2000, **72**, 317.
4 N. Toshima, M. Harada, Y. Yamazaki and K. Asakura, *J. Phys. Chem.*, 1992, **96**, 9927.
5 L. Favre, S. Stanescu, V. Dupuis, E. Bernstein, T. Epicier, P. Melinon and A. Perez, *Appl. Surf. Sci.*, 2004, **226**, 265.
6 H. Zeng, J. Li, Z. L. Wang, J. P. Liu and S. Sun, *Nano Lett.*, 2004, **4**, 187.
7 S. Basu and D. Chakravorty, *J. Non-Cryst. Solids*, 2006, **352**, 380.
8 M. Gaudry, E. Cottancin, M. Pellarin, J. Lerme, L. Arnaud, J. R. Huntzinger, J. L. Vialle, M. Broyer, J. L. Rousset, M. Treilleux and P. Melinon, *Phys. Rev. B: Condens. Matter Mater. Phys.*, 2003, **67**, 155409.
9 J. Zhu, Y. Wang, L. Huang and Y. Lu, *Phys. Lett. A*, 2004, **323**, 455.
10 D. Garcia-Gutierrez, C. Gutierrez-Wing, M. Miki-Yoshida and M. Jose-Yacaman, *Appl. Phys. A: Mater. Sci. Process.*, 2004, **79**, 481.
11 S. Sun, C. B. Murray, D. Weller, L. Folks and A. Moser, *Science*, 2000, **287**, 1989.
12 S. P. Withrow, C. W. White, J. D. Budai, L. A. Boatner, K. D. Sorge, J. R. Thompson and R. Kalyanaraman, *J. Magn. Magn. Mater.*, 2003, **260**, 319.

13 M. Cazayous, C. Langlois, T. Oikawa, C. Ricolleau and A. Sacuto, *Phys. Rev. B*, 2006, **73**, 113402.
14 T. Miyazaki, O. Kitakami, S. Okamoto, Y. Shimada, Z. Akase, Y. Murakami, D. Shindo, Y. K. Takahashi and K. Hono, *Phys. Rev. B*, 2005, **72**, 144419.
15 K. Sato, B. Bian, T. Hanada and Y. Hirotsu, *Scripta Materialia*, 2001, **44**, 1389.
16 L. Castaldi, K. Giannakopoulos, A. Travlos and D. Niarchos, *Appl. Phys. Lett.*, 2004, **85**, 2854.
17 D. B. Chrisey and G. K. Hubler, *Pulsed Laser Deposition of Thin Films*, John Wiley & Sons, London, 1994.
18 D. Alloyeau, C. Langlois, C. Ricolleau, Y. Le Bouar and A. Loiseau, *Nanotechnology*, 2007, **18**(37), 375301.
19 T. W. Trelenberg, L. N. Dinh, B. C. Stuart and M. Balooch, *Appl. Surf. Sci.*, 2004, **229**, 268.
20 L. Vitos, A. V. Ruban, H. L. Skriver and J. Kollar, *Surf. Sci.*, 1998, **411**, 186.
21 J. P. Hirth and J. Lothe, *Theory of Dislocations*, Krieger Publishing Company, Berlin 1982.
22 A. E. Romanov, W. Pompe, S. Mathis, G. E. Beltz and J. S. Speck, *J. Appl. Phys.*, 1999, **85**, 182.
23 C. Mottet, J. Goniakowski, F. Baletto, R. Ferrando and G. Treglia, *Phase Transitions: Multinatl. J.*, 2004, **77**, 101.
24 K. Koga and K. Sugawara, *Surf. Sci.*, 2003, **529**.
25 Y. Le Bouar, A. Loiseau and A. Finel, *Phys. Rev. B*, 2003, **68**, 224203.
26 D. J. Sellmyer, M. Yu and R. D. Kirby, *Nanostruct. Mater.*, 1999, **12**, 1021.
27 M. Yu, Y. Liu and D. J. Sellmyer, *J. Appl. Phys.*, 2000, **87**, 6959.
28 G. Palasantzas, T. Vystavel, S. A. Koch and L. T. D. Hosson, *J. Appl. Phys.*, 2006, **99**, 024307.
29 I. M. Lifshitz and V. V. Slyozov, *Sov. Phys. JETP*, 1959, **35**, 331.
30 C. T. Campbell, *Surf. Sci. Rep.*, 1997, **27**, 1.
31 C. R. Henry, *Surf. Sci. Rep.*, 1998, **31**, 231.
32 D. R. Lide, *CRC Handbook of Chemistry and Physics*, CRC press, London, 2002–2003.
33 B. Behkam, Y. Yang and M. Asheghi, *Int. J. Heat Mass Transfer*, 2005, **48**, 2023.
34 Y. H. Takahashi, T. Ohkubo, M. Ohnuma and K. Hono, *J. Appl. Phys.*, 2003, **93**, 7166.
35 B. F. Mc Ewen and M. Marko, *J. Histochem. Cytochem.*, 2001, **49**, 553.

Binding energy and preferred adsorption sites of CO on gold and silver–gold cluster cations: Adsorption kinetics and quantum chemical calculations

Marco Neumaier,[a] Florian Weigend,[a] Oliver Hampe*[ab] and Manfred M. Kappes[ab]

Received 3rd April 2007, Accepted 27th April 2007
First published as an Advance Article on the web 12th September 2007
DOI: 10.1039/b705043g

We revisit the reactivity of trapped pure gold (Au_n^+, $n < 26$) and silver–gold alloy cluster cations ($Ag_mAu_n^+$, $m + n < 7$) with carbon monoxide as studied in a Fourier transform ion cyclotron resonance (FT-ICR) mass spectrometer. The experimental results are discussed in terms of *ab initio* computations which provide a comprehensive picture of the chemical binding behaviour (like binding energy, adsorption sites, associated vibrational frequencies) of CO to the noble metal as a function of cluster size and composition. Starting from results for pure gold cluster cations for which an overall decrease of CO binding energy with increasing cluster size was experimentally observed—from about 1.09 ± 0.1 eV (for $n = 6$) to below 0.65 ± 0.1 eV (for $n > 26$)—we demonstrate that metal–CO bond energies correlate with the total electron density and with the energy of the lowest unoccupied molecular orbital (LUMO) on the bare metal cluster cation as obtained by density functional theory (DFT) computations. This is a consequence of the predominantly σ-donating character of the CO–M bond. Further support for this concept is found by contrasting the predictions of binding energies to the experimental results for small alloy cluster cations ($Ag_mAu_n^+$, $4 < m + n < 7$) as a function of composition. Here, binding energy drops with increasing silver content, while CO still binds always in a head-on fashion to a gold atom. Finally we show how the CO stretch frequency of $Ag_mAu_nCO^+$ may be used to identify possible adsorption sites and pre-screen favorable isomers.

1. Introduction

The fascinating interplay between the electronic/geometric structure of coinage metal clusters (M_x, M = Cu, Ag and Au) and their chemical properties has given rise to numerous theoretical and experimental studies. Coinage metal atoms have an electronic structure featuring completely filled d-shells and a singly occupied valence s-shell. To first order, this renders them as "simple" s^1-metals. However, whereas d- and s-orbitals are energetically well-separated in silver, relativistic effects

[a] Institut für Nanotechnologie, Forschungszentrum Karlsruhe, P.O. Box 3640, D-76021 Karlsruhe, Germany. E-mail: Oliver.Hampe@int.fzk.de; Fax: +49 7247 826368
[b] Institut für Physikalische Chemie, Universität Karlsruhe, Kaiserstr. 12, D-76128 Karlsruhe, Germany

significantly decrease the energy difference between 6s and 5d orbitals in gold—as reviewed by Pyykkö.[1] For the bulk metal this is most directly manifested in the yellowish appearance of gold surfaces due to the shift of the corresponding interband transitions to lower energies compared to silver.[2] Given that d-electrons participate strongly in gold–gold bonding, an adequate description of associated relativistic effects is also important in any theoretical treatment of gold clusters and of their chemical derivatives. Early computational studies of coinage metal clusters included: (i) the identification of a bent ground-state geometry of mixed coinage metal trimers,[3] (ii) a relativistically corrected *ab initio* description of the ground and excited states of neutral Au_3[4] as well as Ag_3, Cu_3, and mixed bimetallic trimers,[5] (iii) determination of the ground state structures of neutral and anionic coinage metal dimers and trimers,[6] and (iv) prediction of the structural properties of neutral Au_n (n = 3–6) using Møller–Plesset perturbation theory to second order.[7] More recently, considerable theoretical effort has been invested into describing the structural and optical properties of silver clusters by Bonačić-Koutecký and coworkers.[8–11] Similarly, the properties of gold clusters have been intensely studied by various groups [*e.g.* ref. 12–15].

There have also been numerous previous experimental studies of mass-selected coinage metal clusters including negative-ion photoelectron spectroscopy,[16–23] photo-dissociation and -ionization studies,[24–29] collision-induced dissociation probes[30–34] and ion mobility measurements.[35–37] As a consequence of the prevailing relativistic effects, gold clusters favour planar structures up to surprisingly large sizes (planar cations for sizes Au_n^+, $n < 8$[35] and planar anions for sizes Au_n^-, $n \leq 12$[36]) whereas small silver clusters already prefer three-dimensional geometries ($n \geq 5$ for cations[37,38]). More recently, the structures of even larger gold and silver clusters have been assigned in studies utilizing a combination of trapped ion electron diffraction (TIED), density functional theory and/or photoelectron spectroscopy.[39,40]

The chemistry of gold clusters has also been extensively probed. Studies relevant to adsorption/oxidation of CO have been so far phenomenological and/or limited to very small cluster sizes. For example, in a pioneering FTMS study of the reactivity of gold cluster cations with CO under single collision conditions, Smalley *et al.*[41] found strong cluster size dependencies for CO adsorption but did not study the associated kinetics extensively. In a low-pressure guided-ion-beam study Lee and Ervin[42] established absolute rate constants for CO adsorption on Au_n^- (n = 1–7) at room temperature. In another FTMS study Bondybey *et al.*[43] studied the efficiency of CO adsorption on gold cluster anions up to n = 16 and found a pronounced maximum at n = 11. In a combined experimental and theoretical approach Landman, Wöste and coworkers[44,45] have probed the catalytic activity of Au_2^- towards CO oxidation with O_2. Recently, Wallace *et al.* probed the multiple uptake of small Au_n^- (n = 2–5) in a fast-flow reactor which was found rather insensitive towards temperature changes in a range between 220 K and 320 K. Neutral gold clusters Au_n ($n < 68$) from a laser-vaporization source (at room and liquid nitrogen temperature) were studied towards CO absorption in a reaction cell and positioned to mass-selectively probe the reaction products. Observation of certain local maxima in the reactivity was rationalized by electronic shell closings.[46]

In a recent study from our laboratory, the binding energies of carbon monoxide to gold cluster cations, Au_n^+ ($5 \leq n \leq 65$), were established to decrease monotonically from 1.09 ± 0.1 eV (n = 6) to below 0.65 ± 0.1 eV ($n > 26$) with notable exceptions at n = 30, 31 and n = 48, 49, which show local binding energy maxima.[47]

These gas-phase studies as well as investigations on supported gold clusters[48–50] have demonstrated that the cluster's overall charge state and charge distribution play decisive roles in determining chemical behaviour. In order to further explore this issue, it was recently suggested to study binary silver–gold clusters as a means to tune charge density within the gold cluster by means of doping it with silver atoms.[51,52] In general, much less is known concerning binary silver–gold clusters than for the pure metallic systems. A laser-induced fluorescence study on diatomic AgAu established its vibrational constants.[52] The fast intramolecular dynamics of the Ag_2Au ground state prepared in the equilibrium geometry of the monoanion was probed by a negative ion-to-neutral-to-positive ion (NeNePo) pump–probe scheme in conjunction with *ab initio* molecular dynamics (MD) simulations.[53] A photoelectron spectroscopy study by Negishi *et al.* on $Au_nAg_m^-$ ($n + m < 5$) found that electron affinities tend to increase with increasing gold content.[54] The structural properties of mixed Ag–Au-clusters were explored by DFT computations.[55,56] Furthermore, the structure of cationic clusters $Au_nAg_m^+$ ($n + m < 6$) was probed in a complementary ion-mobility and density-functional computational study[57] which also found clear evidence for inhomogeneous ground state charge distributions reflecting electron transfer from silver to the more electronegative gold atoms. A recent DFT computational study yielded propensity rules for the binding of propene to neutral tri- and tetratomic gold–silver clusters with a tendency of weakening the cluster–propene bond upon exchanging certain "active sites" in the cluster with a less electronegative atom (*i.e.* replacing a Au atom by a Ag atom).[58]

In this contribution, we briefly review the binding energies of CO to positively charged gold clusters Au_n^+ ($5 \leq n < 26$) and silver–gold alloy clusters $Ag_mAu_n^+$ ($m + n = 5$ and 6) as experimentally probed by radiative association kinetics measurements in the ion trap of a FT-ICR mass spectrometer at room temperature. Extensive computational studies at various DFT levels are reported to bring about a deeper understanding of possible correlations between binding energy and both total electron densities and LUMO energies of the respective metal cluster. We finally present calculated harmonic frequencies of the CO stretch vibration as a sensitive probe for changes in the strength of the metal–CO bond.

2. Methods

A. Experiment

1. Cluster ion generation and trapping. Experiments were carried out in a Fourier transform ion cyclotron resonance (FT-ICR) mass spectrometer (Bruker Daltonics, Billerica, MA, USA) equipped with a 7.0 T magnet and a cylindrical Infinity ICR trap. Details of the experimental setup are given elsewhere.[47] Briefly, the ions were generated in a homebuilt laser vaporization disc source. Ions were then extracted, collimated and transferred into the ICR cell held at a base pressure of about 2×10^{-10} mbar. The second harmonic (532 nm) of a Nd : YAG laser (Continuum Electrooptics, Surelite SL-II) with typical pulse energies of 20–90 mJ was used as vaporization laser. Synchronization of the external ion source/laser and the mass spectrometer was achieved by using a pulse delay generator (S.M.V., PDG 204) as master trigger to run the ion source at a repetition rate of 5 Hz. In order to improve the duty cycle of the experiment, ion packages from several laser shots were accumulated in the ICR cell prior to thermalization and CO exposure. Upon on-the-fly trapping each ion package was kinetically cooled by a pulse of argon at typically 9×10^{-7} mbar. After the ion injection the trapping electrodes were held symmetrically at $+2$ V. This sequence was repeated up to 10 times leading to ion intensities (and duty cycle) five times better compared to single shot trapping/detection. As target material we used pure gold or a self-made silver–gold alloy foil (0.2 mm $\times$ 50 mm) with a Ag : Au ratio of 20 : 80 wt%. The cluster composition formed under these laser vaporization conditions is significantly enriched in silver compared to the alloy target composition.[57]

2. Kinetic data acquisition. Isolation and thermalization of the cluster ions were found to be crucial in order to obtain reproducible rate constants of CO adsorption. A particular cluster ion size (parent ion) was isolated in the trap by radial ejection of all unwanted ions from the cell by broadband RF excitation. The frequency window around the parent ion's frequency turned out to be crucial and was chosen to be

~2000 Hz corresponding to $\Delta m = 20$ amu (for $m = 1000$ amu). Any sharper isolation was avoided to minimize frequency components that might kinetically excite the parent ion to well-above thermal. However, as residual excitation couldn't be ruled out completely a supplementary thermalization step using room temperature helium was performed.

The trapping (Ar, 99.99%), thermalization (He, 99.9995%) as well as the reaction (CO, 99.97%, Messer Griesheim) gases were admitted into the Penning trap by three pulsed valves (General Valve Corp.) which allow for the desired pressure to be reached within <100 ms. Typical thermalization times were 10 s at He pressures of $\sim 7 \times 10^{-7}$ mbar. After a pump down delay of 3 s, carbon monoxide was allowed into the cell to reach a constant pressure of 2×10^{-7} mbar for a variable reaction time t (of up to 120 s) controlled by the opening time of the valve. After another pump down of typically 5 s mass spectra were recorded. Typically, this sequence was repeated several times for each reaction time to give sufficient signal : noise ratios. Ion intensities extracted from the mass spectra were normalized to the sum of product and educt ions and evaluated as a function of the reaction time t. In order to obtain rate constants the coupled differential rate equations corresponding to the proposed reaction mechanism were integrated numerically by using the fourth-order Runge–Kutta algorithm and fitted to the experimental data points with an iterative non-linear least squares fit as implemented in the DETMECH program.[59]

B. Computations

Quantum chemical computations were performed to describe the ground state structures of mixed $Ag_mAu_n^+$ ($m + n = 4$–6) and $Ag_mAu_n(CO)^+$ ($m + n = 3$–6) cluster ions as well as pure Au_n^+ and Au_nCO^+ ($n = 2$–8, and 20, 21) cations using the TURBOMOLE program package.[60] The GGA level of density functional theory (DFT) was employed with the S-VWN + Becke–Perdew parametrization (BP86)[61–63] and the RI-J method for the Coulomb part of the Fock operator.[64,65] For open shell cases (spin doublets) the unrestricted Kohn–Sham formalism was used. Deviations of the expectation values of the S^2 operator from the eigenvalue 0.75 were found to be always below 0.008 indicating pure doublet states.

For the alloy cluster species the most stable isomers found were also subjected to another geometry relaxation step applying the hybrid functional B3-LYP.[66] From these calculations we derived binding energies of CO to the Au_n^+ and $Ag_mAu_n^+$ clusters as the difference between the total electronic energies (after correcting for the zero-point energy) of the respective ground state isomer of the cluster carbonyl and bare cluster ion (both in their equilibrium states). For Au and Ag, relativistic effects were included by means of effective core potentials.[67] This leaves 19 valence electrons to be treated explicitly for each Au (with configuration $5s^2 5p^6 5d^{10} 6s^1$) and Ag ($4s^2 4p^6 4d^{10} 5s^1$) atom. If not otherwise mentioned, Au was described using a (9s7p5d1f)/[7s5p3d1f] basis set derived from a standard SVP basis[65] by adding optimized polarization and diffuse functions as specified previously[68] and Ag was described using a (8s7p5d1f)/[6s5p3d1f] basis set obtained in a similar treatment.[69] For carbon and oxygen we used the def-TZVP basis set.[70] Vibrational frequencies have been calculated analytically with the BP86 functional also with a basis set of triple zeta quality (as above) without the diffuse f function.

3.   Results and discussion

A.   CO on Au_n^+

1.   Reaction rates and binding energies. Under the well-defined conditions used in the FT-ICR mass spectrometer setup, kinetic data of CO adsorption on mass-selected Au_n^+ yields absolute bimolecular rate constants which can be converted at well-known pressures—here $p_{CO} = 2.3 \times 10^{-7}$ mbar—to quasi-unimolecular rate constants. The rate constants obtained under these experimental conditions are

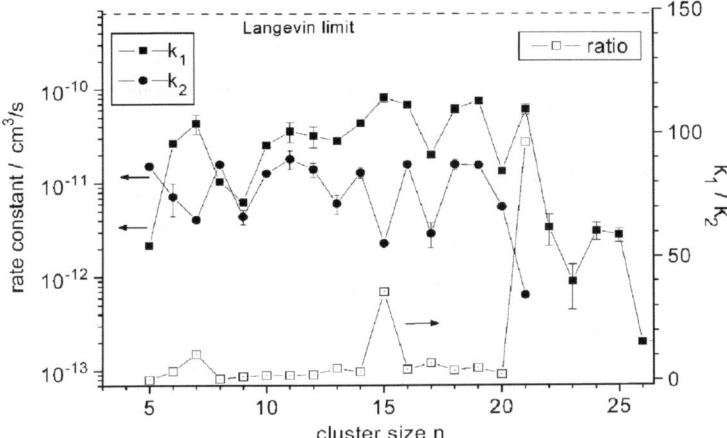

Fig. 1 Experimentally determined absolute rate constants (solid points) for adsorption of the first (k_1) and second (k_2) CO molecule to gold cluster cations Au_n^+; also given is the ratio k_1/k_2 (empty squares) as a function of cluster size n showing local maxima for $n = 7$, 15, and 21.

shown in Fig. 1. In most cases multiple CO uptake is found. The corresponding rate constants for the first two CO molecules (k_1 and k_2) are evaluated and included in Fig. 1. In general, the rate constants stay well below the Langevin limit by at least one order of magnitude even for the most reactive cluster size ($n = 15$). This most likely represents the fact that not all collisions lead to a reaction but re-dissociate. We point out in passing that the relative rate constants show local maxima, particularly pronounced for $n = 21$ and to a lesser extent for $n = 7$ and 15. As going from $n = 20$ to $n = 21$ this ratio of rate constants for absorbing the first and the second CO molecule changes by at least 2 orders of magnitude. We shall return to this finding below.

As outlined recently in detail[47] one can arrive at the binding energy for the first carbon monoxide molecule adsorbed upon modelling such an ion–molecule reaction by the radiative association kinetics model based on the work by Dunbar *et al.*[71,72] Briefly, one assumes that the energized complex formed upon a reactive ion–molecule collision has equipartitioned the released energy, *i.e.* binding energy and center-of-mass collision energy, among all internal degrees of freedom. Here, the efficiency of complex formation is taken as the capture cross-section as described by the Langevin capture model.[73] One then applies statistical rate theories to describe (i) the back reaction leading to the educts, *i.e.* dissociation of the energized complex as formulated in RRKM theory and (ii) the probability of spontaneous emission of an infrared (IR) photon to give a vibrationally stabilized product. For both rate models the molecular constants used are derived from DFT calculations, namely the vibrational frequency spectra and corresponding IR intensities (Einstein coefficients). We note in passing that by far the most intense and therefore only relevant IR-active mode is the CO stretch vibration, which in a fortuitous way simplifies the description of the radiative association model and makes the modelling more robust towards possible inaccuracies in describing the low-frequency modes of metal–metal vibrations.

The results of this radiative association model (and corresponding fits to the experimental kinetic data with the metal–CO binding energy as the only fit parameter) are summarized in Fig. 2. There is an overall trend towards smaller binding energies with increasing cluster size which can be attributed to a charge dilution effect. In Fig. 2 an electrostatic model is overlaid which is calculated as the interaction potential between a point charge and a polarizable molecule with polarizability volume α. The effective distance of interaction is taken as the cluster radius plus half the CO bond length (0.56 Å). This leads essentially to a $n^{-1/3}$ size

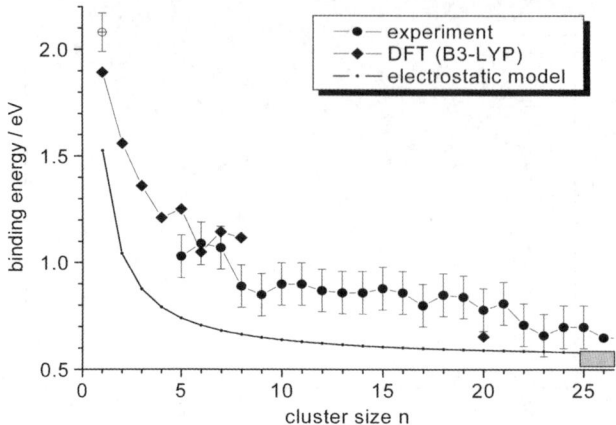

Fig. 2 Binding energies for the first CO molecule to Au_n^+ as derived from the experimental rate constants using the radiative association model. Superimposed is the electrostatic interaction potential between a charged metal sphere and an induced dipole shifted by the bulk heat of adsorption (0.55 eV) of CO on Au(111).

dependence which describes the overall decrease of the bond energy. Clearly, the attainable accuracy in experimental binding energies doesn't allow for structural inferences detailed enough to allow assignment of specific isomers with respect to the bare precursor cluster or preferred adsorption sites. However, identifying particularly reactive cluster sizes (reactive islands) and comparison with possible candidate structures can provide useful complementary data for structure determination methodologies such as TIED.

One might wonder if it is possible to correlate the strength of the metal–CO bond to parameters from quantum chemical computations of the bare cluster cations—thus aiding in understanding the σ-donor type of bonding between CO and metal cluster and/or allowing to predict specific adsorption sites. Two different quantities might reasonably be expected to correlate with the bond energy of μ_1-bound CO. First, the total electron density at the cluster surface is likely to play a role, as σ-electron donation (the dominant bonding mechanism) is most easily feasible for clusters exhibiting bonding sites with low electron density. Secondly, we consider the energy of the lowest unoccupied molecular orbital (LUMO), as electron transfer into an energetically lower lying LUMO will lead to a larger energy gain than into a higher lying one. The latter quantity can be directly taken from the output of the quantum chemical treatment and is plotted in Fig. 3 *versus* the calculated binding energy of CO to the pure gold cluster cations in their respective optimized structure. Note that for the open shell clusters, *i.e.* for even *n*, the first *completely* unoccupied orbital was used (*i.e.* not the unoccupied opposite-spin counterpart of the HOMO).

For obtaining the total electron densities at the relevant positions we proceeded as follows. The surface of each Au_n^+ cluster was generated by putting spheres around each gold atom with a radius of 1.97 Å, which is a typical value for the Au–C distance. Note that the actual choice is not very critical, as the variation of total electron density with distance is similar for the whole surface. Next, the total electron density ρ was calculated for each gold atom on a grid of 900 points per atom on that sphere. The position of the minimum of ρ for each atom, ρ_c, was selected leading to *n* possible bonding sites with *n* corresponding values for ρ_c. Finally the lowest of these values, ρ_{min}, was chosen for each Au_n^+ cluster and plotted *versus* the binding energy of CO in Fig. 4.

The general trend of decreasing bond energies for increasing clusters size is in line with both increasing LUMO energy (Fig. 3) and increasing total electron density (Fig. 4). For example, CO binding energy for Au_{20}^+ is about half as large as for

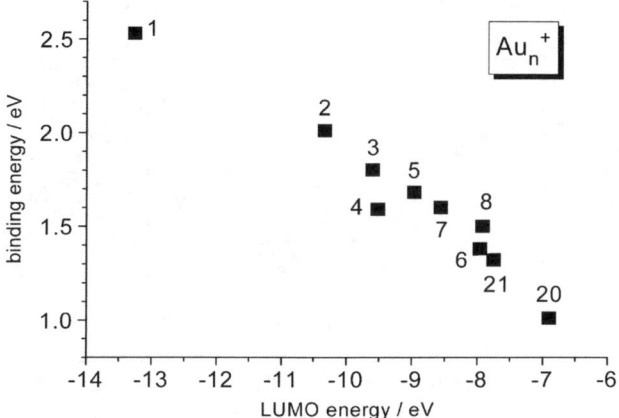

Fig. 3 Binding energies for Au_n^+–CO calculated using the BP86 functional (*i.e.* difference in total energies between the bare and the cluster carbonyl both in their fully relaxed geometry) contrasted to the Kohn–Sham energy of the LUMO of the respective bare gold cluster cation in its relaxed geometry.

Au_2^+ while ρ_{min} increases by a factor of two and the LUMO energy increases from *ca.* −10.5 eV to −7 eV. Even exceptions from the general size trend are reproduced: the CO binding energy for $n = 21$ is significantly larger than for $n = 20$. This reversal is also found for ρ_{min} as well as for the HOMO energy. We also note, that for $n = 7$ and 8 the correlation might be rather fortuitous since CO adsorption leads to rather strong rearrangement of the gold cluster structure.[47]

2. **Preferred adsorption sites.** As a follow-up it was of interest to extend this study to the various atomic sites on a cluster of a given size. Au_{21}^+ was chosen as a model cluster. For this cluster a structure was chosen (as depicted in Fig. 5) that can be deduced from a tetrahedral structure—recently suggested for Au_{20}^-[74]—with an additional atom capping four gold atoms (labelled as #12, 17, 18, 20 in Fig. 5) with bond lengths of 272 pm (#17 and 18), 281 pm (#12) and 293 pm (#20). This structure most probably is a local minimum structure though a vibrational analysis

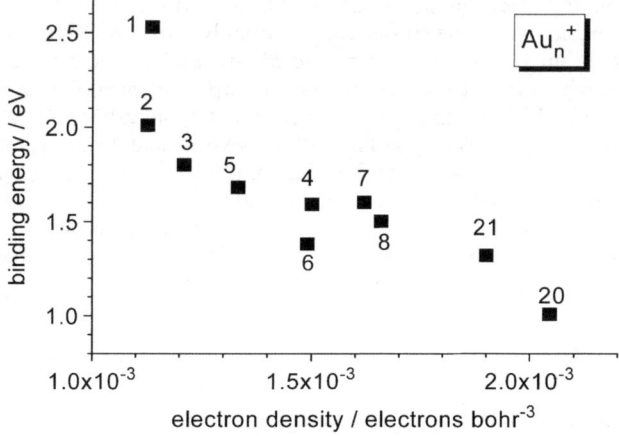

Fig. 4 CO binding energy (BP86) (as in Fig. 3) *versus* the total electron density (given in atomic units) of the Au_n^+ cluster ion taken at the position of the preferred adsorption site on the cluster for various cluster sizes *n*.

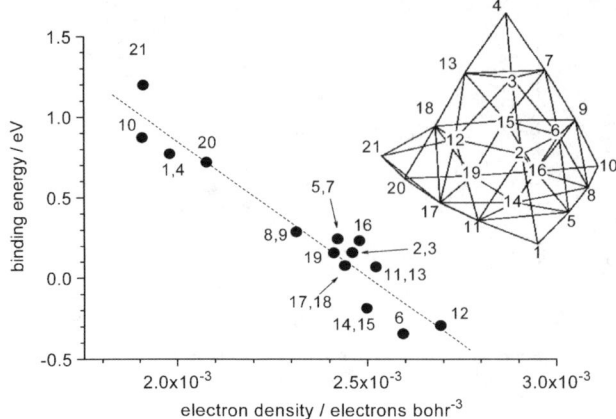

Fig. 5 Computed binding energy (BP86) of CO on Au_{21}^{+} for all possible binding sites. CO is always bound to a single gold atom in a head-on fashion. Note the correlation between bond energy and the local electron density.

was omitted (however the relaxation procedure, when applied in C_1 symmetry, is designed to avoid running into saddle points). Other minima like atom #21 bridging three gold atoms of the quasi-(111) faces were also found and are similar in total energy within 0.05 eV.

Fig. 5 also presents this structure together with the calculated binding energies of CO to all possible μ_1-adsorption sites on the model Au_{21}^{+}. The possible binding sites for CO were obtained by the procedure described in the previous section for the electron density. Binding energies in Fig. 5 are calculated as differences of single-point energies for the individual $Au_{21}CO^{+}$ species and the (optimized) bare Au_{21}^{+} cluster. For $Au_{21}CO^{+}$ the Au–C and C–O distance were chosen as 1.97 Å and 1.13 Å, respectively, with an Au–C–O angle of 180°. The use of non-equilibrium structures explains the occasional occurrence of negative values for the bond energy. In fact, fundamental rearrangement of gold atoms was observed when applying a relaxation procedure to systems with such negative nominal CO bond energies. As can be seen from Fig. 5, high binding energies for the individual bonding positions of CO are well in line with low electron densities ρ_c. The different positions on the cluster can be roughly grouped into three categories: the single ad atom (#21), which is predicted to have a binding energy for CO about 0.35 eV higher than for the second best group of adsorption sites (gold atoms #1, 4, 10 and 20) which are the apex atoms of the Au_{20}-tetrahedron. Note that this description of Au_{21}^{+} is perfectly in line with the experimental observation of a high propensity of Au_{21}^{+} to adsorb the first CO molecule (see above), whereas Au_{20}^{+} was recently found to adsorb four CO molecules with rate constants that differ by no more than a factor of 3.[47] In accordance with chemical intuition the least likely absorption sites are the highly coordinated atoms forming the edges and faces of the tetrahedron having the highest electron density.

B. CO on $Ag_mAu_n^{+}$ ($m + n < 7$)

1. Reaction rates and binding energies. Fig. 6 summarizes the experimentally obtained adsorption rates for carbon monoxide onto the alloy cluster ions $Ag_mAu_n^{+}$ (for $m + n = 5, 6$). There is a clear trend towards smaller reactivity for the first and second CO binding to the cluster ion with increasing silver content.[75]

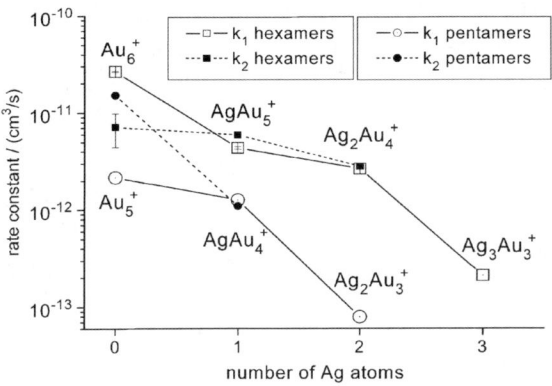

Fig. 6 Experimentally determined absolute rate constants for the CO adsorption on $Ag_mAu_n^+$ ($m + n = 5$ and 6) for several compositions (m, n).

As described in more detail recently[76] an extensive search for the lowest-energy structures has been undertaken and we restrict ourselves to describing the structures for cluster carbonyls of interest in this study. For the pentatomic alloy cluster species the most stable structure found is a twisted X-shaped geometry as proposed recently;[55,57] adding a CO molecule gives rise to only slight distortions— other isomers being considerably higher in energy by more than 0.2 eV, except for one trigonal bipyramid structure for $Ag_2Au_3CO^+$ lying about 0.06 eV above the ground state structure.[76] For the hexamers, a number of different structural motifs are found. As silver content is increased more three-dimensional isomers compete energetically with the prevailing two-dimensional geometries. On the other hand adsorption of a CO molecule seems to reinforce planar motifs again. We have argued that in the case of $Ag_2Au_2^+$ energetic barriers between isomers are small compared to the energy released by CO adsorption.[76] Extrapolating this result to the larger penta- and hexaatomic cluster sizes one might then assume that CO adsorption samples/generates the lowest energy structure. This assumption is implicitly used in the kinetic modelling to evaluate binding energies.

The radiative association model as outlined above is also applied here and the results are summarized in Fig. 7. For the pentamers as well as for the hexamers the general trend is a decrease in the experimentally determined binding energy with increasing silver content of the cluster. This trend is well reproduced by the computed values, however, the B3-LYP energies appear systematically higher by 0.15–0.2 eV (see also Table 1). The values lie well within the limits of binding energies given on the one hand by CO binding to atomic metal ions—where binding is dominated by electrostatic interaction—and on the other by binding to a single-crystal surface of the bulk metal. Note that for both limits the binding energy of CO is roughly twice as large for gold as for silver. The decrease in binding energy of CO to the cluster ion of a given size as a function of increasing silver composition can be qualitatively understood in terms of the charge distribution in the cluster. Due to the higher electronegativity of gold the electron density is preferentially increased on the gold atoms, thereby weakening the binding to CO.

Indeed, when exchanging Au atoms with Ag, one observes an increased total electron density for the surface (see above) of the remaining Au atoms. We note, that for the mixed clusters the correlation of total electron density and binding energy of CO is much less pronounced than for the pure Au clusters investigated previously. Carrying out the same procedure as described before for Au_{21}^+, it turns out that for the present cases only the adsorption site, *i.e.* gold atom, with the lowest electron

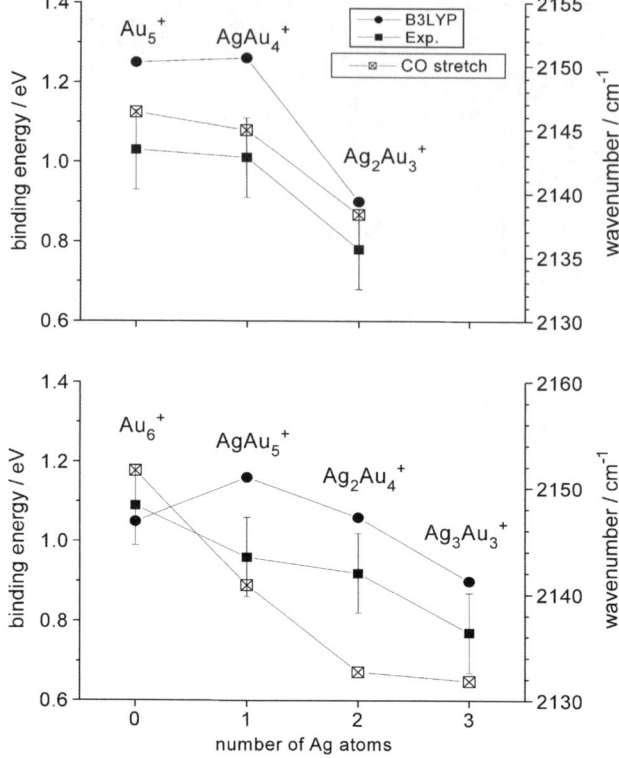

Fig. 7 Binding energies for the first CO adsorbed to pentameric ($m + n = 5$, top panel) and hexameric ($m + n = 6$, lower panel) $Ag_mAu_n^+$ as determined from experimental rate constants *via* radiative association kinetic modelling and from DFT calculations employing the hybrid functional B3LYP (see text for details). Also shown are the calculated vibrational frequencies for the C–O stretch mode for $Ag_mAu_n^+CO$.

density of a given isomer gives rise to the highest binding energy, whereas for the other gold atoms this correlation is much less reliable. We further note, that also the correlation of LUMO energies and CO-bond energies is much less pronounced than for the homoatomic clusters.

We have also computed the harmonic frequencies of the C–O stretch and the Au–C stretch vibration for each $Ag_mAu_nCO^+$ species with $m + n = 5$ and 6 for compositions which were also probed in the experiment. Frequencies and IR intensities are summarized in Table 2. Interestingly, there is a strong correlation between the relative strength of the metal–CO bond and the vibrational stretch frequency of the CO moiety as can also be seen from Fig. 7. The weaker the interaction of the σ-donor (M–CO) bond the more redshifted the CO vibrational frequency becomes. This effect is particularly significant for the hexamers for which changes in experimentally determined binding energies of about 0.2 eV (in going from Au_6^+ to $Ag_2Au_4^+$) lead to a shift of about 20 cm^{-1}. We note that the textbook explanation for a redshift of this vibrational mode with decreasing metal–CO bond strength—stating that the M ← CO σ-donor type bond stems from an *antibonding* σ-orbital of CO—was recently questioned.[77] In fact, a redshift has been observed in IR multi-photon dissociation spectra of gold cluster carbonyl cations with increasing cluster size (*i.e.* decreasing bond energy) and be attributed to charge effects[78] polarizing the CO bond. This means in turn that the frequency shift predicted here for the alloy clusters may be used as a sensitive tool to probe local charge densities.

Table 1 Experimental (E_0) and calculated (D_0) binding energies of CO on $Ag_mAu_n^+$

m, n	E_0/eV^a	D_0/eV^b
0,1	2.08 ± 0.15[c]	1.89
1,0	0.93 ± 0.05[d]	
0,5	1.03 ± 0.1	1.25
1,4	1.01 ± 0.1	1.26
2,3	0.78 ± 0.1	0.9
0,6	1.09 ± 0.1	1.05
1,5	0.96 ± 0.1	1.16
2,4	0.92 ± 0.1	1.06
3,3	0.77 ± 0.1	0.9
0, ∞	0.55 ± 0.05[e]	
∞, 0	0.28 ± 0.02[f]	

[a] E_0 from the radiative association kinetic model (this work, unless otherwise noted). [b] Calculated 0 K binding energies (DFT-B3LYP) corrected for zero-point energies. [c] Value for Au^+–CO from ref. 79. [d] Value for Ag^+–CO from ref. 80. [e] Heat of adsorption of CO on Au(110) from ref. 81. [f] Heat of adsorption of CO on Ag(111) from ref. 82.

Table 2 Calculated harmonic frequencies and IR intensities for the $Ag_mAu_n^+$–CO and $Ag_mAu_n^+$ C–O vibrational stretching modes. ($M^+ = Ag_mAu_n^+$). Note that in all cases the CO is bound to a single gold atom

$Ag_mAu_nCO^+$ m, n	$\nu(M^+–CO)/cm^{-1}$	IR intensity/ km mol^{-1}	$\nu(M^+C–O)/cm^{-1}$	IR intensity/ km mol^{-1}
0,5	391	11.4	2146	574
1,4	396	10.9	2145	591
2,3	365	6.3	2138	614
3,2	363	5.6	2135	645
0,6	373	6.1	2152	637
1,5	379	7.3	2141	722
2,4	372	7.8	2133	790
3,3	357	3.8	2132	701

4. Summary

A detailed study of low-pressure kinetics of CO adsorption on gold and mixed silver–gold cluster cations in a FT-ICR mass spectrometer has been analyzed in terms of the binding energies involved in the gold–CO bond. Overall, the bond strength is found to decrease with increasing cluster size. For alloy clusters of fixed size the CO bond strength is found to decrease with increasing silver content. DFT-based computations employing the BP86 and B3LYP functionals have been performed which support these findings. For the homoatomic clusters the CO bond energy was found to clearly correlate with both the energy of the LUMO and the total electron density on the cluster surface. The latter finding may be used for a screening of possible binding sites of CO, as exemplified for Au_{21}^+. For mixed silver–gold cluster ions these correlations are much less pronounced, and therefore screening of possible bonding sites by searching points of lowest total electron density on the cluster surface is not very reliable, at least for the cases treated. Finally, we show that the computed vibrational frequencies for the CO stretch mode exhibit a distinct redshift with decreasing CO adsorption energy which might potentially serve as an experimental probe to describe the typical μ_1-type of gold–CO bonding.

Acknowledgements

The authors thank Lars Walter for technical support in performing the experiments. This research was supported in part by the Deutsche Forschungsgemeinschaft (DFG-Center for Functional Nanostructures).

References

1   (a) P. Pyykkö, Chem. Rev., 1988, 88, 563; (b) P. Pyykkö, Angew. Chem., Int. Ed., 2004, 43, 4412; (c) P. Pyykkö, Inorg. Chim. Acta, 2005, 358, 4113.
2   N. W. Ashcroft and N. D. Mermin, Solid State Physics, Saunders College, Philadelphia, 1976.
3   S. C. Richt smeier, T. Jagger, J. L. Gole and D. A. Dixon, Chem. Phys. Lett., 1985, 117, 274.
4   K. Balasubramanian and M. Z. Liao, J. Chem. Phys., 1987, 86, 5587.
5   S. P. Walch, C. W. Bauschlicher and S. R. Langhoff, J. Chem. Phys., 1986, 85, 5900.
6   C. W. Bauschlicher, Jr, S. R. Langhoff and H. Partridge, J. Chem. Phys., 1989, 91, 2412.
7   G. Bravo-Peréz, I. L. Garzón and O. Novaro, Chem. Phys. Lett., 1999, 313, 655.
8   V. Bonačić-Koutecký, L. Češpiva, P. Fantucci and J. Koutecký, J. Chem. Phys., 1993, 98, 7981.
9   V. Bonačić-Koutecký, L. Češpiva, P. Fantucci and J. Koutecký, Z. Phys. D, 1993, 26, 287.
10 V. Bonačić-Koutecký, L. Češpiva, P. Fantucci, J. Pittner and J. Koutecký, J. Chem. Phys., 1994, 100, 490.
11 V. Bonačić-Koutecký, J. Pittner, M. Boiron and P. Fantucci, J. Chem. Phys., 1999, 110, 3876.
12 G. Bravo-Perez, I. L. Garzon and O. Novaro, J. Mol. Struct. (THEOCHEM), 1999, 493, 225.
13 H. Grönbeck and W. Andreoni, Chem. Phys., 2000, 262, 1.
14 H. Häkkinen and U. Landman, Phys. Rev. B, 2000, 62, R2287.
15 H. Häkkinen, M. Moseler and U. Landman, Phys. Rev. Lett., 2002, 89, 033401.
16 D. Leopold, J. Ho and W. C. Lineberger, J. Chem. Phys., 1987, 86, 1715.
17 J. Ho, K. Ervin and W. C. Lineberger, J. Chem. Phys., 1990, 93, 6987.
18 G. Ganteför, M. Gausa, K.-H. Meiwes-Broer and H. O. Lutz, J. Chem. Soc., Faraday Trans., 1990, 86, 2483.
19 K. J. Taylor, C. L. Pettiette-Hall, O. Cheshnovski and R. E. Smalley, J. Chem. Phys., 1992, 96, 3319.
20 G. Ganteför, D. Cox and A. Kaldor, J. Chem. Phys., 1992, 96, 4102.
21 H. Handschuh, C.-Y. Cha, P. S. Bechthold, G. Ganteför and W. Eberhardt, J. Chem. Phys., 1995, 102, 6406.
22 J. Li, X. Li, H.-J. Zhai and L.-S. Wang, Science, 2003, 299, 864.
23 H. Häkkinen, B. Yoon, U. Landman, X. Li, H.-J. Zhai and L.-S. Wang, J. Phys. Chem. A, 2003, 107, 6168.
24 B. A. Collings, K. Athanassenas, D. M. Rayner and P. A. Hackett, Z. Phys. D, 1993, 26, 36.
25 B. A. Collings, K. Athanassenas, D. Lacombe, D. M. Rayner and P. Hackett, J. Chem. Phys., 1994, 101, 3506.
26 G. Dietrich, S. Krückeberg, K. Lützenkirchen, L. Schweikhard and C. Walther, J. Chem. Phys., 2000, 112, 752.
27 A. Terasaki, S. Minemoto, M. Iseda and T. Kondow, Eur. Phys. J. D, 1999, 9, 163.

28 D. Schooss, S. Gilb, J. Kaller, M. M. Kappes, F. Furche, A. Köhn, K. May and R. Ahlrichs, *J. Chem. Phys.*, 2000, **113**, 5361.
29 A. Schweizer, J. M. Weber, S. Gilb, H. Schneider, D. Schooss and M. M. Kappes, *J. Chem. Phys.*, 2003, **119**, 3699.
30 S. Krückeberg, L. Schweikhard, G. Dietrich, K. Lützenkirchen, C. Walther and J. Ziegler, *Chem. Phys.*, 2000, **262**, 105.
31 S. Krückeberg, L. Schweikhard, G. Dietrich, K. Lützenkirchen, C. Walther and J. Ziegler, *Eur. Phys. J. D*, 1999, **9**, 145.
32 J. Ziegler, G. Dietrich, S. Krückeberg, K. Lützenkirchen, L. Schweikhard and C. Walther, *Int. J. Mass Spectrom.*, 2000, **202**, 47.
33 V. A. Spasov, T. H. Lee, J. P. Maberry and K. M. Ervin, *J. Chem. Phys.*, 1999, **110**, 5208.
34 V. A. Spasov, Y. Shi and K. M. Ervin, *Chem. Phys.*, 2000, **262**, 75.
35 S. Gilb, P. Weis, F. Furche, R. Ahlrichs and M. M. Kappes, *J. Chem. Phys.*, 2002, **116**, 4094.
36 F. Furche, R. Ahlrichs, P. Weis, C. Jacob, S. Gilb, T. Bierweiler and M. M. Kappes, *J. Chem. Phys.*, 2002, **117**, 6982.
37 P. Weis, *Int. J. Mass Spectrom.*, 2005, **245**, 1, and references therein.
38 E. M. Fernandez, J. M. Soler, I. L. Garzon and L. C. Balbas, *Int. J. Quantum Chem.*, 2005, **101**, 740, and references therein.
39 A. Lechtken, D. Schooss, J. R. Stairs, M. N. Blom, F. Furche, N. Morgner, O. Kostko, B. v. Issendorff and M. M. Kappes, *Angew. Chem., Int. Ed.*, 2007, **46**(16), 2944.
40 D. Schooss, M. N. Blom, J. H. Parks, B. v. Issendorff, H. Haberland and M. M. Kappes, *Nano Lett.*, 2005, **5**(10), 1972.
41 M. A. Nygren, P. E. M. Siegbahn, C. Jin, T. Guo and R. E. Smalley, *J. Chem. Phys.*, 1991, **95**, 6181–6184.
42 T. H. Lee and K. M. Ervin, *J. Chem. Phys.*, 1994, **98**, 10023–10031.
43 I. Balteanu, O. P. Balaj, B. S. Fox, M. K. Beyer, Z. Bastl and V. E. Bondybey, *Phys. Chem. Chem. Phys.*, 2003, **5**, 1213–1218.
44 L. D. Socaciu, J. Hagen, T. M. Bernhardt, L. Wöste, U. Heiz, H. Häkkinen and U. Landman, *J. Am. Chem. Soc.*, 2003, **125**, 10437–10445.
45 T. M. Bernhardt, *Int. J. Mass Spectrom.*, 2005, **243**, 1.
46 N. Veldeman, P. Lievens and M. Andersson, *J. Phys. Chem. A*, 2005, **109**, 11793.
47 M. Neumaier, F. Weigend, O. Hampe and M. M. Kappes, *J. Chem. Phys.*, 2005, **122**, 104702.
48 A. Sanchez, S. Abbet, U. Heiz, W.-D. Schneider, H. Häkkinen, R. N. Barnett and U. Landman, *J. Phys. Chem. A*, 1999, **103**, 9572.
49 U. Heiz, A. Sanchez, S. Abbet and W.-D. Schneider, *Chem. Phys.*, 2000, **262**, 189–200.
50 M. Sterrer, M. Yulikov, E. Fischbach, M. Heyde, H.-P. Rust, G. Pacchioni, T. Risse and H.-J. Freund, *Angew. Chem., Int. Ed.*, 2006, **45**, 2630.
51 R. Mitrić, C. Bürgel, J. Burda, V. Bonačić-Koutecký and P. Fantucci, *Eur. Phys. J. D*, 2003, **24**, 41.
52 J. C. Fabbi, J. D. Langenberg, Q. D. Costello and M. D. Morse, *J. Chem. Phys.*, 2001, **15**, 7543.
53 T. M. Bernhardt, J. Hagne, L. D. Socaciu, R. Mitrić, A. Heidenreich, J. Le Roux, D. Popolan, M. Vaida, L. Wöste, V. Bonačić-Koutecký and J. Jortner, *ChemPhysChem*, 2005, **6**, 243.
54 Y. Negishi, Y. Nakamura and A. Nakajiama, *J. Chem. Phys.*, 2001, **115**, 3657.
55 V. Bonačić-Koutecký, J. Burda, R. Mitrić, M. Ge, G. Zampella and P. Fantucci, *J. Chem. Phys.*, 2002, **117**, 3120.
56 H. M. Lee, M. Ge, B. R. Sahu, P. Tarakeshwar and K. S. Kim, *J. Phys. Chem. B*, 2003, **107**, 9994.
57 P. Weis, O. Welz, E. Vollmer and M. M. Kappes, *J. Chem. Phys.*, 2004, **120**, 677.
58 S. Chretien, M. S. Gordon and H. Metiu, *J. Chem. Phys.*, 2004, **121**, 9931.
59 E. Schumacher, DETMECH, *Chemical Reaction Kinetics Software*, 1997; http://www.chemsoft.ch/.
60 TURBOMOLE V5-9, © Universität Karlsruhe.
61 B. Becke, *Phys. Rev. A*, 1988, **38**, 3098.
62 J. Perdew, *Phys. Rev. B*, 1986, **33**, 8822.
63 O. Treutler and R. Ahlrichs, *J. Chem. Phys.*, 1995, **102**, 346.
64 K. Eichkorn, O. Treutler, H. Öhm, M. Häser and R. Ahlrichs, *Chem. Phys. Lett.*, 1995, **242**, 652.
65 K. Eichkorn, F. Weigend, O. Treutler and R. Ahlrichs, *Theor. Chem. Acc.*, 1997, **97**, 119.
66 C. T. Lee, W. T. Yang and R. G. Parr, *Phys. Rev. B*, 1988, **37**, 785.
67 D. Andrae, U. Haeussermann, M. Dolg, H. Stoll and H. Preuss, *Theor. Chim. Acta*, 1990, **77**, 123.

68 S. Gilb, P. Weis, F. Furche, R. Ahlrichs and M. M. Kappes, *J. Chem. Phys.*, 2002, **16**, 4094–4101.
69 P. Weis, T. Bierweiler, S. Gilb and M. M. Kappes, *Chem.Phys. Lett.*, 2002, **355**, 355.
70 A. Schäfer, H. Horn and R. Ahlrichs, *J. Chem. Phys.*, 1992, **97**, 2571.
71 R. C. Dunbar, T. B. McMahon, D. Thölmann, D. S. Tonner, D. R. Salahub and D. Wie, *J. Am. Chem. Soc.*, 1995, **117**, 12819.
72 V. Ryzov, S. J. Klippenstein and R. C. Dunbar, *J. Am. Chem. Soc.*, 1996, **118**, 5462.
73 G. Gioumousis and D. P. Stevenson, *J. Chem. Phys.*, 1958, **29**, 294.
74 J. Li, X. Li, H.-J. Zhai and L.-S. Wang, *Science*, 2003, **299**, 864.
75 For clusters with an even higher content of silver the reaction rate constant is below our experimentally accessible detection limit of about 8×10^{-14} cm^3 s^{-1}. For smaller cluster sizes the reaction with CO leads to dissociative reactions.
76 M. Neumaier, F. Weigend, O. Hampe and M. M. Kappes, *J. Chem. Phys.*, 2006, **125**, 104308.
77 A. J. Lupinetti, S. Fau, G. Frenking and S. H. Strauss, *J. Phys. Chem. A*, 1997, **101**, 9551.
78 A. Fielicke, G. v. Helden, G. Meijer, B. Simard and D. M. Rayner, *J. Phys. Chem. B*, 2005, **109**, 23935.
79 D. Schröder, H. Schwarz, J. Hrušák and P. Pyykkö, *Inorg. Chem.*, 1998, **37**, 624.
80 F. Meyer, Y.-M. Chen and P. B. Armentrout, *J. Am. Chem. Soc.*, 1995, **117**, 4071.
81 J. M. Gottfried, K. J. Schmidt, S. L. M. Schroeder and K. Christmann, *Surf. Sci.*, 2003, **536**, 206–224.
82 G. McElhiney and J. Pritchard, *Surf. Sci.*, 1976, **54**, 617.

Preparation of regular arrays of bimetallic clusters with independent control of size and chemical composition†

M. Marsault, G. Hamm A. Wörz, G. Sitja, C. Barth and C. R. Henry*

Received 3rd April 2007, Accepted 24th May 2007
First published as an Advance Article on the web 3rd October 2007
DOI: 10.1039/b705083f

Regular arrays of bimetallic clusters have been prepared by atomic deposition, under UHV, on a nanostructured ultrathin alumina film. The alumina films are obtained by oxidation at 1000 K of a Ni_3Al (111) surface. They present two regular hexagonal superstructures with lattice parameters of 2.4 and 4.1 nm. Pd clusters nucleate exclusively on the 4.1 nm superstructure forming regular arrays of clusters extending on the whole (1 cm^2) substrate. Gold deposited on a previously formed Pd clusters array, condenses exclusively on the Pd clusters in forming a regular array of bimetallic AuPd clusters with a narrow size distribution. The size and the composition of the AuPd clusters can be controlled independently. Gold clusters nucleates also on the 4.1 nm superstructure but they can escape from the nucleation sites and coalesce with other gold clusters. By condensing Pd on the preformed Au clusters, PdAu clusters are formed together with pure Pd clusters nucleated on the free sites of the 4.1 nm superstructure of the alumina film.

1. Introduction

Bimetallic nanoparticles supported on oxides have become increasingly used as heterogeneous catalysts. They present better resistance again ageing,[1] less sensitivity to contamination[2] or a better selectivity.[3] However, their potential is not fully utilised because of a lack of knowledge of the properties of alloy nanoparticles. Studies on extended surfaces of alloy single crystals are numerous[4] but these results often cannot be extrapolated to nanoparticles. For example Pd/Ni (110) has shown an increase of reactivity by a factor 35 for butadiene hydrogenation compared to Pd(110)[5] which has the best performance for this reaction whereas Pd/Ni core shell nanoparticles show roughly the same activity as pure Pd particles for this reaction.[6] From the theory side, *ab initio* calculations have recently permitted the prediction of trends for catalytic properties among a series of alloy surfaces[7,8] but again the structure and the composition of bimetallic nanoparticles can be very different from their bulk counterpart.[9] There is then clearly a need for systematic studies on bimetallic nanoparticles. From the experimental point of view it is rather difficult to prepare bimetallic particles with homogeneous composition and controlled size.

CMCN-CNRS, Campus de Luminy, case 913, 13288 Marseille cedex 09 France. E-mail: henry@crmcn.univ-mrs.fr; Fax: +33 491 41 89 16; Tel: +33 662 92 28 32

† The HTML version of this article has been enhanced with colour images.

A good way to control the growth of metallic nanoparticles is to evaporate atoms on an oxide single crystal. By choosing a good combination of atomic flux and substrate temperature, uniform collections of monometallic particles with a unique shape can be obtained.[10] However for bimetallic particles these requirements are very difficult to reach. If the two metals are successively deposited, generally alloy and pure metallic particles are obtained.[11,12] If the two metals are deposited simultaneously, only bimetallic particles can be obtained but their composition evolve with their size.[13] Recently, we have shown that by using nanostructured alumina thin films as a template, it was possible to obtain regular arrays of PdAu bimetallic particles with an independent control of the size and the chemical composition.[14] In the work presented in this paper we have studied this new method in more detail. Firstly, we investigate the nucleation of the pure metals (Pd and Au), then we analyze the effect of the order of deposition of the two metals. Finally we discuss the morphology and the distribution of the two metals inside the particles.

2. Experimental

Experiments were carried out on a multichamber UHV system with a base pressure in the low 10^{-10} mbar range. Alumina thin films are prepared by oxidation of a clean Ni_3Al (111) surface at 1000 K under an oxygen pressure of 5×10^{-8} mbar for an exposure of 45 L. The film was then annealed at 1050 K for 5 min. The quality of the film was controlled *in situ* by STM (Omicron) at room temperature (RT). Fig. 1 shows an STM picture of such an alumina film. At a bias voltage of 3.2 V an hexagonal network with a parameter of 2.4 nm is seen, changing the bias voltage to 2.3 V another hexagonal superstructure is observed with a parameter of 4.1 nm, in agreement with previous studies.[15] In some cases the two superstructures can be imaged simultaneously (see top of Fig. 1) that evidences their $(\sqrt{3} \times \sqrt{3})R30°$ relationship. These superstructures were first investigated by LEED and STM by the group of K. Wandelt.[15] By using dynamic force microscopy we have been able to obtain the first atomic resolution of the surface of the film, which allowed us to

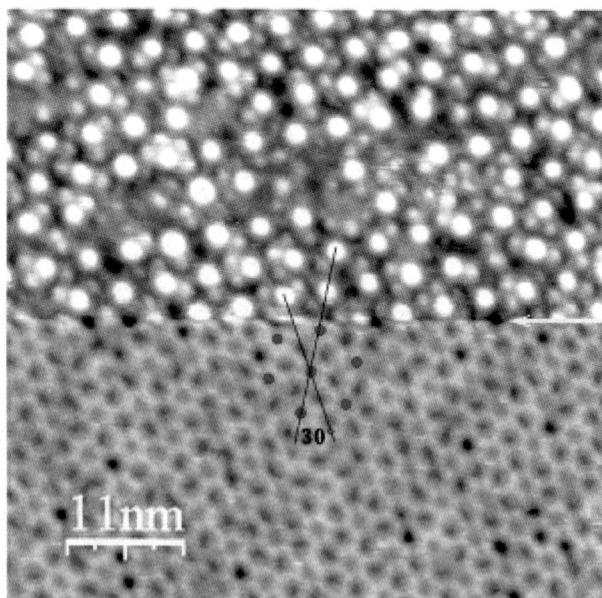

Fig. 1 STM image of an Al_2O_3 film on Ni_3Al (111). The lower part corresponds to a bias voltage of 3.2 V and the upper part to a bias voltage of 2.3 V.

Fig. 2 STM image of an array of Pd clusters on an Al_2O_3 film on Ni_3Al (111). ($V = 0.8$ V, $I = 8$ pA).

understand the origin of the nanostructurization and the existence of two domains rotated by 23°.[16]

The metal is deposited *in situ* at RT on the freshly prepared alumina films using two Knusdsen sources calibrated *in situ* by a quartz microbalance. Just after deposition the samples are imaged by STM. The height and the size distributions of the deposited particles are measured by manually counting at least 300 particles on the STM images.

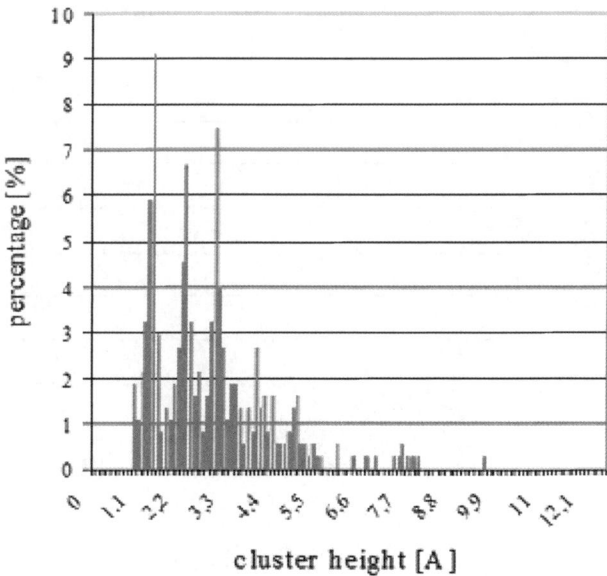

Fig. 3 Height distribution of Pd clusters on an Al_2O_3 film on Ni_3Al (111) obtained from STM images.

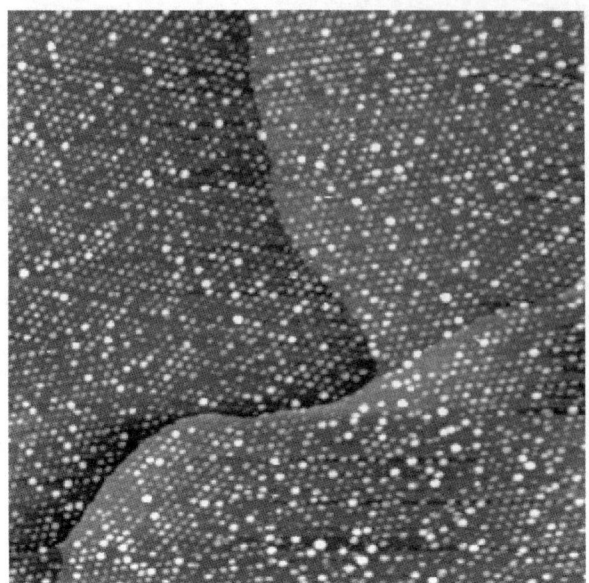

Fig. 4 STM image of an array of AuPd clusters on a Al_2O_3 film on Ni_3Al (111) obtained by successive deposition of Pd and Au (200×200 nm^2, $V = 0.8$ V, $I = 8$ pA).

3. Pd deposition

As previously shown,[14,17] Pd nucleates exclusively on the nodes of the 4.1 nm superstructure of the alumina film. Fig. 2 shows an STM image of a typical Pd deposit. Pd clusters form a regular hexagonal array extending on the whole image. On some pictures the two domains rotated by $24°$ are observed. The long range organisation of the metal clusters has been studied by grazing incidence small angle

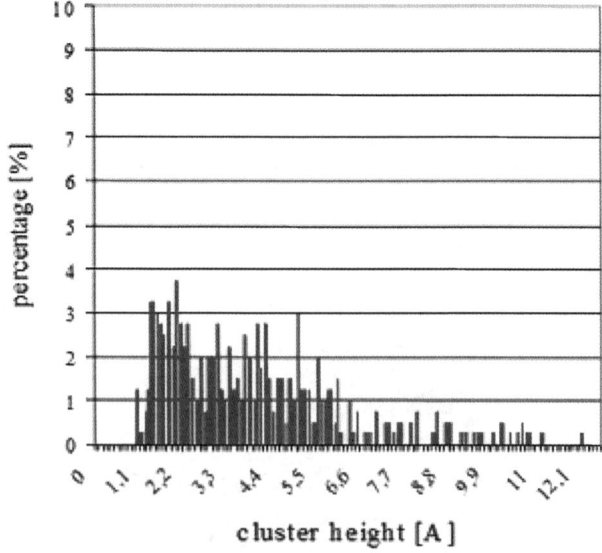

Fig. 5 Height distribution of AuPd clusters on a Al_2O_3 film on Ni_3Al (111) obtained from STM images.

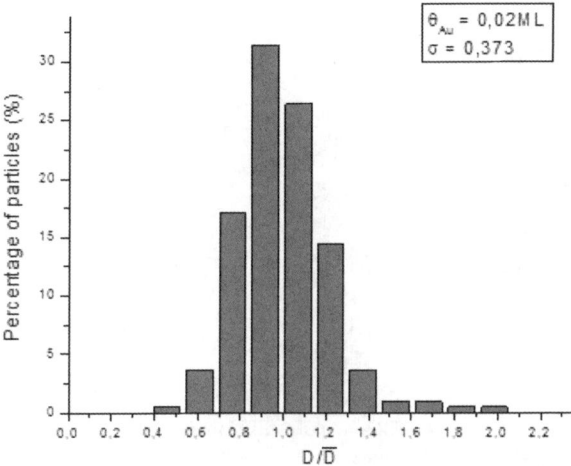

$\theta_{Au} = 0,02 ML$
$\sigma = 0,373$

Fig. 6 STM image and corresponding normalized size distribution of a 0.02 ML gold deposit on an Al_2O_3 film on Ni_3Al (111). (200×200 nm^2, $V = 0.8$ Volt, $I = 9$ pA). σ represents the size dispersion of the small clusters.

X-ray scattering (GISAXS) at ESRF in Grenoble. This study has shown that the cluster array extends on the whole 1 cm^2 sample.[18] The size distribution is narrow but due to the tip-cluster convolution in the STM images it is difficult to give accurate absolute size measurements. From the total amount of deposited Pd we know the average number of atoms per clusters: it has been varied from around 10 atoms to about 140 atoms corresponding to clusters of 2 nm in diameter. For larger clusters coalescence occurs and the organisation in the cluster array starts to decrease. Kinetic Monte Carlo simulation of the growth of the Pd clusters arrays is currently under study. The first results show that the size distribution is very narrow and close to the statistical limit.[19] Height distribution of the clusters is not corrupted by the tip–sample convolution effect. Fig. 3 shows a typical height

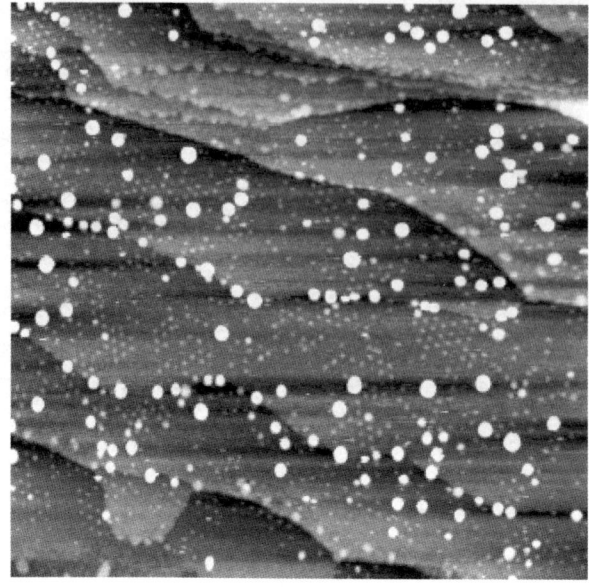

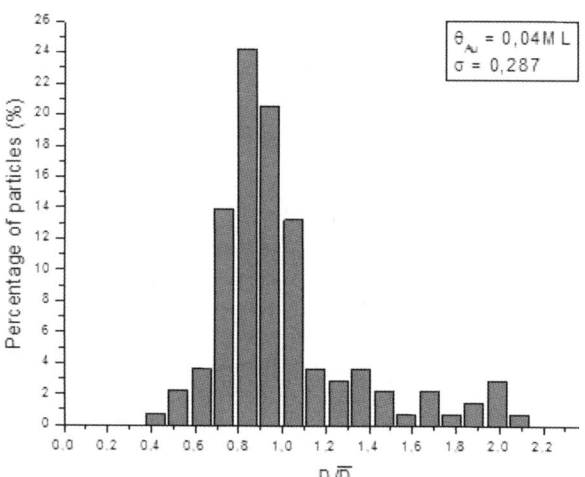

Fig. 7 STM image and corresponding size distribution of a 0.04 ML gold deposit on an Al_2O_3 film on Ni_3Al (111). (200 × 200 nm^2, V = 0.8 Volt, I = 9 pA). σ represents the size dispersion of the small clusters.

distribution of Pd clusters obtained from STM images. It is remarkable that the histogram is not monotonous but shows sharp peaks regularly distributed. They can be attributed to the successive atomic layers. The spacing between the peaks is around 0.11 nm. This value is smaller than the typical interdistance between (111) planes which is 0.24 nm. The measured shorter distance is probably due partly to an imperfect calibration of the z-piezo of the STM but also partly to an electronic effect as previously observed on Pd clusters supported on MoS_2.[20] On thick deposits the shape of the large (around 2 nm) Pd clusters appears clearly (image not shown) and corresponds to hexagonal truncated tetrahedra in agreement with STM observations for Pd clusters grown on alumina films on NiAl (110).[21]

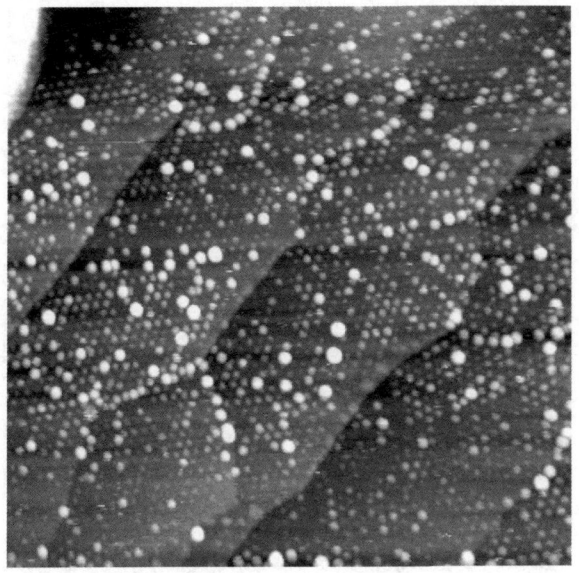

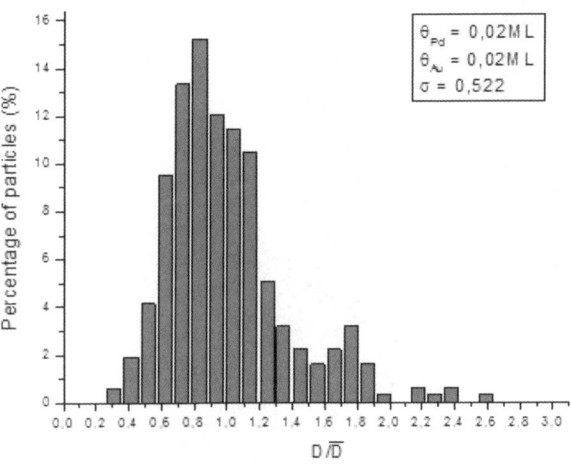

$$\theta_{Pd} = 0,02\,ML$$
$$\theta_{Au} = 0,02\,ML$$
$$\sigma = 0,522$$

Fig. 8 STM image and corresponding size distribution of a deposition of 0.02 ML of Pd on 0.02 ML of Au predeposited on the alumina film. (200×200 nm^2, $V = 1.49$ V, $I = 9$ pA). σ represents the size dispersion of the small clusters.

4. Au deposition on Pd clusters

We have shown in a previous study[14] that Au atoms deposited at RT on an array of Pd clusters go exclusively on the preformed Pd clusters and therefore no pure Au clusters are formed. This is a definite advantage of using a nanostructured substrate over a normal one where the nucleation centres, which are surface defects, are randomly distributed.[22] As long as the mean diffusion length of the gold adatoms is larger than the distance between two Pd clusters (*i.e.* 4.1 nm) all of them are captured by the preformed Pd clusters. Another advantage of this method is that the composition and the size of the clusters can be adjusted independently. Indeed the mean size is given by the total amount of deposited metal divided by the number of clusters while their chemical composition is given by the ratio of the two amounts of

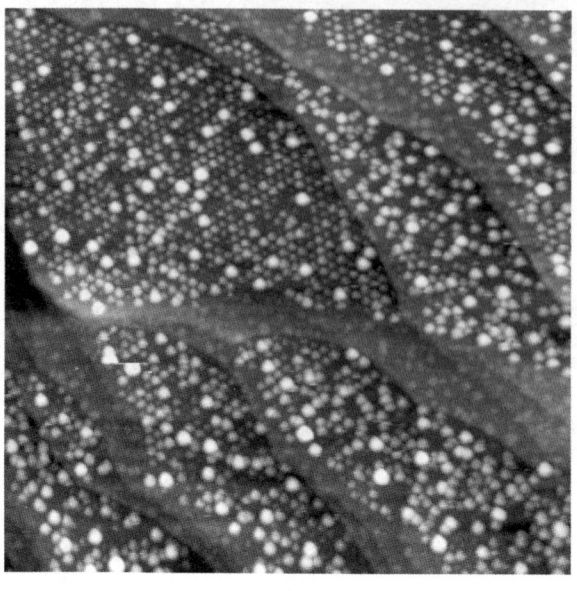

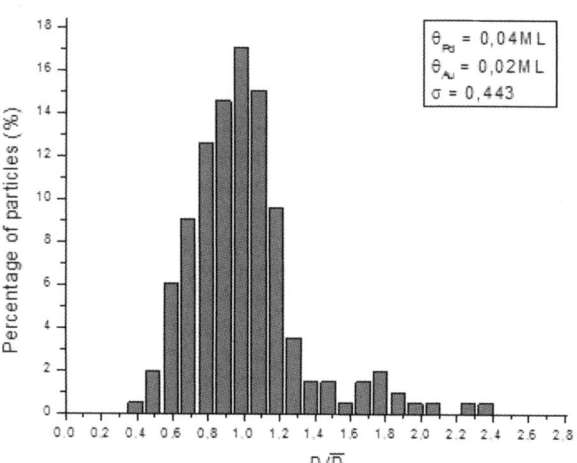

Fig. 9 STM image and corresponding size distribution of a deposition of 0.04 ML of Pd on 0.02 ML of Au predeposited on the alumina film. (200 × 200 nm², V = 1.49 V, I = 9 pA). σ represents the size dispersion of the small clusters.

deposited metals. Fig. 4 shows a typical STM image of an array of AuPd clusters. As in the case of pure Pd the AuPd clusters are regularly distributed forming a hexagonal array of 4.1 nm lattice parameter. We have also studied the long-range organisation by GISAXS.[18] The ordering in the cluster hexagonal array extends on the whole substrate as long as the size of the clusters is not too large to avoid coalescence. Fig. 5 shows a typical height distribution for AuPd clusters. Unlike the case of pure Pd clusters, the histogram is monotonous. If we had a layer-by-layer growth of Au on top of the Pd clusters we would also expect a modulated histogram (like on Fig. 3). The fact that the height distribution is monotonous could mean that the top layer of the clusters is not homogeneous but constituted of mixed Pd and Au atoms. Indeed Au and Pd are miscible in any proportion.[23]

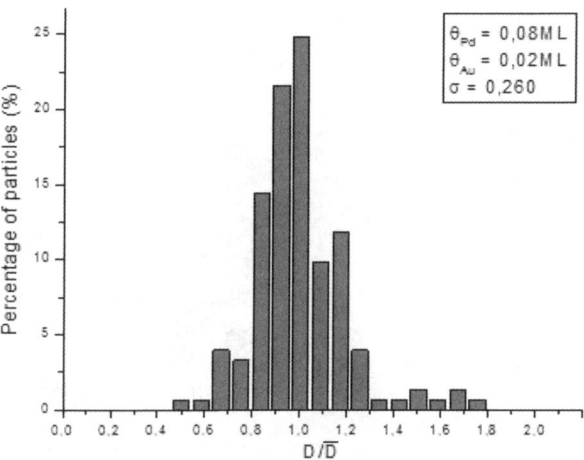

Fig. 10 STM image and corresponding size distribution of a deposition of 0.08 ML of Pd on 0.02 ML of Au predeposited on the alumina film. (200×200 nm², $V = 1.49$ V, $I = 9$ pA). σ represents the size dispersion of the small clusters.

5. Au deposition

We have studied the nucleation of gold on the nanostructured alumina films. Fig. 6 and 7 show STM images and size distributions of two gold deposits. In Fig. 6, corresponding to 0.02 ML of Au, a large majority of small clusters with a few larger clusters are present. A careful analysis of the image shows that all the small clusters sit on the nodes of the 4.1 nm superstructure of the alumina film. Another remarkable feature is that only a fraction of the sites are occupied by gold clusters. By increasing the amount of gold to 0.04 ML (Fig. 7) more sites are occupied but also a bigger proportion of large clusters are present. A detailed study of the nucleation of gold which will be published elsewhere[24] has shown that the proportion of large clusters increases with coverage but saturates at nearly 25%. The proposed explanation for the different behaviour of gold in comparison with Pd is

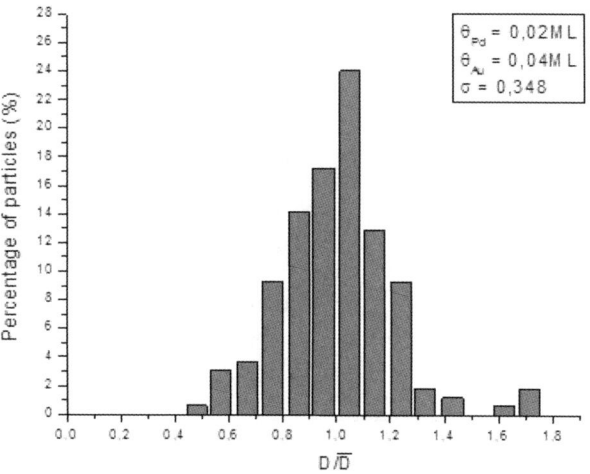

Fig. 11 STM image and corresponding size distribution of a deposition of 0.02 ML of Pd on 0.04 ML of Au predeposited on the alumina film. (200×200 nm^2, $V = 1.49$ V, $I = 9$ pA). σ represents the size dispersion of the small clusters.

based on a weaker interaction for Au relative to Pd on the alumina films. Briefly, Au always nucleates on the nodes of the 4.1 nm superstructure but at a given size the gold clusters can escape from the nucleation centres and diffuse on the substrates before coalescing with other gold clusters. Thus, by increasing the density of clusters the number of coalescence events increases. The size distribution is larger for the thicker deposit but in the two cases (Fig. 6 and 7) the size dispersion for the small clusters remains the same (around 30%).

6. Pd deposition on Au clusters

Using deposition of Pd on preformed Au clusters we have attempted to grow PdAu clusters. Fig. 8, 9 and 10 show a series of STM pictures and size distributions

Fig. 12 STM image of a deposition of 0.04 ML of Pd on 0.04 ML of Au predeposited on the alumina film. (200×200 nm^2, $V = 1.49$ V, $I = 9$ pA).

corresponding to increasing coverage of Pd on a 0.02 ML deposit of Au clusters represented on Fig. 6. After the first Pd deposition (Fig. 8) corresponding to 0.02 ML, it appears that two types of clusters are present, large ones and small ones. This fact is reflected in the size distribution. Despite the fact that the cluster size are not accurately known, due to the tip–cluster convolution in the STM measurements, we can qualitatively follow the evolution of the size distribution with the Pd coverage. By increasing the amount of deposited palladium, the same qualitative behaviour is observed (Fig. 9), but after deposition of 0.08 ML of Pd, the size distribution of the clusters appears more homogeneous (see Fig. 10). In fact on the size distribution the peak at larger sizes almost vanishes and the width of the total size distribution is strongly reduced. We have deposited Pd on a thicker deposit of gold (0.04 ML, Fig. 7). The STM images and size distributions corresponding to increasing amounts of Pd are displayed in Fig. 11 to 14. At low Pd coverage, as in the previous case, a double-peak size distribution is observed (Fig. 11). After a Pd deposition larger than 0.04 ML the second peak disappears and the size distribution narrows (Fig. 13).

Fig. 15 shows the evolution of the density of clusters as a function of the amount of deposited Pd on the two preformed collections of gold corresponding to Fig. 6 and 7. The density of clusters increases with the amount of deposited Pd up to the point where all the nodes of the 4.1 nm superstructure of the alumina film are covered (saturation density: 6.5×10^{12} clusters cm^{-2}). The new formed clusters are in fact pure Pd clusters. At the beginning of the Pd deposition the large clusters correspond to PdAu bimetallic clusters whereas the small ones correspond mainly to pure Pd clusters. Contrary to the case of the deposition of Au on Pd, the deposition of Pd on preformed gold clusters leads to an inhomogeneous collection of clusters: some contain only Pd and the others are bimetallic. This behaviour comes probably from a shorter diffusion length of the Pd clusters resulting in the possibility of nucleation of pure Pd clusters. This is only allowed by the fact that originally a large fraction of the nodes of the superstructure are not occupied by gold clusters. The sharpening of the size distribution after deposition of a large amount of Pd is firstly due to the fact that a large majority of clusters are pure Pd clusters and secondly to the fact that the proportion of gold atoms in the bimetallic clusters is small.

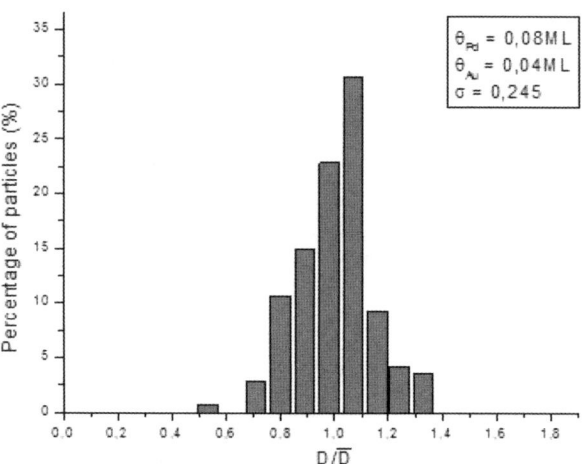

$\theta_{Pd} = 0,08\,ML$
$\theta_{Au} = 0,04\,ML$
$\sigma = 0,245$

Fig. 13 STM image and corresponding size distribution of a deposition of 0.08 ML of Pd on 0.04 ML of Au predeposited on the alumina film. (200 × 200 nm^2, V = 1.49 V, I = 9 pA). σ represents the size dispersion of the small clusters.

7. Conclusion and perspectives

Nearly perfect arrays of AuPd bimetallic clusters supported on alumina have been obtained by successive deposition of Pd and Au on a nanostructured film of alumina obtained by oxidation at high temperature of a Ni$_3$Al (111) surface. Pd nucleates only on the nodes of the hexagonal superstructure of the alumina films with a lattice parameter of 4.1 nm. Due to the regular distribution of the nucleation centres the growth rate of the clusters is uniform and the resulting size distribution is very sharp. During gold deposition, the gold adatoms rapidly diffuse on the substrate to be captured by the preformed Pd clusters. In the growth conditions of the experiments homogeneous nucleation of pure gold clusters is not possible, because two gold adatoms have no chance to collide before they meet a preformed Pd cluster. In that

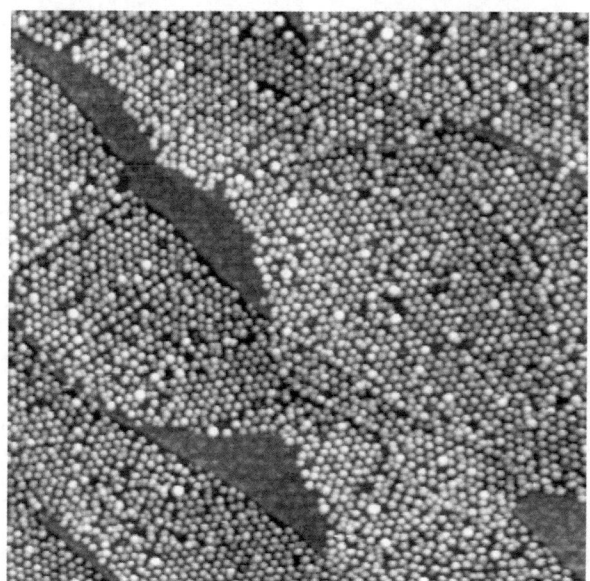

Fig. 14 STM image of a deposition of 0.16 ML of Pd on 0.04 ML of Au predeposited on the alumina film. (200×200 nm^2, V = 1.49 V, I = 9 pA).

way, almost perfect bimetallic model catalysts are obtained. The size and the composition of the bimetallic clusters can be controlled independently by the total amounts of deposited metals and by their ratio, respectively. Then, it becomes possible to study the reactivity of bimetallic clusters as a function of the composition at a given size. The array of the bimetallic clusters is very regular and extends over the whole substrate (1 cm^2). The regular organisation of the clusters is not only important for the homogeneity of their size and composition but also for the study of catalytic reactions. In fact, it is now well established that the supply of reactants through surface diffusion on the catalyst support depends strongly on the cluster spatial distribution and their size, this phenomenon is called reverse-spillover.[25] So in order to study size effect in catalysis it is necessary to quantitatively know this

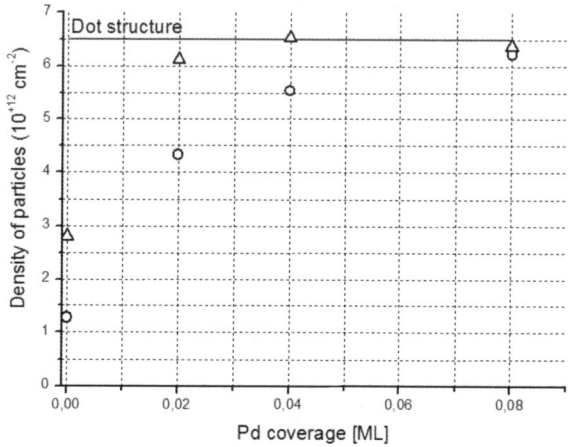

Fig. 15 Variation of the density of clusters as a function of the Pd deposited amount. Circular points correspond to 0.02 ML of predeposited Au and triangular points to 0.04 ML predeposited Au.

effect.[26] In the case of a random distribution of clusters it is difficult to model the diffusion of reactants whereas in the case of a regular array of clusters with a sharp size distribution an exact analytical solution of the problem is available.[27]

The deposition of Pd on preformed gold clusters leads to an inhomogeneous distribution of clusters. Both bimetallic and pure Pd are present. This result is mainly due to the weaker interaction of gold adatoms with the nucleation centres of the alumina films compared to palladium atoms. Then, not all the sites of the nanostructure are occupied by gold clusters even at high coverage because small gold clusters can escape from the nucleation sites and coalesce with other gold clusters. It is expected, from Monte Carlo simulations, that by decreasing temperature all the nucleation centres could be occupied by gold clusters. Experiments in this direction are in progress.

Acknowledgements

M. Dayez and C. Claeys are gratefully acknowledged for the technical support during this work.

References

1   J. H. Sinfelt, *Bimetallic Catalysts: Discoveries, Concepts and Applications*, Wiley, New York, 1983.
2   F. Besenbacher, I. Chorkendorff, B. S. Clausen, B. Hammer, A. M. Molenbroek, J. K. Nøsrkov and I. Stesgaard, *Science*, 1998, **279**, 1913.
3   W. D. Provine, P. Mills and J. J. Lerou, *Stud. Surf. Sci. Catal.*, 1996, **101**, 191.
4   J. C. Bertolini, *Surf. Rev. Lett.*, 1996, **5&6**, 1857.
5   P. Hermann, J. M. Guignier, B. Tardy, Y. Jugnet, D. Simon and J. C. Bertolini, *J. Catal.*, 1996, **163**, 169.
6   (*a*) S. Sao Joao, S. Giorgio, J. M. Penisson, C. Chapon, S. Bourgeois and C. R. Henry, *J. Phys. Chem. B*, 2005, **109**, 342; (*b*) S. Sao-Joao, *PhD Thesis*, University of Aix-Marseille II, 2005.
7   B. Hammer and J. K. Nørskov, *Adv. Catal.*, 2000, **45**, 71.
8   M. Neurock, *J. Catal.*, 2003, **216**, 73.
9   F. Baletto, C. Mottet and R. Ferrando, *Phys. Rev. B*, 2002, **66**, 155420.
10 C. R. Henry, *Prog. Surf. Sci.*, 2005, **80**, 92.
11 A. K. Santra, F. Yang and W. D. Goodman, *Surf. Sci.*, 2004, **548**, 324.
12 M. Heemeier, A. F. Carlsson, M. Naschitzki, M. Schmal, M. Bäumer and H. J. Freund, *Angew. Chem., Int. Ed.*, 2002, **41**, 4073.
13 F. Gimenez, C. Chapon and C. R. Henry, *New J. Chem.*, 1998, **22**, 1289.
14 G. Hamm, C. Becker and C. R. Henry, *Nanotechnology*, 2006, **17**, 1943.
15 S. Degen, A. Krupski, M. Krajl, A. Langner, C. Becker and K. Wandelt, *Surf. Sci.*, 2005, **576**, L57.
16 G. Hamm, C. Barth and C. R. Henry, *Phys. Rev. Lett.*, 2006, **97**, 126106.
17 S. Degen, C. Becker and K. Wandelt, *Faraday Discuss.*, 2003, **125**, 343.
18 Unpublished results.
19 G. Sitja, R. O. Unac and C. R. Henry, to be published.
20 A. Piednoir, E. Perrot, S. Granjeaud, A. Humbert, C. Chapon and C. R. Henry, *Surf. Sci.*, 1997, **391**, 19.
21 K. Højrup-Hansen, T. Worren, E. Laegsgaard, M. Bäumer, H. J. Freund, F. Besenbacher and I. Stensgaard, *Phys. Rev. Lett.*, 1999, **83**, 4120.
22 C. R. Henry, *Surf. Sci. Rep.*, 1998, **31**, 231.
23 *Binary alloy Phase Diagrams*, ed. H. Okamoto, R. O. Subramanian and L. Kacpraz, ASM International, Materials Park, Ohio, 2nd edn, 1992, vol. 1.
24 A. Wörz, M. Marsault, G. Hamm and C. R. Henry, submitted.
25 C. R. Henry, in *The Chemical Physics of Solid Surfaces Surface Dynamics*, ed. D. P. Woodruff, Elsevier, 2003, vol. 11, ch. 9, p. 247.
26 L. Piccolo and C. R. Henry, *J. Mol. Catal. A: Chem.*, 2001, **167**, 181.
27 C. R. Henry, *Surf. Sci.*, 1989, **223**, 519.

General Discussion

Professor Henry opened the discussion of Professor Kiely's paper: What is the limit of sensitivity of the chemical analysis? In other words, is it possible to detect the presence of one monolayer segregated on an homogeneous nanoparticle alloy?

Professor Kiely answered: To date, the best spatial resolution we have achieved with the STEM-XEDS spectrum imaging technique on samples having spherical nanoparticle geometry is to detect a 1.5 nm thick-shell on an approximately 7 nm diameter particle. Below this particle size we just tend to observe spatially co-incident X-ray emission signals from both metallic components and we lose the ability to resolve any surface segregation effects. However, on planar geometry specimens (*i.e.* grain boundaries in thin metal foils) we have recently demonstrated a STEM-XEDS spatial resolution down to 0.4 nm in our aberration corrected VG HB603 instrument.[1]

1   M. Watanabe, D. W. Ackland, A. Burrows, C. J. Kiely, D. B. Williams, O. L. Krivanek, N. Dellby, M. F. Murfitt and Z. Szilagyi, *Microsc. Microanal.*, 2006, **12**, 515–526.

Professor Sir Thomas asked: Your EDS work is very impressive. Yesterday you drew our attention to the advantages of EELS chemical analysis. Could you comment on whether EELS would give you even greater sensitivity in studying the Au–Ag or Au–Pd core–shell structures?

Professor Kiely answered: Thank you for the kind comments regarding our STEM-XEDS spectrum imaging experiments. We have indeed considered doing complementary STEM-EELS spectrum imaging experiments on the Au–Ag and Au–Pd bimetallic particles. Unfortunately it turns out to be extremely difficult because the various EELS transitions for Au fall at very inconvenient positions in the EELS spectrum. The low energy-loss O2,3 and N6,7 edges for Au occur in the 65–85 eV range and lie on the trailing edge of the strong plasmon peak. The M4,5 and M3 edges for Au lie between 2200 and 2800 eV which is beyond the practical operation range of our EELS spectrometer system due to the very low signal generated at such large energy-loss values. Hence for Au–Ag and Au–Pd systems, the potential higher sensitivity and improved spatial resolution offered by EELS cannot be capitalized upon because of the unfortunate locations of the characteristic Au absorption edges. However, it is well worth noting that in other non Au-based systems where the absorption edges are more conveniently located, STEM-EELS spectrum imaging experiments can be carried out. Furthermore the STEM-EELS data cube acquired can be analyzed and manipulated using the same MSA methodology employed for STEM-XEDS spectrum images.[1] Indeed in my laboratory we are currently attempting to simultaneously acquire STEM-XEDS and STEM-EELS spectrum images from other types of materials.

1   M. Watanabe, M. Kanno, D. W. Ackland, C. J. Kiely and D. B. Williams, *Microsc. Microanal.*, 2007, **13**(2), 1264 CD.

Professor Sir Thomas commented: It was remarkable to observe your result that you never get Au–Pd core–shell structures when you use carbon as support (as distinct from TiO_2 or Al_2O_3). There are very many different kinds of carbon *e.g.* graphite, carbon black, coconut charcoal *etc.* I would be surprised if graphite behaved in the same way (as a support) as, say, more amorphous variants.

Professor Kiely answered: It was indeed surprising to discover that the Au–Pd core shell morphology only developed upon calcination on oxide supports and not on carbon. To date we have only prepared the AuPd alloy particles on amorphous activated carbon (*e.g.* G60 type) supports. It is an excellent suggestion to repeat the preparation on a graphite support, to see if the development of the core–shell morphology is affected by the degree of crystallinity of the underlying carbon.

Dr Russell commented: Do you find the same Au : Pd ratio in individual particles on each support? (*e.g.* comparing the composition of the core/shell particles on the oxide supports and the mixed/alloy particles on the carbon).

Professor Kiely replied: In general we do a find similar a composition trend on all the supports irrespective of whether they are oxides or carbon. The largest particles always tend to be Au-rich whereas the smallest ones tend primarily to be Pd-rich. Only those particles in the intermediate size range (*i.e.* about 10 nm) exhibit the nominal 1 : 1 Au : Pd ratio. It should be noted however that the resultant particle size distribution observed can be strongly influenced by the precise identity of the support[1] and indeed the chemical treatment of a support by acid washing prior to metal deposition.[2]

1   J. K. Edwards, A. F. Carley, A. A. Herzing, C. J. Kiely and G. J. Hutchings, *Faraday Discuss.*, 2008, **138**, DOI: 10.1039/b705915a
2   J. K. Edwards, B. Solsana, T. Garcia, P. Landon, A. F. Carley, G. J. Hutchings, A. A. Herzing and C. J. Kiely, "Switching Off Hydrogen Peroxide Hydrogenation in the Direct Synthesis Process", 2007, submitted for publication.

Professor Ricolleau remarked: Has the Pd enrichment of Au–Pd nanoparticles been observed on the as-grown particles? What is the influence of the beam on the composition? Is evaporation of Pd of the smallest nanoparticles under beam irradiation possible or not?

Professor Kiely responded: The Au–Pd bimetallic particles which have only been dried at 120 °C never exhibit core–shell morphologies irrespective of their support material. However, upon subsequent calcination we find that those AuPd particles resting on oxide supports do have a strong tendency to form the Pd-rich shell/Au-rich core morphology.[1] We have not encountered any significant problems with preferential loss of Au or Pd from our bimetallic particles during irradiation under the electron beam. A more significant problem is the gradual loss of the lighter support material during exposure to the intense aberration corrected probe. It is possible however to induce structural re-configurations of the smaller metallic particle under conditions of prolonged electron beam irradiation.

1   A. A. Herzing, A. F. Carley, J. K. Edwards, G. J. Hutchings and C. J. Kiely, "Microstructural Development and Catalytic Performance of Au–Pd Nanoparticles on Al$_2$O$_3$ Supports: The Effect of Heat Treatment, Temperature and Atmosphere", submitted for publication.

Professor Bond remarked: It is possible to envisage mechanisms to explain the dependence of composition on the particle size observed with PdAu/C catalysts. First, the breadth of the size distribution (Fig.3(a)) reflects the size distribution of pores in the support, because, for catalysts made by impregnation where interaction between the metal ions and the surface is weak, each particle results from a volume of solution residing in an isolated pore.[1] It is therefore of interest to examine the location of metal or alloy particles within the support grain, because normally the pore diameter decreases on passing from the surface of the grain towards the interior. The largest particles are therefore expected to be close to the surface and the smallest in the centre. It is therefore relevant to measure the pore size distribution of the support to see if it can provide a model for the alloy size distribution.

Now it is also possible with bimetallic systems for there to be a chromatographic separation of the ions as they diffuse through the liquid-filled pores, so that at the stage of drying where independent microvolumes of solution become isolated the relative concentrations of the two components may depend on the location within the pore structure. Such separation is well known to occur with oxide supports,[2] and is easily detected by scanning across a support grain using EDX; it may be controlled by adjusting the pH or by other competing ions. It would however appear from Fig. 3(a) that most of the particles will have compositions close to the nominal. The particle size distribution for the $PdAu/TiO_2$ is much narrower because the support is non-porous and a different deposition mechanism operates.

The control of the location of the active component within the support grain can be extremely important, depending on the phase in which the catalyst exists during reaction, because diffusion within liquid-filled is relatively slow, so rates are improved and secondary reactions are minimised if the active phase is located close to the exterior surface of the support granule.

1   G. C. Bond, *Heterogeneous catalysis, Principles and Applications*, 2nd edn., Oxford University Press, Oxford, 1987, p. 79
2   G. C. Bond and V. Ponec, *Catalysis by Metals and Alloys*, Elsevier, Amsterdam 1996, ch. 7.

Professor Kiely replied: Professor Bond has proposed some very interesting and plausible mechanisms for explaining the systematic particle size/composition variations we have identified with our STEM-XEDS spectrum imaging experiments. We can in principle test the proposition that the larger Au-rich particles are preferentially located at the surface of the oxide particle aggregates, whereas the smaller Pd-rich particle are trapped in the interior interparticle voids by using the technique from 3-D electron tomography. This method has been recently used to great effect by Dr Midgely and Professor Sir Thomas to characterize the spatial distribution of metal particles within zeolitic matrix grains.[1] Attempting to correlate tomographic information with void size distribution measurements and our STEM-XEDS particle size/composition data would indeed be a valuable and fruitful exercise to perform. Some evidence in support of the possibility that the pore size distribution of the support plays a key role in determining the eventual size of the Au–Pd nanoparticles can be provided in our earlier paper.[2] The particle size distribution for the Au–Pd/TiO$_2$ catalyst is considerably broader and also displays much larger particles than that of the Au–Pd/carbon catalyst. This is consistent with the surface areas of the two supports since the carbon has a surface area that is an order of magnitude higher that of the TiO$_2$ utilized, and hence has considerably smaller pores present.

1   P. A. Midgley, J. M. Thomas, L. Laffont, M. Weyland, R. Raja, B. F. F. Johnson, and T. Khimyak, *J. Phys. Chem. B*, 2004, **108**(15), 4590.
2   J. K. Edwards, A. F. Carley, A. A. Herzing, C. J. Kiely and G. J. Hutchings, *Faraday Discuss.*, 2008, **138**, DOI: 10.1039/b705915a

Dr Geng asked: What is the reason for larger particles being Au rich but for smaller ones being Pd rich?

Is it possible that this compositional difference resulted from separate nucleations? This is to say that, the larger particles nucleated from Au but smaller ones from Pd?

Professor Kiely replied: This is a very intriguing question which we are currently trying to resolve. As noted in my response to Dr Russell's question we do a find similar a size-composition trend on all the supports irrespective of whether they are oxides or carbon. The largest particles always tend to be Au-rich whereas the smallest ones tend primarily to be Pd-rich. We believe that this trend is driven by the initial metal element dispersion profile and subsequent particle coarsening phenomena. Recent experiments[1] on uncalcined (*i.e.* just dried) specimens of AuPd on Al$_2$O$_3$

suggest that much of the Pd component is highly (maybe even atomically) dispersed over the support, whereas the Au component does not wet the support so well and has a strong tendency to nucleate as Au-rich particles from the outset. Subsequent calcination causes the highly dispersed Pd component to coarsen and form very small particles that just contain a trace amount of Au, whereas the larger Au-rich particles coarsen and incorporate any directly adjacent Pd.

1 A. A. Herzing, A. F. Carley, J. K. Edwards, G. J. Hutchings and C. J. Kiely, "Microstructural Development and Catalytic Performance of Au–Pd Nanoparticles on Al_2O_3 Supports: The Effect of Heat Treatment, Temperature and Atmosphere", submitted for publication.

Professor Mattei asked:

(1) When considering clusters quite large in size (about 50 nm, as the ones shown in the paper), is it possible that effects like X-ray absorption within the nanocluster can affect the composition quantification due to a preferential absorption of the lighter element X-ray emission?

(2) Regarding the quantification of the alloy composition, are the measured EDS spectra pre-processed (background subtraction, absorption correction...) before the MSA procedure is started?

Professor Kiely answered:

(1) Indeed it is probable that the larger bimetallic particles do exhibit some absorption of the Pd L series X-rays by the Au component. The particle compositions we quote in the paper have been determined using a standardless Cliff–Lorimer type quantification procedure using theoretical k-factor values and ignoring the effects of absorption and fluorescence. This method does reveal the broad trends in particle composition with size, but the composition results should not by any means be treated as absolute determinations of particle composition. It should be noted that a much better approach to XEDS quantification, which we now routinely employ, is the new zeta-factor method introduced by Watanabe and Williams,[1] which requires the microscopist to measure the sample thickness and probe current at the time of the experiment.

(2) The MSA analysis described in our paper was performed on the spectrum image data cube without any prior background subtraction or absorption correction. Indeed, one of the great advantages of the technique is that no *a priori* knowledge of the sample or data on the part of the microscopist is necessary. The processed data cube can be analyzed after MSA reconstruction and further data improvement such as background subtraction, absorption correction, and compositional determination can then be performed.

1 M. Watanabe M. and D. B. Williams, *J. Microsc.*, 2006, **221**, 89–109.

Professor Johnston opened the discussion of Dr Mejia's paper: In the paper, apparently the composition of the particle shown in Fig. 7 is approximately 20% Au, 80% Pd. However, your STEM image and profile simulations (Fig. 6) appear to be for 50 : 50 composition. Is it certain that the experimental particle is alloyed, or could it have a core of one metal surrounded by an alloyed shell?

Dr Mejia answered: By the studies made with HAADF-STEM in single particles we had (maybe not conclusive) evidence of alloying in the whole of the particle. Later on, ultra high-resolution HAADF-STEM images (not included in the paper but shown in the discussion) confirmed this observation. It is not possible to discard the possibility of a core of one metal surrounded by an alloy, at least not from the STEM simulations, but this situation is unlikely, given the synthesis process.

Professor Lievens asked: How was the composition of atoms (after cluster preparation and deposition) investigated, and how does this relate to the composition of the targets that were used for the synthesis?

Dr Mejia responded: The EDX measurements were performed more in a qualitative fashion, with the purpose of verifying the presence of both atomic species and to check the possible presence of contaminants. We believe that it is necessary to perform more quantitatively reliable measurements to have a fair comparison with the composition of the target.

Professor Ferrando asked: Does the icosahedral shape of your AuPd clusters prevail for all compositions you considered?

Do you know whether coalescence is occurring during the formation stage of clusters in your chamber?

Dr Mejia responded: Response to the first part of the question: With the method described in this paper the prevailing shape of the particles was the icosahedral. As far as we could resolve with STEM methods, all the particles have I_h symmetry including the ones with diameter ~ 1 nm. This is in sharp contrast with colloidal methods which give cuboctahedral particles. In the paper we just included results for particles generated from a 50 : 50 Au/Pd target, and the production of particles with other compositions is a work in progress.

Response to the second part of the question: The production of nanoparticles can be explained by the occurrence of three mechanisms: aggregation or attachment of atoms around small clusters (dimers, trimers, *etc.*); coagulation of nanoparticles; and a pure coalescence process (which takes place preferentially at high temperatures, see ref. 1). The competition between these three mechanisms will define the final size and shape distributions. In our case it is likely that the first mechanism dominates over the other two, given the synthesis conditions.

1   I. Shyjumon, Ph.D. Thesis, University of Greifswald, 2005.

Professor Henry asked: Comment to the former question about coalescence on the substrate: As the size distribution obtained by mass spectrometry in the beam is even larger than the size distribution obtained after deposition by TEM, coalescence is unexpected?

Dr Mejia answered: The deposition was made at a rate sufficiently low to allow the landing of isolated nanoparticles, minimizing the coalescence in the substrate. Nevertheless, there were some small regions in the substrate with aggregates of particles. These aggregates were not considered in the statistics. The size distribution obtained by TEM is narrower than that of the mass spectrometer mostly because the TEM measurements are more exact than those of the spectrometer; the main purpose of the mass spectrometer in the system is to act as a mass filter, and the size estimates were intended to serve as a guideline for the control of the deposition parameters.

Professor Leiva remarked: You have shown in your contribution that the Au/Pd nanoparticles fabricated by inert-gas condensation on a sputtering reactor show no evidence of a core–shell structure, but are rather formed by an alloy which has a homogeneous composition. In this respect, it is interesting to point out that the cluster collision method that we simulated yielded (see ref. 1), for the collision of Au and Pd particles, a Pd/Au alloyed cluster in a wide range of initial velocities (10 m s^{-1} and 500 m s^{-1}). Furthermore, according to the results of Table 1 of our contribution, it appears that Pd and Au atoms have a strong tendency to mix with each other.

1   M. M. Mariscal, N. A. Oldani, S. A. Dassie and E. P. M. Leiva, *Faraday Discuss.*, 2008, **138**, DOI: 10.1039/b706149h

Dr Mejia replied: It is interesting that your results support the idea of a strong tendency of Au and Pd to form alloys. On the other hand, we have made simulations of the melting of Au–Pd nanoparticles, where we found that, if there is enough thermal energy available, Au tends to migrate to the surface of the particle.[1,2] I think this is not contrary to your results, but it only shows that the energy considerations are not the only important issues to predict the kind of mixing (core–shell A–B *vs.* core–shell B–A *vs.* homogeneous alloy), but we have to consider the kinetic details of the synthesis process.

1   Sergio J. Mejía-Rosales, Carlos Fernández-Navarro, Eduardo Pérez-Tijerina, Juan Martín Montejano-Carrizales, and Miguel José-Yacamán, *J. Phys. Chem. B*, 2006, **110**(26), 12884–12889.
2   Sergio J. Mejía-Rosales, Carlos Fernández-Navarro, Eduardo Pérez-Tijerina, Douglas A. Blom, Lawrence F. Allard, and Miguel José-Yacamán, *J. Phys. Chem. C*, 2007, **111**, 1256–1260.

Professor El-Shall asked:

(1) Knowing the conditions of your inert condensation source such as temperature, degree of supersaturation and carrier gas pressure, it may be possible to use nucleation theory to compare the behaviors of Au and Pd and to check if the Au–Pd nanoparticles are formed by a binary nucleation mechanism or by heterogeneous nucleation of one component over a core nucleus of the other component. Have you examined this possibility?

(2) According to your conclusions, the method allows a strict control of the size, composition and shape of the particles. What are the experimental parameters that control the size and shape? In other words, what are the conditions needed to produce small particles or non-spherical shapes?

Dr Mejia answered:

(1) The short answer is not yet. On the other hand, we don't expect to find the nucleation of one of the species before the other one, since both elements are being sputtered at the same rate from a target composed by an homogeneous alloy, and, as Dr Leiva points out, the mixing of these two species is energetically favourable, which under these synthesis conditions must be specially important.

(2) The size is controlled by an interplay of magnetron power, the gas flow, and, the condensation zone length, the latter being the most important. It is not that the shape is being controlled, but the size of the particle, along with the relative concentration of the atomic species, determines the shape of the particles. Particles in the range of 1 to 5 nm have a tendency to be icosahedron in shape, at least for the composition of the target that we used. Apparently, by varying the temperature on the condensation zone it is possible to change the shape of the particles. We are doing a set of experimental runs to investigate this dependency.

Professor Kiely asked: I would like to add a remark about why Pd segregates to the surface in AuPd bimetallic particles. Theoretical predictions on isolated particles (in a vacuum) suggest that Au should move to the surface, and that the Pd should stay in the core. We have evidence from XPS[1] and energy filtered transmission electron microscopy (EFTEM)[2] that the driving force for the observed surface segregation of Pd is related to the fact that the bimetallic particles can interact with the oxygen present in the air. We detect the presence of Pd^{2+} from XPS and can image a shell of oxygen in EFTEM that spatially correlates with the Pd signal. Both of these observations suggest that Pd has a strong propensity to react with oxygen to generate a thin layer of PdO on the surface of the AuPd bimetallic particle.

1 A. A. Herzing, A. F. Carley, J. K. Edwards, G. J. Hutchings and C. J. Kiely, "Micro-structural Development and Catalytic Performance of Au–Pd Nanoparticles on Al₂O₃ Supports: The Effect of Heat Treatment, Temperature and Atmosphere", submitted for publication.
2 A. A. Herzing, Ph.D. Thesis, Lehigh University, 2007.

Professor Fortunelli asked: The use of plasmon resonance to force particles into thermodynamically more stable structures. Is that possible or sensible?

Dr Mejia responded: Yes, it is possible to use plasmon resonance to send the particles to lower energy configurations, using for example a high energy laser. This is something that actually we are planning to include in our sputtering system.[1,2]

1 Cecilia Noguez, *J. Phys. Chem. C*, 2007, **111**(10), 3806–3819.
2 Iván O. Sosa, Cecila Noguez, and Rubén G. Barrera, *J. Phys. Chem. B*, 2003, **107**(26), 6269–6275.

Professor Broyer commented: What is the phase diagram of the Pd–Au system in the bulk? Are these alloys existing in the bulk? If alloys exist, this may explain the particle composition and the absence of core shell.

Dr Mejia replied: There indeed exists a phase diagram for Pd–Au, and the alloy exists in the bulk, but this is not enough to explain the alloy nanoparticles, since it is possible to produce core–shell structures with other methods.

Professor Leiva opened the discussion of Dr Li's paper: Could you deposit your nanoparticles on a Ag substrate?

(The answer was yes) Comment: Then, you could take advantage of the so-called "underpotential deposition" electrochemical phenomenon in order to identify the nature of your core–shell particles. If you deposit your particles on a Ag substrate and then polarize it in a Cu^{2+} containing solution, you could get Cu decoration of the Ag(core)–Au(shell) nanoparticles at potentials more positive than the Cu/Cu^{2+} Nernst reversible potential. This would not be the case for the Au(core)–Ag(shell) nanoparticles, since Cu is known to deposit on Ag only at overpotentials. Furthermore, from the charge related to the underpotential deposition of Cu on the Ag(core)–Au(shell) nanoparticles you could even quantify the Au area exposed to the electrolyte.

Dr Russell communicated: Follow up to comment by Professor Leiva to Dr Li in discussion of her paper:

The electrochemical characterisation by upd (underpotential deposition) of Cu onto the AuAg particles is not likely to work because the particles in this case are protected by the surfactant used to prepare the micelles from which the particles are formed.

Professor Jellinek commented: Photoemission spectra of core–shell Ag/Au nano-particles[1] display two peaks. Your photoabsorption spectra of core–shell Ag/Au nanoparticles of approximately the same size have one peak. It would be of interest to identify the possible reasons for this difference.

1 W. Benten, N. Nilius, N. Ernst and H.-J. Freund, *Phys. Rev. B*, **72**, 2005, 045403.

Professor Ferrando opened the discussion of Dr Langlois' paper: For large AgCu nanoparticles, which are of the fcc structure, you observe that Ag does not completely cover Cu, so that your clusters are not really core–shell. You attribute this finding to a kinetic effect involving the different growth rates of Ag on the facets of a truncated octahedron. I expect that the fast growing facets of a truncated octahedron are the (001) facets.

Therefore, you should observe the formation of complete octahedra.[1] Are these structures present in your samples?

1. F. Baletto, C. Mottet, R. Ferrando, *Surf. Sci.*, 2000, **446**, 31.

Dr Langlois responded: According to your simulations, with a temperature of 700 K, we should observe complete octahedra for copper. However, except the icosahedra and decahedra, the shapes that we observe for the defect-free fcc Cu particles are not very well defined or clearly truncated octahedra. After the deposition of silver, we do not observe a significant difference concerning the morphologies, and truncated octahedral do not turn into complete octahedra, even after the growth of silver on the Cu cores. Maybe the size of the nanoparticles as well as the fact that they are supported by a substrate can explain the differences between our results and your simulations?

Dr Baletto asked:
(1) Did you try the deposition of Cu on Ag?
(2) Did you observe diffusion properties of simple species and of clusters?
(3) What is the substrate temperature used?

Dr Langlois answered: We did not try the sequential deposition of first Ag and then Cu. It would be interesting to see if the low surface energy of Ag is a large enough driving force to allow a segregation of Ag onto the surface of the nanoparticle...and if at the substrate temperature used for the deposition (400 °C), Ag and Cu metallic species diffuse very well on the surface. Diffusion of a particle already formed on the substrate has also been observed under the electron beam.

Dr Geng asked: Could you explain the structure of the bimetallic nanoparticle based on your HRTEM image?

Dr Langlois answered: Some examples of the use of the HRTEM images to determine the structure of the nanoparticles are presented in the paper. The crystallographic structure of the nanoparticles can be determined from the HRTEM images. However, chemical information is more difficult to obtain from HRTEM images. Moiré patterns and energy-filtered imaging are more efficient for determining the chemical arrangement (core/shell or not) of the two metallic species. Other electron microscopy techniques like STEM and EDX are very useful to gain insight on the chemical configurations.

Professor Henry remarked: From the Moiré patterns of Ag/Cu particles, can you tell what is the lattice parameter of the Ag shell?

Dr Langlois responded: A Moiré pattern appears when two lattices are super-imposed. The periodicity of the Moiré is related to the difference of the two lattice parameters, and also to the rotation between the two lattices. In most of the cases that we identified on the HRTEM images, the Ag lattice and the Cu lattice are in a perfect epitaxial relationship, without rotation. In this case, and when the two lattices are in the [110] zone axis, the periodicity of the Moiré that appears due to the d_{111} Ag and Cu interplane distances is expressed by this formula : $d_{Moiré} = 1/\Delta g = d_{Ag111} \times d_{Cu111}/(d_{Ag111} - d_{Cu111})$ From the micrographs, we know $d_{Moiré}$. If ones wants to obtain, say, d_{Ag111}, an assumption must be made on the value of d_{Cu111}. If we give d_{Cu111} its bulk value, we also find the bulk value for d_{Ag111} using the formula. In such cases, we say that both Cu and Ag keep their own lattice parameters. In other words, if we know the epitaxial relationship between the two lattices, and if we make an hypothesis concerning one of the two lattice parameters, we can estimate the second lattice parameter.

Professor Hou said: The comparison between your experimental results and those of the Lyon group (see my comment about the discussion paper by Rossi *et al.*) suggests that the CoPt particle size may be a key factor governing the formation of a $L1_0$ phase. Can you deduce from your observations what is the smallest particle diameter and the smallest particle height at which the $L1_0$ phase is formed?

Dr Langlois replied: From our experiment, the $L1_0$ phase cannot be obtained in nanoparticles with a mean diameter smaller than 3 nm (measured in the plane of the substrate).

From simple thermodynamic considerations, we can consider that the particle shape is very near to the equilibrium morphology so that they have more or less the same height as their size in the plane of the substrate (confirmation by TEM tomography experiments).

Professor Kiely remarked: In this morning's paper by Herzing *et al.*[1] I explained the significant advantages that the new generation of aberration corrected electron microscopes hold for the compositional analysis of bimetallic nanoalloys *via* the STEM-XEDS and STEM-EELS spectrum imaging techniques. I would also like to draw your attention to the potential benefits that aberration correction offers for the imaging of supported bimetallic particles *via* the high angle annular dark field (HAADF) imaging technique. Firstly, the more traditionally used phase contrast (HREM) technique is not very good at resolving the atomic structure of clusters that contain less than about 30 atoms. In contrast, the aberration corrected HAADF technique is now allowing us to clearly image clusters of just 3–4 atoms and also even atomic dispersions of heavy metal atoms on light oxide supports.[2] This advance is invaluable in allowing us to directly observe for the first time all of the various metal species present in supported metal catalyst systems, not just particles greater than about 1 nm in dimension.

Secondly, the atomic columns resolved in a crystalline structure using the HAADF technique show atomic mass (Z) contrast. In ordered alloy systems such as Ni_3Al it is possible to directly distinguish the Al and Ni sub-lattices by HAADF imaging provided a suitable crystal projection (*i.e.* [100]) is chosen.[3] Furthermore the atomic number discrimination can be as little as 4 (*i.e.* Fe columns ($Z = 26$) can be distinguished from Ti columns ($Z = 22$) in $FeTiO_3$–Fe_2O_3).[4] Atomic resolution Z-contrast HAADF imaging could therefore potentially be a very useful way of unambiguously determining whether non core/shell morphology particles are random alloys or ordered alloys. This application however still needs some development since the projected thickness variations across the diameter of the particle would need to be taken into account in the analysis.

1 Andrew A. Herzing, Masashi Watanabe, Jennifer K. Edwards, Marco Conte, Zi-Rong Tang, Graham J. Hutchings and Christopher J. Kiely, *Faraday Discuss.*, 2008, **138**, DOI: 10.1039/b706293c
2 A. A. Herzing, C. J. Kiely, A. F. Carley, P. Landon and G. J. Hutchings, *Identification of Active Gold Nanoclusters for CO Oxidation in Iron Oxide Supported Catalysts*, 2007, submitted for publication.
3 M. Watanabe, D. W. Ackland, C. J. Kiely, D .B. Williams, M. Kanno, R. Hynes and H. Sawada, *JEOL News*, 2006, **41**(1), 2–7.
4 M. Watanabe, H. Hojo, D. W. Ackland, C. J. Kiely and D. B. Williams, *Microsc. Microanal.*, 2007, **13**(2), 1198 CD.

Professor Sir Thomas opened the discussion of Dr Hampe's paper: Very nice work. May I urge you to look not just at CO adsorption on these charged (or neutral) Au clusters but also to place oxygen (either as O_2 or O) onto the cluster also and then carry your calculation right through to investigate a possible reaction path for the Au_n catalyzed $CO + H_2O_2 \rightarrow CO_2$ reaction.

Dr Hampe responded: That is what we have started studying and will continue to look at both experimentally and theoretically.

Professor Johnson remarked: I note with pleasure that your results may be taken to suggest that the atom with the lowest connectivity preferentially bonds the CO. This is consistent with the role of individual sites or roughness of the surface rather than the overall structure.

Dr Hampe responded: I am afraid the story is not quite that simple (at least for gold): it doesn't seem that connectivity alone tells you to which gold atom the CO will bind. For example, to a rhombic Au_4^+ cluster the CO prefers to bind to the more highly connected gold atom.

Our computations suggest that electron density might be a good measure to predict preferred adsorption sites as outlined in our paper.

Professor Fortunelli commented: There are structural effects on top of even–odd alternation: see *e.g.* $N = 6$ and 8 (2D $\rightarrow$ 3D transition) Is Au_{20}^+ Td? And what about Au_{21}^+?

Connectivity is not the entire story: Au_{20} vertexes have coordination 3, but have the lowest BF in the series. The choice of the xc functional can make a difference: DFT-GC usually underestimates Au–Au bond energy appreciably more than Ag–Ag.

Dr Hampe answered: I agree that connectivity is not the entire story. Note, that we do not claim from our study that Au_{20}^+ or Au_{21}^+ have Td structure, though they are local minima. The point we are making is that there is a correlation between binding energy (Au–CO) and the local electron density on a given gold atom of a given structure.

Professor Henry remarked: You compare CO binding energy on the Au_n cations with CO adsorption on bulk Au. So, I would expect that the charge plays an important role for the adsorption energy of CO?

Dr Hampe answered: In fact there is an overall contribution from the charge on the binding energy of CO as we have shown for pure gold cluster cations.[1]

1   M. Neumaier, F. Weigend, O. Hampe, and M. M. Kappes, *J. Chem. Phys.*, 2005, **122**, 104702.

However, there is also a strong size effect superimposed on the relative binding energies that one would like to understand and to correlate with the individual structure.

Dr Ellis remarked: From your work on CO adsorption on gold clusters, is there anything you have found which could help in developing a chemisorption method to measure gold dispersion in supported catalysts?

Dr Hampe responded: It has been shown that it is possible to gain structural information of unsupported metal clusters by complete coverage using weakly bound molecules. I think it will be of importance to further our understanding of the interplay between structure and charge (distribution) on a gold particle to understand its adsorption/desorption and ultimately the catalytic properties. This will also further our understanding of supported catalysts.

Professor El-Shall asked:

(1) Do you expect the same trend for binding energies of CO to Au_n^+ as to the neutral clusters?

(2) Why would the binding energies of CO to the neutral clusters decrease gradually with n?

(3) Knowing the ionization energies of Au_n and $(CO)Au_n$ clusters, is it possible to estimate the binding energies of the neutral clusters?

Dr Hampe replied: I think that every cluster size may exhibit a different chemical behaviour that is not scalable with size. In that sense neutral clusters may behave very differently. We know already for some cluster sizes that the charge has a crucial effect on the structure. Gold cluster anions for instance are much less reactive towards CO than the corresponding cations.

Of course, applying a Born–Haber cycle (given you have the IPs as you are suggesting) to estimate neutral binding energies.

Dr Schofield remarked: What you have been describing in this paper appears to be the chemisorption of CO onto gold, including a dimension of site discrimination. Chemisorption is a technique regularly used for catalyst characterisation with other metals—Pt, Pd, Ru—but so far no-one has been able to come up with a technique for adsorption onto gold. Can this technique be applied to particles on a support, bearing in mind:

—in the paper you say that as the cluster increases in size the Au–CO bond weakens—that the support may also adsorb?

Dr Hampe responded: There are a number of things we need to keep in mind when going from the gas-phase to supported particles as structural rearrangement, partial charging (and others) as induced by the support. They may all play a crucial effect and are definitely worth investigating.

Professor Johnston commented: In recent DFT calculations on Ag–Au clusters we have found significant charge transfer from the more electropositive Ag to the more electronegative Au atoms.[1] This may explain the trend in Au–CO binding energy as a function of Ag doping. Also, I would expect the Au–CO interaction and, hence, the CO stretching frequency to change as a function of the net charge on the cluster.

1   L. O. Paz-Borbon, R. L. Johnston, G. Barcaro and A. Fortunelli, submitted for publication.

Dr Hampe responded: There is indeed a partial charge transfer from silver to the more electronegative gold as has been recently pointed out by several groups (*i.e.* predicted by *ab initio* computations) and also confirmed experimentally.[1,2]

1   See *e.g.* V. Bonačić-Koutecký, J. Burda, R. Mitrić, M. Ge, G. Zampella, and P. Fantucci, *J. Chem. Phys.*, 2002, **117**, 3120.
2   P. Weis, O. Welz, E. Vollmer, and M. M. Kappes, *J. Chem. Phys.*, 2004, **120**, 677.

Professor Lievens remarked: Concerning the lower panel of Fig. 7. There seems to be a controversy between the DFT results, predicting a higher binding energy for CO upon replacing one Au atom in Au_6 by Ag (as opposed to the general trends observed), while the experiments yield a lower value. What could be the reason for this?

Dr Hampe replied: We don't really know, but I suppose it is conceivable that in this case we missed one isomer that is realized in the experiment.

Professor Hutchings remarked: Returning to the points raised earlier by Professor J. M. Thomas and Professor B. F. G. Johnson concerning the reaction of CO with O_2 that you have indicated you will carry out. Will you be able to obtain experimental rate data for the reaction? This information will be crucial, as you will get information concerning unsupported gold clusters in the gas phase. Your clusters have no support, all the current models concerning the reaction between CO and O_2 to form CO_2 using gold catalysts involve the support in the activation of dioxygen. If the rate you obtain is comparable to the rate observed in real supported catalysts then this would clearly demonstrate that the support is not crucial; conversely, if the rate is negligible then this would demonstrate the support is crucial. So your data would help resolve a very important catalysis question.

Professor Henry responded: All the experiments on CO oxidation performed in my group were on supported gold clusters either under UHV by molecular beams or in mbar pressure range by environmental HRTEM, then it is not possible in these conditions to rule out a possible effect of the support.

Professor Johnson remarked: The Au_6^+ system will be planar and therefore contain three Au atoms with a connectivity of two. That being the case and according to our earlier discussion, I would have assumed that these gold atoms would be prime targets for CO absorption. This is not the case, and I must conclude that there is no direct relationship between absorption and connectivity.

Professor Pastor asked: Could you comment in general on the dependence of your results (structure, binding energy *etc.*) on the choice of the xc functional? In particular, could the choice and inaccuracies of the xc functional be a reason for the discrepancy in the trends in the binding energy of $Au_{6-m}Ag_m$ as a function of the m for $0 \leq m \leq 2$?

Dr Hampe responded: To answer this question would be worth a complete paper in its own right. What we did is to carefully check the consistency between structural parameters (like from ion mobility data for gold cluster sizes up to Au_8^+) and the computed structures: here BP86 gives the better results compared to B3LYP. On the other hand for describing the metal–CO bond the hybrid functional B3LYP does a significantly better job.

Dr Mejia opened the discussion of Mr Marsault's paper:
(1) To what temperature are the Pd clusters stable?
(2) Are these clusters (and the film) stable under a reaction environment? What if the atmosphere was reducing not oxidising?

Mr Marsault answered: The Pd clusters are stable at least up to 400 °C. The stability of the clusters under CO and O_2 atmospheres are under investigation.

Professor Johnson asked: We are conducting very similar chemistry. In our core we deposit a thin layer of support *e.g.* silicate, on Al_2O_3, TiO_2 to provide a template for the organisation of our active nanoparticles which we prepare by wet chemical methods. They are catalytically active!

Dr Baletto asked: What is the role of the thickness of the alumina film on the formation of clusters after deposition of Au and Pd?

Mr Marsault responded: The alumina films are 5 Å thick. If we use higher oxygen exposure or air exposures the films become thicker but the superstructure formed by the defects disappears.

Concluding remarks

Brian Johnson

Received 6th November 2007, Accepted 6th November 2007
First published as an Advance Article on the web 5th February 2008
DOI: 10.1039/b717215j

From the onset, it was clear that a meeting devoted to this topic was essential. Given the nature of the subject and its diversity, it was clearly time to take stock and begin to make an attempt at a rationalisation of this intriguing but highly complex subject.

The meeting began with a discussion of the fundamental nature of nano-structured materials. The importance of shape, 'magic numbers', *etc.* was developed in detail. A call was made for a new approach and search for some fundamental property, which would allow a better understanding of the chemical and physical properties these materials possess. Shape dominates much of these properties but in itself provides problems both in its determination by experimental means and prediction by theory. For very small numbers of atoms, clearly shape can be more easily determined by both means and its importance in defining orbital distribution and electronic effects more easily appreciated. For much larger particles it could be argued that the use of simple geometrical forms does not serve its purpose. It might be argued that for these more highly complex systems the type of atom within the structure was more important *e.g.* plateau, edge, or corner. Clearly, this is of relevance to those interested in catalysis and reactivity, which by its nature is largely governed by surface properties and to some measured less by atoms contained within the interior. In the section devoted to catalysis this issue was raised—with some arguing that activity is a function of size—the smaller the better—and others that larger particles were better. The problem is of course that insufficient studies of actual processes have been conducted and in almost all instances the nature of the particle actually during—as opposed to before and after—the catalytic reaction not established.

The term 'magic number' was debated. Again the problem revolves around this simple concept in which key and fundamental structures are governed by a given electronic count. As with elsewhere within chemistry the problem is in deciding the appropriate magic number for a given system. As with organometallic chemistry the number will change according to shape and the nature of the metal atom or ion under consideration. For example it may not be the same for gold as copper, and so on. In organometallic systems this leads to magic number systems with, for example 18-, 16- and 14-electrons. Much evidence for the use of the magic number comes from mass-spectrometric evidence, and there is little doubt that to a large degree stability appears to correlate with a magic number of electrons. It must be remembered however, that in mass spectroscopy one is dealing with kinetic rather than ThD stability, and this may lead to erroneous conclusions. Frequently it would appear that ions also appear that do not correlate with given magic numbers and thereby tend to be ignored. There is much work in this area and no doubt rules will be established and the structure magic number relationship more firmly established. In this connection the section within the meeting dedicated to the theoretical understanding of nano-structures was highly impressive and it is clear that in some systems at least properties and structures are clearly understood and there relationships to electronic structure established.

An area, which was impressive in its achievements, was developed in the session devoted to the important magnetic and optical properties of new nano-materials.

University of Cambridge, UK

Here theory and experiment appeared to be progressing hand-in-hand. The nature of discovery was still led predominantly by experiment, but many remarkable properties were unveiled, and this must remain an area of high interest. This study was, by the nature of the subject, coupled with synthetic approaches. Here again the record was impressive. The need for new and advanced methods of structural study was apparent. However, it was also clear that to a large measure this need was being met.

In conclusion, an excellent meeting which clearly met the needs of a wide group of researchers. Of major importance are the very special—in some cases unique—optical and magnetic properties of the special materials. It is also apparent that they show very special catalytic properties. The field is at an exciting stage of continuous discovery and has much to offer the experimentalist and the theoretician.

Poster titles

Grafting of metal clusters on chemically functionalized active carbon for the synthesis of supported nanoalloys, **Christopher Willocq, Deborah Vidick, Sophie Hermans, Arnaud Delcorte, Patrick Bertrand** and **Michel Devillers**, *Université Catholique de Louvain, Belgium*

Study of small Ag/MgO(100) and AgPd/MgO(100) clusters, **Z. Kuntova, G. Rossi** and **R. Ferrando**, *Università di Genova and CNR/INFM, Italy*

MD simulation of the melting of AgCo and NiCo nanoclusters, **Z. Kuntova, G. Rossi** and **R. Ferrando**, *Università di Genova and CNR/INFM, Italy*

Mackay and anti-Mackay motifs in AgCo clusters, **G. Schiappelli, G. Rossi, F. Nita** and **R. Ferrando**, *Università degli studi di Genova and INFM/CNR, Italy*

DFT studies of 38-atom bimetallic clusters, **Lauro Oliver Paz-Borbón, Roy L. Johnston, Giovanni Barcaro** and **Alessandro Fortunelli**, *University of Birmingham, UK*

Computational studies of Pd–Au clusters, **Faye Pittaway, Lauro Oliver Paz-Borbón** and **Roy L. Johnston**, *University of Birmingham, UK*

Structures and stabilities of Pt–Au clusters, **Andrew J. Logsdail, Lauro Oliver Paz-Borbón** and **Roy L. Johnston**, *University of Birmingham, UK*

Computational studies of the structures and energetics of PdPt nanoalloy clusters of 98, **Alvaro Posada-Amarillas, Lauro Oliver Paz-Borbón, Thomas V. Mortimer-Jones, Roy L. Johnston, Giovanni Barcaro** and **Alessandro Fortunelli**, *Universidad de Sonora, Mexico*

Metal clusters on a defected MgO(100) surface: a magic Pd–Ag structure grown on an oxygen vacancy (F_5 center), **Giovanni Barcaro** and **Alessandro Fortunelli**, *Instituto per i Processi Chimico-Fisici (IPCF)—CNR, Italy*

FePt nanoparticles—can we promote $L1_0$ ordering by ion irradiation, **T. T. Järvi, A. Kuronen, K. Nordlund, M. Müller** and **K. Albe**, *University of Helsinki, Finland*

Thermal response of hollow nanometer-sized systems, **Giuseppe Manai, F. Delogu** and **I. V. Shvets**, *Trinity College Dublin, Ireland*

Bimetallic nanocluster in silica formed by sequential ion implantation, **G. Mattei, V. Bello, C. de Julián Fernández, P. Mazzoldi, G. Pellegrini** and **C. Maurizo**, *University of Padova, Italy*

Nb-doped TiO_2 layers as optical windows for solar cells, **B. Neumann, F. Bierau, K. Ellmer** and **H. Tributsch**, *Hahn-Meitner-Institut, Germany*

Alloying analysis of Ag–Au bimetallic core–shell structures by molecular dynamics simulation, **Fuyi Chen** and **Roy L. Johnston**, *University of Birmingham, UK*

Structure and stability of scandium doped copper clusters: a combined experimental and computational study, **Nele Veldeman, Tibor Höltzl, Sven Neukermans, Tamás Veszprémi, Minh Tho Nguyen** and **P. Lievens**, *K. U. Leuven, Belgium*

Argon complex formation as a structural probe for endohedral transition metal doped silicon clusters, **E. Janssens, P. Grüne, G. Meijer, L. Wöste, A. Fielicke** and **P. Lievens**, *K. U. Leuven, Belgium*

Hexadecagold backbone for a tuneable sub-nanometer oxidation and reduction agent, **Michael Walter** and **Hannu Häkkinen**, *University of Jyväskylä, Finland*

PdRu alloy nanoparticles as anode catalysts in direct methanol fuel cells, **Noelia Cabello, Elvis Christian, Janet Fisher** and **David Thompsett**, *Johnson Matthey Technology Centre, UK*

Characterisation and activity of Pt_3Co nanoparticle alloys for oxygen reduction in PEM FC fuel cells, **S. C. Ball, S. L. Hudson, A. E. Russell, D. Thompsett** and **B. Theobald**, *Johnson Matthey Technology Centre, UK*

Production of metal wires from salt-containing zeolites "A" & "X", **Alvaro Mayoral** and **P. A. Anderson**, *University of Birmingham, UK*

Hydrogen chloride on ice, **U. F. T. Ndongmouo, M.-S. Lee, R. Rousseau, F. Baletto** and **S. Scandolo**, *King's College London, UK*

Density Functional Theory study of Pt and Pd pseudomorphic monolayer alloy catalysts for NO_x storage reduction applications, **Jelena Jelic** and **Randall Meyer**, *University of Illinois, USA*

The Skinner Poster Prize was split between two winners as they were both excellent, but not really comparable, as one is theoretical and one experimental. The prize winners were Giovanni Barcaro for his poster on metal clusters on a defected MgO(100) surface: a magic Pd–Ag structure grown on an oxygen vacancy (F_5 center) and Alvaro Mayoral for his poster on the production of metal wires from salt-containing zeolites "A" & "X".

List of participants

Dr F. Baletto, *King's College London, UK*
Dr G. Barcaro, *IPCF, Italy*
Dr M. Binhazzaa, *King Saud University, Saudi Arabia*
Professor G. Bond, *UK*
Professor M. Broyer, *Université Claude Bernard Lyon 1, France*
Miss N. Cabello, *Johnson Matthey Plc., UK*
Dr F. Calvo, *LASIM University of Lyon, France*
Dr V. Caps., *IRCELYON, France*
Dr M. Chapman, *Royal Society of Chemistry, UK*
Dr F. Chen, *University of Birmingham, UK*
Dr J. Cookson, *Johnson Matthey Plc., UK*
Dr E. Crabb, *The Open University, UK*
Mr B. Curley, *University of Birmingham, UK*
Dr P. Ellis, *Johnson Matthey Plc., UK*
Professor S. El-Shall, *Virginia Commonwealth University, USA*
Professor J. Evans, *University of Southampton, UK*
Professor R. Ferrando, *University of Genoa, Italy*
Professor A. Fortunelli, *Instituto per i Processi Chimico-Fisici (IPCF), Italy*
Dr M. Fox, *Royal Society of Chemistry, UK*
Dr J. Geng, *University of Cambridge, UK*
Ms M. Gilbert, *Royal Society of Chemistry, UK*
Dr H. Grönbeck, *Chalmers University of Technology, Sweden*
Professor H. Hakkinen, *University of Jyväskylä, Finland*
Dr O. Hampe, *Forschungszentrum Karlsruhe GmbH, Germany*
Professor C. Henry, *CRMCN-CNRS, France*
Dr P. Hinde, *Johnson Matthey Plc., UK*
Dr S. Horswell, *University of Birmingham, UK*
Professor M. Hou, *Université Libre de Bruxelles, Belgium*
Professor G. Hutchings, *Cardiff University, UK*
Mr T. Jarvi, *University of Helsinki, Finland*
Professor J. Jellinek, *Argonne National Laboratory, USA*
Professor B. Johnson, *University of Cambridge, UK*
Professor R. Johnston, *University of Birmingham, UK*
Mr C. Kiely, *Lehigh University, USA*
Dr A. King, *Renishaw Plc., UK*
Dr Z. Kuntova, *Università di Genova, Italy*
Dr C. Langlois, *Université Paris 7/MPQ, France*
Professor E. Leiva, *Universidad Nacional de Cordoba, Argentina*
Ms F. Lequien, *LEMHE/ICMMO, France*
Dr Z. Li, NPRL, *University of Birmingham, UK*
Dr N. Lidgi-Guigui, *NPRL University of Birmingham, UK*
Professor P. Lievens, *University of Leuven, Belgium*
Mr A. Logsdail, *University of Birmingham, UK*
Dr H. Lunn, *Royal Society of Chemistry, UK*
Dr G. Manai, *Trinity College Dublin, UK*
Mr M. Marsault, *CNRS, France*
Professor G. Mattei, *University of Padova, Italy*
Mr A. Mayoral-Garcia, *University of Birmingham, UK*
Dr S. Mejía, *Universidad Autonoma de Nuevo Leon, Mexico*
Professor R. Meyer, *University of Illinois at Chicago, USA*
Dr C. Mottet, *CRMCN-CNRS, France*
Mr B. Neumann, *Hahn-Meitner-Institut, Germany*

Professor R. Palmer, *University of Birmingham, UK*
Professor G. Pastor, *Universität Kassel, Germany*
Mr O. Paz-borbon, *University of Birmingham, UK*
Miss F. Pittaway, *University of Birmingham, UK*
Professor C. Ricolleau, *University Paris 7, France*
Dr G. Rossi, *INFM-CNR, Italy*
Dr A. Russell, *University of Southampton, UK*
Dr E. Schofield, *Johnson Matthey Plc., UK*
Mr A. Scott, *Royal Society of Chemistry, UK*
Professor P. Sermon, *University of Surrey, UK*
Miss E. Shiells, *Royal Society of Chemistry, UK*
Dr B. Theobald, *Johnson Matthey Technology Centre, UK*
Prof Sir J. Thomas, *University of Cambridge, UK*
Dr M. Tromp, *University of Southampton, UK*
Miss N. Veldeman, *Katholieke Universiteit Leuven, Belgium*
Dr P. P. Wells, *University of Southampton, UK*
Mr C. Willocq, *Université Catholique de Louvain, Belgium*
Dr R. Wiltshire, *University of Birmingham, UK*
Miss Y. Zhang, *University of Birmingham, UK*

Index of contributors*

* The page numbers in **bold** type indicate papers submitted for discussions.